An Introduction to Metamaterials and Waves in Composites

Biswajit Banerjee

CRC Press
Taylor & Francis Group
Boca Raton London New York

CRC Press is an imprint of the
Taylor & Francis Group, an **informa** business
A TAYLOR & FRANCIS BOOK

First published 2011 by Taylor & Francis

Published 2021 by CRC Press
Taylor & Francis Group
6000 Broken Sound Parkway NW, Suite 300
Boca Raton, FL 33487-2742

International Standard Book Number: 978-1-4398-4157-0 (Hardback)

Visit the Taylor & Francis Web site at
http://www.taylorandfrancis.com

and the CRC Press Web site at
http://www.crcpress.com

Contents

Preface vii

Introduction ix

1 Elastodynamics, Acoustics, and Electrodynamics 1

1.1 A note about notation . . . 2
1.2 Elastodynamics . . . 3
1.2.1 Elastodynamic potentials . . . 6
1.2.2 Cartesian coordinates . . . 7
1.2.3 Curvilinear coordinates . . . 8
1.2.4 Anisotropic and frequency-dependent mass density . . . 9
1.2.5 Elastodynamics at a fixed frequency . . . 10
1.2.6 Antiplane shear strain . . . 13
1.2.7 Elastodynamic power and dissipation . . . 13
1.3 Acoustics . . . 16
1.3.1 Acoustic potential . . . 19
1.3.2 Cartesian and curvilinear coordinates . . . 19
1.3.3 Anisotropic and frequency-dependent mass density . . . 20
1.3.4 Acoustics at a fixed frequency . . . 21
1.3.5 Acoustic equation and antiplane shear . . . 22
1.3.6 Acoustic power and dissipation . . . 23
1.4 Electrodynamics . . . 24
1.4.1 Governing equations . . . 25
1.4.2 Electrodynamic potentials . . . 29
1.4.3 Cartesian and curvilinear coordinates . . . 31
1.4.4 A slight simplification . . . 33
1.4.5 Maxwell's equations at fixed frequency . . . 34
1.4.6 TE and TM modes in electromagnetism . . . 37
1.4.7 Maxwell's equations in elasticity form . . . 39
1.4.8 Electrodynamic power and dissipation . . . 40
1.4.9 The electrical conductivity equation . . . 41

2 Plane Waves and Interfaces 47

2.1 Plane wave solutions . . . 48
2.2 Plane wave solutions in elastodynamics . . . 49
2.2.1 Elastodynamic potentials for plane waves . . . 51

2.2.2 Reflection of plane P-waves at a free surface 53
2.2.3 Reflection of P-waves at the interface between two solids . 58
2.2.4 Reflection of SH-waves at the interface between two solids 59
2.2.5 Negative refractive index and SH-waves 60
2.3 Plane wave solutions in acoustics 61
2.3.1 Reflection of an acoustic wave at a plane interface 62
2.3.2 Acoustic impedance 65
2.3.3 Reflection at a liquid-solid interface 66
2.3.4 Negative refractive index 68
2.4 Plane wave solutions in electrodynamics 69
2.4.1 Reflection of a plane wave at an interface 70
2.4.2 Wave polarized parallel to the plane of incidence 71
2.4.3 Wave polarized perpendicular to the plane of incidence . . 75
2.4.4 Electromagnetic impedance 77
2.4.5 Negative refractive index 78
2.5 Wave propagation through a slab 80
2.5.1 Acoustic waves in a slab 80
2.5.2 Impedance tube measurements 83
2.5.3 Electromagnetic waves in a slab 85

3 Sources and Scattering **93**
3.1 Plane wave expansion of sources 93
3.1.1 Two-dimensional scalar wave equation 94
3.1.2 Three-dimensional scalar wave equation 98
3.1.3 Expansion of electromagnetic dipole sources 102
3.2 Single scattering from spheres 104
3.2.1 Scattering of acoustic waves from a liquid sphere 105
3.2.2 Acoustic scattering cross-section 108
3.2.3 Scattering by an elastic sphere in an elastic medium 110
3.2.4 Scattering of electromagnetic radiation from a sphere . . . 113
3.3 Multiple scattering 117
3.3.1 The self-consistent method 117
3.3.2 Acoustic scattering from multiple circular cylinders 120

4 Electrodynamic Metamaterials **125**
4.1 A model for the permittivity of a medium 126
4.2 Negative permittivity materials 129
4.3 Artificial magnetic metamaterials 133
4.4 Negative refraction and perfect lenses 139
4.4.1 Pendry's argument 143

5 Acoustic and Elastodynamic Metamaterials **147**
5.1 Dynamic mass density 150
5.1.1 A simple model with negative mass 150
5.1.2 Complex mass density 154
5.1.3 Anisotropic mass 161
5.1.4 General expression for frequency-dependent mass 163
5.2 Frequency-dependent moment of inertia 167
5.3 Negative elastic moduli 173
5.3.1 A Helmholtz resonator model 173
5.3.2 Negative moduli and viscoelasticity 178
5.4 Band gaps, negative index and lenses 183
5.5 Anisotropic density 186
5.6 Willis materials in elastodynamics 188
5.6.1 Ensemble averaging 188
5.6.2 The Willis equations 190
5.6.3 Derivation of the Willis equations 192
5.6.4 Extension to composites with voids 194
5.6.5 Willis equations for electromagnetism 196
5.7 A Milton-Willis model material 197
5.8 Extremal materials 208
5.8.1 Pentamode materials 208

6 Transformation-based Methods and Cloaking **217**
6.1 Transformations 218
6.2 Cloaking of electrical conductivity 221
6.2.1 Coordinate transformation for electrical conductivity . . . 222
6.2.2 Electrical tomography 223
6.2.3 Examples of transformation-based cloaking 225
6.3 Cloaking for electromagnetism 225
6.3.1 The perfect lens 231
6.3.2 Magnification 233
6.4 Inertial acoustic cloaking 234
6.4.1 The infinite mass requirement 239
6.5 Transformation-based cloaking in elastodynamics 241
6.5.1 Energy considerations 244
6.6 Acoustic metafluids and pentamode materials 245
6.6.1 A pentamode to pentamode transformation 247

7 Waves in Periodic Media **251**
7.1 Periodic media and the Bloch condition 253
7.2 Elastodynamics in the quasistatic limit 256
7.2.1 Effective stiffness and dispersion relation 260
7.2.2 Hashin's relations 261
7.3 Electromagnetic waves in the quasistatic limit 264
7.3.1 Effective properties and dispersion relations 266

7.4 Band gap phenomena in periodic composites 268
7.4.1 One-dimensional lattice models 269
7.4.2 Two-dimensional lattice models 272
7.4.3 Dispersion in continuous elastic composites 277

8 Waves in Layered Media 283
8.1 Wave equations in layered media 283
8.1.1 Antiplane shear elastic waves 284
8.1.2 Acoustic waves 286
8.1.3 Electromagnetic waves 287
8.2 Piecewise-constant multilayered media 289
8.3 Smoothly varying layered media 292
8.3.1 The thin layer limit 292
8.3.2 The Airy function solution 295
8.3.3 The Wentzel-Kramers-Brillouin-Jeffreys method 297
8.4 Propagator matrix approach 302
8.4.1 A three-layered sandwich medium 306
8.4.2 Anisotropic electromagnetic layered media 308
8.4.3 Anisotropic elastic layered media 309
8.5 Periodic layered media 310
8.5.1 The Schoenberg-Sen model 310
8.5.2 Low-frequency effective properties 315
8.6 Quasistatic homogenization of layered media 318
8.6.1 Quasistatic effective properties of elastic laminates 318
8.6.2 Quasistatic effective properties of electromagnetic laminates 322
8.6.3 Laminates with arbitrary lamination direction 324
8.6.4 Tartar-Murat-Lurie-Cherkaev formula 326
8.6.5 Hierarchical laminates 331

Epilogue 337

References 339

Index 355

Preface

In the fall of 2006 I read a couple of articles on transformation-based cloaking in *Science* and heard a talk by Prof. Graeme Milton on cloaking due to anomalous resonances. I was intrigued by these new ideas; were they applicable to shocks, vibrations, seismic waves? An opportunity to learn more came early in 2007, when Prof. Milton offered a course called "Waves in Composites and Metamaterials." I soon realized that the contents of the course were worth preserving and disseminating to a broader audience. I started to work out the details of each lecture and posting them on my web-page at the University of Utah. The next three months were a juggling act between my research on computational plasticity and trying to understand metamaterial concepts. Later that year, I converted my notes into Wiki form and gave them a more permanent home at Wikiversity.* This book has grown out of those lecture notes.

The book has been written with beginning graduate students and advanced undergraduates in mind. The book will also be of use to engineers and researchers who wish to understand recent developments in the dynamics of composites. A background of elementary calculus and differential equations, linear algebra, complex analysis, and vector and tensor analysis is assumed. The reader unfamiliar with these topics can find excellent sources on the web that are sufficient to understand most of the book. The notation in the book has been kept as independent of coordinate system as possible. Components of vectors and tensors are introduced only when necessary and the Einstein summation convention is used extensively.

Special emphasis is placed on elastic media and acoustics, in part because of my background in solid mechanics but also because there is a large gap between the literature on the dynamics of elastic/acoustic composites and that on electromagnetic composite media. Previous knowledge of elasticity and electrodynamics will help in the navigation of this book, but is not essential. It is crucial that the reader develop a familiarity with electrodynamics during the reading of this book; many developments in the fields of phononic crystals and acoustic metamaterials have been based on previous developments in electromagnetic materials.

I have adopted a "direct," deductive, approach in this book in the sense that results do not appear magically. This, of course, has required that an amount of detail be included in the text. I suggest that the student work through some of the detail to gain an understanding but spend more time analyzing the results and developing concepts. The problems at the end of each chapter are designed to consolidate the

*The web page is http://en.wikiversity.org/wiki/Waves_in_composites_and_metamaterials.

understanding of techniques used in the book. Beyond general ideas, the book does not discuss the design of metamaterials and artificial crystals; that task is left to the readers of this book.

The style and content of this book have been influenced by several people, particularly Prof. Graeme Milton, Prof. William Pariseau, Prof. Robert Smith, and Prof. Rebecca Brannon of the University of Utah, and Prof. Andrew Norris of Rutgers University. Special thanks are due to my wife Champa who suffered through a year of lost weekends and evenings while this book was being written. I would like to thank Dr. Emilio Calius for writing part of the introduction, Dr. Eric Wester for showing me his approach to impedance tube calculations, Dr. Bryan Smith for helping with cloaking simulations, and our librarian Alison Speakman for giving me quick access to the older literature on the dynamics of composites. Thanks also to my editor, Dr. John Navas, who persisted and made this book possible.

All errors in this book are, of course, mine and I will be grateful if readers inform me of any errors that they find.

Biswajit Banerjee
Auckland, New Zealand

Introduction

> Traditionally, the answer to a given problem is obtained by copying suitable equations, submodels, and boundary conditions with their appropriate solution techniques from available sources. This is called "matching" and may result in a good first-step learning experience; however, it should be augmented later on by more independent work ...
>
> C. KLEINSTREUER, *Two-phase flow: Theory and applications*, 2003.

Let us consider the propagation of light through glass, clear water, or any other reasonably transparent media; the propagation of sound through plasterboard walls or steel bulkheads; the propagation of vibration through aluminum airframes; or the propagation of radio waves through the earth's atmosphere. What all of these situations have in common is that they involve the interaction of waves with a medium, be it solid, fluid, or gas. If the structure of the medium is just right, these interactions can have interesting effects. The colors in a peacock's feathers are caused by light interacting with a periodic solid material; a rainbow is produced by the interaction of light with spherical water droplets; a mirage is caused by the interaction of light with our layered atmosphere. If we have special instruments we can observe similar effects for sound and elastic vibrations. But to what extent can we control these phenomena?

That light can be bent must have been known since soon after humans started fishing, and even the controlled bending of light through lenses has been known for a long time. Translucent crystals of various kinds have been found among both Neanderthal and Cro-Magnon remains that are tens of thousands of years old. The oldest polished lens artifacts seem to be plano-convex (flat on one side and convex on the other) rock crystal lenses that may have been used as magnifying glasses, or as burning-glasses to start fires by concentrating sunlight. The earliest written records of lenses date to Ancient Greece, with Aristophanes' play *The Clouds* (424 BC) mentioning a burning-glass (a biconvex lens used to focus the sun's rays to produce fire). The oldest lens artifact is the Nimrud lens, which is over three thousand years old, dating back to ancient Assyria. Clearly, attempts to control the propagation of waves has a long history.

Ordinary matter is fundamentally discrete. All naturally occurring media known to man are composed of discrete molecules, which in turn are made up of atoms. An atom consists of electrons, positively charged protons, and neutrons and, even further down in the length scale, quarks. A central assumption in classical mechanics and electrodynamics is that the discrete nature of matter can be overlooked, provided the length scales of interest are large compared to the length scales of discrete molecular structure. Thus, matter at sufficiently large length scales can be treated as a continuum in which all physical quantities of interest, including density, stiffness, magnetic permeability, and electric permittivity are continuously differentiable.

Wave propagation in periodic structures such as natural crystals and in layered media has been studied since the pioneering work of Kelvin, Rayleigh, and Maxwell in the late 1800s. But, since the geometries of these structures could be varied only slightly, there was limited control over the effects that could be achieved. In 1987, the work of Yablonovitch and John triggered an interest in the creation of artificial crystals that could be designed to exhibit particular wave phenomena. However, most of these artificial photonic or phononic crystals were designed to operate at wavelengths of the order of the lattice parameter until the development of metamaterials.

Metamaterials

The name metamaterials emerged in the late 1990s and was first used officially by the Defense Advanced Research Projects Agency (DARPA) Symposium on Meta-Materials in 1999. It was coined by the pioneers in the field by using the prefix *meta*, which can be translated from the Greek as beyond, to imply beyond conventional materials. Metamaterials are a new class of complex composite materials that have created considerable excitement because they can be engineered to exhibit any desired electromagnetic, acoustic, or mechanical effective properties up to and including such exotic behaviors as negative refraction, negative bulk modulus, or negative mass under certain excitations. The physics of metamaterials and their interaction with waves of all kinds can be extremely counter-intuitive, causing strong criticism and debate as well as an explosion of research.

Metamaterials have a long prehistory, dating back at least as far as the Lycurgus cup from the 4th century AD that uses metallic nanoparticle colloids embedded in glass to dramatically change its color as a function of the illumination angle. But scientific research in this area only started in the late 19th century, as Floquet, Rayleigh, Bose, and others investigated waves in periodic systems of various kinds. Although materials that exhibited reversed physical characteristics were first described theoretically by Veselago in 1967, it was not until the late 1990s that practical designs of such materials were discovered.

In our view metamaterials involve inclusions and inter-inclusion distances that are much smaller than a wavelength. As long as this condition is met, the dimensions of the metamaterials' internal structure are independent of the wavelength they interact with, and determined mainly by practical considerations. Consequently such media can be described by homogenization and effective media concepts. They typically involve coupling of the waves with resonances of some kind. On the other hand, the phononic and photonic crystals (also called band-gap materials) involve length scales that are on the order of half a wavelength or more and are described by Bragg reflection and other periodic media concepts. The key dimensions of a band gap material are directly linked to the wavelengths that it will strongly interact with. However, there is often no clear distinction between metamaterials and photonic/phononic crystals and the same structure may behave as one or the other, depending on the wavelength and frequency of the incident waves.

Modern developments

Metamaterials have a prehistory dating back many centuries; but modern development really started in the mid-1990s when it was realized that split-ring resonators and thin wire structures provided the means of constructing electromagnetic metamaterials with a negative refractive index, which was first demonstrated experimentally by Smith et al. at the University of California in 2000 (Smith et al., 2000). Research in this area, although active, remained something of a niche until Pendry's seminal paper (Pendry, 2000), published that same year, in which he described how metamaterials could enable the perfect lens, one whose imaging resolution is not limited by diffraction, which would have profound consequences for microscopy, spectroscopy, and micro-fabrication. Also in 2000, a group of researchers in Hong Kong reported the first elastic metamaterial and its ability to greatly affect acoustic transmission in a narrow frequency band (Liu et al., 2000). The composite material was later shown to exhibit negative density at those frequencies (Liu et al., 2005).

By 2006 both Leonhardt (Leonhardt, 2006b) and Pendry (Pendry et al., 2006) had developed theories of electromagnetic cloaking using metamaterials, and later that year Schurig and Smith, now at Duke University, experimentally demonstrated cloaking of a small region at one microwave frequency (Schurig et al., 2006a). Theoretical concepts for acoustic cloaking were first published in 2007 (Cummer and Schurig, 2007, Chen and Chan, 2007a), but experimental results were not available until 2010, elastic structures being more complex as they can support shear as well as longitudinal wave modes. Cloaks to protect marine structures from waves (Farhat et al., 2008) and buildings from earthquakes (Brun et al., 2009) have also been proposed recently. Progress in cloaking is continuing, with the first optical frequency metamaterial being produced in 2008 and the first cloak that operates over a range of frequencies in the microwave spectrum being demonstrated in 2010. There are still major challenges, such as the issue of losses, which is driving interest in superconducting metamaterials (Kurter et al., 2010) and how to achieve the desired effects over an usefully wide range of frequencies.

Recent developments and future directions

Although cloaking and super lensing continue to be major foci of metamaterials research, the field has begun to broaden significantly over the past couple of years. A team at Caltech is designing metamaterial coatings to improve the effectiveness of solar cells; several groups are looking at applications involving THz focusing and imaging, others at wireless communications in the 100 Mhz to 10 GHz range, some at controlling heat transmission through phonons; yet others are using metamaterials to simulate black holes and other aspects of the structure of the universe by exploiting the analogy between metamaterials' ability to distort electromagnetic space and gravity's distortions of spacetime. A team at Nanjing's Southeast University reported the first microwave black hole in 2009 (Cheng et al., 2009, Cheng et al., 2010).

Research worldwide has focused mainly on electromagnetic metamaterials because there are many kinds of modern devices that require interactions with electromagnetic waves. Applications range from health care to communications, from power generation and conversion to semiconductor manufacturing, from stealth applications for the military to security. Electromagnetic metamaterial antennas that offer improved performance in smaller sizes appeared in 2009 and many other commercial applications will undoubtedly follow.

But there is a gap in published research on elastic and acoustic metamaterials when compared with that on electrodynamic metamaterials. Though deep sub-wavelength acoustic imaging has been demonstrated (Zhu et al., 2010), there is a paucity of experimental realizations of elastic metamaterial designs and applications. For instance, there is a need in the construction industry for new approaches to sound insulation and for earthquake protection, vibration cloaking of sensitive components (perhaps for protecting electrodynamic metamaterial components), and mechanical sensors and actuators by controlling the vibrations of beams and plates. As our understanding of the capabilities of metamaterials improves, we can expect the opportunities for exciting new research and applications to multiply.

The book

This book deals mainly with theoretical aspects of metamaterials, periodic composites, and layered composites. The first chapter introduces the reader to elasticity, acoustics, and electrodynamics in media. Concepts such as an anisotropic, tensorial mass density, frequency-dependent material properties, and dissipation and constraints imposed by causality are introduced from the beginning. The second chapter deals with plane wave solutions to the wave equations that describe elastic, acoustic, and electromagnetic waves. Reflection and refraction at plane interfaces are explored for various situations and transmission through slabs is discussed. The third chapter deals with the plane wave expansion of sources and with scattering from curved interfaces, specifically spheres and cylinders. Multiple scattering is also explained in brief.

Electrodynamic metamaterials are covered in the fourth chapter. Particular emphasis is placed on homogenization of metamaterials as proposed by Pendry and co-workers and an effort is made to give some physical insight into metamaterials using the Drude model. Perfect lensing and negative refractions are also discussed.

The fifth chapter is on elastodynamic and acoustic metamaterials. Simple spring-mass models are used to motivate the possibility of negative and anisotropic mass density. The Milton-Willis theory of frequency-dependent mass is explained and a gyrocontinuum model with a frequency-dependent moment of inertia is discussed. Helmholtz resonator models are used to show that effective dynamic elastic moduli can be negative. The Willis equations, which appear to be a general descriptor of composites with microstructure, are discussed in detail and a spring-mass model that exhibits Willis behavior is examined. The last section of the chapter deals with extremal materials such as negative Poisson's ratio and pentamode materials.

Chapter 6 deals with transformation-based cloaking. The invariance of the conductivity equations and of Maxwell's equation under coordinate transformations is derived. Examples of transformations are discussed and aspects of acoustic cloaking are introduced. We then show that the equations of elastodynamics transform into the Willis equations under coordinate transformations. The special behavior of pentamode materials and their application to acoustic cloaking is also examined.

The next chapter deals with periodic composites. The Bloch-Floquet theorem is introduced and applied to the problems of finding the effective behavior of composites in the quasistatic limit. The quasistatic equations are found to be identical to the equations of elastostatics and electrostatics, but with complex parameters. The Hashin effective medium solution is used to show how results from elastostatics can be extended to dynamic problems using analytic continuation. Finally, Brillouin zones and band gaps in periodic structures are discussed in the context of lattice models of elastic structures.

The final chapter involves layered media. Wave propagation in smoothly varying layered media and approximate solutions to problems involving such media are examined. The propagator matrix is introduced and used to show that a periodic layered medium can exhibit anisotropic density. Quasistatic homogenization of laminates is explored and hierarchical laminates are shown to possess remarkable properties.

Many of the ideas in this book are yet to be realized experimentally and we hope that the readers of this book will explore some of these ideas and bring them to technological maturity.

1

Elastodynamics, Acoustics, and Electrodynamics

> We, however, working as we do to advance a single department of science, can devote but little of our time to the simultaneous study of other branches. As soon as we enter upon any investigation, all our powers have to be concentrated on a field of narrowed limit.
>
> HERMANN VON HELMHOLTZ, On the aim and progress of physical science, 1869.

Vibrations, sound, and light involve the propagation of waves. In the case of vibrations, these waves carry information about small changes in the shape of an elastic body. Vibrations are elastic waves. Our sense of touch can be used to track these elastic changes in shape as a function of time. Sound is the propagation of small changes in pressure through a fluid and our ears may be used to track these waves. Light is the propagation of small disturbances in electric and magnetic fields and we can use our eyes to track these changes. But for certain frequencies our senses are no longer adequate and specialized instruments are needed to sense vibrations, sound, and light.

Vibrations and seismic waves are special types of elastic waves. These waves are governed by the equations of elastodynamics. Acoustics deals with various types of sound waves. As you can guess, elastodynamics and acoustics are closely related. Light is a special type of electromagnetic wave and its propagation is described by Maxwell's equations of electrodynamics. Electromagnetic waves are quite different from elastic and acoustic waves in that they can travel through vacuum. But, remarkably, all three types of waves can be described by a similar set of equations and hence these disparate phenomena can be studied simultaneously.

The governing equations of linear elastodynamics, acoustics, and electrodynamics are the starting point of our study of waves in heterogeneous media and composites.* We will explore these governing equations and look at ways in which these equations

*Detailed derivations of the equations of linear elasticity can be found in Atkin and Fox (1980) and Slaughter (2002). Expositions on elastodynamics can be found in Achenbach (1973) and Harris (2001). An accessible introduction to the applications of elastodynamics in the study of vibrations can be found in Volterra and Zachmanoglou (1965). For seismic waves and problems related to seismology a good starting point is Aki and Richards (1980). Excellent introductions to acoustics can be found in Reynolds (1981) and Skudrzyk (1971), and numerous solved problems are given in Morse and Ingard (1986). The governing equations of electrodynamics are discussed in detail in Feynman et al. (1964) and Jackson (1999).

can be simplified. We will start with direct notation and then explore specialized coordinate systems. Fourier transforms will be used to convert the time-dependent equations into frequency-dependent form. We will discuss dissipation and the constraints placed by causality on the solutions of the governing equations. We will also identify the similarities between the equations of elastodynamics, acoustics, and electrodynamics. These similarities will allow us to use some of the same techniques to solve problems in these seemingly disparate fields.

1.1 A note about notation

The notation used in this book is based on Ogden (1997). Scalar quantities are denoted by italic letters (a) and Greek letters (ϕ). Vector quantities are represented with bold font lower-case letters ($\mathbf{v}$) and in some cases as upper-case bold letters ($\mathbf{E}$). Second-order tensor quantities are written with bold italic ($\boldsymbol{E}$) or with bold Greek ($\boldsymbol{\sigma}$). Third-order tensors are written with calligraphic fonts ($\mathcal{S}$) and fourth-order tensors are written in bold sans-serif fonts ($\mathsf{\mathbf{C}}$).

The scalar product of two vectors is indicated by $\mathbf{u}\cdot\mathbf{v}$, the vector product by $\mathbf{u}\times\mathbf{v}$, and the tensor product by $\mathbf{u}\otimes\mathbf{v}$. The tensor product is a second-order tensor. The action of a second-order tensor on a vector is represented by $\boldsymbol{S}\cdot\mathbf{v}$. The definition of the tensor product is

$$(\mathbf{u}\otimes\mathbf{v})\cdot\mathbf{w} = (\mathbf{v}\cdot\mathbf{w})\mathbf{u}\ .$$

The transpose of a second-order tensor is denoted by $\boldsymbol{S}^T$ and is defined via the relation

$$\mathbf{v}\cdot(\boldsymbol{S}^T\cdot\mathbf{u}) = \mathbf{u}\cdot(\boldsymbol{S}\cdot\mathbf{v})\ .$$

The trace of a second-order tensor is an invariant and can be defined with respect to an orthonormal basis $(\mathbf{e}_1, \mathbf{e}_2, \mathbf{e}_3)$ as

$$\mathrm{tr}\boldsymbol{S} = S_{ii} = \mathbf{e}_i\cdot(\boldsymbol{S}\cdot\mathbf{e}_i)$$

where summation over repeated indices is implied. The determinant of a second-order tensor is likewise an invariant and can be expressed as

$$\det(\boldsymbol{S}) = e_{ijk}\, S_{i1}\, S_{j2}\, S_{k3}$$

where e_{ijk} is the permutation tensor which is defined as

$$e_{ijk} = \begin{cases} 1 & \text{for even permutations, i.e., 123, 231, 312} \\ -1 & \text{for odd permutations, i.e., 132, 321, 213} \\ 0 & \text{otherwise.} \end{cases}$$

The inner product of two second-order tensors is a second-order tensor given by $\boldsymbol{S}\cdot\boldsymbol{T}$, in index notation $S_{ij}T_{jk}$. If $\det\boldsymbol{S} \neq 0$ then there exists a unique tensor called the inverse of $\boldsymbol{S}$ and denoted by $\boldsymbol{S}^{-1}$ such that

$$\boldsymbol{S}\cdot\boldsymbol{S}^{-1} = \boldsymbol{S}^{-1}\cdot\boldsymbol{S} = \boldsymbol{1}$$

where $\boldsymbol{1}$ is the identity tensor that takes a second-order tensor to itself. The contraction of two indices between two tensors is represented as $\boldsymbol{S} : \boldsymbol{T}$ and $\mathbf{C} : \boldsymbol{S}$, in index notation, $S_{ij}T_{ij}$ and $C_{ijk\ell}S_{k\ell}$. The magnitude of a vector is $\|\mathbf{v}\| = \sqrt{\mathbf{v}\cdot\mathbf{v}}$ and the magnitude of a second-order tensor is $\|\boldsymbol{S}\| = \sqrt{\boldsymbol{S}^T : \boldsymbol{S}}$. We use the notations $\mathbf{v}\cdot\boldsymbol{S}$ and $\boldsymbol{S}^T\cdot\mathbf{v}$ interchangeably. For higher-order tensors, the quantity $\mathbf{v}\cdot\mathbf{C}\cdot\mathbf{u}$ is equivalent in index notation to $C_{ijk\ell}v_i u_\ell$.

The notation used for differential operators is similar to that used in vector calculus. However, the definitions are slightly different. The gradient of a tensor field of any order is defined by its action on a vector $\mathbf{a}$ as

$$[\boldsymbol{\nabla}\boldsymbol{S}(\mathbf{x})]\cdot\mathbf{a} = \left.\frac{d}{d\alpha}\boldsymbol{S}(\mathbf{x}+\alpha\mathbf{a})\right|_{\alpha=0} = (\mathbf{a}\cdot\boldsymbol{\nabla})\boldsymbol{S}(\mathbf{x})\,.$$

The divergence of a vector field is defined as $\boldsymbol{\nabla}\cdot\mathbf{v} = \text{tr}(\boldsymbol{\nabla}\mathbf{v})$ and the divergence of a second-order tensor field is defined as

$$[\boldsymbol{\nabla}\cdot\boldsymbol{S}(\mathbf{x})]\cdot\mathbf{a} = \boldsymbol{\nabla}\cdot[\boldsymbol{S}(\mathbf{x})\cdot\mathbf{a}]\,.$$

The curl of a vector field is defined by

$$[\boldsymbol{\nabla}\times\mathbf{v}(\mathbf{x})]\cdot\mathbf{a} = \boldsymbol{\nabla}\cdot[\mathbf{v}(\mathbf{x})\times\mathbf{a}]\,.$$

The Laplacian of a scalar field is denoted by ∇^2 and is defined as

$$\nabla^2\phi = \boldsymbol{\nabla}\cdot(\boldsymbol{\nabla}\phi) = (\boldsymbol{\nabla}\cdot\boldsymbol{\nabla})\phi\,.$$

The Laplacian of a vector field is defined as

$$\nabla^2\mathbf{v} = \boldsymbol{\nabla}\cdot(\boldsymbol{\nabla}\mathbf{v})^T = (\boldsymbol{\nabla}\cdot\boldsymbol{\nabla})\mathbf{v}\,.$$

We use the symbols $a := b$ and $b =: a$ to indicate that the quantity a is being defined to be equal to b.

1.2 Elastodynamics

Consider the body Ω_0 with boundary Γ_0 shown in Figure 1.1. Points in the body can be located using the position vector $\mathbf{x}$. Let Ω be a subpart of the body with boundary Γ. Let the unit vector $\mathbf{n}$ be the outward normal to the surface Γ. The region Ω can be in the interior of the body or share a part of the surface of the body. Let there be

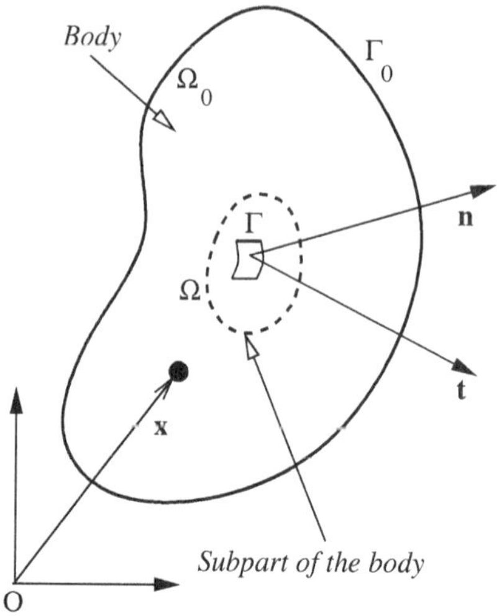

FIGURE 1.1

Illustration of the concept of stress in part of a body. The stress is defined via the surface traction $\mathbf{t}$ on an internal surface Γ with outward unit normal $\mathbf{n}$.

a force $\mathbf{t}(\mathbf{x},\mathbf{n})$ per unit area on the surface Γ. The balance of forces requires that $\mathbf{t}$ is the force exerted on Γ by the material outside Γ or by surface tractions on Γ_0. From the balance of forces on a small tetrahedron,[†] we can show that $\mathbf{t}(\mathbf{x},\mathbf{n})$ is linear in $\mathbf{n}$ as the size of the tetrahedron tends to zero while keeping $\mathbf{t}$ finite. Therefore,

$$\mathbf{t} = \boldsymbol{\sigma}^T \cdot \mathbf{n} = \mathbf{n} \cdot \boldsymbol{\sigma} \tag{1.1}$$

where $\boldsymbol{\sigma}(\mathbf{x})$ is a second-order tensor called the *Cauchy stress tensor*. The vector $\mathbf{t}$ is also called the *traction vector*. Since the tetrahedron cannot rotate at infinite velocity as its size goes to zero,[‡] we can show that the stress tensor is symmetric, i.e.,

$$\boldsymbol{\sigma} = \boldsymbol{\sigma}^T \,. \tag{1.2}$$

For dynamic problems the Cauchy stress is also a function of time, t. In that case the equation of motion of the body is given by the conservation of linear momentum and can be written as

$$\boxed{\boldsymbol{\nabla} \cdot \boldsymbol{\sigma} + \rho(\mathbf{x},t)\,\mathbf{b}(\mathbf{x},t) = \rho(\mathbf{x},t)\,\frac{\partial \mathbf{v}}{\partial t}\,; \quad \mathbf{v} := \frac{\partial \mathbf{u}}{\partial t}} \tag{1.3}$$

where ρ is the mass density, $\boldsymbol{\nabla} \cdot \boldsymbol{\sigma}$ is the internal force per unit volume, $\mathbf{b}$ is the body force per unit volume, $\mathbf{u}(\mathbf{x},t)$ is the displacement field in the body, and $\partial \mathbf{v}/\partial t$

[†]See Slaughter (2002, p. 175) for a detailed explanation.

[‡]This is a way of conceptualizing the principle of conservation of angular momentum.

is the spatial acceleration. It is implicit in the above equation that mass is conserved because the equation for the conservation of mass has been used to simplify the expression for the rate of change of linear momentum on the right-hand side.

Let us assume that the stress depends only on the strain (and not on strain gradients or strain rates), where the strain tensor is defined as[§]

$$\boldsymbol{\varepsilon} = \tfrac{1}{2}\left[\boldsymbol{\nabla}\mathbf{u} + (\boldsymbol{\nabla}\mathbf{u})^T\right] . \tag{1.4}$$

We need a constitutive relation between the stress and the strain to complete the system of equations of elastodynamics. Assume that $\boldsymbol{\sigma}$ depends linearly on $\boldsymbol{\varepsilon}$ so that we can use the principle of superposition to get the constitutive relation[¶]

$$\boldsymbol{\sigma}(\mathbf{x},t) = \int_{\Omega} \mathrm{d}\mathbf{y} \left[\int_{-\infty}^{\infty} \mathbf{K}_{\varepsilon}(\mathbf{x},\mathbf{y},t-t') : \boldsymbol{\varepsilon}(\mathbf{x},t')\, \mathrm{d}t'\right] \tag{1.5}$$

where $\mathbf{K}_{\varepsilon}$ is a fourth-order kernel function called the stiffness tensor. This assumption ignores preexisting internal stresses such as those found in pre-stressed concrete. The kernel function may be singular, a delta function for example, and in such situations the integrals are interpreted in the sense of measure theory. If the material can be approximated as being local in space, i.e., the stresses at a point are determined solely by the strains at that point, then

$$\boxed{\boldsymbol{\sigma}(\mathbf{x},t) = \int_{-\infty}^{\infty} \mathbf{K}_{\varepsilon}(\mathbf{x},t-t') : \boldsymbol{\varepsilon}(\mathbf{x},t')\, \mathrm{d}t' .} \tag{1.6}$$

Causality implies that stresses at time t can only depend on strains from previous times, i.e., if $t \geq t'$. Therefore

$$\mathbf{K}_{\varepsilon}(\mathbf{x},\tau) = \mathbf{0} \qquad \text{if} \qquad \tau = t - t' < 0 .$$

In addition, if we assume that the material is local in time (i.e., the stresses at a particular point in time do not depend on strains at earlier times), then

$$\boldsymbol{\sigma}(\mathbf{x},t) = \mathbf{K}_{\varepsilon}(\mathbf{x},t) : \boldsymbol{\varepsilon}(\mathbf{x},t) =: \mathbf{C}(\mathbf{x},t) : \boldsymbol{\varepsilon}(\mathbf{x},t) . \tag{1.7}$$

In that case, the governing equation (1.3) becomes

$$\boxed{\boldsymbol{\nabla}\cdot[\mathbf{C}(\mathbf{x},t) : \boldsymbol{\nabla}\mathbf{u}] + \rho(\mathbf{x},t)\, \mathbf{b}(\mathbf{x},t) = \rho(\mathbf{x},t)\, \frac{\partial^2 \mathbf{u}}{\partial t^2} .} \tag{1.8}$$

[§]Note that the gradient of the displacement field is used to define the strain because rigid body motions should not affect $\boldsymbol{\sigma}$ and a rigid body rotation gives zero strains (for small displacements).

[¶]The principle of linear superposition implies that the constitutive relation should have the form

$$\boldsymbol{\sigma}(\mathbf{x},t) = \int_{\Omega} \mathrm{d}\mathbf{y} \left[\int_{-\infty}^{t} \mathbf{K}_{\varepsilon}(\mathbf{x},\mathbf{y},t-t') : \boldsymbol{\varepsilon}(\mathbf{x},t')\, \mathrm{d}t'\right] .$$

We have extended the limit to $+\infty$ in order to make it easier to Fourier transform the equation. Causality requires that $\mathbf{K}_{\varepsilon} = \mathbf{0}$ if $t' > t$. Hence the extension of the limit is valid.

For an isotropic material, the stiffness tensor has the form

$$\mathsf{C}(\mathbf{x},t) = \lambda(\mathbf{x},t)\,\boldsymbol{1}\otimes\boldsymbol{1} + 2\mu(\mathbf{x},t)\,\mathsf{I}$$

where λ is the Lamé elastic modulus, μ is the shear modulus, $\boldsymbol{1}$ is the second-order identity tensor, and I is the symmetric part of the fourth-order identity tensor. If the material is also homogeneous, then equation (1.8) can be expressed as

$$(\lambda+\mu)\,\boldsymbol{\nabla}(\boldsymbol{\nabla}\cdot\mathbf{u}) + \mu\boldsymbol{\nabla}\cdot(\boldsymbol{\nabla}\mathbf{u}^T) + \rho\,\mathbf{b} = \rho\frac{\partial^2\mathbf{u}}{\partial t^2}\,. \tag{1.9}$$

Using the identity

$$\boldsymbol{\nabla}\times(\boldsymbol{\nabla}\times\mathbf{u}) = \boldsymbol{\nabla}(\boldsymbol{\nabla}\cdot\mathbf{u}) - \boldsymbol{\nabla}\cdot(\boldsymbol{\nabla}\mathbf{u}^T)$$

we can write (1.9) in the form

$$\boxed{(\lambda+2\mu)\,\boldsymbol{\nabla}(\boldsymbol{\nabla}\cdot\mathbf{u}) - \mu\boldsymbol{\nabla}\times(\boldsymbol{\nabla}\times\mathbf{u}) + \rho\,\mathbf{b} = \rho\,\frac{\partial^2\mathbf{u}}{\partial t^2}\,.} \tag{1.10}$$

Let us define

$$c_p^2 := \frac{\lambda+2\mu}{\rho}\,;\quad c_s^2 := \frac{\mu}{\rho}\,. \tag{1.11}$$

Then equation (1.10) can be written as

$$c_p^2\,\boldsymbol{\nabla}(\boldsymbol{\nabla}\cdot\mathbf{u}) - c_s^2\,\boldsymbol{\nabla}\times(\boldsymbol{\nabla}\times\mathbf{u}) + \mathbf{b} = \frac{\partial^2\mathbf{u}}{\partial t^2}\,. \tag{1.12}$$

Note that the volumetric strain in given by the divergence of $\mathbf{u}$ and the infinitesimal rotation vector (which is the axial vector of the skew-symmetric part of the displacement gradient tensor $\boldsymbol{\nabla}\mathbf{u}$) is given by the curl of $\mathbf{u}$, i.e.,

$$e := \operatorname{tr}(\boldsymbol{\varepsilon}) = \boldsymbol{\nabla}\cdot\mathbf{u}\,;\quad \boldsymbol{\theta} := \tfrac{1}{2}\,\boldsymbol{\nabla}\times\mathbf{u}\,. \tag{1.13}$$

1.2.1 Elastodynamic potentials

It is common to invoke Lamé's theorem and express the above equation in terms of a scalar potential ϕ and a vector potential $\boldsymbol{\psi}$ for $\mathbf{u}$ such that[‖]

$$\mathbf{u} = \boldsymbol{\nabla}\phi + \boldsymbol{\nabla}\times\boldsymbol{\psi}\,;\quad \boldsymbol{\nabla}\cdot\boldsymbol{\psi} = 0\,. \tag{1.14}$$

Then $e = \nabla^2\phi$ and $\boldsymbol{\theta} = -\frac{1}{2}\boldsymbol{\nabla}\cdot(\boldsymbol{\nabla}\boldsymbol{\psi})$ and we can show that

$$\boxed{\frac{\partial^2\phi}{\partial t^2} = \frac{\Phi}{\rho} + c_p^2\,\nabla^2\phi\,;\quad \frac{\partial^2\boldsymbol{\psi}}{\partial t^2} = \frac{\boldsymbol{\Psi}}{\rho} + c_s^2\,\boldsymbol{\nabla}\cdot(\boldsymbol{\nabla}\boldsymbol{\psi})^T} \tag{1.15}$$

where the Helmholtz potential representation of the body force is $\mathbf{b} = \boldsymbol{\nabla}\Phi + \boldsymbol{\nabla}\times\boldsymbol{\Psi}$. The first of equations (1.15) is called the equation for the "P-wave" while the second is the equation for the "S-wave." The quantities c_p and c_s are the wave speeds of the P-wave and the S-wave, respectively.

[‖] See Aki and Richards (1980) for more details.

1.2.2 Cartesian coordinates

So far we have represented the equations of elastodynamics in a form that is independent of the coordinate system. It is often convenient to express these equations in terms of components with respect to a particular coordinate system. In a rectangular Cartesian coordinate system a vector can be expressed as $\mathbf{v} = v_i \mathbf{e}_i$ where v_i are the components of the vector in terms of the basis vector set $\{\mathbf{e}_i\}$. Summation over repeated indices is implied. A second-order tensor such as the stress can be expressed as $\boldsymbol{S} = S_{ij} \mathbf{e}_i \otimes \mathbf{e}_j$ where S_{ij} are the components of the tensor. Then the traction-stress relation can be written as

$$t_i \mathbf{e}_i = (\sigma^T_{km} \mathbf{e}_k \otimes \mathbf{e}_m) \cdot (n_j \mathbf{e}_j).$$

From the definition of the dyadic product, $(\mathbf{a} \otimes \mathbf{b}) \cdot \mathbf{c} = (\mathbf{b} \cdot \mathbf{c})\mathbf{a}$, we have $(\mathbf{e}_k \otimes \mathbf{e}_m) \cdot \mathbf{e}_j = \delta_{mj} \mathbf{e}_k$ where δ_{mj} is the *Kronecker delta* which has the value 1 when $m = j$ and 0 otherwise. Therefore,

$$t_i \mathbf{e}_i = n_j \sigma^T_{km} \delta_{mj} \mathbf{e}_k = n_j \sigma^T_{kj} \mathbf{e}_k = n_j \sigma^T_{ij} \mathbf{e}_i = n_j \sigma_{ji} \mathbf{e}_i \quad \Leftrightarrow \quad t_i = n_j \sigma_{ji} \; .$$

Note that the *dummy index k* can be replaced has been replaced with i in the above equation.

We can proceed in a similar manner with the conservation of linear momentum, equation (1.3). The divergence of a second-order tensor is defined as

$$\boldsymbol{\nabla} \cdot \boldsymbol{S} = \frac{\partial S_{ij}}{\partial x_i} \, \mathbf{e}_j = S_{ij,i} \, \mathbf{e}_j \; .$$

With this definition, equation (1.3) can be expressed as

$$S_{ij,i} + \rho \, b_j = \rho \, \frac{\partial v_j}{\partial t} = \rho \dot{v}_j \; .$$

The gradient of a vector is defined as

$$\boldsymbol{\nabla} \mathbf{v} = \frac{\partial v_i}{\partial x_j} \, \mathbf{e}_i \otimes \mathbf{e}_j = v_{i,j} \, \mathbf{e}_i \otimes \mathbf{e}_j \; .$$

This definition can be use to obtain the component form of the strain-displacement relation which is

$$\varepsilon_{ij} = \tfrac{1}{2} \left(u_{i,j} + u_{j,i} \right) \; .$$

The stress-strain relation for an elastic material in terms of components can be expressed in components as

$$\sigma_{ij} = C_{ijk\ell} \, \varepsilon_{k\ell}$$

where the symmetries of the stiffness tensor are

$$C_{ijk\ell} = C_{jik\ell} = C_{ij\ell k} = C_{k\ell ij} \; .$$

Therefore the governing equation for an inhomogeneous but local elastic body can be written in the form

$$\left[C_{ijk\ell}\varepsilon_{k\ell}\right]_{,i} + \rho\, b_j = \rho\, \ddot{u}_j \,.$$

If we recognize that we can express the gradient and divergence terms in the momentum balance for an isotropic and homogeneous elastic solid as

$$\boldsymbol{\nabla}\cdot\mathbf{u} = u_{j,j}\,;\;\; \boldsymbol{\nabla}(\boldsymbol{\nabla}\cdot\mathbf{u}) = u_{j,ji}\mathbf{e}_i\,;\;\; \nabla^2\mathbf{u} := \boldsymbol{\nabla}\cdot(\boldsymbol{\nabla}\mathbf{u})^T = u_{i,jj}\,\mathbf{e}_i$$

we can write the balance of linear momentum (1.9) in component form as

$$(\lambda+\mu)u_{j,ji} + \mu u_{i,jj} + \rho b_i = \rho\ddot{u}_i \,.$$

For the component form of equation (1.10), we use the definition of the curl of a vector

$$\boldsymbol{\nabla}\times\mathbf{v} = e_{mnq}\, v_{n,m}\,\mathbf{e}_q \quad\Longrightarrow\quad \boldsymbol{\nabla}\times(\boldsymbol{\nabla}\times\mathbf{v}) = e_{pqi}e_{mnq}\, v_{n,mp}\,\mathbf{e}_i$$

where e_{ijk} is the permutation tensor which equals 1 for even permutations of ijk (123, 231, 312), has a value of –1 for odd permutations (132, 321, 213), and equals zero otherwise. Then equation (1.10) can be expressed as

$$(\lambda+2\mu)u_{j,ji} - e_{pqi}e_{mnq}\,\mu\, u_{n,mp} + \rho b_i = \rho\ddot{u}_i \,.$$

The component forms of the wave equations for the elastodynamic potentials ϕ and $\boldsymbol{\psi}$ can be found in a similar manner. These are

$$\ddot{\phi} = \frac{\Phi}{\rho} + c_p^2\,\phi_{,ii}\,;\;\; \ddot{\psi}_i = \frac{\Psi_i}{\rho} + c_s^2\,\psi_{i,jj}\,.$$

1.2.3 Curvilinear coordinates

Though rectangular Cartesian coordinates are the easiest to work with, it is often convenient to express the equations of elastodynamics in general curvilinear coordinates. In that case, if an invertible mapping exists between the Cartesian components ($\{x_i\}$) and the curvilinear coordinates $\{\theta_i\}$, then the natural basis vectors $\mathbf{g}_i$ and the metric tensor components g_{ij} are given by

$$\mathbf{g}_i = \frac{\partial x_j}{\partial\theta^i}\mathbf{e}_j\,;\;\; g_{ij} = \mathbf{g}_i\cdot\mathbf{g}_j\,;\;\; \mathbf{g}^i\cdot\mathbf{g}_j = \delta^i_j\,;\;\; g^{ij} = \mathbf{g}^i\cdot\mathbf{g}^j\,.$$

A vector $\mathbf{v}$ can be expressed in terms of components with respect to the natural basis as $\mathbf{v} = v^k\mathbf{g}_k = v_k\mathbf{g}^k$. A second-order tensor $\boldsymbol{S}$ may be expressed in the form $\boldsymbol{S} = S^{ij}\,\mathbf{g}_i\otimes\mathbf{g}_j = S^{i}_{\cdot j}\,\mathbf{g}_i\otimes\mathbf{g}^j = S_i^{\cdot j}\,\mathbf{g}^i\otimes\mathbf{g}_j = S_{ij}\,\mathbf{g}^i\otimes\mathbf{g}^j$. Using these definitions we can show that the relation between the Cauchy stress and the traction vector can be expressed as**

$$t^i = \sigma^{mi}n^p g_{ip}\,.$$

**The most appropriate representation of the stress tensor depends on the operational use to which it is to be put. The traction vector is often thought of as a 1-form in which case the right form of the stress tensor will be one with mixed components. See Marsden and Hughes (1993) for further details.

The divergence of a second-order tensor in curvilinear coordinates is given by

$$\nabla \cdot \boldsymbol{S} = \left[\frac{\partial S^{ij}}{\partial \theta^i} + \Gamma^i_{i\ell}\, S^{\ell j} + \Gamma^j_{i\ell}\, S^{i\ell}\right] \mathbf{g}_j = \left[\frac{\partial S^i{}_j}{\partial \theta^i} + \Gamma^i_{i\ell}\, S^\ell{}_j - \Gamma^\ell_{ij}\, S^i{}_\ell\right] \mathbf{g}^j$$

$$= \left[\frac{\partial S_{ij}}{\partial \theta^k} - \Gamma^\ell_{ki}\, S_{\ell j} - \Gamma^\ell_{kj}\, S_{i\ell}\right] g^{ik}\, \mathbf{g}^j = \left[\frac{\partial S_i{}^j}{\partial \theta^k} - \Gamma^\ell_{ik}\, S_\ell{}^j + \Gamma^j_{k\ell}\, S_i{}^\ell\right] g^{ik}\, \mathbf{g}_j$$

where the *Christoffel symbols* of the second kind are given by

$$\Gamma^k_{ij} = \frac{g^{km}}{2}\left[\frac{\partial g_{mi}}{\partial \theta^j} + \frac{\partial g_{mj}}{\partial \theta^i} - \frac{\partial g_{ij}}{\partial \theta^m}\right] .$$

Therefore, for small deformations, equation (1.3) can be expressed in curvilinear coordinates as

$$\frac{\partial S^{ij}}{\partial \theta^i} + \Gamma^i_{i\ell}\, S_{\ell j} + \Gamma^j_{i\ell}\, S_{i\ell} + \rho b^j = \rho \ddot{u}^j .$$

The gradient and the divergence of a vector are expressed in curvilinear coordinates as

$$\nabla \mathbf{v} = \left[\frac{\partial v^i}{\partial \theta^k} + \Gamma^i_{\ell k}\, v^\ell\right] \mathbf{g}_i \otimes \mathbf{g}^k = \left[\frac{\partial v_i}{\partial \theta^k} - \Gamma^\ell_{ki}\, v_\ell\right] \mathbf{g}^i \otimes \mathbf{g}^k$$

$$\nabla \cdot \mathbf{v} = \frac{\partial v^i}{\partial \theta^i} + \Gamma^i_{\ell i}\, v^\ell = \left[\frac{\partial v_i}{\partial \theta^j} - \Gamma^\ell_{ji}\, v_\ell\right] g^{ij} = \frac{1}{\sqrt{g}} \frac{\partial}{\partial \theta^i}\left(v^i \sqrt{g}\right)$$

where $g := \det([g_{ij}])$. These can be used, in conjunction with the definition of the divergence of a second-order tensor, to find an expression for the Laplacian of a vector, $\nabla^2 \mathbf{u} := \nabla \cdot (\nabla \mathbf{u}^T)$ and the form of equation (1.9) in curvilinear coordinates. However, that form is rarely used because of its complexity.

It is of value to express the first of equations (1.15) in curvilinear coordinates. The Laplacian (also called the Laplace-Beltrami operator) of a scalar field (ϕ) in curvilinear coordinates is given by

$$\nabla^2 \phi = \frac{1}{\sqrt{g}} \frac{\partial}{\partial \theta^i}\left[\sqrt{g}\, g^{ji}\, \frac{\partial \phi}{\partial \theta^j}\right] .$$

Therefore the wave equation for the P-wave in terms of the scalar potential ϕ can be written as

$$\ddot{\phi} = \frac{\Phi}{\rho} + \frac{c_p^2}{\sqrt{g}} \frac{\partial}{\partial \theta^i}\left[\sqrt{g}\, g^{ji}\, \frac{\partial \phi}{\partial \theta^j}\right] . \tag{1.16}$$

1.2.4 Anisotropic and frequency-dependent mass density

In Chapter 5 we will see that the macroscopic dynamic mass density of a composite containing masses and springs can be both anisotropic and frequency dependent. If the mass density is anisotropic we can represent it as a second-order tensor ($\boldsymbol{\rho}$) and equation (1.3) can then be expressed in the form

$$\boldsymbol{\nabla}\cdot\boldsymbol{\sigma}+\boldsymbol{\rho}(\mathbf{x})\cdot\mathbf{b}(\mathbf{x})=\boldsymbol{\rho}(\mathbf{x})\cdot\frac{\partial^2\mathbf{u}}{\partial t^2}\,. \tag{1.17}$$

Additionally, if the mass density is a function of the frequency, we can express equation (1.17) as

$$\boxed{\boldsymbol{\nabla}\cdot\boldsymbol{\sigma}+\int_{-\infty}^{\infty}\boldsymbol{\rho}(\mathbf{x},t-t')\cdot\mathbf{b}(\mathbf{x})\,\mathrm{d}t'=\int_{-\infty}^{\infty}\boldsymbol{\rho}(\mathbf{x},t-t')\cdot\frac{\partial^2\mathbf{u}}{\partial t'^2}\,\mathrm{d}t'\,.} \tag{1.18}$$

Causality requires that the dynamic mass density $\boldsymbol{\rho}$ at time t should not depend on the dynamic mass density at times $t'\geq t$. Hence, $\boldsymbol{\rho}=\mathbf{0}$ if $\tau=t-t'<0$. It is easier to explore the above equations in the frequency domain and we shall see why in the next section.

1.2.5 Elastodynamics at a fixed frequency

In the study of waves in composites it is often convenient to examine the behavior of elastic materials and fixed frequencies. This is because we can think of the solutions at fixed frequencies as the individual Fourier components of the time-dependent solution. After the relevant Fourier components have been computed, an inverse Fourier transform can be performed to obtain the solution in the time domain. This is allowed because of the linearity of the problem and the applicability of the principle of linear superposition.

Let us define the Fourier transform of a function $f(t)$ as

$$\widehat{f}(\omega)=\mathcal{F}[f(t)]=\int_{-\infty}^{\infty}f(t)\,e^{i\omega t}\,\mathrm{d}t\,.$$

Then the Fourier transform of a convolution of two functions $f(t)$ and $g(t)$ is given by

$$\mathcal{F}[f\star g]=\int_{-\infty}^{\infty}\left[\int_{-\infty}^{\infty}f(t-t')\,g(t')\,\mathrm{d}t'\right]\,e^{i\omega t}\,\mathrm{d}t\,.$$

Making the substitution $\tau=t-t'$ leads to

$$\begin{aligned}\mathcal{F}[f\star g]&=\int_{-\infty}^{\infty}\left[\int_{-\infty}^{\infty}f(\tau)\,g(t')\,\mathrm{d}t'\right]\,e^{i\omega(\tau+t')}\,\mathrm{d}\tau\\&=\left[\int_{-\infty}^{\infty}f(\tau)\,e^{i\omega\tau}\,\mathrm{d}\tau\right]\left[\int_{-\infty}^{\infty}g(t')\,e^{i\omega t'}\,\mathrm{d}t'\right]=\left[\int_{-\infty}^{\infty}f(\tau)\,e^{-i\omega\tau}\,\mathrm{d}\tau\right]\,\widehat{g}(\omega)\,.\end{aligned}$$

Equation (1.6) expressed in rectangular Cartesian components is

$$\sigma_{ij}(\mathbf{x},t)=\int_{-\infty}^{\infty}K_{ijk\ell}(\mathbf{x},t-t')\varepsilon_{k\ell}(\mathbf{x},t')\,\mathrm{d}t'\,.$$

Since the right-hand side is a sum of integrals we can transform each component and sum the results to get the Fourier transformed equation which is

$$\widehat{\sigma}_{ij}(\mathbf{x},\omega) = \left[\int_{-\infty}^{\infty} K_{ijk\ell}(\mathbf{x},\tau)\, e^{i\omega\tau}\, \mathrm{d}\tau\right] \widehat{\varepsilon}_{k\ell}(\mathbf{x},\omega) = C_{ijk\ell}(\mathbf{x},\omega)\, \widehat{\varepsilon}_{k\ell}(\mathbf{x},\omega) .$$

In direct tensor notation,

$$\boxed{\widehat{\boldsymbol{\sigma}}(\mathbf{x},\omega) = \mathbf{C}(\mathbf{x},\omega) : \widehat{\boldsymbol{\varepsilon}}(\mathbf{x},\omega)} \tag{1.19}$$

where

$$\mathbf{C}(\mathbf{x},\omega) = \int_{-\infty}^{\infty} \mathbf{K}_{\varepsilon}(\mathbf{x},\tau)\, e^{i\omega\tau}\, \mathrm{d}\tau \quad \text{and} \quad \tau := t - t' . \tag{1.20}$$

Note the similarity between this equation and equation (1.7). Causality implies that stresses at time t cannot depend on strains developed at times $t' > t$. Hence $\mathbf{K}_{\varepsilon}(\mathbf{x},\tau) = \mathbf{0}$ if $\tau = t - t < 0$. Now consider a complex frequency $\omega = \omega_r + i\,\omega_i$ where ω_r is the real part and ω_i is the imaginary part of the frequency. Then

$$\mathbf{C}(\mathbf{x},\omega) = \int_{-\infty}^{\infty} \mathbf{K}_{\varepsilon}(\mathbf{x},\tau)\, e^{i\omega_r\tau} e^{-\omega_i\tau}\, \mathrm{d}\tau = \int_{-\infty}^{\infty} \mathbf{K}_{\varepsilon}(\mathbf{x},\tau)\, [\cos(\omega_r\tau) + i\sin(\omega_r\tau)]\, e^{-\omega_i\tau}\, \mathrm{d}\tau .$$

Nonzero values of the integral are obtained only when $\tau \geq 0$. When $\omega_i < 0$ the product $-\omega_i\tau > 0$ and the integral diverges. This implies that the integral converges only if $\omega_i = \mathrm{Im}(\omega) \geq 0$, i.e., $\mathbf{C}(\mathbf{x},\omega)$ is analytic only in the upper-half of the complex ω plane.[††]

Let us now consider the momentum balance equation (1.18) for a material with anisotropic and frequency-dependent mass density in the absence of a body force. In terms of components with respect to a Cartesian basis the balance of momentum reads

$$\sigma_{ji,j} = \int_{-\infty}^{\infty} \rho_{ik}(\mathbf{x}, t - t')\, \frac{\partial^2 u_k}{\partial t'^2}\, \mathrm{d}t' .$$

Fourier transformation of both sides of the above equation leads to

$$\int_{-\infty}^{\infty} \sigma_{ji,j} e^{i\omega t} \mathrm{d}t = \int_{-\infty}^{\infty} \left[\int_{-\infty}^{\infty} \rho_{ik}(\mathbf{x}, t - t')\, \frac{\partial^2 u_k}{\partial t'^2}\, \mathrm{d}t'\right] e^{i\omega t} \mathrm{d}t .$$

As we have seen previously, we can use a change of variable, $\tau = t - t'$, to get

$$\int_{-\infty}^{\infty} \sigma_{ji,j} e^{i\omega t} \mathrm{d}t = \left[\int_{-\infty}^{\infty} \rho_{ik}(\mathbf{x},\tau)\, e^{i\omega\tau} \mathrm{d}\tau\right] \left[\int_{-\infty}^{\infty} \frac{\partial^2 u_k}{\partial t'^2}\, e^{i\omega t'}\, \mathrm{d}t'\right] .$$

[††]The analytic nature of $\mathbf{C}$ arises from the observation that $\exp(i\omega\tau)$ is an analytic function of ω. Since a sum of analytic functions is analytic and a convergent integral of analytic functions is also analytic, the function $\mathbf{C}(\mathbf{x},\omega)$ is an analytic function of ω in the upper half of the ω-plane, $\mathrm{Im}(\omega) > 0$.

We can use integration by parts to find the Fourier transform of the derivative on the right-hand side, i.e.,

$$\begin{aligned}\int_{-\infty}^{\infty} \frac{\partial^2 u_k}{\partial t'^2}\, e^{i\omega t'}\, \mathrm{d}t' &= \frac{\partial u_k}{\partial t'}\, e^{i\omega t'}\Big|_{-\infty}^{\infty} - \int_{-\infty}^{\infty} i\omega \frac{\partial u_k}{\partial t'}\, e^{i\omega t'}\, \mathrm{d}t' \\ &= \frac{\partial u_k}{\partial t'}\, e^{i\omega t'}\Big|_{-\infty}^{\infty} - i\omega u_k\, e^{i\omega t'}\Big|_{-\infty}^{\infty} - \int_{-\infty}^{\infty} \omega^2 u_k\, e^{i\omega t'}\, \mathrm{d}t'\end{aligned}$$

If $\mathrm{Im}(\omega) \geq 0$ the boundary terms vanish and we have

$$\int_{-\infty}^{\infty} \sigma_{ji,j} e^{i\omega t}\, dt - -\left[\int_{-\infty}^{\infty} \rho_{lk}(\mathbf{x},\tau)\, e^{i\omega\tau} \mathrm{d}\tau\right]\left[\int_{-\infty}^{\infty} \omega^2 u_k\, e^{i\omega t'}\, \mathrm{d}t'\right] .$$

Therefore the Fourier transformed form of equation (1.18) is

$$\boxed{\nabla \cdot \widehat{\boldsymbol{\sigma}}(\mathbf{x},\omega) = -\omega^2\, \widehat{\boldsymbol{\rho}}(\mathbf{x},\omega) \cdot \widehat{\mathbf{u}}(\mathbf{x},\omega)\, .} \tag{1.21}$$

Substituting equation (1.19) into equation (1.21) we get

$$\nabla \cdot (\mathbf{C} : \widehat{\boldsymbol{\varepsilon}}) + \omega^2\, \widehat{\boldsymbol{\rho}} \cdot \widehat{\mathbf{u}} = 0\, . \tag{1.22}$$

Also, taking the Fourier transform of equation (1.4), we have

$$\widehat{\boldsymbol{\varepsilon}} = \tfrac{1}{2}\left[\nabla \widehat{\mathbf{u}} + (\nabla \widehat{\mathbf{u}})^T\right] . \tag{1.23}$$

Since $\widehat{\boldsymbol{\sigma}}$ and $\widehat{\boldsymbol{\varepsilon}}$ are symmetric, we must have $C_{ijk\ell} = C_{jik\ell} = C_{ij\ell k}$. Because of this symmetry, we can replace $\widehat{\boldsymbol{\varepsilon}}$ by $\nabla\widehat{\mathbf{u}}$ in equation (1.22) to get

$$\nabla \cdot (\mathbf{C} : \nabla \widehat{\mathbf{u}}) + \omega^2\, \widehat{\boldsymbol{\rho}} \cdot \widehat{\mathbf{u}} = 0\, .$$

Dropping the hats, we have the *wave equation for elastodynamics* for fixed frequency

$$\boxed{\nabla \cdot (\mathbf{C} : \nabla \mathbf{u}) + \omega^2\, \boldsymbol{\rho} \cdot \mathbf{u} = 0\, .} \tag{1.24}$$

The above equation is the fixed frequency form of equation (1.8) in the absence of body force and for the situation where the mass density is anisotropic and frequency dependent. For an isotropic material

$$\mathbf{C} : \nabla \mathbf{u} = \mu \left[\nabla \mathbf{u} + (\nabla \mathbf{u})^T\right] + \lambda\, \mathrm{tr}(\nabla \mathbf{u})\, \boldsymbol{1} \tag{1.25}$$

where μ is the shear modulus and λ is the Lamé modulus. For a homogeneous and isotropic medium we can split the elastodynamic wave equation into equations for the p-wave and the s-wave components. In that case, the wave equations at fixed frequency in the absence of body forces take the form

$$\nabla^2 \phi + \frac{\omega^2}{c_p^2}\, \phi = 0\, ; \quad \nabla \cdot (\nabla \boldsymbol{\psi})^T + \frac{\omega^2}{c_s^2}\, \boldsymbol{\psi} = \mathbf{0}\, . \tag{1.26}$$

The quantities c_p and c_s are phase velocities and must be real for waves to propagate through the elastic medium.

1.2.6 Antiplane shear strain

Let us now consider the case of antiplane shear strain which is a special state of deformation where the components of the displacement ($\mathbf{u}$) with respect to an orthonormal basis $\{\mathbf{e}_1, \mathbf{e}_2, \mathbf{e}_3\}$ are given by

$$u_1 = u_1(x_2, x_3)\ , \quad u_2 = 0\ , \quad u_3 = 0\ .$$

This is an out-of-plane mode of deformation. Let us also assume that C is isotropic and that μ and λ are independent of x_1, i.e., $\mu \equiv \mu(x_2, x_3)$ and $\lambda \equiv \lambda(x_2, x_3)$. Since $\text{tr}(\boldsymbol{\nabla}\mathbf{u}) = \boldsymbol{\nabla}\cdot\mathbf{u} = u_{i,i} = 0$ equation (1.25) can be written as

$$[\mathsf{C} : \boldsymbol{\nabla}\mathbf{u}]_{ij} = \mu\,(u_{i,j} + u_{j,i}) \Leftrightarrow [\mathsf{C} : \boldsymbol{\nabla}\mathbf{u}] = \begin{bmatrix} 0 & \mu u_{1,2} & \mu u_{1,3} \\ \mu u_{1,2} & 0 & 0 \\ \mu u_{1,3} & 0 & 0 \end{bmatrix} .$$

Therefore,

$$[\boldsymbol{\nabla}\cdot(\mathsf{C} : \boldsymbol{\nabla}\mathbf{u})]_j = [\mu\,(u_{i,j} + u_{j,i})]_{,i} \Leftrightarrow [\boldsymbol{\nabla}\cdot(\mathsf{C} : \boldsymbol{\nabla}\mathbf{u})] = \begin{bmatrix} (\mu\, u_{1,2})_{,2} + (\mu\, u_{1,3})_{,3} \\ 0 \\ 0 \end{bmatrix} .$$

Plugging into the wave equation (1.24) we get

$$(\mu\, u_{1,2})_{,2} + (\mu\, u_{1,3})_{,3} + \omega^2\, \rho_{11}\, u_1 = 0\,.$$

Let us define the two-dimensional gradient operator ($\overline{\boldsymbol{\nabla}}$) such that the gradient of a scalar (ϕ) and the divergence of a vector ($\mathbf{v}$) are given by

$$\overline{\boldsymbol{\nabla}}\phi = \phi_{,\alpha}\mathbf{e}_\alpha\ ; \quad \overline{\boldsymbol{\nabla}}\cdot\mathbf{v} = v_{\alpha,\alpha}\ ; \qquad \alpha, \beta = 2, 3\ .$$

Then we can write the wave equation for antiplane shear as

$$\boxed{\overline{\boldsymbol{\nabla}}\cdot(\mu\,\overline{\boldsymbol{\nabla}}u_1) + \omega^2\,\rho\,u_1 = 0\,.} \tag{1.27}$$

Shear waves that satisfy the above equation are also called SH-waves (shear horizontal waves), particularly in seismology. Later we shall see that this equation is analogous to those for plane acoustic wave propagation and transverse electromagnetic wave propagation.

1.2.7 Elastodynamic power and dissipation

The rate of work done on an elastic body by external surface tractions and body forces is

$$P = \int_\Gamma \mathbf{t}\cdot\dot{\mathbf{u}}\,\text{dA} + \int_\Omega \mathbf{b}\cdot\dot{\mathbf{u}}\,\text{dV}$$

where $\mathbf{t}$ is the surface traction, $\mathbf{b}$ is the body force, and $\mathbf{u}$ is the displacement. If $\mathbf{n}$ is the unit outward normal to the surface Γ, we can use the definition of Cauchy stress to get

$$P = \int_\Gamma (\mathbf{n}\cdot\boldsymbol{\sigma})\cdot\dot{\mathbf{u}}\,\mathrm{dA} + \int_\Omega \mathbf{b}\cdot\dot{\mathbf{u}}\,\mathrm{dV}\,.$$

In the absence of jump discontinuities in the body, the divergence theorem[‡‡] and the symmetry of the Cauchy stress gives us

$$P = \int_\Omega [\boldsymbol{\nabla}\cdot(\dot{\mathbf{u}}\cdot\boldsymbol{\sigma}) + \mathbf{b}\cdot\dot{\mathbf{u}}]\ \mathrm{dV} = \int_\Omega [\boldsymbol{\nabla}\cdot(\boldsymbol{\sigma}\cdot\dot{\mathbf{u}}) + \mathbf{b}\cdot\dot{\mathbf{u}}]\ \mathrm{dV}\,.$$

Expanding the divergence term leads to

$$P = \int_\Omega [(\boldsymbol{\nabla}\cdot\boldsymbol{\sigma})\cdot\dot{\mathbf{u}} + \boldsymbol{\sigma}:\boldsymbol{\nabla}\dot{\mathbf{u}} + \mathbf{b}\cdot\dot{\mathbf{u}}]\ \mathrm{dV}\,.$$

We can express equation (1.3) for the balance of linear momentum in the form

$$\boldsymbol{\nabla}\cdot\boldsymbol{\sigma} + \mathbf{b} = \dot{\mathbf{p}} \quad \text{where} \quad \dot{\mathbf{p}} := \int_{-\infty}^{\infty} \rho(\mathbf{x}, t-t')\cdot\frac{\partial^2\mathbf{u}}{\partial t'^2}\,\mathrm{d}t'\,.$$

Therefore,

$$\boxed{P = \int_\Omega [\dot{\mathbf{p}}\cdot\dot{\mathbf{u}} + \boldsymbol{\sigma}:\boldsymbol{\nabla}\dot{\mathbf{u}}]\ \mathrm{dV}\,.} \tag{1.28}$$

Then the *average* work done over a time period, T, is given by

$$W_{\text{cycle}} = \int_\Omega \left[\frac{1}{T}\int_{-T/2}^{T/2} (\dot{\mathbf{p}}\cdot\dot{\mathbf{u}} + \boldsymbol{\sigma}:\boldsymbol{\nabla}\dot{\mathbf{u}})\,\mathrm{d}t\right]\ \mathrm{dV}\,. \tag{1.29}$$

It is convenient to use the notation $\mathbf{v} = \dot{\mathbf{u}}$ and to take advantage of the symmetry of the Cauchy stress to express the power in terms of the velocity and strain rate as

$$P = \int_\Omega [\dot{\mathbf{p}}\cdot\mathbf{v} + \boldsymbol{\sigma}:\dot{\boldsymbol{\varepsilon}}]\ \mathrm{dV} \quad \text{where} \quad \dot{\boldsymbol{\varepsilon}} := \tfrac{1}{2}\left[\boldsymbol{\nabla}\dot{\mathbf{u}} + (\boldsymbol{\nabla}\dot{\mathbf{u}})^T\right]. \tag{1.30}$$

Recall that the stress is related to the strain by

$$\boldsymbol{\sigma}(\mathbf{x},t) = \int_{-\infty}^{\infty} \mathsf{K}(\mathbf{x}, t-t') : \boldsymbol{\varepsilon}(\mathbf{x}, t')\,\mathrm{d}t'\,.$$

[‡‡] The divergence theorem for a tensor field can be symbolically written as

$$\int_\Omega \boldsymbol{\nabla}\cdot\boldsymbol{T}\,\mathrm{dV} = \int_\Gamma \boldsymbol{T}\otimes\mathbf{n}\,\mathrm{dA}$$

where $\mathbf{n}$ is the unit normal to the surface Γ. For a vector field $\mathbf{v}$ and a second-order tensor field $\boldsymbol{S}$, we have

$$\int_\Omega \boldsymbol{\nabla}\cdot\mathbf{v}\,\mathrm{dV} = \int_\Gamma \mathbf{v}\cdot\mathbf{n}\,\mathrm{dA} \quad \text{and} \quad \int_\Omega \boldsymbol{\nabla}\cdot\boldsymbol{S}\,\mathrm{dV} = \int_\Gamma \boldsymbol{S}^T\cdot\mathbf{n}\,\mathrm{dA}\,.$$

If we consider time harmonic displacements and strains of the form

$$\mathbf{u}(\mathbf{x},t) = \mathrm{Re}[\widehat{\mathbf{u}}(\mathbf{x})\,e^{-i\omega t}] \qquad \text{and} \qquad \boldsymbol{\varepsilon}(\mathbf{x},t) = \mathrm{Re}[\widehat{\boldsymbol{\varepsilon}}(\mathbf{x})\,e^{-i\omega t}]$$

we have

$$\mathbf{v} = \mathrm{Re}[-i\omega\widehat{\mathbf{u}}\,e^{-i\omega t}] \; ; \; \dot{\boldsymbol{\varepsilon}} = \mathrm{Re}[-i\omega\widehat{\boldsymbol{\varepsilon}}\,e^{-i\omega t}]$$

and

$$\boldsymbol{\sigma} = \mathrm{Re}[\mathbf{C} : \widehat{\boldsymbol{\varepsilon}} e^{-i\omega t}] \; ; \; \dot{\mathbf{p}} = \mathrm{Re}[-\omega^2 \widehat{\boldsymbol{\rho}} \cdot \widehat{\mathbf{u}}\, e^{-i\omega t}] \, .$$

Let us break up the above quantities into their real and imaginary parts, i.e.,

$$\widehat{\mathbf{u}} = \widehat{\mathbf{u}}_r + i\widehat{\mathbf{u}}_i \; ; \; \widehat{\boldsymbol{\varepsilon}} = \widehat{\boldsymbol{\varepsilon}}_r + i\widehat{\boldsymbol{\varepsilon}}_i \; ; \; \widehat{\boldsymbol{\rho}} = \widehat{\boldsymbol{\rho}}_r + i\widehat{\boldsymbol{\rho}}_i \; \text{ and } \; \mathbf{C} = \mathbf{C}_r + i\mathbf{C}_i \, .$$

Then we have

$$\begin{aligned}
\mathbf{v} &= \omega[\widehat{\mathbf{u}}_i \cos(\omega t) - \widehat{\mathbf{u}}_r \sin(\omega t)] \; ; \; \dot{\boldsymbol{\varepsilon}} = \omega[\widehat{\boldsymbol{\varepsilon}}_i \cos(\omega t) - \widehat{\boldsymbol{\varepsilon}}_r \sin(\omega t)] \\
\boldsymbol{\sigma} &= (\mathbf{C}_r : \widehat{\boldsymbol{\varepsilon}}_r - \mathbf{C}_i : \widehat{\boldsymbol{\varepsilon}}_i)\cos(\omega t) + (\mathbf{C}_r : \widehat{\boldsymbol{\varepsilon}}_i + \mathbf{C}_i : \widehat{\boldsymbol{\varepsilon}}_r)\sin(\omega t) \\
\dot{\mathbf{p}} &= \omega^2 \left[(\widehat{\boldsymbol{\rho}}_i \cdot \widehat{\mathbf{u}}_i - \widehat{\boldsymbol{\rho}}_r \cdot \widehat{\mathbf{u}}_r)\cos(\omega t) - (\widehat{\boldsymbol{\rho}}_r \cdot \widehat{\mathbf{u}}_i + \widehat{\boldsymbol{\rho}}_i \cdot \widehat{\mathbf{u}}_r)\sin(\omega t) \right] .
\end{aligned}$$

Plugging these into equation (1.30) gives us

$$\begin{aligned}
P = \int_\Omega \Big[& \omega \left[\widehat{\boldsymbol{\varepsilon}}_i \cos\omega t - \widehat{\boldsymbol{\varepsilon}}_r \sin\omega t \right] : \\
& \left[(-\mathbf{C}_i : \widehat{\boldsymbol{\varepsilon}}_i + \mathbf{C}_r : \widehat{\boldsymbol{\varepsilon}}_r)\cos\omega t + (\mathbf{C}_r : \widehat{\boldsymbol{\varepsilon}}_i + \mathbf{C}_i : \widehat{\boldsymbol{\varepsilon}}_r)\sin\omega t \right] + \\
& \omega^3 \left[\widehat{\mathbf{u}}_i \cos\omega t - \widehat{\mathbf{u}}_r \sin\omega t \right] \cdot \\
& \left[(\widehat{\boldsymbol{\rho}}_i \cdot \widehat{\mathbf{u}}_i - \widehat{\boldsymbol{\rho}}_r \cdot \widehat{\mathbf{u}}_r)\cos\omega t - (\widehat{\boldsymbol{\rho}}_r \cdot \widehat{\mathbf{u}}_i + \widehat{\boldsymbol{\rho}}_i \cdot \widehat{\mathbf{u}}_r)\sin\omega t \right] \Big] \, \mathrm{dV}.
\end{aligned}$$

Therefore, using (1.29), the average work done in a cycle of oscillation will be

$$\begin{aligned}
W_{\text{cycle}} &= \int_\Omega \left[\frac{\omega}{2\pi} \int_{-\pi/\omega}^{\pi/\omega} P \, \mathrm{d}t \right] \mathrm{dV} \\
&= \int_\Omega \frac{\omega}{2} \left[- \left(\widehat{\boldsymbol{\varepsilon}}_r : \mathbf{C}_i : \widehat{\boldsymbol{\varepsilon}}_r + \widehat{\boldsymbol{\varepsilon}}_i : \mathbf{C}_i : \widehat{\boldsymbol{\varepsilon}}_r \right) + \omega^2 \left(\widehat{\mathbf{u}}_i \cdot \widehat{\boldsymbol{\rho}}_i \cdot \widehat{\mathbf{u}}_i + \widehat{\mathbf{u}}_r \cdot \widehat{\boldsymbol{\rho}}_i \cdot \widehat{\mathbf{u}}_r \right) \right] \mathrm{dV} \, .
\end{aligned}$$

This quadratic form will be non-negative for all choices of $\widehat{\mathbf{u}}$ and $\widehat{\boldsymbol{\varepsilon}}$ if and only if $\boldsymbol{\varepsilon} : (-\mathbf{C}_i) : \boldsymbol{\varepsilon} > 0$ and $\mathbf{u} \cdot \widehat{\boldsymbol{\rho}}_i \cdot \mathbf{u} > 0$ for all real $\omega > 0$, i.e, $-\mathbf{C}_i$ and $\widehat{\boldsymbol{\rho}}_i$ are positive definite. Note that the quadratic forms do not contain the real parts of $\mathbf{C}$ and $\widehat{\boldsymbol{\rho}}$. Since the work done in a cycle should be zero in the absence of dissipation, this implies that the imaginary parts of the elastic stiffness tensor and the density are connected to energy dissipation.

1.3 Acoustics

The equations of acoustics describe the propagation of sound waves in a medium. These equations consist of simplified forms of the conservation of mass and the Navier-Stokes equations for the conservation of linear momentum in fluids. In this section we consider only the linear equations (also called the "small signal" equations) for acoustic waves.

The equations of elastodynamics implicitly assume that mass is conserved within the body Ω and there is no mass flow in and out of the body. Such a description of the dynamics of a continuum is called Lagrangian. However, a Lagrangian description usually proves inconvenient for acoustics and a spatial or Eulerian description is preferred. In the Eulerian description of a continuum, mass is allowed to flow in and out of a region which is assumed fixed in time.

The equation for the conservation of mass in a fluid volume (without any mass sources or sinks) is given by

$$\frac{\partial \rho}{\partial t} + \boldsymbol{\nabla} \cdot (\rho \mathbf{v}) = 0 \tag{1.31}$$

where $\rho(\mathbf{x},t)$ is the mass density of the fluid and $\mathbf{v}(\mathbf{x},t)$ is the fluid velocity.

In linear acoustic theory we assume that the amplitudes of *disturbances in the fluid are small*. This assumption allows us to express the field variables as the sum of a time-averaged mean spatial field ($\langle\cdot\rangle$) and a small fluctuating field ($\tilde{\cdot}$) that varies in space and time. Therefore,

$$p = \langle p \rangle + \tilde{p}; \quad \rho = \langle \rho \rangle + \tilde{\rho}; \quad \mathbf{v} = \langle \mathbf{v} \rangle + \tilde{\mathbf{v}}$$

and

$$\frac{\partial \langle p \rangle}{\partial t} = 0; \quad \frac{\partial \langle \rho \rangle}{\partial t} = 0; \quad \frac{\partial \langle \mathbf{v} \rangle}{\partial t} = \mathbf{0}.$$

Plugging these expressions into the mass balance equation (1.31) and neglecting terms higher than first in the fluctuations gives

$$\frac{\partial \tilde{\rho}}{\partial t} + [\langle \rho \rangle + \tilde{\rho}]\, \boldsymbol{\nabla} \cdot \langle \mathbf{v} \rangle + \langle \rho \rangle \boldsymbol{\nabla} \cdot \tilde{\mathbf{v}} + \boldsymbol{\nabla} [\langle \rho \rangle + \tilde{\rho}] \cdot \langle \mathbf{v} \rangle + \boldsymbol{\nabla} \langle \rho \rangle \cdot \tilde{\mathbf{v}} = 0.$$

If, in addition, the medium is *homogeneous*, i.e., $\boldsymbol{\nabla} \langle \rho \rangle = 0$ then we have

$$\frac{\partial \tilde{\rho}}{\partial t} + [\langle \rho \rangle + \tilde{\rho}]\, \boldsymbol{\nabla} \cdot \langle \mathbf{v} \rangle + \langle \rho \rangle \boldsymbol{\nabla} \cdot \tilde{\mathbf{v}} + \boldsymbol{\nabla} \tilde{\rho} \cdot \langle \mathbf{v} \rangle = 0.$$

If the disturbances are such that the *medium can be assumed to be at rest*, i.e., $\langle \mathbf{v} \rangle = 0$ the mass balance equation reduces to

$$\frac{\partial \tilde{\rho}}{\partial t} + \langle \rho \rangle \boldsymbol{\nabla} \cdot \tilde{\mathbf{v}} = 0.$$

The conservation of mass in an acoustic fluid is usually expressed in terms of the pressure. To get that relation we need an equation of state for the pressure. In linear acoustics it is usually assumed that acoustic waves compress the medium in an *adiabatic and reversible* manner. The equation of state of such a medium is defined as the relation between the pressure and the density, $p = f(\rho)$ and can have several forms. In linear acoustics these relations are assumed to be given by

$$\frac{dp}{d\rho} = c^2 = \begin{cases} \gamma p/\rho & \text{for gases} \\ \kappa/\rho & \text{for liquids} \end{cases}$$

where c is the speed of sound in the medium, γ is the adiabatic index for ideal gases, and κ is the bulk modulus of the liquid. These assumptions lead, after dropping the tildes and defining $\rho_0 := \langle\rho\rangle$, to the commonly used expression for the balance of mass in an acoustic medium

$$\boxed{\frac{\partial p}{\partial t} + \rho_0\, c_0^2\, \boldsymbol{\nabla} \cdot \mathbf{v} = 0} \tag{1.32}$$

where p is the pressure and c_0 is the speed of sound in the medium.

We can derive the above equation in an alternative manner from the constitutive relations of linear elastodynamics given in equation (1.7). For an isotropic material the stress is given by

$$\boldsymbol{\sigma} = \mathsf{C} : \boldsymbol{\nabla}\mathbf{u} = \mu\,[\boldsymbol{\nabla}\mathbf{u} + (\boldsymbol{\nabla}\mathbf{u})^T] + (\lambda\, \boldsymbol{\nabla} \cdot \mathbf{u})\, \mathit{1}\,.$$

If the material is an inviscid fluid, the shear modulus is zero ($\mu = 0$) and the Lamé constant equals the bulk modulus (κ) since $\lambda = \kappa - 2/3\,\mu = \kappa$. Therefore, for fluids,

$$\boldsymbol{\sigma} = \mathsf{C} : \boldsymbol{\nabla}\mathbf{u} = (\kappa\, \boldsymbol{\nabla} \cdot \mathbf{u})\, \mathit{1} = -p\, \mathit{1}\,.$$

The quantity

$$\boxed{p = -\kappa\, \boldsymbol{\nabla} \cdot \mathbf{u}} \tag{1.33}$$

is the pressure in the fluid. For small disturbances we can assume that κ does not change with time, i.e., $\kappa = \kappa_0 = \rho_0 c_0^2$. Then we can take the time derivative of the pressure equation (1.33) to get

$$\frac{\partial p}{\partial t} + \kappa_0\, \boldsymbol{\nabla} \cdot \mathbf{v} = 0 \quad \Leftrightarrow \quad \frac{\partial p}{\partial t} + \rho_0 c_0^2\, \boldsymbol{\nabla} \cdot \mathbf{v} = 0\,.$$

If we also assume that the bulk modulus is a function of time, then we can use the principle of linear superposition to write equation (1.33) as

$$p(\mathbf{x},t) = -\int_{-\infty}^{\infty} \kappa(\mathbf{x}, t - t')\, \boldsymbol{\nabla} \cdot \mathbf{u}(\mathbf{x},t')\, dt'\,. \tag{1.34}$$

Causality implies that the pressure at time t depends only on velocities from previous times and hence $\kappa(\mathbf{x},\tau) = 0$ if $\tau = t - t' < 0$. We can write the above in terms of velocity as

$$\boxed{\frac{\partial p(\mathbf{x},t)}{\partial t} = -\int_{-\infty}^{\infty} \kappa(\mathbf{x}, t - t')\, \boldsymbol{\nabla} \cdot \mathbf{v}(\mathbf{x},t')\, dt'\,.} \tag{1.35}$$

The local equations for the conservation of linear momentum in a fluid can be deduced from equation (1.3) where the acceleration on the right-hand side now takes the form of a total derivative, i.e.,

$$\boldsymbol{\nabla}\cdot\boldsymbol{\sigma}+\rho(\mathbf{x})\,\mathbf{b}(\mathbf{x})=\rho(\mathbf{x})\,\frac{D\mathbf{v}}{Dt};\quad \frac{D\mathbf{v}}{Dt}:=\frac{\partial\mathbf{v}}{\partial t}+\mathbf{v}\cdot(\boldsymbol{\nabla}\mathbf{v})^T. \tag{1.36}$$

The stress tensor $\boldsymbol{\sigma}$ can be split into volumetric and deviatoric parts

$$\sigma=-p\,\mathit{1}+\mathbf{s};\quad p:=-\tfrac{1}{3}\,\mathrm{tr}(\sigma)$$

and we can write equation (1.36) as

$$-\boldsymbol{\nabla}p+\boldsymbol{\nabla}\cdot\mathbf{s}+\rho(\mathbf{x})\,\mathbf{b}(\mathbf{x})=\rho(\mathbf{x})\left(\frac{\partial\mathbf{v}}{\partial t}+\mathbf{v}(\mathbf{x})\cdot(\boldsymbol{\nabla}\mathbf{v})^T\right).$$

The constitutive behavior of the fluid is assumed to be Newtonian, i.e., the deviatoric stress depends on the symmetric part of the velocity gradient which is defined as

$$\dot{\boldsymbol{\varepsilon}}:=\tfrac{1}{2}\left(\boldsymbol{\nabla}\mathbf{v}+\boldsymbol{\nabla}\mathbf{v}^T\right). \tag{1.37}$$

Then, assuming that the fluid is local in space, we have

$$\mathbf{s}(\mathbf{x},t)=\int\mathbf{K}_\varepsilon(\mathbf{x},t-t'):\dot{\boldsymbol{\varepsilon}}(\mathbf{x},t')\,\mathrm{d}t'. \tag{1.38}$$

If the fluid is also local in time and isotropic, we have

$$\mathbf{s}(\mathbf{x},t)=\lambda(\mathbf{x})\;(\boldsymbol{\nabla}\cdot\mathbf{v})\;\mathit{1}+\mu(\mathbf{x})\left(\boldsymbol{\nabla}\mathbf{v}+\boldsymbol{\nabla}\mathbf{v}^T\right).$$

where μ is the *shear viscosity* and $\mu_b=\lambda+2/3\mu$ is the *bulk viscosity*. If we repeat the process used to arrive at equation (1.10), the balance of momentum for a *homogeneous medium* becomes

$$-\boldsymbol{\nabla}p+(\lambda+2\mu)\,\boldsymbol{\nabla}(\boldsymbol{\nabla}\cdot\mathbf{v})-\mu\boldsymbol{\nabla}\times(\boldsymbol{\nabla}\times\mathbf{v})+\rho\,\mathbf{b}(\mathbf{x})=\rho\left(\frac{\partial\mathbf{v}}{\partial t}+\mathbf{v}\cdot(\boldsymbol{\nabla}\mathbf{v})^T\right). \tag{1.39}$$

The flow is assumed to be *irrotational* and the *body force is neglected* in linear acoustics, i.e., $\boldsymbol{\nabla}\times\mathbf{v}=0$ and $\mathbf{b}(\mathbf{x})=\mathbf{0}$. Therefore the momentum equation reduces to

$$-\boldsymbol{\nabla}p+(\lambda+2\mu)\,\boldsymbol{\nabla}(\boldsymbol{\nabla}\cdot\mathbf{v})=\rho\left(\frac{\partial\mathbf{v}}{\partial t}+\mathbf{v}\cdot(\boldsymbol{\nabla}\mathbf{v})^T\right). \tag{1.40}$$

If we also assume that there are *no viscous forces in the medium*, i.e., the bulk and shear viscosities are zero, we get

$$-\boldsymbol{\nabla}p=\rho\left(\frac{\partial\mathbf{v}}{\partial t}+\mathbf{v}\cdot(\boldsymbol{\nabla}\mathbf{v})^T\right). \tag{1.41}$$

If we introduce the assumption of *small disturbances* that we had used in the derivation of the mass balance equation and neglect products of fluctuation terms, we can write

$$-\boldsymbol{\nabla}\left[\langle p\rangle+\tilde{p}\right]=\langle\rho\rangle\,\frac{\partial\tilde{\mathbf{v}}}{\partial t}+\left[\langle\rho\rangle+\tilde{\rho}\right]\left[\langle\mathbf{v}\rangle\cdot\boldsymbol{\nabla}\langle\mathbf{v}\rangle\right]$$
$$+\langle\rho\rangle\left[\langle\mathbf{v}\rangle\cdot(\boldsymbol{\nabla}\tilde{\mathbf{v}})^T+\tilde{\mathbf{v}}\cdot(\boldsymbol{\nabla}\langle\mathbf{v}\rangle)^T\right].$$

If the medium is *homogeneous* in the sense that the time-averaged variables $\langle p\rangle$ and $\langle\rho\rangle$ have zero spatial gradients, i.e., $\boldsymbol{\nabla}\langle p\rangle=0;\ \ \boldsymbol{\nabla}\langle\rho\rangle=0$, the momentum balance equation takes the form

$$-\boldsymbol{\nabla}\tilde{p}=\langle\rho\rangle\,\frac{\partial\tilde{\mathbf{v}}}{\partial t}+\left[\langle\rho\rangle+\tilde{\rho}\right]\left[\langle\mathbf{v}\rangle\cdot(\boldsymbol{\nabla}\langle\mathbf{v}\rangle)^T\right]+\langle\rho\rangle\left[\langle\mathbf{v}\rangle\cdot(\boldsymbol{\nabla}\tilde{\mathbf{v}})^T+\tilde{\mathbf{v}}\cdot(\boldsymbol{\nabla}\langle\mathbf{v}\rangle)^T\right].$$

An acoustic medium *is assumed to be at rest*, i.e., the mean velocity ($\langle\mathbf{v}\rangle$) is zero. With this assumption, after dropping the tildes and using $\rho_0:=\langle\rho\rangle$, we get the momentum equation for linear acoustics

$$\boxed{\rho_0\,\frac{\partial\mathbf{v}}{\partial t}+\boldsymbol{\nabla}p=0\,.} \tag{1.42}$$

Equations (1.32) and (1.42) are frequently combined into the forms

$$\frac{\partial^2\mathbf{v}}{\partial t^2}-c_0^2\,\boldsymbol{\nabla}\cdot(\boldsymbol{\nabla}\mathbf{v})=0;\quad \frac{\partial^2 p}{\partial t^2}-c_0^2\,\nabla^2 p=0\,. \tag{1.43}$$

1.3.1 Acoustic potential

Recall that we had assumed that the flow is irrotational. For irrotational motions the velocity $\mathbf{v}$ can be derived from a scalar potential $\mathbf{v}=-\boldsymbol{\nabla}\phi$ since $\boldsymbol{\nabla}\times\mathbf{v}=0$. Plugging the expression for $\mathbf{v}$ into equation (1.42) gives

$$\boldsymbol{\nabla}\left(\rho_0\frac{\partial\phi}{\partial t}\right)=\boldsymbol{\nabla}p\quad\Longrightarrow\quad p=\rho_0\frac{\partial\phi}{\partial t}+\text{constant}\,.$$

If we replace p in equation (1.32) with the above expression and $\mathbf{v}$ with the gradient of ϕ, we get

$$\boxed{\nabla^2\phi=\frac{1}{c_0^2}\,\frac{\partial^2\phi}{\partial t^2}\,.} \tag{1.44}$$

This is the acoustic wave equation expressed in terms of a scalar potential.

1.3.2 Cartesian and curvilinear coordinates

In terms of components with respect to a rectangular Cartesian basis, the equations of linear acoustics are

$$p=-\kappa\,\frac{\partial u_i}{\partial x_i}\,;\quad \frac{\partial p}{\partial t}+\rho_0 c_0^2\frac{\partial v_i}{\partial x_i}=0\,;\quad \rho_0\,\frac{\partial v_i}{\partial t}+\frac{\partial p}{\partial x_i}=0\,.$$

The acoustic wave equations in terms of the velocity and the pressure are

$$\frac{\partial^2 v_i}{\partial t^2} - c_0^2 \frac{\partial^2 v_j}{\partial x_i \partial x_j} = 0 \; ; \quad \frac{\partial^2 p}{\partial t^2} - c_0^2 \frac{\partial^2 p}{\partial x_j \partial x_j} = 0 \, .$$

The acoustic wave equation in terms of the scalar potential ϕ is

$$\frac{\partial^2 \phi}{\partial x_j \partial x_j} = \frac{1}{c_0^2} \frac{\partial^2 \phi}{\partial t^2} \, .$$

In curvilinear spatial coordinates, the governing equations of linear acoustics have the form

$$p = -\frac{\kappa}{\sqrt{g}} \frac{\partial}{\partial \theta^i} \left(u^i \sqrt{g}\right) \; ; \quad \frac{\partial p}{\partial t} + \frac{\rho_0 c_0^2}{\sqrt{g}} \frac{\partial}{\partial \theta^i} \left(v^i \sqrt{g}\right) = 0 \; ; \quad \rho_0 \frac{\partial}{\partial t} \left(v_i \mathbf{g}^i\right) + \frac{\partial p}{\partial \theta^i} \mathbf{g}^i = 0 \, .$$

The acoustic wave equation for pressure is

$$\frac{\partial^2 p}{\partial t^2} - \frac{c_0^2}{\sqrt{g}} \frac{\partial}{\partial \theta^i} \left[\sqrt{g} g^{ji} \frac{\partial p}{\partial \theta^j} \right] = 0 \, . \tag{1.45}$$

Similarly the acoustic potential equation in curvilinear coordinates is

$$\frac{1}{\sqrt{g}} \frac{\partial}{\partial \theta^i} \left[\sqrt{g} g^{ji} \frac{\partial \phi}{\partial \theta^j} \right] = \frac{1}{c_0^2} \frac{\partial^2 \phi}{\partial t^2} \, .$$

1.3.3 Anisotropic and frequency-dependent mass density

In Section 8.5.1 we will show that periodic and stratified structure can have an anisotropic dynamic density. Assuming that such a situation is possible, we can rewrite equation (1.42) in the form

$$\boldsymbol{\rho}_0 \cdot \frac{\partial \mathbf{v}}{\partial t} + \boldsymbol{\nabla} p = 0$$

where $\boldsymbol{\rho}_0$ is now a second-order anisotropic mass density tensor. Also, from equation (1.32) we have

$$\frac{\partial p}{\partial t} + \kappa \, \boldsymbol{\nabla} \cdot \mathbf{v} = 0 \, .$$

Combining the two equations gives us a single equation for p which is

$$\boxed{\frac{\partial^2 p}{\partial t^2} - \kappa \, \boldsymbol{\nabla} \cdot \left(\boldsymbol{\rho}_0^{-1} \cdot \boldsymbol{\nabla} p\right) = 0 \, .} \tag{1.46}$$

This is equation $(1.43)_2$ for a medium with anisotropic density.

If we also allow the mass density to be frequency dependent we can write

$$\boxed{\int_{-\infty}^{\infty} \boldsymbol{\rho}_0(\mathbf{x}, t - t') \cdot \frac{\partial \mathbf{v}}{\partial t'} \, \mathrm{d}t' + \boldsymbol{\nabla} p = 0 \, .} \tag{1.47}$$

This is a significant departure from classical linear acoustics. As we have seen before, causality implies that the pressure at time t depends only on velocities from previous times and hence $\boldsymbol{\rho}_0(\mathbf{x}, \tau) = 0$ if $\tau = t - t' < 0$.

1.3.4 Acoustics at a fixed frequency

In section 1.2.5 we saw how Fourier transformation of the equations of elastodynamics could be used to find equations that are valid for a given frequency ω. We can use the same process on equations (1.35) and (1.47) to get the acoustics equations at a fixed frequency. Fourier transforming these two equations leads to

$$-i\omega\widehat{p} = -\widehat{\kappa}\boldsymbol{\nabla}\cdot\widehat{\mathbf{v}} = i\omega\widehat{\kappa}\boldsymbol{\nabla}\cdot\widehat{\mathbf{u}} \qquad \Leftrightarrow \qquad \boldsymbol{\nabla}\cdot\widehat{\mathbf{u}} = -\frac{\widehat{p}}{\widehat{\kappa}} \tag{1.48}$$

$$-\boldsymbol{\nabla}\widehat{p} = -i\omega\widehat{\boldsymbol{\rho}}_0\cdot\widehat{\mathbf{v}} = -\omega^2\widehat{\boldsymbol{\rho}}_0\cdot\widehat{\mathbf{u}} \qquad \Leftrightarrow \qquad \widehat{\mathbf{u}} = \frac{1}{\omega^2}\widehat{\boldsymbol{\rho}}_0^{-1}\cdot\boldsymbol{\nabla}\widehat{p} \tag{1.49}$$

where we have used $\mathbf{v}(\mathbf{x},t) = \dot{\mathbf{u}}$ and

$$\widehat{p}(\mathbf{x},\omega) = \int_{-\infty}^{\infty} p(\mathbf{x},t)\, e^{i\omega t}\, dt\ ;\ \widehat{\kappa}(\mathbf{x},\omega) = \int_{-\infty}^{\infty} \kappa(\mathbf{x},\tau)\, e^{i\omega\tau}\, d\tau$$

$$\widehat{\mathbf{u}}(\mathbf{x},\omega) = \int_{-\infty}^{\infty} \mathbf{u}(\mathbf{x},t')\, e^{i\omega t'}\, dt'\ ;\ \widehat{\boldsymbol{\rho}}_0(\mathbf{x},\omega) = \int_{-\infty}^{\infty} \boldsymbol{\rho}(\mathbf{x},\tau)\, e^{i\omega\tau}\, d\tau\ ;\ \tau = t - t'\ .$$

Taking the divergence of equation (1.49) gives

$$\boldsymbol{\nabla}\cdot\widehat{\mathbf{u}} = \frac{1}{\omega^2}\boldsymbol{\nabla}\cdot\left(\widehat{\boldsymbol{\rho}}_0^{-1}\cdot\boldsymbol{\nabla}\widehat{p}\right)\ .$$

Replacing the divergence of $\mathbf{u}$ with the right side of equation (1.48), we get the *acoustic wave equation for fixed frequency*,

$$\boxed{\boldsymbol{\nabla}\cdot\left(\widehat{\boldsymbol{\rho}}_0^{-1}\cdot\boldsymbol{\nabla}\widehat{p}\right) + \frac{\omega^2}{\widehat{\kappa}}\,\widehat{p} = 0\,.} \tag{1.50}$$

If the mass density is isotropic, i.e., $\widehat{\boldsymbol{\rho}}_0 = \widehat{\rho}_0\,\boldsymbol{1}$, then the acoustic wave equation at fixed frequency is

$$\boldsymbol{\nabla}\cdot\left(\frac{1}{\widehat{\rho}_0}\boldsymbol{\nabla}\widehat{p}\right) + \frac{\omega^2}{\kappa}\,\widehat{p} = 0\,.$$

For a homogeneous medium we can write the above equation as

$$\boxed{\nabla^2\widehat{p} + k^2\,\widehat{p} = 0} \tag{1.51}$$

where $k^2 = \omega^2/c_0^2$ and $c_0^2 = \kappa/\rho$. Here c_0 is the phase velocity of the wave and must be real for a wave to propagate. This will be the case when κ and ρ are real and both positive or both negative as represented in Figure 1.2.

Proceeding in a similar manner, the acoustic equations in terms of a scalar potential (ϕ) can be expressed as

$$\widehat{\mathbf{v}} = -\boldsymbol{\nabla}\widehat{\varphi} \quad \text{and} \quad \widehat{p} = -i\omega\rho_0\widehat{\varphi}\,.$$

The acoustic wave equation becomes the Helmholtz equation

$$\nabla^2\widehat{\varphi} + \frac{\omega^2}{c_0^2}\widehat{\varphi} = 0 \quad \text{or} \quad \boxed{\nabla^2\widehat{\varphi} + k^2\widehat{\varphi} = 0}\,. \tag{1.52}$$

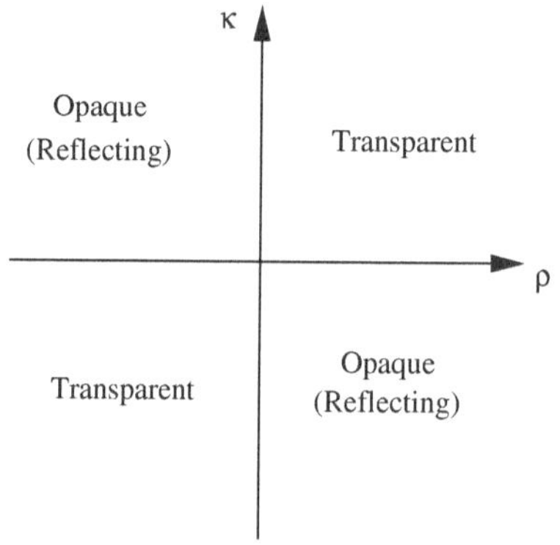

FIGURE 1.2
Acoustic transparency and opacity of a material as a function of ρ and κ.

1.3.5 Acoustic equation and antiplane shear

Let us now consider the case of two-dimensional acoustic wave propagation in an isotropic medium. We can express the acoustic equation for that situation in terms of components with respect to a rectangular Cartesian basis $(\mathbf{e}_1, \mathbf{e}_2, \mathbf{e}_3)$ as (after dropping the hats for convenience)

$$\frac{\partial}{\partial x_i}\left(\frac{1}{\rho_0}\frac{\partial p}{\partial x_i}\right) + \frac{\omega^2}{\kappa} p = 0 .$$

In the two-dimensional case p, ρ_0 and κ depend only on x_2 and x_3 and the acoustic wave equation can be written as

$$\frac{\partial}{\partial x_2}\left(\frac{1}{\rho_0}\frac{\partial p}{\partial x_2}\right) + \frac{\partial}{\partial x_3}\left(\frac{1}{\rho_0}\frac{\partial p}{\partial x_3}\right) + \frac{\omega^2}{\kappa} p = 0 .$$

That means that the three-dimensional gradient operator $\boldsymbol{\nabla}$ can be replaced with the two-dimensional gradient operator $\overline{\boldsymbol{\nabla}}$, and we get

$$\overline{\boldsymbol{\nabla}} \cdot \left(\frac{1}{\rho_0}\overline{\boldsymbol{\nabla}} p\right) + \frac{\omega^2}{\kappa}\, p = 0 \,; \quad \mathbf{u} = \frac{1}{\omega^2\,\rho_0}\,\overline{\boldsymbol{\nabla}} p . \tag{1.53}$$

This equation is similar to the elastodynamic wave equation for antiplane shear with the substitutions

$$p \leftrightarrow u_1 \,; \quad \frac{1}{\rho} \leftrightarrow \mu \,; \quad \frac{1}{\kappa} \leftrightarrow \rho .$$

Therefore, the solutions of the two problems will lead to similar conclusions if we make the appropriate interpretations of the quantities involved.

1.3.6 Acoustic power and dissipation

The time-averaged *intensity* of an acoustic wave is given by

$$\mathbf{I}(\mathbf{x}) = \frac{1}{T}\int_{-T/2}^{T/2} p(\mathbf{x},t)\,\mathbf{v}(\mathbf{x},t)\mathrm{d}t \tag{1.54}$$

where T is the period of the acoustic wave for a single frequency or a suitably long period for random fluctuations that have reached a steady state. The *acoustic intensity level* in a particular direction is calculated with respect to a reference intensity (I_0) and is defined as

$$L_I := 10\log_{10}(I/I_0)\ \mathrm{dB}\ .$$

Acoustic power over a surface area Γ with outward unit normal $\mathbf{n}$ is then given by

$$P = \int_\Gamma \mathbf{I}(\mathbf{x})\cdot\mathbf{n}\ \mathrm{dA} = \frac{1}{T}\int_{-T/2}^{T/2}\mathrm{d}t\int_\Gamma \mathrm{dA}\ [p(\mathbf{x},t)\,\mathbf{v}(\mathbf{x},t)]\cdot\mathbf{n}\ .$$

For a simply connected region Ω of which Γ is the boundary, the divergence theorem yields

$$P = \frac{1}{T}\int_{-T/2}^{T/2}\mathrm{d}t\int_\Omega \mathrm{dV}\ \boldsymbol{\nabla}\cdot[p(\mathbf{x},t)\,\mathbf{v}(\mathbf{x},t)]$$

or

$$P = \int_\Omega \mathrm{dV}\left[\frac{1}{T}\int_{-T/2}^{T/2}\mathrm{d}t(\boldsymbol{\nabla}p\cdot\mathbf{v}+p\boldsymbol{\nabla}\cdot\mathbf{v})\right]\ . \tag{1.55}$$

Now consider harmonically varying pressure $p(t)$ and particle velocity $\mathbf{v}(t)$ given by

$$p(t) = \mathrm{Re}(\widehat{p}e^{-i\omega t}) \quad \text{and} \quad \mathbf{v}(t) = \mathrm{Re}(\widehat{\mathbf{v}}e^{-i\omega t}) \tag{1.56}$$

where the dependence on $\mathbf{x}$ is implicit. If $\widehat{p} = \widehat{p}_r + i\widehat{p}_i$ and $\widehat{\mathbf{v}} = \widehat{\mathbf{v}}_r + i\widehat{\mathbf{v}}_i$, we may write these as

$$p(t) = \widehat{p}_r\cos(\omega t) + \widehat{p}_i\sin(\omega t) \quad \text{and} \quad \mathbf{v}(t) = \widehat{\mathbf{v}}_r\cos(\omega t) + \widehat{\mathbf{v}}_i\sin(\omega t)\ .$$

Recall from (1.35) and (1.47) that, for isotropic media,

$$\frac{\partial p}{\partial t} = -\int_{-\infty}^{\infty}\kappa(t-t')\,\boldsymbol{\nabla}\cdot\mathbf{v}(t')\,\mathrm{d}t' \quad \text{and} \quad \boldsymbol{\nabla}p = -\int_{-\infty}^{\infty}\rho(t-t')\frac{\partial\mathbf{v}}{\partial t'}\mathrm{d}t'.$$

Substitution of the expressions for p and $\mathbf{v}$ into the above equations gives

$$-i\omega\widehat{p}e^{-i\omega t} = -\int_{-\infty}^{\infty}\kappa(t-t')\,\boldsymbol{\nabla}\cdot\widehat{\mathbf{v}}e^{-i\omega t'}\,\mathrm{d}t'$$

$$\boldsymbol{\nabla}\widehat{p}e^{-i\omega t} = i\omega\int_{-\infty}^{\infty}\rho(t-t')\widehat{\mathbf{v}}e^{-i\omega t'}\mathrm{d}t'\ .$$

Using the replacement $\tau = t - t'$ in the equations above, we get

$$-i\omega\widehat{p}\,e^{-i\omega t} = -\left[\int_{-\infty}^{\infty} \kappa(\tau)\,e^{i\omega\tau}\mathrm{d}\tau\right] \boldsymbol{\nabla}\cdot\widehat{\mathbf{v}}\,e^{-i\omega t} = -\widehat{\kappa}\boldsymbol{\nabla}\cdot\widehat{\mathbf{v}}\,e^{-i\omega t}$$

$$\boldsymbol{\nabla}\widehat{p}\,e^{-i\omega t} = i\omega\left[\int_{-\infty}^{\infty} \rho(\tau)\,e^{i\omega\tau}\mathrm{d}\tau\right] \widehat{\mathbf{v}}\,e^{-i\omega t} = i\omega\widehat{\rho}\,\widehat{\mathbf{v}}\,e^{-i\omega t}$$

or

$$\widehat{\kappa}\boldsymbol{\nabla}\cdot\widehat{\mathbf{v}} = i\omega\widehat{p} \quad \text{and} \quad \boldsymbol{\nabla}\widehat{p} = i\omega\widehat{\rho}\widehat{\mathbf{v}}\,. \tag{1.57}$$

Therefore,

$$(\boldsymbol{\nabla}\cdot\widehat{\mathbf{v}})_r = \frac{\omega}{\widehat{\kappa}_i^2+\widehat{\kappa}_r^2}(\widehat{p}_r\widehat{\kappa}_i - \widehat{p}_i\widehat{\kappa}_r)\ ;\ \ (\boldsymbol{\nabla}\cdot\widehat{\mathbf{v}})_i = \frac{\omega}{\widehat{\kappa}_i^2+\widehat{\kappa}_r^2}(\widehat{p}_r\widehat{\kappa}_r + \widehat{p}_i\widehat{\kappa}_i)$$

and

$$(\boldsymbol{\nabla}\widehat{p})_r = -\omega(\widehat{\rho}_r\widehat{\mathbf{v}}_i + \widehat{\rho}_i\widehat{\mathbf{v}}_r)\ ;\ \ (\boldsymbol{\nabla}\widehat{p})_i = \omega(\widehat{\rho}_r\widehat{\mathbf{v}}_r - \widehat{\rho}_i\widehat{\mathbf{v}}_i)\,.$$

Also, from (1.56) we have

$$\begin{aligned}\boldsymbol{\nabla}p &= \mathrm{Re}(\boldsymbol{\nabla}\widehat{p}\,e^{-i\omega t}) = (\boldsymbol{\nabla}\widehat{p})_r\cos(\omega t) + (\boldsymbol{\nabla}\widehat{p})_i\sin(\omega t) \quad \text{and}\\ \boldsymbol{\nabla}\cdot\mathbf{v} &= \mathrm{Re}(\boldsymbol{\nabla}\cdot\widehat{\mathbf{v}}\,e^{-i\omega t}) = (\boldsymbol{\nabla}\cdot\widehat{\mathbf{v}})_r\cos(\omega t) + (\boldsymbol{\nabla}\cdot\widehat{\mathbf{v}})_i\sin(\omega t)\,.\end{aligned}$$

Therefore, from (1.55), the average acoustic power *density* in a cycle of oscillation will be

$$\begin{aligned}P_{\mathrm{den}} &= \frac{\omega}{2\pi}\int_{-\pi/\omega}^{\pi/\omega}(\boldsymbol{\nabla}p\cdot\mathbf{v} + p\boldsymbol{\nabla}\cdot\mathbf{v})\,\mathrm{d}t\\ &= \frac{\omega}{2}\left[\frac{\widehat{\kappa}_i}{\widehat{\kappa}_i^2+\widehat{\kappa}_r^2}(\widehat{p}_i^2+\widehat{p}_r^2) - \widehat{\rho}_i(\widehat{\mathbf{v}}_r\cdot\widehat{\mathbf{v}}_r + \widehat{\mathbf{v}}_i\cdot\widehat{\mathbf{v}}_i)\right].\end{aligned}$$

This quadratic form will be non-negative for all choices of $\widehat{p}$ and $\widehat{\mathbf{v}}$ if and only if $\widehat{\kappa}_i = \mathrm{Im}(\widehat{\kappa}) > 0$ and $\widehat{\rho}_i = \mathrm{Im}(\widehat{\rho}) < 0$ for all real $\omega > 0$. Note that the quadratic form does not contain the real part of $\widehat{\rho}$. Since the work done in a cycle should be zero in the absence of dissipation, this implies that the imaginary parts of the bulk modulus and the density are connected to energy dissipation.

1.4 Electrodynamics

Electrodynamics describes the forces between moving electrical and magnetic fields. The equations that are used to relate these fields are called Maxwell's equations. The equations of Maxwell were derived from earlier discoveries in electromagnetism by Coulomb, Ampere, and Faraday. These laws were initially developed for isolated charges and currents in vacuum and later extended to macroscopic media. In this section we will start of with a discussion of the laws of electrostatics in vacuum and extend them to macroscopic media. That will be followed by a discussion of Faraday's law of induction and an extension of Ampere's law to the dynamics case to arrive at Maxwell's equations. Our approach is based on Jackson (1999) and a detailed exposition of the subject can be found in that text.

1.4.1 Governing equations

Coulomb's law of electrostatics is an equation for the force between point charges at rest. If q_x and q_y are two charges located at the points $\mathbf{x}$ and $\mathbf{y}$, the force between the two charges is given by

$$\mathbf{f}_{xy} = k\, q_x\, q_y\, \frac{\mathbf{x}-\mathbf{y}}{\|\mathbf{x}-\mathbf{y}\|^3}$$

where k is a constant. The *electric field* is defined at the force per unit charge. Therefore the electric field at a point $\mathbf{x}$ due to a charge q_y at point $\mathbf{y}$ can be expressed as

$$\mathbf{E}(\mathbf{x}) = k\, q_y\, \frac{\mathbf{x}-\mathbf{y}}{\|\mathbf{x}-\mathbf{y}\|^3}\,.$$

In the SI system, q_y has units of coulombs, $\mathbf{E}$ has units of volts/m, and $k = 1/(4\pi\varepsilon_0)$ where ε_0 is the *permittivity of free space* and has a value 8.854×10^{-12} farads/m.

Electric fields due to a number of charges in free space can be linearly superposed. If the point charges can be located in a region Ω and if a charge density ρ can be assigned to each point in the region, the electric field at point $\mathbf{x}$ due to the charges inside Ω is given by

$$\mathbf{E}(\mathbf{x}) = \frac{1}{4\pi\varepsilon_0}\int_\Omega \rho(\mathbf{y})\, \frac{\mathbf{x}-\mathbf{y}}{\|\mathbf{x}-\mathbf{y}\|^3}\, d\mathbf{y} = -\frac{1}{4\pi\varepsilon_0}\int_\Omega \rho(\mathbf{y})\, \boldsymbol{\nabla}_x \left(\frac{1}{\|\mathbf{x}-\mathbf{y}\|}\right) d\mathbf{y}\,.$$

Since the curl of a gradient is zero, the above relation leads directly to

$$\boldsymbol{\nabla} \times \mathbf{E} = \mathbf{0}\,. \tag{1.58}$$

Coulomb's law is more conveniently expressed in terms of an integral over the surface of the region Ω. The resulting relation is called Gauss's law and can be written as

$$\int_{\partial\Omega} \mathbf{E}\cdot\mathbf{n}\; dA = \frac{1}{\varepsilon_0}\int_\Omega \rho(\mathbf{x})\; dV\,.$$

Application of the divergence theorem leads to the differential form of Gauss's law

$$\boldsymbol{\nabla}\cdot\mathbf{E} = \frac{\rho}{\varepsilon_0}\,. \tag{1.59}$$

Another set of relations in electromagnetism are obtained from magnetostatics. Isolated electric charges can be observed in nature. However, the equivalent situation of isolated magnetic charges has never been observed. An important assumption in electromagnetism is that there are no magnetic monopoles in nature. In magnetism, the quantity that is analogous to the electrical charge is the magnetic dipole. The orientation of the magnetic dipole is given by the direction of the *magnetic flux density* (**B**), also called the *magnetic induction*.

Let us now consider the magnetic flux density produced by a current. A current is a motion of charges and is described by a current density (**J**). Consider an element of length $d\mathbf{l}$ carrying a current J. The Biot-Savart law relates the elemental magnetic flux density ($d\mathbf{B}$) produced at a point **x** due to the current in the element located at point **y** and can be expressed as

$$d\mathbf{B} = k\,J\,\frac{d\mathbf{l}\times(\mathbf{x}-\mathbf{y})}{\|\mathbf{x}-\mathbf{y}\|^3}$$

where k is a constant. In SI units $k = \mu_0/(4\pi)$ where μ_0 is the *magnetic permeability of free space* and has units of henry/m (H/m). Linear superposition also applies for many magnetic phenomena. If we assume that the magnetic induction produced by elementary current elements can be superposed, the total magnetic induction produced at a point **x** by a current density **J** located in a region Ω is given by

$$\mathbf{B}(\mathbf{x}) = \frac{\mu_0}{4\pi}\int_\Omega \mathbf{J}(\mathbf{y})\times\frac{\mathbf{x}-\mathbf{y}}{\|\mathbf{x}-\mathbf{y}\|^3}\,d\mathbf{y}\;. = -\frac{\mu_0}{4\pi}\int_\Omega \mathbf{J}(\mathbf{y})\times\boldsymbol{\nabla}_x\left(\frac{1}{\|\mathbf{x}-\mathbf{y}\|}\right)\,d\mathbf{y}\;.$$

Since the divergence of a curl is zero we have

$$\boldsymbol{\nabla}\cdot\mathbf{B} = 0\;. \tag{1.60}$$

The above states that there are no free magnetic monopoles. If we take the curl of **B** instead, we have

$$\boldsymbol{\nabla}\times\mathbf{B} = \frac{\mu_0}{4\pi}\,\boldsymbol{\nabla}\times\left[\boldsymbol{\nabla}\times\int_\Omega\left(\frac{\mathbf{J}(\mathbf{y})}{\|\mathbf{x}-\mathbf{y}\|}\right)\,d\mathbf{y}\right]\;.$$

A small amount of algebraic manipulation leads to

$$\boldsymbol{\nabla}\times\mathbf{B} = \mu_0\,\mathbf{J} + \frac{\mu_0}{4\pi}\,\boldsymbol{\nabla}\left[\int_\Omega\left(\frac{\boldsymbol{\nabla}_y\cdot\mathbf{J}(\mathbf{y})}{\|\mathbf{x}-\mathbf{y}\|}\right)\,d\mathbf{y}\right]\;.$$

In magnetostatics $\boldsymbol{\nabla}\cdot\mathbf{J} = 0$ and we get the differential form of Ampere's law

$$\boldsymbol{\nabla}\times\mathbf{B} = \mu_0\,\mathbf{J}\;. \tag{1.61}$$

The first relations between dynamic electric and magnetic fields were discovered by Faraday who found that a changing magnetic flux induces an electric field in a circuit. Let us consider a loop C in a magnetic induction field **B**. Let the loop bound an open surface $\partial\Omega$ with unit outward normal **n**. If **E** is the electric field at the element $d\mathbf{l}$

of the loop, the relation between the electromotive force around the loop and the magnetic induction is given by Faraday's law:

$$\oint_C \mathbf{E}\cdot d\mathbf{l} = -k\,\frac{D}{Dt}\left[\int_{\partial\Omega} \mathbf{B}\cdot\mathbf{n}\; dA\right]$$

where k is a constant which has the value 1 in SI units. If the fields **E** and **B** are defined in the same reference frame, Stokes' theorem can be used to get the differential form of Faraday's law

$$\boldsymbol{\nabla}\times\mathbf{E} + \frac{\partial\mathbf{B}}{\partial t} = \mathbf{0}\,. \tag{1.62}$$

Note that equation (1.62) extends the expression in equation (1.58) beyond electrostatics to include the case of time-dependent magnetic induction fields.

For the macroscopic media that are the subject of this book, the application of an electric field **E** causes the bound charges to realign themselves in response to the field and cause an *electric polarization* (**P**) which can be thought of as a charge dipole moment per unit volume. The effective charge density in a medium is reduced due to this polarization and the macroscopic form of equation (1.59) becomes

$$\boldsymbol{\nabla}\cdot\mathbf{E} = \frac{1}{\varepsilon_0}(\rho - \boldsymbol{\nabla}\cdot\mathbf{P})\,.$$

If we define the *electrical displacement* as $\mathbf{D} := \varepsilon_0\,\mathbf{E} + \mathbf{P}$, Gauss's law takes the form

$$\boldsymbol{\nabla}\cdot\mathbf{D} = \rho\,. \tag{1.63}$$

For linear media* the electric polarization is linear in **E**, i.e.,

$$\mathbf{P}(\mathbf{x},t) = \int_\Omega d\mathbf{y}\,\left[\int_{-\infty}^{\infty} \varepsilon_0\,\boldsymbol{\chi}_E(\mathbf{x},\mathbf{y},t-t')\cdot\mathbf{E}(\mathbf{x},t')\,dt'\right]$$

where $\boldsymbol{\chi}_\varepsilon$ is a second-order tensor called the *electrical susceptibility*. Therefore, we have the following relationship between **D** and **E** (where we have added the assumption that the material is local in space)*

$$\mathbf{D}(\mathbf{x},t) = \int_{-\infty}^{\infty} \varepsilon_0\,\{\mathbf{1} + \boldsymbol{\chi}_E(\mathbf{x},t-t')\}\cdot\mathbf{E}(\mathbf{x},t')\,dt' = \int_{-\infty}^{\infty} \boldsymbol{\varepsilon}(\mathbf{x},t-t')\cdot\mathbf{E}(\mathbf{x},t')\,dt'\,. \tag{1.64}$$

The quantity $\boldsymbol{\varepsilon}$ is called the *electrical permittivity* of the medium and the ratio $\boldsymbol{\varepsilon}/\varepsilon_0$ is called the *dielectric constant*. For a medium that does not have a time-dependent response

$$\mathbf{D}(\mathbf{x}) = \boldsymbol{\varepsilon}(\mathbf{x})\cdot\mathbf{E}(\mathbf{x})\,.$$

*Linearity applies as long as the fields do not become too large. This can be an issue for some metamaterials where large fields are predicted close to resonance.

*Note that the assumption that the material is local in space is reasonable for poor conductors but may fail for good conductors or small-scale structures due to Debye screening.

And for an isotropic medium the constitutive relation between **D** and **E** is $\mathbf{D} = \varepsilon\,\mathbf{E}$.

A similar set of constitutive relations can be obtained for the magnetic induction field in macroscopic media by considering an average *magnetic moment density* **M** (also called the *magnetization vector*) which contributes an effective current density $\mathbf{J}_M = \boldsymbol{\nabla} \times \mathbf{M}$ to equation (1.61). We can then write

$$\boldsymbol{\nabla} \times \mathbf{B} = \mu_0\ (\mathbf{J} + \boldsymbol{\nabla} \times \mathbf{M})\ .$$

If we define the *magnetic field* as $\mathbf{H} := \mu_0^{-1}\ \mathbf{B} - \mathbf{M}$, we get

$$\boldsymbol{\nabla} \times \mathbf{H} = \mathbf{J}\,. \tag{1.65}$$

For linear magnetic materials the relationship between **H** and **B** can be deduced in a manner similar to that for electric fields and electric displacements. Thus we can write

$$\mathbf{H}(\mathbf{x},t) = \int_{-\infty}^{\infty} \mu_0^{-1}\,\{\mathbf{1} - \boldsymbol{\chi}_H(\mathbf{x},t-t')\} \cdot \mathbf{B}(\mathbf{x},t')\ \mathrm{d}t' = \int_{-\infty}^{\infty} \boldsymbol{\mu}^{-1}(\mathbf{x},t-t') \cdot \mathbf{B}(\mathbf{x},t')\ \mathrm{d}t' \tag{1.66}$$

where $\boldsymbol{\mu}$ is the *magnetic permeability* of the medium. For time-independent media the above relation becomes

$$\mathbf{H}(\mathbf{x}) = \boldsymbol{\mu}^{-1}(\mathbf{x}) \cdot \mathbf{B}(\mathbf{x}) \quad \Longrightarrow \quad \mathbf{B}(\mathbf{x}) = \boldsymbol{\mu}(\mathbf{x}) \cdot \mathbf{H}(\mathbf{x})\ .$$

A relation between the electric and the magnetic field can be obtained with the help of the generalized Ohm's law which predicts the current density generated by an electric field,

$$\mathbf{J}(\mathbf{x}) = \boldsymbol{\sigma}(\mathbf{x}) \cdot \mathbf{E}(\mathbf{x})$$

where $\boldsymbol{\sigma}$ is the second-tensor *electrical conductivity tensor.*

We had assumed that $\boldsymbol{\nabla} \cdot \mathbf{J} = 0$ while determining the differential form of Ampere's law given in equation (1.61). However, if there is a flux of charge in and out of the domain Ω, we have to consider the contribution of that charge flux to the current in the domain. The equation for the conservation of charge is

$$\boxed{\frac{\partial \rho}{\partial t} + \boldsymbol{\nabla} \cdot \mathbf{J} = 0\,.} \tag{1.67}$$

The above equation can be derived in a manner analogous to that used for the conservation of mass in acoustics and the conservation of linear momentum in elastodynamics. Maxwell derived a relation between the magnetic field **H**, the current **J** and the electrical displacement **D** from equation (1.63) and the conservation of charge. Substitution of equation (1.63) into (1.67) gives

$$\frac{\partial}{\partial t}(\boldsymbol{\nabla} \cdot \mathbf{D}) + \boldsymbol{\nabla} \cdot \mathbf{J} = 0 \quad \Longrightarrow \quad \boldsymbol{\nabla} \cdot \left(\mathbf{J} + \frac{\partial \mathbf{D}}{\partial t}\right) = 0\,.$$

The above equation is analogous to $\boldsymbol{\nabla}\cdot\mathbf{J}=0$ but with an addition component to the current density from the bound charges. Adding the extra term to the current density in equation (1.65) leads to

$$\boldsymbol{\nabla}\times\mathbf{H}=\mathbf{J}+\frac{\partial\mathbf{D}}{\partial t}\,. \tag{1.68}$$

We now have the time-dependent Maxwell's equations in media in the absence of any internal sources of magnetic induction in the region Ω. These equations are (1.62), (1.68), (1.60), (1.63) and can be written as

$$\boxed{\begin{aligned}\boldsymbol{\nabla}\times\mathbf{E}&=-\frac{\partial\mathbf{B}}{\partial t}\,; &\quad \boldsymbol{\nabla}\times\mathbf{H}&=\frac{\partial\mathbf{D}}{\partial t}+\mathbf{J}_f\\ \boldsymbol{\nabla}\cdot\mathbf{B}&=0\,; &\quad \boldsymbol{\nabla}\cdot\mathbf{D}&=\rho_f\end{aligned}} \tag{1.69}$$

where $\mathbf{E}(\mathbf{x},t)$ is electric field, $\mathbf{B}(\mathbf{x},t)$ is the magnetic induction, $\mathbf{H}(\mathbf{x},t)$ is the magnetic field intensity, $\mathbf{D}(\mathbf{x},t)$ is the electric displacement field due to the movement of bound charges, $\mathbf{J}_f(\mathbf{x},t)$ is the free current density, and $\rho_f(\mathbf{x},t)$ is the free charge density. The primary variables in the above equations are **E** and **B**. The quantities **H** and **D** are obtained through the constitutive relations

$$\boxed{\begin{aligned}\mathbf{H}&=\boldsymbol{\mu}_0^{-1}\cdot\mathbf{B}-\mathbf{M}=\boldsymbol{\mu}^{-1}\cdot\mathbf{B}\\ \mathbf{D}&=\boldsymbol{\varepsilon}_0\cdot\mathbf{E}+\mathbf{P}=\boldsymbol{\varepsilon}\cdot\mathbf{E}\end{aligned}} \tag{1.70}$$

where $\boldsymbol{\mu}_0(\mathbf{x},t)$ is the magnetic permeability tensor of free space, $\boldsymbol{\mu}$ is the magnetic permeability of the medium, $\boldsymbol{\varepsilon}_0(\mathbf{x},t)$ is the permittivity tensor of free space, $\boldsymbol{\varepsilon}$ is the permittivity of the medium, $\mathbf{M}(\mathbf{x},t)$ is the magnetization vector, and $\mathbf{P}(\mathbf{x},t)$ is the polarization vector. The magnetization vector **M** measures the net magnetic dipole moment per unit volume. This dipole is associated with electron or nuclear spins. The polarization vector **P** measures the net electric dipole moment per unit volume and is caused by the close proximity of two charges of opposite sign. A point electric dipole is obtained when the distance between two charges tends to zero.

1.4.2 Electrodynamic potentials

Let us now examine some vector potentials that are useful in solving problems in electromagnetism. Since $\boldsymbol{\nabla}\cdot\mathbf{B}=0$, there exists a vector potential **A** such that $\mathbf{B}=\boldsymbol{\nabla}\times\mathbf{A}$. Hence,

$$\boxed{\mathbf{H}=\boldsymbol{\mu}^{-1}\cdot(\boldsymbol{\nabla}\times\mathbf{A})\,.} \tag{1.71}$$

Plugging $\mathbf{B}=\boldsymbol{\nabla}\times\mathbf{A}$ into Faraday's equation (1.62) gives

$$\boldsymbol{\nabla}\times\left(\mathbf{E}+\frac{\partial\mathbf{A}}{\partial t}\right)=0\,.$$

Therefore, there exists a scalar potential ϕ such that

$$\boxed{\mathbf{E}=-\frac{\partial\mathbf{A}}{\partial t}-\boldsymbol{\nabla}\phi \quad\Longrightarrow\quad \mathbf{D}=-\boldsymbol{\varepsilon}\cdot\left(\frac{\partial\mathbf{A}}{\partial t}+\boldsymbol{\nabla}\phi\right)\,.} \tag{1.72}$$

Let us assume that the material is isotropic. Then, for all rotations $\boldsymbol{R}^T \cdot \boldsymbol{R} = \boldsymbol{R} \cdot \boldsymbol{R}^T = \boldsymbol{1}$ we have

$$\boldsymbol{R}^T \cdot \boldsymbol{\varepsilon} \cdot \boldsymbol{R} = \boldsymbol{\varepsilon} \qquad \text{and} \qquad \boldsymbol{R}^T \cdot \boldsymbol{\mu} \cdot \boldsymbol{R} = \boldsymbol{\mu}$$

and we can write $\boldsymbol{\varepsilon} = \varepsilon\, \boldsymbol{1}$ and $\boldsymbol{\mu} = \mu\, \boldsymbol{1}$. If the medium is also homogeneous, plugging equations (1.71) and (1.72) into the two remaining Maxwell's equations leads to

$$\frac{\partial}{\partial t}(\boldsymbol{\nabla} \cdot \mathbf{A}) + \nabla^2 \phi = -\frac{\rho}{\varepsilon}\,; \quad \varepsilon\mu\, \frac{\partial^2 \mathbf{A}}{\partial t^2} - \nabla^2 \mathbf{A} + \boldsymbol{\nabla}\left[\boldsymbol{\nabla} \cdot \mathbf{A} + \varepsilon\mu\, \frac{\partial \phi}{\partial t}\right] = \mu\, \mathbf{J}\,.$$

At this stage there is some flexibility in the choice of $\mathbf{A}$ and ϕ because the magnetic induction is not changed if we substitute $\mathbf{A}$ with $\mathbf{A} - \boldsymbol{\nabla}\psi$ where ψ is a scalar field. To keep the electric field consistent with this definition of $\mathbf{A}$ we have to adjust ϕ to $\phi + \partial\psi/\partial t$. This flexibility in the allowable scalar and vector potentials allows us to choose our potentials such that they satisfy the *Lorenz condition* (Lorenz, 1867) which is equivalent to requiring that the charge be conserved, i.e.,

$$\boldsymbol{\nabla} \cdot \mathbf{A} + \varepsilon\,\mu\, \frac{\partial \phi}{\partial t} = 0\,. \tag{1.73}$$

Then, in the absence of free charges and currents in an isotropic homogeneous medium, both potentials satisfy the wave equation, i.e.,

$$\boxed{\nabla^2 \phi - \varepsilon\,\mu\, \frac{\partial^2 \phi}{\partial t^2} = 0\,; \;\; \nabla^2 \mathbf{A} - \varepsilon\,\mu\, \frac{\partial^2 \mathbf{A}}{\partial t^2} = \mathbf{0}\,.} \tag{1.74}$$

Even after these restrictions the potentials are not uniquely defined and one is free to make the gauge transformations

$$\phi' = \phi + \frac{\partial f}{\partial t}\,; \;\; \mathbf{A}' = \mathbf{A} - \boldsymbol{\nabla} f \quad \text{where} \quad \nabla^2 f - \varepsilon\,\mu\, \frac{\partial^2 f}{\partial t^2} = 0$$

to obtain new potentials ϕ', $\mathbf{A}'$. The preceding potentials are well known. However, one can go one step further and define *superpotentials* (see, for example, Bowman et al. (1969)). The most widely used superpotentials are the electric and magnetic *Hertz vector potentials* $\boldsymbol{\Pi}_e$ and $\boldsymbol{\Pi}_m$ (also known as polarization potentials). In terms of these potentials, the fields $\mathbf{E}$ and $\mathbf{H}$ can be expressed as

$$\boxed{\begin{aligned} \mathbf{E} &= \boldsymbol{\nabla} \times \boldsymbol{\nabla} \times \boldsymbol{\Pi}_e - \mu\, \boldsymbol{\nabla} \times \frac{\partial \boldsymbol{\Pi}_m}{\partial t} \\ \mathbf{H} &= \boldsymbol{\nabla} \times \boldsymbol{\nabla} \times \boldsymbol{\Pi}_m + \varepsilon\, \boldsymbol{\nabla} \times \frac{\partial \boldsymbol{\Pi}_e}{\partial t}\,. \end{aligned}} \tag{1.75}$$

Comparing equations (1.75) with (1.72) and (1.71) one sees that the superpotentials lead to symmetric representations of $\mathbf{E}$ and $\mathbf{H}$ unlike when standard vector and scalar potentials are used.

Of course, the superpotentials $\mathbf{\Pi}_e$ and $\mathbf{\Pi}_m$ are not uniquely defined and one is free to make gauge transformations

$$\mathbf{\Pi}'_e = \mathbf{\Pi}_e + \nabla g_e(\mathbf{x},t)$$
$$\mathbf{\Pi}'_m = \mathbf{\Pi}_m + \nabla g_m(\mathbf{x},t)$$

where $g_e(\mathbf{x},t)$ and $g_m(\mathbf{x},t)$ are arbitrary scalar potential functions. Plugging these definitions into the Maxwell equations leads to

$$\boxed{\begin{aligned} \nabla\times\nabla\times\mathbf{\Pi}_e + \varepsilon\,\mu\,\frac{\partial^2\mathbf{\Pi}_e}{\partial t^2} &= \nabla f \\ \nabla\times\nabla\times\mathbf{\Pi}_m + \varepsilon\,\mu\,\frac{\partial^2\mathbf{\Pi}_m}{\partial t^2} &= \nabla f \end{aligned}} \tag{1.76}$$

where f is an arbitrary scalar potential which is a function of position and time. The Lorenz condition is satisfied if $f = \nabla\cdot\mathbf{\Pi}_e$. The potentials $\mathbf{A}$ and ϕ can also be expressed in terms of $\mathbf{\Pi}_e$ and $\mathbf{\Pi}_m$ as

$$\phi = -\nabla\cdot\mathbf{\Pi}_e\ ;\quad \mathbf{A} = \varepsilon\,\mu\,\frac{\partial\mathbf{\Pi}_e}{\partial t} + \mu\,\nabla\times\mathbf{\Pi}_m\ .$$

1.4.3 Cartesian and curvilinear coordinates

Recall that the curl of a vector expressed in components with respect to a Cartesian basis is given by

$$\nabla\times\mathbf{v} = e_{ijk}\,\frac{\partial v_j}{\partial x_i}\,\mathbf{e}_k = e_{kij}\,v_{j,i}\,\mathbf{e}_k$$

where $e_{ijk} = e_{kij} = e_{jki} = -e_{kji} = -e_{jik} = -e_{ikj} = 1(i \neq j \neq k)$ is the permutation symbol. Maxwell's equations (1.69) for media can be expressed as

$$\begin{aligned} e_{kij}\,E_{j,i} &= -\frac{\partial B_k}{\partial t}\ ; & e_{kij}\,H_{j,i} &= -\frac{\partial D_k}{\partial t} + J_k \\ B_{i,i} &= 0\ ; & D_{i,i} &= \rho\ . \end{aligned} \tag{1.77}$$

Similarly, the constitutive relations (1.70) are

$$H_i = \left[\mu^{-1}\right]_{ij}\,B_j\ ;\quad D_i = \varepsilon_{ij}\,E_j\ . \tag{1.78}$$

The expression for the curl of the curl of a vector in Cartesian coordinates is

$$\nabla\times\nabla\times\mathbf{v} = e_{mpq}e_{qij}v_{j,ip}\,\mathbf{e}_m\ .$$

Therefore Maxwell's equations (1.75) in terms of the Hertz superpotentials are

$$e_{rpq}e_{qij}\Pi^e_{j,ip} + \varepsilon\mu\frac{\partial^2\Pi^e_r}{\partial t^2} = f_{,r}\ ;\quad e_{rpq}e_{qij}\Pi^m_{j,ip} + \varepsilon\mu\frac{\partial^2\Pi^m_r}{\partial t^2} = f_{,r}\ .$$

To express Maxwell's equations in terms of spatial curvilinear coordinates we start with the definition of the curl of a vector ($\mathbf{v}$),

$$(\boldsymbol{\nabla} \times \mathbf{v}) \cdot \mathbf{a} = \boldsymbol{\nabla} \cdot (\mathbf{v} \times \mathbf{a})$$

where $\mathbf{a}$ is an arbitrary constant vector. Let us express the vectors $\mathbf{v}$ and $\mathbf{a}$ in the form $\mathbf{v} = v_i \mathbf{g}^i$ and $\mathbf{a} = a_i \mathbf{g}^i$. Then

$$\mathbf{v} \times \mathbf{a} = v_i a_j \left(\mathbf{g}^i \times \mathbf{g}^j\right) \,.$$

We can show that

$$\mathbf{g}^i \times \mathbf{g}^j = \frac{1}{\sqrt{g}}\, e^{kij}\, \mathbf{g}_k \quad \text{where } g = \det g_{ij} \quad \text{and } e^{ijk} = e_{ijk} \,.$$

Therefore, we get the expression for the vector product of two vectors in curvilinear coordinates,

$$\mathbf{b} := \mathbf{v} \times \mathbf{a} = \frac{1}{\sqrt{g}}\, e^{kij}\, v_i a_j\, \mathbf{g}_k = b^k \mathbf{g}_k \quad \Leftrightarrow \quad b^k = \frac{1}{\sqrt{g}}\, e^{kij}\, v_i a_j \,.$$

Now, the divergence of the vector $(\mathbf{v} \times \mathbf{a})$ can be expressed in curvilinear coordinates as

$$\boldsymbol{\nabla} \cdot (\mathbf{v} \times \mathbf{a}) = \boldsymbol{\nabla} \cdot \mathbf{b} = \frac{1}{\sqrt{g}} \frac{\partial}{\partial \theta^k} \left(b^k \sqrt{g}\right) = \frac{1}{\sqrt{g}} \frac{\partial}{\partial \theta^k} \left(e^{kij}\, v_i a_j\right) \,.$$

Hence we have, using the definition of the curl of a vector,

$$(\boldsymbol{\nabla} \times \mathbf{v}) \cdot \mathbf{a} = \left([\boldsymbol{\nabla} \times \mathbf{v}]^p \mathbf{g}_p\right) \cdot \left(a_q \mathbf{g}^q\right) = \frac{e^{kij}}{\sqrt{g}} \frac{\partial v_i}{\partial \theta^k}\, a_j$$

or, invoking the arbitrariness of $\mathbf{a}$,

$$[\boldsymbol{\nabla} \times \mathbf{v}]^j = \frac{e^{kij}}{\sqrt{g}} \frac{\partial v_i}{\partial \theta^k} \quad \Leftrightarrow \quad [\boldsymbol{\nabla} \times \mathbf{v}]_\ell = g_{\ell j} [\boldsymbol{\nabla} \times \mathbf{v}]^j = \frac{e^{kij}}{\sqrt{g}}\, g_{\ell j}\, \frac{\partial v_i}{\partial \theta^k} \,.$$

We therefore get an expression for the curl of a vector in curvilinear coordinates,

$$\boldsymbol{\nabla} \times \mathbf{v} = e^{kij} \frac{g_{\ell j}}{\sqrt{g}} \frac{\partial v_i}{\partial \theta^k}\, \mathbf{g}^\ell \,.$$

Using the above expression, Maxwell's equations (1.69) for media in spatial curvilinear coordinates for stationary media are

$$\begin{aligned} & e^{kij} \frac{g_{\ell j}}{\sqrt{g}} \frac{\partial E_i}{\partial \theta^k} = -\frac{\partial B_\ell}{\partial t} \,; \quad e^{kij} \frac{g_{\ell j}}{\sqrt{g}} \frac{\partial H_i}{\partial \theta^k} = -\frac{\partial D_\ell}{\partial t} + J_\ell \\ & \frac{1}{\sqrt{g}} \frac{\partial}{\partial \theta^k} \left(B^k \sqrt{g}\right) = 0 \,; \quad \frac{1}{\sqrt{g}} \frac{\partial}{\partial \theta^k} \left(D^k \sqrt{g}\right) = \rho \end{aligned} \tag{1.79}$$

and the constitutive relations (1.70) have the form

$$H_i = g^{j\ell}\left[\mu^{-1}\right]_{ij} B_\ell\,; \quad D_i = g^{j\ell}\varepsilon_{ij}\, E_\ell\,. \tag{1.80}$$

It is also convenient in some situations to express the above equations in terms of curved space-time coordinates and some of the literature on metamaterials based on coordinate transformations use these equations. However, we will not explore these forms of Maxwell's equations.

1.4.4 A slight simplification

To simplify our discussion of electrodynamics in much of the rest of the book let us make some assumptions:

1. The free current density ($\mathbf{J}_f$) arises only from conduction currents arising from the response of the medium and not from beams of charged particles.
2. In the far distant past ($t = -\infty$) all fields are zero.
3. The media are at rest.

Define

$$\widetilde{\mathbf{D}}(\mathbf{x},t) := \mathbf{D}(\mathbf{x},t) + \int_{-\infty}^{t} \mathbf{J}_f(\mathbf{x},\tau)\,\mathrm{d}\tau\,. \tag{1.81}$$

Then taking the time derivative of (1.81) and using Maxwell's equation connecting the magnetic field to the total current, we have

$$\frac{\partial \widetilde{\mathbf{D}}}{\partial t} = \frac{\partial \mathbf{D}}{\partial t} + \mathbf{J}_f = \boldsymbol{\nabla}\times\mathbf{H}\,. \tag{1.82}$$

Taking the divergence of (1.81) and using the conservation of charge (1.67) and Maxwell's equation relating the electric displacement to the free charge density, we get

$$\boldsymbol{\nabla}\cdot\widetilde{\mathbf{D}} = \boldsymbol{\nabla}\cdot\mathbf{D} + \int_{-\infty}^{t} \boldsymbol{\nabla}\cdot\mathbf{J}_f\,\mathrm{d}\tau = \boldsymbol{\nabla}\cdot\mathbf{D} - \int_{-\infty}^{t} \frac{\partial \rho_f}{\partial t}\,\mathrm{d}\tau = \boldsymbol{\nabla}\cdot\mathbf{D} - \rho_f = 0\,. \tag{1.83}$$

Therefore, we can write Maxwell's equations in terms of $\widetilde{\mathbf{D}}$ as

$$\boxed{\boldsymbol{\nabla}\times\mathbf{E} = -\frac{\partial \mathbf{B}}{\partial t}\,; \quad \boldsymbol{\nabla}\cdot\mathbf{B} = 0\,; \quad \boldsymbol{\nabla}\times\mathbf{H} = \frac{\partial \widetilde{\mathbf{D}}}{\partial t}\,; \quad \boldsymbol{\nabla}\cdot\widetilde{\mathbf{D}} = 0\,.} \tag{1.84}$$

This reduction reflects the fact that it is difficult to distinguish the free current density $\mathbf{J}_f$ from currents arising from the electric displacement field through $\partial\mathbf{D}/\partial t$. To complete the system of equations (1.84), we need relations between the fields $\mathbf{E}$, $\widetilde{\mathbf{D}}$, $\mathbf{B}$, and $\mathbf{H}$. We reiterate some further assumptions that are made regarding these relations.

1. We assume that $\widetilde{\mathbf{D}}$ is only coupled with $\mathbf{E}$ and that $\mathbf{H}$ is only coupled with $\mathbf{B}$. This is a good approximation for many stationary materials. But more generally there is cross coupling between these quantities, for example, in bi-isotropic and bi-anisotropic materials.
2. We assume that the relations between $\widetilde{\mathbf{D}}$ and $\mathbf{E}$, and $\mathbf{H}$ and $\mathbf{B}$ are linear.
3. Causality applies, i.e., the net magnetic dipole moment $\mathbf{M}$ (and hence the magnetic field $\mathbf{H}$) does not depend on future values of $\mathbf{B}$. For the same reason the free current density $\mathbf{J}_f$ (and hence the electric displacement field $\mathbf{D}$) cannot depend on future values of $\mathbf{E}$.
4. We further assume that the media can be approximated as being local in space (this is true for poor conductors but may fail for good conductors due to Debye screening).

Recall the relation between the magnetic field and the electric displacement in equation (1.66),

$$\mathbf{H}(\mathbf{x},t) = \int_{-\infty}^{\infty} \boldsymbol{\mu}^{-1}(\mathbf{x},t-t') \cdot \mathbf{B}(\mathbf{x},t')\, \mathrm{d}t' \,.$$

Linearity implies that we have a similar relationship between $\widetilde{\mathbf{D}}$ and $\mathbf{E}$. For simplicity of notation let us write these constitutive relations in terms of two second-order tensor valued kernel functions $\bar{\boldsymbol{K}}_B$ and $\bar{\boldsymbol{K}}_E$. Recall that such kernel functions may be singular (such as delta functions) and the integrals should be interpreted in the sense of measure theory under such conditions. Therefore, we have

$$\boxed{\begin{aligned} \mathbf{H}(\mathbf{x},t) &= \int_{-\infty}^{\infty} \bar{\boldsymbol{K}}_B(\mathbf{x},t-t') \cdot \mathbf{B}(\mathbf{x},t')\, \mathrm{d}t' \\ \widetilde{\mathbf{D}}(\mathbf{x},t) &= \int_{-\infty}^{\infty} \bar{\boldsymbol{K}}_E(\mathbf{x},t-t') \cdot \mathbf{E}(\mathbf{x},t')\, \mathrm{d}t' \end{aligned}} \tag{1.85}$$

where causality requires that $\bar{\boldsymbol{K}}_B(\mathbf{x},\tau) = \bar{\boldsymbol{K}}_E(\mathbf{x},\tau) = 0$ if $\tau = t - t' < 0$.

1.4.5 Maxwell's equations at fixed frequency

To obtain Maxwell's equations at fixed frequency (ω) we Fourier transform the equations. This is equivalent to assuming that all fields depend harmonically on time. Taking the first of Maxwell's equations (1.84), we have

$$\int_{-\infty}^{\infty} \boldsymbol{\nabla} \times \mathbf{E}\, e^{i\omega t}\, \mathrm{d}t = -\int_{-\infty}^{\infty} \frac{\partial \mathbf{B}}{\partial t} e^{i\omega t}\, \mathrm{d}t$$

or

$$\boldsymbol{\nabla} \times \left[\int_{-\infty}^{\infty} \mathbf{E} e^{i\omega t}\, \mathrm{d}t\right] = -\, \mathbf{B} e^{i\omega t}\Big|_{-\infty}^{\infty} + i\omega \int_{-\infty}^{\infty} \mathbf{B} e^{i\omega t}\, \mathrm{d}t \,.$$

If we assume that the magnetic induction $\mathbf{B}$ is zero at $\pm\infty$ we get, using the usual hat notation for the Fourier transformed quantities,

$$\boldsymbol{\nabla} \times \widehat{\mathbf{E}}(\mathbf{x},\omega) = i\omega \widehat{\mathbf{B}}(\mathbf{x},\omega) \,.$$

In a similar manner we get $\nabla \cdot \widehat{\mathbf{B}} = 0$, $\nabla \times \widehat{\mathbf{H}} = -i\omega \widehat{\mathbf{D}}$, and $\nabla \cdot \widehat{\mathbf{D}} = 0$. Therefore the fixed frequency version of Maxwell's equations for media are

$$\nabla \times \widehat{\mathbf{E}} = i\omega\, \widehat{\mathbf{B}}(\mathbf{x}) \; ; \quad \nabla \cdot \widehat{\mathbf{B}} = 0 \; ; \quad \nabla \times \widehat{\mathbf{H}} = -i\omega\, \widehat{\mathbf{D}}(\mathbf{x}) \; ; \quad \nabla \cdot \widehat{\mathbf{D}} = 0 \qquad (1.86)$$

where

$$\widehat{\mathbf{E}}(\mathbf{x},\omega) = \int_{-\infty}^{\infty} \mathbf{E}(\mathbf{x},t) e^{i\omega t}\, dt \; ; \quad \widehat{\mathbf{B}}(\mathbf{x},\omega) = \int_{-\infty}^{\infty} \mathbf{B}(\mathbf{x},t) e^{i\omega t}\, dt$$
$$\widehat{\mathbf{H}}(\mathbf{x},\omega) = \int_{-\infty}^{\infty} \mathbf{H}(\mathbf{x},t) e^{i\omega t}\, dt \; ; \quad \widehat{\mathbf{D}}(\mathbf{x},\omega) = \int_{-\infty}^{\infty} \widetilde{\mathbf{D}}(\mathbf{x},t) e^{i\omega t}\, dt \; .$$

The convolutions in the constitutive relations (1.85) turn into products when Fourier transformed. Therefore, these relations may be written in terms of the magnetic permeability ($\boldsymbol{\mu}$) and the electric permittivity ($\boldsymbol{\varepsilon}$) at fixed frequency as

$$\widehat{\mathbf{H}}(\mathbf{x},\omega) = [\boldsymbol{\mu}(\mathbf{x},\omega)]^{-1} \cdot \widehat{\mathbf{B}}(\mathbf{x},\omega) \; ; \quad \widehat{\mathbf{D}}(\mathbf{x},\omega) = \boldsymbol{\varepsilon}(\mathbf{x},\omega) \cdot \widehat{\mathbf{E}}(\mathbf{x},\omega) \qquad (1.87)$$

where

$$[\boldsymbol{\mu}(\mathbf{x},\omega)]^{-1} = \int_{-\infty}^{\infty} \bar{\boldsymbol{K}}_B(\mathbf{x},\tau)\, e^{i\omega\tau}\, d\tau \; ; \quad \boldsymbol{\varepsilon}(\mathbf{x},\omega) = \int_{-\infty}^{\infty} \bar{\boldsymbol{K}}_E(\mathbf{x},\tau)\, e^{i\omega\tau}\, d\tau \; .$$

Substituting equations (1.87) into equations (1.86), and dropping the hats, we get

$$\boxed{\nabla \times \mathbf{E} = i\omega\, \boldsymbol{\mu} \cdot \mathbf{H} \; ; \; \nabla \times \mathbf{H} = -i\omega\, \boldsymbol{\varepsilon} \cdot \mathbf{E} \; ; \; \nabla \cdot \mathbf{B} = 0 \; ; \; \nabla \cdot \mathbf{D} = 0 \, .} \qquad (1.88)$$

These are Maxwell's equations for media at a fixed frequency.*

If the medium is isotropic we have $\boldsymbol{\mu} = \mu \boldsymbol{1}$ and $\boldsymbol{\varepsilon} = \varepsilon \boldsymbol{1}$ where $\boldsymbol{1}$ is the second-order identity tensor. In that case Maxwell's equations at fixed frequency can be written as

$$\nabla \times \mathbf{E} = i\omega\, \mu\, \mathbf{H} \; ; \; \nabla \times \mathbf{H} = -i\omega\, \varepsilon\, \mathbf{E} \; ; \; \nabla \cdot \mathbf{B} = 0 \; ; \; \nabla \cdot \mathbf{D} = 0 \, . \qquad (1.89)$$

If we also assume that μ and ε do not depend upon position, i.e., $\mu \equiv \mu(\omega)$ and $\varepsilon \equiv \varepsilon(\omega)$, and take the curl of equations (1.89), we get

$$\nabla \times (\nabla \times \mathbf{E}) = i\omega\, \mu\, \nabla \times \mathbf{H} = \omega^2\, \varepsilon\, \mu\, \mathbf{E}(\mathbf{x})$$
$$\nabla \times (\nabla \times \mathbf{H}) = -i\omega\, \varepsilon\, \nabla \times \mathbf{E} = \omega^2\, \varepsilon\, \mu\, \mathbf{H}(\mathbf{x}) \, .$$

Replacing $\nabla \times (\nabla \times \mathbf{A})$ with $\nabla(\nabla \cdot \mathbf{A}) - \nabla^2 \mathbf{A}$ in the above equations gives

$$\begin{aligned} \nabla(\nabla \cdot \mathbf{E}) - \nabla^2 \mathbf{E} &= \omega^2\, \varepsilon\, \mu\, \mathbf{E}(\mathbf{x}) \\ \nabla(\nabla \cdot \mathbf{H}) - \nabla^2 \mathbf{H} &= \omega^2\, \varepsilon\, \mu\, \mathbf{H}(\mathbf{x}) \, . \end{aligned} \qquad (1.90)$$

Recall from equations (1.87) that $\mathbf{H} = \mu^{-1}\, \mathbf{B}$ and $\mathbf{D} = \varepsilon\, \mathbf{E}$. We also know from Maxwell's equations that $\nabla \cdot \mathbf{B} = 0$ and $\nabla \cdot \mathbf{D} = 0$. Therefore,

$$\nabla \cdot \mathbf{H} = \mu^{-1}\, \nabla \cdot \mathbf{B} = 0 \qquad \text{and} \qquad \nabla \cdot \mathbf{E} = \varepsilon^{-1}\, \nabla \cdot \mathbf{D} = 0 \, .$$

*For time harmonic problems $\mathbf{E} \equiv \mathbf{E}(\mathbf{x})$, $\mathbf{H} \equiv \mathbf{H}(\mathbf{x})$, $\mathbf{B} \equiv \mathbf{B}(\mathbf{x})$, and $\mathbf{D} \equiv \mathbf{D}(\mathbf{x})$.

Plugging the above results into equation (1.90) gives us

$$\nabla^2\mathbf{E} + \omega^2\,\varepsilon\,\mu\,\mathbf{E} = \mathbf{0}\,;\;\; \nabla^2\mathbf{H} + \omega^2\,\varepsilon\,\mu\,\mathbf{H} = \mathbf{0}\,.$$

These wave equations are often expressed in the form

$$\boxed{\nabla^2\mathbf{E} + \kappa^2\,\mathbf{E}(\mathbf{x}) = \mathbf{0}\,;\quad \nabla^2\mathbf{H} + \kappa^2\,\mathbf{H}(\mathbf{x}) = \mathbf{0}} \tag{1.91}$$

where $\kappa^2 = \omega^2/c^2$ and $c^2 = 1/(\varepsilon\,\mu)$. The quantity c is the phase velocity (the velocity at which the wave crests travel). To have a propagating wave, c must be real. This will be the case when ε and μ are both positive or both negative as shown in the schematic in Figure 1.3.

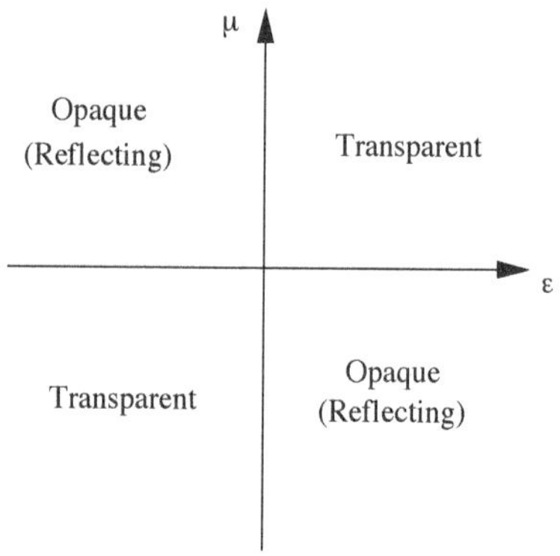

FIGURE 1.3
Electromagnetic transparency and opacity of a material as a function of μ and ε.

For some conductors, at low frequencies, the permittivity tensor is expressed in terms of the conductivity $\boldsymbol{\sigma}$ as

$$\boxed{\boldsymbol{\varepsilon} = \boldsymbol{\varepsilon}_0 + \frac{i}{\omega}\,\boldsymbol{\sigma}} \tag{1.92}$$

where $\boldsymbol{\varepsilon}_0$ is the real part of the permittivity tensor and $\boldsymbol{\sigma}$ is the electrical conductivity tensor. The above relation for the permittivity tensor can be obtained as follows. Recall that

$$\widetilde{\mathbf{D}}(\mathbf{x},t) = \mathbf{D}(\mathbf{x},t) + \int_{-\infty}^{t} \mathbf{J}_f(\mathbf{x},\tau)\;d\tau\,.$$

Differentiation with respect to time gives

$$\frac{\partial\widetilde{\mathbf{D}}}{\partial t} = \frac{\partial\mathbf{D}}{\partial t} + \mathbf{J}_f(\mathbf{x},t)\,.$$

Assuming harmonic solutions of the form

$$\widetilde{\mathbf{D}}(\mathbf{x},t) = \widehat{\widetilde{\mathbf{D}}}(\mathbf{x})\ \exp(-i\omega t)\ ;\ \ \mathbf{D}(\mathbf{x},t) = \widehat{\mathbf{D}}(\mathbf{x})\ \exp(-i\omega t)$$
$$\mathbf{J}_f(\mathbf{x},t) = \widehat{\mathbf{J}}_f(\mathbf{x})\ \exp(-i\omega t)\ ;\ \ \mathbf{E}(\mathbf{x},t) = \widehat{\mathbf{E}}(\mathbf{x})\ \exp(-i\omega t)$$

and plugging into the differential equation, we get

$$(-i\omega)\widehat{\widetilde{\mathbf{D}}}(\mathbf{x}) = (-i\omega)\widehat{\mathbf{D}}(\mathbf{x}) + \widehat{\mathbf{J}}_f(\mathbf{x})\ .$$

If we assume that the medium obeys Ohm's law and has a "reference" dielectric constant $\boldsymbol{\varepsilon}_0$, the free current density $\mathbf{J}_f$ and the electric displacement $\mathbf{D}$ are related to the electric field $\mathbf{E}$ by

$$\mathbf{J}_f = \boldsymbol{\sigma}\cdot\mathbf{E}\ ;\ \ \mathbf{D} = \boldsymbol{\varepsilon}_0\cdot\mathbf{E}\ .$$

Therefore,

$$\widehat{\widetilde{\mathbf{D}}}(\mathbf{x}) = \boldsymbol{\varepsilon}_0\cdot\widehat{\mathbf{E}}(\mathbf{x}) + \frac{i}{\omega}\boldsymbol{\sigma}\cdot\widehat{\mathbf{E}}(\mathbf{x}) = \boldsymbol{\varepsilon}\cdot\widehat{\mathbf{E}}(\mathbf{x}) \quad \text{with} \quad \boldsymbol{\varepsilon} := \boldsymbol{\varepsilon}_0 + \frac{i}{\omega}\boldsymbol{\sigma}\ .$$

When we discuss electromagnetic metamaterials in Chapter 4 we will see that we get a different relation for the permittivity at high frequencies.

1.4.6 TE and TM modes in electromagnetism

Let us now consider two important modes of electromagnetic wave propagation: the transverse electric field and transverse magnetic field. These modes are closely analogous to antiplane shear in elastodynamics and to acoustic waves.

A transverse electric (TE) field (also called an 's'-polarized wave) is a special two-dimensional solution of Maxwell's equations that has the form $E_1 \equiv E_1(x_2,x_3)$ and $E_2 = E_3 = 0$ in rectangular Cartesian coordinates. Let us assume that μ and ε are scalars which are independent of x_1, i.e., $\mu \equiv \mu(x_2,x_3)$ and $\varepsilon \equiv \varepsilon(x_2,x_3)$. Then, from Maxwell's equations (1.89) at a fixed frequency (ω) in isotropic media, we have

$$\boldsymbol{\nabla}\times\mathbf{E} = e_{kij}E_{j,i}\mathbf{e}_k = E_{1,3}\ \mathbf{e}_2 - E_{1,2}\ \mathbf{e}_3\ .$$

Since $\boldsymbol{\nabla}\times\mathbf{E} = i\omega\mu\mathbf{H}$, we have

$$\mathbf{H} = \frac{1}{i\omega\mu}E_{1,3}\ \mathbf{e}_2 - \frac{1}{i\omega\mu}E_{1,2}\ \mathbf{e}_3\ .$$

Therefore,

$$\boldsymbol{\nabla}\times\mathbf{H} = e_{kij}H_{j,i}\mathbf{e}_k = (H_{3,2} - H_{2,3})\mathbf{e}_1 = \frac{i}{\omega}\left[\left(\mu^{-1}\ E_{1,2}\right)_{,2} + \left(\mu^{-1}\ E_{1,3}\right)_{,3}\right]\mathbf{e}_1\ .$$

From (1.89) we also have $\boldsymbol{\nabla}\times\mathbf{H} = -i\omega\varepsilon\mathbf{E}$ which leads to

$$\left[\left(\mu^{-1}\ E_{1,2}\right)_{,2} + \left(\mu^{-1}\ E_{1,3}\right)_{,3}\right]\mathbf{e}_1 = -\omega^2\varepsilon E_1\mathbf{e}_1\ .$$

or

$$(\mu^{-1}\,E_{1,2})_{,2} + (\mu^{-1}\,E_{1,3})_{,3} + \omega^2 \varepsilon E_1 = 0\,.$$

In terms of the two-dimensional gradient operator discussed in Section 1.2.6 we can write the above equation as

$$\boxed{\overline{\boldsymbol{\nabla}} \cdot \left(\frac{1}{\mu(x_2,x_3)}\overline{\boldsymbol{\nabla}} E_1\right) + \omega^2\,\varepsilon(x_2,x_3)\,E_1 = 0\,.} \tag{1.93}$$

This is the governing equation for the TE mode and has the same form as equation (1.27) for antiplane shear with the substitutions

$$E_1 \leftrightarrow u_1\,,\quad \frac{1}{\mu} \leftrightarrow \mu\,,\quad \varepsilon \leftrightarrow \rho\,.$$

Let us now consider the case where the permeability and permittivity are anisotropic. In that case we can write these second-order tensors in matrix form as

$$\boldsymbol{\mu} = \boldsymbol{\mu}(x_2,x_3) \equiv \begin{bmatrix} \mu_{11} & 0 & 0 \\ 0 & \mu_{22} & \mu_{23} \\ 0 & \mu_{23} & \mu_{33} \end{bmatrix} \quad \text{and} \quad \boldsymbol{\varepsilon} = \boldsymbol{\varepsilon}(x_2,x_3) \equiv \begin{bmatrix} \varepsilon_{11} & 0 & 0 \\ 0 & \varepsilon_{22} & \varepsilon_{23} \\ 0 & \varepsilon_{23} & \varepsilon_{33} \end{bmatrix}\,. \tag{1.94}$$

We can show that the TE equation for this situation can be written as

$$\boxed{\overline{\boldsymbol{\nabla}} \cdot \left[(\boldsymbol{R}_\perp \cdot \boldsymbol{M}^{-1} \cdot \boldsymbol{R}_\perp^T) \cdot \overline{\boldsymbol{\nabla}} E_1\right] + \omega^2\,\varepsilon_{11}\,E_1 = 0} \tag{1.95}$$

where

$$\boldsymbol{R}_\perp \equiv \begin{bmatrix} 0 & 1 \\ -1 & 0 \end{bmatrix} \quad \text{and} \quad \boldsymbol{M} \equiv \begin{bmatrix} \mu_{22} & \mu_{23} \\ \mu_{23} & \mu_{33} \end{bmatrix}\,.$$

A transverse magnetic (TM) field, also called a 'p'-polarized wave, is another special two-dimensional solution of Maxwell's equations. For a TM field, $H_1 \equiv H_1(x_2,x_3)$ and $H_2 = H_3 = 0$ in rectangular Cartesian coordinates. If we again assume that the permeability and permittivity are scalars that are independent of x_1, we have

$$\mathbf{E} = \frac{i}{\omega\varepsilon}\boldsymbol{\nabla}\times\mathbf{H} = \frac{i}{\omega}\left[\frac{H_{1,3}}{\varepsilon}\mathbf{e}_2 - \frac{H_{1,2}}{\varepsilon}\mathbf{e}_3\right]\,.$$

Hence,

$$\mathbf{H} = -\frac{i}{\omega\mu}\boldsymbol{\nabla}\times\mathbf{E} = -\frac{i}{\omega\mu}(E_{3,2}-E_{2,3})\mathbf{e}_1 = \frac{1}{\omega^2\mu}\left[\left(-\frac{H_{1,2}}{\varepsilon}\right)_{,2} - \left(\frac{H_{1,3}}{\varepsilon}\right)_{,3}\right]\mathbf{e}_1\,.$$

Therefore the TM wave equation for the isotropic case has the form

$$\boxed{\overline{\boldsymbol{\nabla}} \cdot \left(\frac{1}{\varepsilon(x_2,x_3)}\overline{\boldsymbol{\nabla}} H_1\right) + \omega^2\,\mu(x_2,x_3)\,H_1 = 0\,.} \tag{1.96}$$

For anisotropic media with permittivity and permeability of the form given in equation (1.94), the TM wave equation can be expressed as

$$\boxed{\overline{\boldsymbol{\nabla}}\cdot\left[\left(\boldsymbol{R}_\perp\cdot\boldsymbol{E}^{-1}\cdot\boldsymbol{R}_\perp^T\right)\overline{\boldsymbol{\nabla}}H_1\right]+\omega^2\,\mu_{11}\,H_1=0} \tag{1.97}$$

where

$$\boldsymbol{R}_\perp \equiv \begin{bmatrix} 0 & 1 \\ -1 & 0 \end{bmatrix} \quad \text{and} \quad \boldsymbol{E} \equiv \begin{bmatrix} \varepsilon_{22} & \varepsilon_{23} \\ \varepsilon_{23} & \varepsilon_{33} \end{bmatrix}.$$

The general solution independent of x_1 is a superposition of the TE and TM solutions. This can be seen by observing that the Maxwell equations decouple under these conditions and a general solution can be written as

$$(E_1,E_2,E_3) = (E_1,0,0) + (0,E_2,E_3) \tag{1.98}$$

where the first term represents the TE solution. We can see that the second term represents the TM solution because

$$\boldsymbol{\nabla}\times(0,E_2,E_3) = [E_{3,2}-E_{2,3},0,0] \tag{1.99}$$

implying that $H_2 = H_3 = 0$ which is the TM solution.

1.4.7 Maxwell's equations in elasticity form

Interestingly, Maxwell's equations can be reduced to the form of the equations of elastodynamics at fixed frequency (Milton, 2006). Recall that, at a fixed ω, Maxwell's equations take the form

$$\boldsymbol{\nabla}\times\mathbf{E} = i\omega\boldsymbol{\mu}(\mathbf{x})\cdot\mathbf{H}(\mathbf{x})\ ;\ \ \boldsymbol{\nabla}\times\mathbf{H} = -i\omega\boldsymbol{\varepsilon}(\mathbf{x})\cdot\mathbf{E}(\mathbf{x})\,.$$

Therefore,

$$\mathbf{H} = \frac{1}{i\omega}\boldsymbol{\mu}^{-1}\cdot(\boldsymbol{\nabla}\times\mathbf{E}) \quad\Longrightarrow\quad \boldsymbol{\nabla}\times\mathbf{H} = \frac{1}{i\omega}\boldsymbol{\nabla}\times\left[\boldsymbol{\mu}^{-1}\cdot(\boldsymbol{\nabla}\times\mathbf{E})\right]$$

or

$$\boldsymbol{\nabla}\times\left[\boldsymbol{\mu}^{-1}\cdot(\boldsymbol{\nabla}\times\mathbf{E})\right] = \omega^2\,\boldsymbol{\varepsilon}\cdot\mathbf{E}\,.$$

Recall that the curl of a vector can be written in orthonormal coordinates as $\boldsymbol{\nabla}\times\mathbf{a} = e_{kij}\,a_{j,i}\,\mathbf{e}_k$ where e_{ijk} is the permutation symbol. Therefore,

$$\begin{aligned}\left[\omega^2\,\boldsymbol{\varepsilon}\cdot\mathbf{E}\right]_j &= e_{jmn}\,[\boldsymbol{\mu}^{-1}\cdot(\boldsymbol{\nabla}\times\mathbf{E})]_{n,m} = e_{jmn}\,[(\boldsymbol{\mu}^{-1})_{np}\,(\boldsymbol{\nabla}\times\mathbf{E})_p]_{,m} \\ &= e_{jmn}\,[(\boldsymbol{\mu}^{-1})_{np}\,e_{pk\ell}\,E_{\ell,k}]_{,m} = [e_{jmn}\,e_{pk\ell}\,(\boldsymbol{\mu}^{-1})_{np}\,E_{\ell,k}]_{,m} \\ &= [C_{mj\ell k}E_{\ell,k}]_{,m} \quad \text{where} \quad C_{mj\ell k} := e_{mnj}\,e_{\ell pk}\,[\boldsymbol{\mu}^{-1}]_{np}\end{aligned}$$

or

$$\boxed{\omega^2\,\boldsymbol{\varepsilon}\cdot\mathbf{E} = \boldsymbol{\nabla}\cdot(\mathbf{C}:\boldsymbol{\nabla}\mathbf{E})\,.} \tag{1.100}$$

This is very similar to the elasticity equation (1.24)

$$-\omega^2 \rho \, \mathbf{u} = \nabla \cdot (\mathbf{C} : \nabla \mathbf{u}) \, . \tag{1.101}$$

The permittivity is similar to a *negative density* and the electric field is similar to the displacement. The equations also hint at an anisotropic tensorial density. However, continuity conditions are different for the two equations, i.e., at an interface, $\mathbf{u}$ is continuous while only the tangential component of $\mathbf{E}$ is continuous. Also, the tensor $\mathbf{C}$ has different symmetries for the two situations. Interestingly, for Maxwell's equations

$$C_{ijk\ell} = -C_{jik\ell} = -C_{ij\ell k} = C_{k\ell ij} \, . \tag{1.102}$$

1.4.8 Electrodynamic power and dissipation

For electrodynamics, the average rate of work done in a harmonic closed cycle is given by

$$P = \frac{\omega}{2\,\pi} \int_0^{2\pi/\omega} \left[\mathbf{E}(t) \cdot \frac{\partial \mathbf{D}(t)}{\partial t} + \mathbf{H}(t) \cdot \frac{\partial \mathbf{B}(t)}{\partial t} \right] dt \, . \tag{1.103}$$

The quantity $\mathbf{E}$ is equivalent to the voltage and the quantity rate of change of electrical displacement $\partial \mathbf{D}/\partial t$ is equivalent to the current (recall that in electrostatics the power is given by $P = V\,I$). In addition, we also have a contribution due to magnetic induction.

Let us assume that the fields can be expressed in harmonic form, i.e.,

$$\mathbf{E}(t) = \text{Re}[\widehat{\mathbf{E}}\, e^{-i\omega t}] \; ; \;\; \mathbf{H}(t) = \text{Re}[\widehat{\mathbf{H}}\, e^{-i\omega t}]$$

or equivalently as

$$\mathbf{E}(t) = \widehat{\mathbf{E}}_r \, \cos(\omega t) + \widehat{\mathbf{E}}_i \, \sin(\omega t) \; ; \;\; \mathbf{H}(t) = \widehat{\mathbf{H}}_r \, \cos(\omega t) + \widehat{\mathbf{H}}_i \, \sin(\omega t)$$

where we have used $\widehat{\mathbf{E}} = \widehat{\mathbf{E}}_r + i\widehat{\mathbf{E}}_i$ and $\widehat{\mathbf{H}} = \widehat{\mathbf{H}}_r + i\widehat{\mathbf{H}}_i$. Also, recall that

$$\frac{\partial \mathbf{D}}{\partial t} = \nabla \times \mathbf{H} = -i\omega\, \boldsymbol{\varepsilon} \cdot \mathbf{E}(t) \qquad \text{and} \qquad \frac{\partial \mathbf{B}}{\partial t} = -\nabla \times \mathbf{E} = -i\omega\, \boldsymbol{\mu} \cdot \mathbf{H}(t) \, .$$

Therefore, for real ω and real P, we can write equation (1.103) as (with the substitution $z = \omega t$),

$$P = \frac{\omega}{2\,\pi} \int_0^{2\pi} \left[\widehat{\mathbf{E}}_r \, \cos z + \widehat{\mathbf{E}}_i \, \sin z \right] \cdot \left\{ \boldsymbol{\varepsilon}_i \cdot \left[\widehat{\mathbf{E}}_r \, \cos z + \widehat{\mathbf{E}}_i \, \sin z \right] \right\} + \left[\widehat{\mathbf{H}}_r \, \cos z + \widehat{\mathbf{H}}_i \, \sin z \right] \cdot \left\{ \boldsymbol{\mu}_i \cdot \left[\widehat{\mathbf{H}}_r \, \cos z + \widehat{\mathbf{H}}_i \, \sin z \right] \right\} \, dz$$

where $\boldsymbol{\mu} = \boldsymbol{\mu}_r + i\boldsymbol{\mu}_i$ and $\boldsymbol{\varepsilon} = \boldsymbol{\varepsilon}_r + i\boldsymbol{\varepsilon}_i$. Expanding out, and using the fact that

$$\int_0^{2\pi} \cos^2 z \, dz = \int_0^{2\pi} \sin^2 z \, dz = \pi \quad \text{and} \quad \int_0^{2\pi} \sin(2z) \, dz = 0$$

we have

$$P = \frac{\omega}{2}\left[\widehat{\mathbf{E}}_r \cdot \boldsymbol{\varepsilon}_i \cdot \widehat{\mathbf{E}}_r + \widehat{\mathbf{E}}_i \cdot \boldsymbol{\varepsilon}_i \cdot \widehat{\mathbf{E}}_i + \widehat{\mathbf{H}}_r \cdot \boldsymbol{\mu}_i \cdot \widehat{\mathbf{H}}_r + \widehat{\mathbf{H}}_i \cdot \boldsymbol{\mu}_i \cdot \widehat{\mathbf{H}}_i\right] . \tag{1.104}$$

If there is no dissipation, then the work done in a closed cycle is zero and $P = 0$. However, when there is dissipation $P > 0$. Since $\omega \geq 0$ and the power $P \geq 0$, the quadratic forms in equation (1.104) require that

$$\boldsymbol{\varepsilon}_i = \text{Im}(\boldsymbol{\varepsilon}) > 0 \qquad \text{and} \qquad \boldsymbol{\mu}_i = \text{Im}(\boldsymbol{\mu}) > 0 . \tag{1.105}$$

Therefore, in dissipative media, the imaginary parts of the permittivity and magnetic permeability must be greater than zero.

Note that if the permittivity is expressed as

$$\boldsymbol{\varepsilon} = \boldsymbol{\varepsilon}_0 + i\frac{\boldsymbol{\sigma}}{\omega}$$

the requirement $\text{Im}(\boldsymbol{\varepsilon}) > 0$ implies that the conductivity $\boldsymbol{\sigma} > 0$. Therefore, if the conductivity is greater than zero, there will be dissipation. These restrictions on the values of the material properties are important in situations where the causality of a given result is not obvious.

1.4.9 The electrical conductivity equation

In our discussion so far we have assumed that the conductivity equation $\boldsymbol{\nabla} \cdot \mathbf{J} = 0$ holds in many situations. Because of the importance of that equation it is worthwhile to examine how we can derive it.

If we assume that the electric field conserves energy, then the vector field $\mathbf{E}$ can be expressed as the gradient of a scalar potential ϕ, i.e.,

$$\mathbf{E} = -\boldsymbol{\nabla}\phi.$$

The constitutive relation between the current $\mathbf{J}$ and the electric field is

$$\mathbf{J}(\mathbf{x}) = \boldsymbol{\sigma}(\mathbf{x}) \cdot \mathbf{E}(\mathbf{x}) = -\boldsymbol{\sigma}(\mathbf{x}) \cdot \boldsymbol{\nabla}\phi$$

where $\boldsymbol{\sigma}$ is a second-order electrical conductivity tensor. The Onsager reciprocity relations (Onsager, 1931) lead to the conclusion that the electrical conductivity $\boldsymbol{\sigma}(\mathbf{x})$ tensor is symmetric. Let us also assume that $\boldsymbol{\sigma}$ is real and positive-definite such that $\alpha\, \boldsymbol{1} \geq \boldsymbol{\sigma}(\mathbf{x}) \geq \beta\, \boldsymbol{1}$ for all $\alpha, \beta > 0$.

Consider the region (Ω) with boundary (Γ) shown in Figure 1.4. From the principle of conservation of charge we know that the flux of positive charge through the boundary $\partial\Omega$ is balanced by a decrease of positive charge within Ω.

Therefore,

$$\int_\Gamma \mathbf{J} \cdot \mathbf{n}\, \mathrm{d}\Gamma = -\frac{dq_{\text{tot}}}{dt} = -\frac{d}{dt}\int_\Omega q\, \mathrm{d}\Omega$$

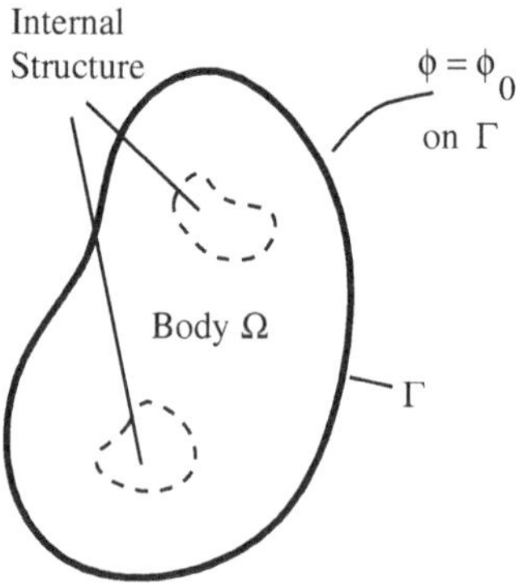

FIGURE 1.4
An electrically conducting region Ω with boundary Γ. A potential $\phi = \phi_0$ has been specified on the boundary.

where $\mathbf{n}$ is the outward unit normal to the surface Γ, q_{tot} is the charge inside Ω, q is the charge density, and t is the time. Assuming continuity of the quantities involved, we can use the divergence theorem and the Leibniz rule to get

$$\int_\Omega \boldsymbol{\nabla}\cdot\mathbf{J}\,\mathrm{d}\Omega = -\int_\Omega \frac{\partial q}{\partial t}\,\mathrm{d}\Omega.$$

Since the volume under consideration is arbitrary, we can use the continuum hypothesis to get the differential form of the conduction equation

$$\boldsymbol{\nabla}\cdot\mathbf{J}(\mathbf{x}) = \boldsymbol{\nabla}\cdot(\boldsymbol{\sigma}\cdot\boldsymbol{\nabla}\phi) = -\frac{\partial q}{\partial t}\,. \tag{1.106}$$

In the absence of sources or sinks of charge inside the body, the charge density remains constant and we have

$$\boxed{\boldsymbol{\nabla}\cdot\mathbf{J} = \boldsymbol{\nabla}\cdot(\boldsymbol{\sigma}\cdot\boldsymbol{\nabla}\phi) = 0\,.} \tag{1.107}$$

Alternatively, we can use Onsager's principle of the least dissipation of energy to arrive at equation (1.107). In the context of electrical conductivity, the principle of minimal power dissipation into heat inside the body can be expressed as

$$\min_{\substack{\phi \\ \phi=\phi_0 \text{ on } \Gamma}} \quad W(\phi) \quad \text{where} \quad W(\phi) := \int_\Omega \boldsymbol{\nabla}\phi\cdot\boldsymbol{\sigma}\cdot\boldsymbol{\nabla}\phi\,\mathrm{d}\Omega\,. \tag{1.108}$$

Now consider a variation ψ where $\psi = 0$ on Γ and let δ be a small parameter. Then

$$W(\phi+\delta\psi) = \int_\Omega \boldsymbol{\nabla}\phi\cdot\boldsymbol{\sigma}\cdot\boldsymbol{\nabla}\phi\,\mathrm{d}\Omega + 2\delta\int_\Omega \boldsymbol{\nabla}\psi\cdot\boldsymbol{\sigma}\cdot\boldsymbol{\nabla}\phi\,\mathrm{d}\Omega + \delta^2\int_\Omega \boldsymbol{\nabla}\psi\cdot\boldsymbol{\sigma}\cdot\boldsymbol{\nabla}\psi\,\mathrm{d}\Omega\,. \tag{1.109}$$

Note that $W(\phi+\delta\psi) > 0$. Using the identity $\mathbf{a}\cdot\boldsymbol{\nabla}b = \boldsymbol{\nabla}\cdot(b\,\mathbf{a}) - b\,\boldsymbol{\nabla}\cdot\mathbf{a}$ in the second term on the right-hand side of (1.109) leads to

$$W(\phi+\delta\psi) = \int_\Omega \boldsymbol{\nabla}\phi\cdot\boldsymbol{\sigma}\cdot\boldsymbol{\nabla}\phi\,\mathrm{d}\Omega + 2\delta\int_\Omega \boldsymbol{\nabla}\cdot(\psi\boldsymbol{\sigma}\cdot\boldsymbol{\nabla}\phi)\,\mathrm{d}\Omega \\ -2\delta\int_\Omega \psi\boldsymbol{\nabla}\cdot(\boldsymbol{\sigma}\cdot\boldsymbol{\nabla}\phi)\,\mathrm{d}\Omega + \delta^2\int_\Omega \boldsymbol{\nabla}\psi\cdot\boldsymbol{\sigma}\cdot\boldsymbol{\nabla}\psi\,\mathrm{d}\Omega\,.$$

From the divergence theorem, we have

$$\int_\Omega \boldsymbol{\nabla}\cdot(\psi\boldsymbol{\sigma}\cdot\boldsymbol{\nabla}\phi)\,\mathrm{d}\Omega = \int_\Gamma \mathbf{n}\cdot(\psi\boldsymbol{\sigma}\cdot\boldsymbol{\nabla}\phi)\,\mathrm{d}\Gamma$$

where $\mathbf{n}$ is the outward unit normal to the surface Γ. Since $\psi = 0$ on Γ, we have

$$\int_\Omega \boldsymbol{\nabla}\cdot(\psi\boldsymbol{\sigma}\cdot\boldsymbol{\nabla}\phi)\,\mathrm{d}\Omega = 0\,.$$

Therefore,

$$W(\phi+\delta\,\psi) = \int_\Omega \boldsymbol{\nabla}\phi\cdot\boldsymbol{\sigma}\cdot\boldsymbol{\nabla}\phi\,\mathrm{d}\Omega - 2\,\delta\int_\Omega \psi\,\boldsymbol{\nabla}\cdot(\boldsymbol{\sigma}\cdot\boldsymbol{\nabla}\phi)\,\mathrm{d}\Omega + \delta^2\int_\Omega \boldsymbol{\nabla}\psi\cdot\boldsymbol{\sigma}\cdot\boldsymbol{\nabla}\psi\,\mathrm{d}\Omega\,.$$

For $W(\phi+\delta\psi)$ to be positive for all ψ, it is sufficient to have

$$\int_\Omega \psi\,\boldsymbol{\nabla}\cdot(\boldsymbol{\sigma}\cdot\boldsymbol{\nabla}\phi)\,\mathrm{d}\Omega = 0\,.$$

If this is to be true for all ψ, then

$$\boxed{\boldsymbol{\nabla}\cdot(\boldsymbol{\sigma}\cdot\boldsymbol{\nabla}\phi) = 0\,.}$$

If we define the flux as $\mathbf{J}(\mathbf{x}) := \boldsymbol{\sigma}(\mathbf{x})\cdot\boldsymbol{\nabla}\phi$ then we have $\boldsymbol{\nabla}\cdot\mathbf{J}(\mathbf{x}) = 0$. If there are sources inside the region Ω we get the modified equation $\boldsymbol{\nabla}\cdot\mathbf{J} = -f$, where f is a source term. The conductivity equation will be used when we discuss transformation-based cloaking.

Exercises

Problem 1.1 The equation of balance of the flow of a physical quantity $f(\mathbf{x},t)$ through a body Ω with surface $\partial\Omega$ can be expressed in the form

$$\frac{d}{dt}\left[\int_\Omega f(\mathbf{x},t)\,\mathrm{dV}\right] = \int_{\partial\Omega}[g(\mathbf{x},t) + f(\mathbf{x},t)[u_n(\mathbf{x},t) - \mathbf{v}(\mathbf{x},t)\cdot\mathbf{n}(\mathbf{x},t)]]\,\mathrm{dA} \\ + \int_\Omega h(\mathbf{x},t)\,\mathrm{dV}$$

where $g(\mathbf{x},t)$ be sources on the surface of the body, $h(\mathbf{x},t)$ are sources inside the body, $\mathbf{n}(\mathbf{x},t)$ is the outward unit normal to the surface $\partial\Omega$, $\mathbf{v}(\mathbf{x},t)$ is the velocity of the physical particles that carry the physical quantity that is flowing, and u_n is the speed at which the bounding surface $\partial\Omega$ is moving in the direction $\mathbf{n}$. Using the above balance relation show that the balance of linear momentum of a linear elastic body can be expressed as

$$\rho\,\dot{\mathbf{v}} - \boldsymbol{\nabla}\cdot\boldsymbol{\sigma} - \rho\,\mathbf{b} = 0$$

where $\rho(\mathbf{x},t)$ is the mass density, $\mathbf{v}(\mathbf{x},t)$ is the spatial velocity, $\boldsymbol{\sigma}(\mathbf{x},t)$ is the Cauchy stress, and $\rho\,\mathbf{b}$ is the body force density. *Hint:* Use the balance of mass $\dot{\rho} + \rho\,\boldsymbol{\nabla}\cdot\mathbf{v} = 0$.

Problem 1.2 Assume that there are no surface couples on $\partial\Omega$ or body couples in Ω. Use the angular momentum density as the conserved quantity in the general balance relation from Problem 2.1, i.e., $f = \mathbf{x}\times(\rho\,\mathbf{v})$, to show that the balance of angular momentum can be expressed as

$$\boldsymbol{\sigma} = \boldsymbol{\sigma}^T\ .$$

Problem 1.3 Show that if $\boldsymbol{\sigma} = \mathsf{C} : \boldsymbol{\varepsilon}$

$$\boldsymbol{\nabla}\cdot\boldsymbol{\sigma} = \boldsymbol{\nabla}\cdot(\mathsf{C} : \boldsymbol{\nabla}\mathbf{u})\ .$$

Problem 1.4 Show that for an isotropic, homogeneous, linear elastic material with stiffness tensor

$$\mathsf{C} = \lambda\boldsymbol{1}\otimes\boldsymbol{1} + 2\mu\mathsf{I}$$

the divergence of the stress can be expressed as

$$\boldsymbol{\nabla}\cdot(\mathsf{C} : \boldsymbol{\nabla}\mathbf{u}) = (\lambda+\mu)\,\boldsymbol{\nabla}(\boldsymbol{\nabla}\cdot\mathbf{u}) + \mu\boldsymbol{\nabla}\cdot(\boldsymbol{\nabla}\mathbf{u}^T)\ .$$

Problem 1.5 Verify that the volumetric strain e is given by $\boldsymbol{\nabla}\cdot\mathbf{u}$ and that the infinitesimal rotation vector is given by $\boldsymbol{\nabla}\times\mathbf{u}$.

Problem 1.6 Show that, in the absence of a body force, the displacement potentials ϕ and $\boldsymbol{\psi}$ satisfy the wave equations

$$\frac{\partial^2\phi}{\partial t^2} = c_p^2\,\nabla^2\phi; \quad \frac{\partial^2\boldsymbol{\psi}}{\partial t^2} = c_s^2\,\boldsymbol{\nabla}\cdot(\boldsymbol{\nabla}\boldsymbol{\psi})^T\ .$$

Problem 1.7 Show, using spatial curvilinear coordinates, that the gradient of a vector field $\mathbf{v}$ can be expressed in spherical coordinates as

$$\boldsymbol{\nabla}\mathbf{v} = \begin{bmatrix} \dfrac{\partial v_r}{\partial r} & \dfrac{1}{r\sin\phi}\dfrac{\partial v_r}{\partial\theta} - \dfrac{v_\theta}{r} & \dfrac{1}{r}\dfrac{\partial v_r}{\partial\phi} - \dfrac{v_\phi}{r} \\ \dfrac{\partial v_\theta}{\partial r} & \dfrac{1}{r\sin\phi}\dfrac{\partial v_\theta}{\partial\theta} + \dfrac{v_r}{r} + \dfrac{\cot\phi}{r}v_\phi & \dfrac{1}{r}\dfrac{\partial v_\theta}{\partial\phi} \\ \dfrac{\partial v_\phi}{\partial r} & \dfrac{1}{r\sin\phi}\dfrac{\partial v_\phi}{\partial\theta} - \dfrac{\cot\phi}{r}v_\theta & \dfrac{1}{r}\dfrac{\partial v_\phi}{\partial\phi} + \dfrac{v_r}{r} \end{bmatrix}$$

where $x_1 = \theta^1\cos\theta^2\sin\theta^3$, $x_2 = \theta^1\sin\theta^2\sin\theta^3$, $x_3 = \theta^1\cos\theta^3$, and $(\theta^1,\theta^2,\theta^3) \equiv (r,\theta,\phi)$.

Problem 1.8 The equation of state for an adiabatic, reversible, ideal gas is given by

$$\frac{dp}{d\rho} = \frac{\gamma p}{\rho}; \quad \gamma := \frac{c_p}{c_v}; \quad c^2 = \frac{\gamma p}{\rho}$$

where c_p is the specific heat at constant pressure, c_v is the specific heat at constant volume, and c is the wave speed. Find the value of γ in air and water. Show that for small disturbances the balance of mass in the gas can be expressed as

$$\frac{\partial p}{\partial t} + \rho_0\, c_0^2\, \boldsymbol{\nabla} \cdot \mathbf{v} = 0$$

where c_0 is the speed of sound in the medium.

Problem 1.9 The total magnetic induction produced at a point $\mathbf{x}$ by a current density $\mathbf{J}$ located in a region Ω is given by

$$\mathbf{B}(\mathbf{x}) = \frac{\mu_0}{4\pi} \int_\Omega \mathbf{J}(\mathbf{y}) \times \frac{\mathbf{x}-\mathbf{y}}{\|\mathbf{x}-\mathbf{y}\|^3}\, d\mathbf{y}\,.$$

Starting from the above equation show that

$$\boldsymbol{\nabla} \times \mathbf{B} = \mu_0\, \mathbf{J} + \frac{\mu_0}{4\pi} \boldsymbol{\nabla} \left[\int_\Omega \left(\frac{\boldsymbol{\nabla}_y \cdot \mathbf{J}(\mathbf{y})}{\|\mathbf{x}-\mathbf{y}\|} \right) d\mathbf{y} \right].$$

Problem 1.10 The relation between the electromotive force (emf) around a loop and the magnetic induction is given by the integral form of Faraday's law:

$$\oint_C \mathbf{E} \cdot d\mathbf{l} = -k\, \frac{D}{Dt} \left[\int_{\partial\Omega} \mathbf{B} \cdot \mathbf{n}\, dA \right].$$

Derive the differential form of Faraday's law,

$$\boldsymbol{\nabla} \times \mathbf{E} + \frac{\partial \mathbf{B}}{\partial t} = \mathbf{0}\,.$$

Problem 1.11 Assume that the electric field, electric displacement, magnetic field, and magnetic inductions depend harmonically of time, i.e.,

$$\mathbf{E}(\mathbf{x},t) = \text{Re}\{\widehat{\mathbf{E}}(\mathbf{x})\, e^{-i\omega t}\}\,; \quad \mathbf{B}(\mathbf{x},t) = \text{Re}\{\widehat{\mathbf{B}}(\mathbf{x})\, e^{-i\omega t}\}$$
$$\widetilde{\mathbf{D}}(\mathbf{x},t) = \text{Re}\{\widehat{\mathbf{D}}(\mathbf{x})\, e^{-i\omega t}\}\,; \quad \mathbf{H}(\mathbf{x},t) = \text{Re}\{\widehat{\mathbf{H}}(\mathbf{x})\, e^{-i\omega t}\}$$

where ω has an infinitesimally small imaginary part. Show that Maxwell's equations under these conditions can be written as

$$\boldsymbol{\nabla} \times \mathbf{E} = i\omega\, \boldsymbol{\mu}(\mathbf{x},\omega) \cdot \mathbf{H}(\mathbf{x})\,; \quad \boldsymbol{\nabla} \times \mathbf{H} = -i\omega\, \boldsymbol{\varepsilon}(\mathbf{x},\omega) \cdot \mathbf{E}(\mathbf{x})$$
$$\boldsymbol{\nabla} \cdot \mathbf{B} = 0\,; \quad \boldsymbol{\nabla} \cdot \mathbf{D} = 0\,.$$

Why does ω need a small imaginary part? Compare these equations with the Fourier transformed form of Maxwell's equations.

Problem 1.12 Show that the transverse electric (TE) wave equation for a material with anisotropic permeability and permittivity

$$\boldsymbol{\mu} = \boldsymbol{\mu}(x_2, x_3) \equiv \begin{bmatrix} \mu_{11} & 0 & 0 \\ 0 & \mu_{22} & \mu_{23} \\ 0 & \mu_{23} & \mu_{33} \end{bmatrix} \quad \text{and} \quad \boldsymbol{\varepsilon} = \boldsymbol{\varepsilon}(x_2, x_3) \equiv \begin{bmatrix} \varepsilon_{11} & 0 & 0 \\ 0 & \varepsilon_{22} & \varepsilon_{23} \\ 0 & \varepsilon_{23} & \varepsilon_{33} \end{bmatrix}$$

is given by

$$\overline{\boldsymbol{\nabla}} \cdot \left[\left(\boldsymbol{R}_{\perp} \cdot \boldsymbol{M}^{-1} \cdot \boldsymbol{R}_{\perp}^{T} \right) \cdot \overline{\boldsymbol{\nabla}} E_1 \right] + \omega^2 \, \varepsilon_{11} \, E_1 = 0$$

where $\overline{\boldsymbol{\nabla}}$ indicates the two-dimensional gradient, and

$$\boldsymbol{M} \equiv \begin{bmatrix} \mu_{22} & \mu_{23} \\ \mu_{23} & \mu_{33} \end{bmatrix} \; ; \; \boldsymbol{R}_{\perp} \equiv \begin{bmatrix} 0 & 1 \\ -1 & 0 \end{bmatrix} .$$

Problem 1.13 Maxwell's equations can be expressed in a form similar to the equations of elastodynamics at a fixed frequency, i.e.,

$$\omega^2 \, \boldsymbol{\varepsilon} \cdot \mathbf{E} = \boldsymbol{\nabla} \cdot (\mathbf{C} : \boldsymbol{\nabla} \mathbf{E}) .$$

Show that the tensor $\mathbf{C}$ has the symmetries

$$C_{ijk\ell} = -C_{jik\ell} = -C_{ij\ell k} = C_{k\ell ij} .$$

2

Plane Waves and Interfaces

Owing to the extraordinary complexity of the investigation when written out in Cartesian form (which I began doing, but gave up aghast) some abbreviated method of expression becomes desirable.

OLIVER HEAVISIDE, On the electromagnetic wave-surface, 1885.

In the previous chapter we have seen that the dynamics of elastic, acoustic, and electromagnetic waves can be expressed in the form of wave equations. In classical linear elastodynamics without body forces we have the wave equations

$$\boldsymbol{\nabla}\cdot[\mathsf{C}:\boldsymbol{\nabla}\mathbf{u}] = \rho\,\ddot{\mathbf{u}}\,; \quad c_p^2\,\nabla^2\phi = \ddot{\phi}\,; \; c_s^2\,\boldsymbol{\nabla}\cdot(\boldsymbol{\nabla}\boldsymbol{\psi})^T = \ddot{\boldsymbol{\psi}}\,.$$

For a fixed frequency (ω) and in the absence of body forces the wave equations of elastodynamics have the forms

$$\rho^{-1}\,\boldsymbol{\nabla}\cdot[\mathsf{C}:\boldsymbol{\nabla}\mathbf{u}] + \omega^2\,\mathbf{u} = 0\,; \quad c_p^2\,\nabla^2\phi + \omega^2\,\phi = 0\,; \; c_s^2\,\boldsymbol{\nabla}\cdot(\boldsymbol{\nabla}\boldsymbol{\psi})^T + \omega^2\,\boldsymbol{\psi} = \mathbf{0}\,.$$

The wave equations in linear acoustics are

$$c_0^2\,\boldsymbol{\nabla}\cdot(\boldsymbol{\nabla}\mathbf{v}) = \ddot{\mathbf{v}}\,; \quad c_0^2\,\nabla^2 p = \ddot{p}\,; \quad c_0^2\,\nabla^2\phi = \ddot{\phi}\,.$$

The acoustic wave equations at fixed frequency have the forms

$$\kappa\,\boldsymbol{\nabla}\cdot\left[\rho_0^{-1}\,\boldsymbol{\nabla}p\right] + \omega^2\,p = 0\,; \; c_0^2\,\nabla^2 p + \omega^2\,p = 0\,; \quad c_0^2\,\nabla^2\phi + \omega^2\,\phi = 0\,.$$

We have also seen that the electromagnetic wave equations can be written as

$$\begin{aligned} &c^2\,\nabla^2\phi = \ddot{\phi}\,; \quad c^2\,\nabla^2\mathbf{A} = \ddot{\mathbf{A}} \\ &c^2\,(\boldsymbol{\nabla}f - \boldsymbol{\nabla}\times\boldsymbol{\nabla}\times\boldsymbol{\Pi}_e) = \ddot{\boldsymbol{\Pi}}_e\,; \quad c^2\,(\boldsymbol{\nabla}f - \boldsymbol{\nabla}\times\boldsymbol{\nabla}\times\boldsymbol{\Pi}_m) = \ddot{\boldsymbol{\Pi}}_m\,. \end{aligned}$$

At fixed frequency we have the electrodynamic wave equations

$$\begin{aligned} &c^2\,\nabla^2\mathbf{E} + \omega^2\,\mathbf{E} = 0\,; \quad c^2\,\nabla^2\mathbf{H} + \omega^2\,\mathbf{H} = 0 \\ &c^2\,\nabla^2\phi + \omega^2\,\phi = 0\,; \quad c^2\,\nabla^2\mathbf{A} + \omega^2\,\mathbf{A} = \mathbf{0}\,. \end{aligned}$$

Notice that, at a fixed frequency ω, most of these equations are of the form of the Helmholtz equation.

In this chapter we will explore plane wave solutions of the wave equation. We will also examine plane wave solutions for problems involving reflection and refraction of waves at plane interfaces. We will look at the reflection of P-waves at a free surface and at the interface between two elastic solids. We will examine the simpler case of acoustic waves at the interface between two inviscid fluids and discuss the idea of acoustic impedance. We will then look at the more complicated fluid-solid interface problem. The possibility of negative refractive index and its mirroring of acoustic pressure gradients will be discussed. Next we will discuss TE- and TM-waves at the interface between two materials with different electromagnetic properties, define the idea of electromagnetic impedance, and explore negative refraction in electromagnetism. We will then proceed to calculate the reflection and transmission through flat slabs, first for acoustics (and the associated impedance tube calculations) and then for electromagnetism.

2.1 Plane wave solutions

Ideally we would to find the most general solutions $\mathbf{u}(\mathbf{x},t)$ to the wave equations

$$\nabla^2\mathbf{u} - \frac{1}{c^2}\ddot{\mathbf{u}} = \mathbf{0} .$$

However, that seems to be rather difficult and a simpler problem is the special situation where $\mathbf{u}(\mathbf{x},t) \equiv \mathbf{u}(x_1,t)$, i.e., the solution varies in the x_1-direction but not in the x_2 and x_3 directions. Waves that have this property are called *plane waves*. The general solution of the equation for a plane wave propagating in the x_1-direction is

$$\mathbf{u}(\mathbf{x},t) = \mathbf{f}_1(x_1 - ct) + \mathbf{f}_2(x_1 + ct) .$$

The function $\mathbf{f}_1$ represents a wave traveling in the positive x_1-direction with a velocity c while the function $\mathbf{f}_2$ represents a wave traveling in the negative x_1 direction. In general, the direction of plane wave will not be oriented with one of the coordinate axes. Let $\widehat{\mathbf{k}}$ be a unit vector in the direction of propagation of the wave, also called the *wave vector*. Then the general solution of the wave equation is

$$\mathbf{u}(\mathbf{x},t) = \mathbf{f}_1(\widehat{\mathbf{k}}\cdot\mathbf{x} - ct) + \mathbf{f}_2(\widehat{\mathbf{k}}\cdot\mathbf{x} + ct) .$$

This solution is not particularly useful because the functions $\mathbf{f}_1$ and $\mathbf{f}_2$ remain to be found. The problem that we face can be resolved by appealing to the Fourier transform, finding a particular solution for a fixed frequency, and summing up the solutions at all frequencies to arrive at the general solution. Such a solution at fixed frequency (ω) is called a *monochromatic plane wave* or *harmonic plane wave* and has the form (for the forward wave)

$$\mathbf{u}(\mathbf{x},t) = \mathbf{A}\, e^{i(\mathbf{k}\cdot\mathbf{x} - \omega\, t)} \tag{2.1}$$

where $\mathbf{A}$ is the amplitude of the wave, $\widehat{\mathbf{k}} = \mathbf{k}/\,\|\mathbf{k}\|$ and $c = \omega/\,\|\mathbf{k}\|$. The magnitude of the wave vector ($\|\mathbf{k}\|$) is called the *wave number* and is related to the wavelength (λ) by $\|\mathbf{k}\| = 2\pi/\lambda$. If the problem is linear as we have assumed in the previous chapter, we can find plane wave solutions corresponding to waves in any number of linearly independent directions, add them up, and get a full three-dimensional solution to the wave equation. Each component plane wave in such a solution travels with the speed of its *phase* (θ) where $\theta := \mathbf{k}\cdot\mathbf{x} - \omega\, t$. The plane wave has a constant amplitude on planes perpendicular to the wave vector. Hence that phase must also be constant on each of these planes and the rate of change of phase for a plane of constant amplitude is $\dot{\varphi} = \mathbf{k}\cdot\dot{\mathbf{x}} - \omega = 0$. The *phase velocity* ($v_p$) is defined as

$$v_p = \widehat{\mathbf{k}}\cdot\dot{\mathbf{x}} = \frac{\omega}{\|\mathbf{k}\|} = c\,.$$

For a wave composed of a number of waves with different frequencies, the value of v_p varies with frequency and the harmonic plane wave components travel at different speeds. This phenomenon is known as *dispersion* and the relation $\omega \equiv \omega(\|\mathbf{k}\|)$ is called a *dispersion relation*. The rate of change of ω as a function of the wave number is called the *group velocity* (v_g) of a superposition of plane waves,

$$v_g = \frac{d\omega}{d\,\|\mathbf{k}\|}\,.$$

Note that we can express the plane wave solution (2.1) as

$$\mathbf{u}(\mathbf{x},t) = \widehat{\mathbf{A}}(\mathbf{x})\; e^{-i\omega t} \quad \text{where} \quad \widehat{\mathbf{A}}(\mathbf{x}) = \mathbf{A}\; e^{i\mathbf{k}\cdot\mathbf{x}}\,.$$

You can check that $\widehat{\mathbf{A}}(\mathbf{x})$ is the plane wave solution at a fixed frequency ω by plugging it into the wave equations at fixed frequency. We shall use this solution frequently in the rest of this chapter.

2.2 Plane wave solutions in elastodynamics

In elastodynamics,* our aim is to find the displacement field $\mathbf{u}(\mathbf{x})$ that satisfies, subject to certain initial and boundary conditions, the equations

$$\boldsymbol{\nabla}\cdot[\mathbf{C} : \boldsymbol{\nabla}\mathbf{u}] = \rho\,\ddot{\mathbf{u}}$$

where $\mathbf{C}$ is the stiffness tensor and ρ is the mass density. For an isotropic material $\mathbf{C} = \lambda\boldsymbol{1}\otimes\boldsymbol{1} + 2\mu\mathbf{I}$ and we have

$$\mathbf{C} : \boldsymbol{\nabla}\mathbf{u} = \lambda(\mathbf{x})(\boldsymbol{\nabla}\cdot\mathbf{u})\boldsymbol{1} + \mu(\mathbf{x})[\boldsymbol{\nabla}\mathbf{u} + (\boldsymbol{\nabla}\mathbf{u})^T]\,.$$

*Our discussion of plane wave solutions in elastodynamics is based on Aki and Richards (1980).

If the material is also homogeneous, the governing equations become

$$\boldsymbol{\nabla}\cdot[\mathbf{C}:\boldsymbol{\nabla}\mathbf{u}] = (\lambda+\mu)\boldsymbol{\nabla}(\boldsymbol{\nabla}\cdot\mathbf{u})+\mu\boldsymbol{\nabla}\cdot(\boldsymbol{\nabla}\mathbf{u})^T = \rho\,\ddot{\mathbf{u}}\,.$$

Let us look for a forward-propagating plane wave solution to the above equation of the form

$$\mathbf{u}(\mathbf{x},t) = \mathbf{u}(\widehat{\mathbf{k}}\cdot\mathbf{x}-ct)\,.$$

Researchers in elastodynamics prefer to work with a quantity called the "slowness vector," defined as $\mathbf{s} := \widehat{\mathbf{k}}/c$, instead of the wave vector $\widehat{\mathbf{k}}$. Note that $\mathbf{s}\cdot\mathbf{s} = 1/c^2$ and that the general plane wave solution in terms of the slowness vector is

$$\mathbf{u}(\mathbf{x},t) = \mathbf{u}(\mathbf{s}\cdot\mathbf{x}-t)\,.$$

Since $c = \omega/\left\lVert\mathbf{k}\right\rVert$, the time-harmonic plane wave solution can be written as

$$\mathbf{u}(\mathbf{x},t) = \widehat{\mathbf{u}}\,e^{i\omega(\mathbf{s}\cdot\mathbf{x}-t)}\,.$$

If we plug in this solution into the governing equations we get

$$(\lambda+\mu)\,(\mathbf{s}\otimes\mathbf{s})\cdot\widehat{\mathbf{u}}+\mu\,(\mathbf{s}\cdot\mathbf{s})\,\widehat{\mathbf{u}} = \rho\,\widehat{\mathbf{u}}\,.$$

Now $[(\mathbf{s}\otimes\mathbf{s})\cdot\widehat{\mathbf{u}}]\times\mathbf{s} = \mathbf{0}$ and $[(\mathbf{s}\otimes\mathbf{s})\cdot\widehat{\mathbf{u}}]\cdot\mathbf{s} = (\mathbf{s}\cdot\mathbf{s})(\mathbf{u}\cdot\mathbf{s})$. Taking the vector and scalar products of the above equation with $\mathbf{s}$ leads to

$$\left(\rho-\frac{\mu}{c^2}\right)(\widehat{\mathbf{u}}\times\mathbf{s}) = \mathbf{0} \quad \text{and} \quad \left(\rho-\frac{\lambda+2\mu}{c^2}\right)(\widehat{\mathbf{u}}\cdot\mathbf{s}) = 0$$

where we have used the relation $\mathbf{s}\cdot\mathbf{s} = 1/c^2$. Therefore, the plane wave decouples into a P-wave with $\widehat{\mathbf{u}}\times\mathbf{s} = \mathbf{0}$ (i.e., $\widehat{\mathbf{u}}$ and $\mathbf{s}$ are in the same direction) and with phase velocity $c_p = \sqrt{(\lambda+2\mu)/\rho}$; and an S-wave with $\widehat{\mathbf{u}}\cdot\mathbf{s} = 0$ (i.e., $\widehat{\mathbf{u}}$ and $\mathbf{s}$ are perpendicular) and with phase velocity $c_s = \sqrt{\mu/\rho}$. *An elastodynamic plane wave in a homogeneous and isotropic medium cannot have both P- and S-wave components.*

If the material is anisotropic and inhomogeneous, we have

$$\boldsymbol{\nabla}\cdot[\mathbf{C}:\boldsymbol{\nabla}(\widehat{\mathbf{u}}\,e^{i\omega(\mathbf{s}\cdot\mathbf{x}-t)})] = -\omega^2\rho\widehat{\mathbf{u}}\,e^{i\omega(\mathbf{s}\cdot\mathbf{x}-t)}$$

or

$$i\omega\boldsymbol{\nabla}\cdot[\mathbf{C}:(\widehat{\mathbf{u}}\otimes\mathbf{s})\,e^{i\omega\mathbf{s}\cdot\mathbf{x}}] = -\omega^2\rho\widehat{\mathbf{u}}\,e^{i\omega\mathbf{s}\cdot\mathbf{x}}\,.$$

If the material is homogeneous the above equation simplifies to

$$-\omega^2\mathbf{s}\cdot[\mathbf{C}:(\widehat{\mathbf{u}}\otimes\mathbf{s})]\,e^{i\omega\mathbf{s}\cdot\mathbf{x}} = -\omega^2\rho\widehat{\mathbf{u}}\,e^{i\omega\mathbf{s}\cdot\mathbf{x}}$$

or

$$[\mathbf{s}\cdot\mathbf{C}\cdot\mathbf{s}-\rho\,\boldsymbol{1}]\cdot\widehat{\mathbf{u}} = \mathbf{0} \tag{2.2}$$

where $\boldsymbol{1}$ is the second-order identity tensor. The second-order tensor quantity $\mathbf{s}\cdot\mathbf{C}\cdot\mathbf{s}$ is called the *acoustic tensor* and the problem has a nontrivial solution only if

$$\det[\mathbf{s}\cdot\mathbf{C}\cdot\mathbf{s}-\rho\,\boldsymbol{1}] = 0\,.$$

Since $\mathbf{s} = \mathbf{k}/\omega$, we can also write the above condition as

$$\det[\mathbf{k}\cdot\mathbf{C}\cdot\mathbf{k}-\omega^2\rho\,\boldsymbol{1}] = 0\,.$$

2.2.1 Elastodynamic potentials for plane waves

Recall that the displacement field can be derived from a P-wave potential (ϕ) and an S-wave potential $\boldsymbol{\psi}$ as

$$\mathbf{u} = \boldsymbol{\nabla}\phi + \boldsymbol{\nabla} \times \boldsymbol{\psi} \quad \text{with} \quad \boldsymbol{\nabla} \cdot \boldsymbol{\psi} = 0 \, .$$

Consider the situation where a plane wave propagates in the $x_1 - x_2$ plane as shown in Figure 2.1. We have chosen a Cartesian coordinate system for simplicity. The coordinate system has been chosen such that the slowness vector $\mathbf{s}$ is perpendicular to the $\mathbf{e}_3$ basis vector.

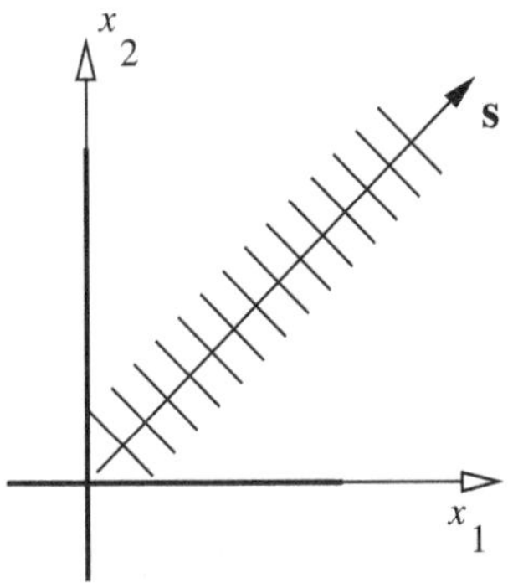

FIGURE 2.1
Schematic of a plane wave with slowness vector $\mathbf{s}$ in the $x_1 - x_2$ plane.

First consider a harmonic P-wave, in the chosen coordinate system, that is derived from a potential of the form

$$\phi(\mathbf{x},t) = \Phi_0 e^{i\omega(\mathbf{s}\cdot\mathbf{x}-t)} \, .$$

For a fixed frequency we have

$$\phi(\mathbf{x}) = \Phi_0 e^{i\omega\mathbf{s}\cdot\mathbf{x}} \, .$$

Since $\mathbf{s} = s_1\mathbf{e}_1 + s_2\mathbf{e}_2 + 0\mathbf{e}_3$ we have $\phi \equiv \phi(x_1,x_2)$, and the displacement field $\mathbf{u} = \boldsymbol{\nabla}\phi$ has components $u_1 \equiv u_1(x_1,x_2)$, $u_2 \equiv u_2(x_1,x_2)$, and $u_3 = 0$. The wave equation for the P-wave at fixed frequency can then be written as

$$c_p^2 \, \overline{\nabla}^2 \phi + \omega^2 \phi = 0 \tag{2.3}$$

where $\overline{\nabla}^2(\cdot)$ is the two-dimensional Laplacian.

Let us next consider a harmonic plane shear wave (S-wave). In that case the potential can be expressed as

$$\boldsymbol{\psi}(\mathbf{x},t) = \boldsymbol{\Psi}_0 e^{i\omega(\mathbf{s}\cdot\mathbf{x}-t)}$$

where $\boldsymbol{\Psi}_0$ is the amplitude. For a fixed frequency we can write

$$\boldsymbol{\psi}(\mathbf{x}) = \boldsymbol{\Psi}_0 e^{i\omega\mathbf{s}\cdot\mathbf{x}} .$$

Once again, since $\mathbf{s} = s_1\mathbf{e}_1 + s_2\mathbf{e}_2 + 0\mathbf{e}_3$ we see that $\boldsymbol{\psi} \equiv \boldsymbol{\psi}(x_1,x_2)$, i.e., the problem is essentially two dimensional. Therefore the requirement $\boldsymbol{\nabla}\cdot\boldsymbol{\psi} = 0$ implies that $\psi_{1,1} + \psi_{2,2} = 0$.

A plane shear wave can be split up into an in-plane shear wave called the SV-wave and an antiplane shear wave called the SH-wave. In the coordinates that we have chosen, an SV-polarized wave has the displacements $u_1 \equiv u_1(x_1,x_2)$, $u_2 \equiv u_2(x_1,x_2)$, and $u_3 = 0$. Therefore the displacements produced by an SV-wave are, in terms of the potential $\boldsymbol{\psi}$,

$$\mathbf{u} = \boldsymbol{\nabla}\times\boldsymbol{\psi} = \psi_{3,2}\,\mathbf{e}_1 - \psi_{3,1}\,\mathbf{e}_2 + (\psi_{2,1} - \psi_{1,2})\mathbf{e}_3 \quad \text{with} \quad \psi_{2,1} - \psi_{1,2} = 0 .$$

If we think in terms of complex variables (x,ψ) where $z = x_1 + ix_2$ and $\psi = \psi_1 + i\psi_2$, then ψ is an analytic function of x for SH-waves.[†] For a plane wave, ψ is bounded and analytic for all z and hence constant. We can take this constant to be zero without loss of generality, i.e., $\psi_1 = \psi_2 = 0$. Therefore $\boldsymbol{\psi} = \psi_3\mathbf{e}_3$ and an SV-wave potential can be expressed as

$$\psi_3(\mathbf{x}) = \Psi_3 e^{i\omega\mathbf{s}\cdot\mathbf{x}} .$$

Recall that the S-wave equation at fixed frequency has the form

$$c_s^2\,\boldsymbol{\nabla}\cdot(\boldsymbol{\nabla}\boldsymbol{\psi})^T + \omega^2\boldsymbol{\psi} = \mathbf{0} .$$

Because $\boldsymbol{\psi}$ has only one component, the vector S-wave equation becomes a scalar wave equation for an SV-wave and has the form

$$c_s^2\,\overline{\nabla}^2\psi_3 + \omega^2\psi_3 = 0 . \tag{2.4}$$

For an antiplane shear wave (SH-wave), the displacement field is $u_1 = u_2 = 0$ and $u_3 \equiv u_3(x_1,x_2)$. It is easier to deal directly with the displacement field in this case because if we plug the displacements into the elastodynamic wave equation at fixed frequency for isotropic, homogeneous elasticity we get

$$c_s^2\,\overline{\nabla}^2 u_3 + \omega^2 u_3 = 0 . \tag{2.5}$$

[†]Recall the Cauchy-Riemann analyticity condition which states that if $f = u + iv$ is analytic is a domain, then

$$\frac{\partial u}{\partial x} - \frac{\partial v}{\partial y} = 0 \quad \text{and} \quad \frac{\partial u}{\partial y} + \frac{\partial v}{\partial x} = 0$$

where $u \equiv u(x+iy)$ and $v \equiv v(x+iy)$.

Note that the components of the slowness vector (and therefore the wave vector) are not all independent. To see this substitute the plane wave expansion of ψ_3 into equation (2.4) to get

$$\mathbf{s}\cdot\mathbf{s} = \frac{1}{c_s^2} \quad \Longrightarrow \quad \mathbf{k}\cdot\mathbf{k} = \frac{\omega^2}{c_s^2}\,.$$

Similar relations can be found for P- and SV-waves. We will now try to solve these equations for the simple case of a P-wave incident upon a free surface.

2.2.2 Reflection of plane P-waves at a free surface

Consider the situation shown in Figure 2.2 where a P-wave in incident upon a free surface from within a solid. The medium on the other side of the free surface can be vacuum or air without changing our conclusions significantly. Let the slowness vector of the incident P-wave be $\mathbf{s}_{ip}$ and let the angle of incidence be θ_{ip}. Then the slowness vector can be expressed in components as

$$\mathbf{s}_{ip} = \|\mathbf{s}_{ip}\|\,(\sin\theta_{ip}\mathbf{e}_1 - \cos\theta_{ip}\mathbf{e}_2) = (\sin\theta_{ip}\mathbf{e}_1 - \cos\theta_{ip}\mathbf{e}_2)/c_p\,.$$

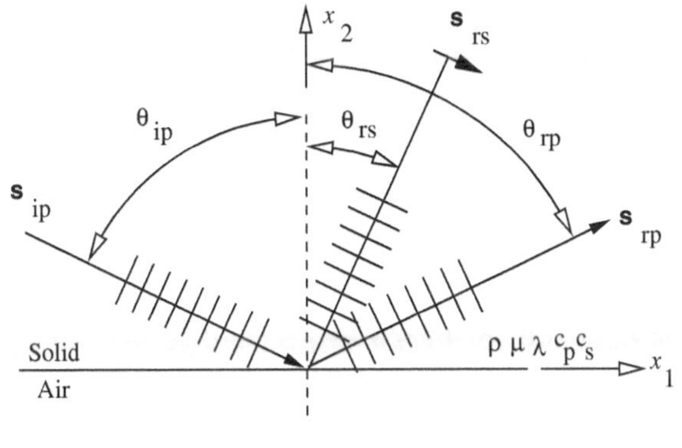

FIGURE 2.2

Reflection of a P-wave at a free interface. There is a reflected P-wave with slowness $\mathbf{s}_{\rm rp}$ and a reflected SV-wave with slowness $\mathbf{s}_{\rm rs}$.

Let us consider a fixed-frequency, plane harmonic, incident P-wave that is described by a potential ϕ_i of the form

$$\phi_i(\mathbf{x}) = \Phi_i\, e^{i\omega\mathbf{s}_{ip}\cdot\mathbf{x}} = \Phi_i\, e^{i\kappa_p\,(x_1\sin\theta_{ip} - x_2\cos\theta_{ip})}$$

where we have used $\kappa_p := \omega/c_p$. The displacement produced by the incident wave is given by

$$\mathbf{u}_{ip} = \boldsymbol{\nabla}\phi_i = i\kappa_p\Phi_i(\sin\theta_{ip}\mathbf{e}_1 - \cos\theta_{ip}\mathbf{e}_2)e^{i\kappa_p\,(x_1\sin\theta_{ip} - x_2\cos\theta_{ip})}\,.$$

The traction at the interface is given by $\mathbf{t} = \boldsymbol{\sigma}^T \cdot \mathbf{n}$ where $\mathbf{n} = -\mathbf{e}_2$ is the outward normal to the free surface. The components of the traction vector are $t_1 = -\sigma_{21}$, $t_2 = -\sigma_{22}$, and $t_3 = -\sigma_{23}$. Therefore the components of the stress tensor that can potentially contribute to the traction are σ_{12}, σ_{22}, and σ_{23}.

Recall that the stress tensor is given by

$$\boldsymbol{\sigma} = \mathbf{C} : \boldsymbol{\nabla}\mathbf{u} = \lambda(\boldsymbol{\nabla}\cdot\mathbf{u})\boldsymbol{1} + \mu[\boldsymbol{\nabla}\mathbf{u} + (\boldsymbol{\nabla}\mathbf{u})^T] .$$

Therefore the relevant stress components are

$$\sigma_{12} = \mu(u_{1,2} + u_{2,1}) \; ; \; \sigma_{22} = \lambda(u_{1,1} + u_{2,2}) + 2\mu u_{2,2} \; ; \; \sigma_{23} = 0$$

since all displacements are functions only of x_1 and x_2 for plane waves in the chosen coordinate system and continuity of displacements requires that $u_3 = 0$. These stress components indicate that there can be two reflected waves from the interface, a P-wave and an SV-wave. However, an SH-wave cannot be produced because $u_3 = 0$.

Since the displacements produced by the incident and reflected P-waves can be derived as the gradient of a potential ($u_i = \phi_{,i}$), the relevant stress components due to the P-waves have the form

$$\sigma_{12} = \mu(\phi_{,12} + \phi_{,21}) = 2\mu\phi_{,12} \quad \text{and} \quad \sigma_{22} = \lambda(\phi_{,11} + \phi_{,22}) + 2\mu\phi_{,22} \, . \tag{2.6}$$

For the SV-wave we have $u_1 = \psi_{,2}$ and $u_2 = -\psi_{,1}$ and the stress components are

$$\sigma_{12} = \mu(\psi_{,22} - \psi_{,11}) \quad \text{and} \quad \sigma_{22} = \lambda(\psi_{,21} - \psi_{,12}) - 2\mu\psi_{,12} = -2\mu\psi_{,12} \, . \tag{2.7}$$

Let the reflected P-wave have a slowness vector $\mathbf{s}_{rp}$ and let the angle of reflection be θ_{rp}. Also, let the reflected SV-wave have a slowness vector $\mathbf{s}_{rs}$ and an angle of reflection θ_{rs}. Then

$$\begin{aligned} \mathbf{s}_{rp} &= \|\mathbf{s}_{rp}\| \, (\sin\theta_{rp}\mathbf{e}_1 + \cos\theta_{rp}\mathbf{e}_2) = (\sin\theta_{rp}\mathbf{e}_1 + \cos\theta_{rp}\mathbf{e}_2)/c_p \\ \mathbf{s}_{rs} &= \|\mathbf{s}_{rs}\| \, (\sin\theta_{rs}\mathbf{e}_1 + \cos\theta_{rs}\mathbf{e}_2) = (\sin\theta_{rs}\mathbf{e}_1 + \cos\theta_{rs}\mathbf{e}_2)/c_s . \end{aligned}$$

Let the corresponding displacement potentials be

$$\begin{aligned} \phi_r(\mathbf{x}) &= \Phi_r e^{i\omega \mathbf{s}_{rp}\cdot\mathbf{x}} = \Phi_r e^{i\kappa_p (x_1 \sin\theta_{rp} + x_2 \cos\theta_{rp})} \\ \psi_r(\mathbf{x}) &= \Psi_r e^{i\omega \mathbf{s}_{rs}\cdot\mathbf{x}} = \Psi_r e^{i\kappa_s (x_1 \sin\theta_{rs} + x_2 \cos\theta_{rs})} \end{aligned}$$

where we have used $\kappa_s := \omega/c_s$. The displacement produced by the reflected P-wave is given by

$$\mathbf{u}_{rp} = \boldsymbol{\nabla}\phi_r = i\kappa_p \Phi_r (\sin\theta_{rp}\mathbf{e}_1 + \cos\theta_{rp}\mathbf{e}_2) e^{i\kappa_p (x_1 \sin\theta_{rp} + x_2 \cos\theta_{rp})} \, .$$

while that produced by the reflected SV-wave is

$$\mathbf{u}_{rs} = \boldsymbol{\nabla}\times\boldsymbol{\psi} = \psi_{r,2}\mathbf{e}_1 - \psi_{r,1}\mathbf{e}_2 = i\kappa_s \Psi_r (\cos\theta_{rs}\mathbf{e}_1 - \sin\theta_{rs}\mathbf{e}_2) e^{i\kappa_s (x_1 \sin\theta_{rs} + x_2 \cos\theta_{rs})} \, .$$

Now we need to find the tractions at the interface that correspond to the superposition of the three plane wave displacement fields. For the incident P-wave

$$\begin{aligned}
u_{1,1} &= -\kappa_p^2 \sin^2\theta_{ip}\Phi_i\, e^{i\kappa_p\,(x_1\sin\theta_{ip}-x_2\cos\theta_{ip})}\\
u_{1,2} &= +\kappa_p^2 \sin\theta_{ip}\cos\theta_{ip}\Phi_i\, e^{i\kappa_p\,(x_1\sin\theta_{ip}-x_2\cos\theta_{ip})}\\
u_{2,1} &= +\kappa_p^2 \sin\theta_{ip}\cos\theta_{ip}\Phi_i\, e^{i\kappa_p\,(x_1\sin\theta_{ip}-x_2\cos\theta_{ip})}\\
u_{2,2} &= -\kappa_p^2 \cos^2\theta_{ip}\Phi_i\, e^{i\kappa_p\,(x_1\sin\theta_{ip}-x_2\cos\theta_{ip})}\,.
\end{aligned}$$

For the reflected P-wave we have

$$\begin{aligned}
u_{1,1} &= -\kappa_p^2 \sin^2\theta_{rp}R_{pp}\Phi_i\, e^{i\kappa_p\,(x_1\sin\theta_{rp}+x_2\cos\theta_{rp})}\\
u_{1,2} &= -\kappa_p^2 \sin\theta_{rp}\cos\theta_{rp}R_{pp}\Phi_i\, e^{i\kappa_p\,(x_1\sin\theta_{rp}+x_2\cos\theta_{rp})}\\
u_{2,1} &= -\kappa_p^2 \sin\theta_{rp}\cos\theta_{rp}R_{pp}\Phi_i\, e^{i\kappa_p\,(x_1\sin\theta_{rp}+x_2\cos\theta_{rp})}\\
u_{2,2} &= -\kappa_p^2 \cos^2\theta_{rp}R_{pp}\Phi_i\, e^{i\kappa_p\,(x_1\sin\theta_{rp}+x_2\cos\theta_{rp})}
\end{aligned}$$

where $R_{pp} = \Phi_r/\Phi_i$ is the *reflection coefficient* for the P-P reflection. Similarly, for the reflected SV-wave we have

$$\begin{aligned}
u_{1,1} &= -\kappa_s^2 \sin\theta_{rs}\cos\theta_{rs}R_{ps}\Phi_i\, e^{i\kappa_s\,(x_1\sin\theta_{rs}+x_2\cos\theta_{rs})}\\
u_{1,2} &= -\kappa_s^2 \cos^2\theta_{rs}R_{ps}\Phi_i\, e^{i\kappa_s\,(x_1\sin\theta_{rs}+x_2\cos\theta_{rs})}\\
u_{2,1} &= +\kappa_s^2 \sin^2\theta_{rs}R_{ps}\Phi_i\, e^{i\kappa_s\,(x_1\sin\theta_{rs}+x_2\cos\theta_{rs})}\\
u_{2,2} &= +\kappa_s^2 \sin\theta_{rs}\cos\theta_{rs}R_{ps}\Phi_i\, e^{i\kappa_s\,(x_1\sin\theta_{rs}+x_2\cos\theta_{rs})}
\end{aligned}$$

where $R_{ps} = \Psi_r/\Phi_i$ is the reflection coefficient for the P-SV reflection. Since the interface is free, the tractions at the interface are zero, i.e., $\sigma_{12} = \mu(u_{1,2}+u_{2,1}) = 0$ and $\sigma_{22} = \lambda u_{1,1} + (\lambda+2\mu)u_{2,2} = 0$. From the shear traction condition we have, at $x_2 = 0$,

$$\begin{aligned}
\kappa_p^2 &\left[\sin(2\theta_{ip})\, e^{i\kappa_p x_1\sin\theta_{ip}} - \sin(2\theta_{rp})R_{pp}\, e^{i\kappa_p x_1\sin\theta_{rp}}\right]\\
&- \kappa_s^2\cos(2\theta_{rs})R_{ps}\, e^{i\kappa_s x_1\sin\theta_{rs}} = 0
\end{aligned}$$

or

$$\kappa_p^2\left[\frac{\sin(2\theta_{ip})}{\sin(2\theta_{rp})}e^{ix_1(\kappa_p\sin\theta_{ip}-\kappa_s\sin\theta_{rs})} - R_{pp}\, e^{ix_1(\kappa_p\sin\theta_{rp}-\kappa_s\sin\theta_{rs})}\right] = \kappa_s^2 R_{ps}\frac{\cos(2\theta_{rs})}{\sin(2\theta_{rp})}\,.$$

Since the right-hand side is not a function of x_1, this relation can hold only if $\kappa_p\sin\theta_{ip} - \kappa_s\sin\theta_{rs} = 0$ and $\kappa_p\sin\theta_{rp} - \kappa_s\sin\theta_{rs} = 0$. Hence we have the relations ‡

$$\boxed{\theta_{ip} = \theta_{rp}} \quad\text{and}\quad \boxed{\frac{\sin\theta_{ip}}{\kappa_s} = \frac{\sin\theta_{rs}}{\kappa_p} \quad\Longrightarrow\quad \frac{\sin\theta_{ip}}{\sin\theta_{rs}} = \frac{c_p}{c_s}} \tag{2.8}$$

‡Since $c_s < c_p$, we have $\theta_{rs} < \theta_{ip}$ for most materials.

and

$$\boxed{\kappa_p^2 \sin(2\theta_{ip})(1 - R_{pp}) = \kappa_s^2 \cos(2\theta_{rs}) R_{ps}\,.} \tag{2.9}$$

The two equations in Equation (2.8) are the equivalent of Snell's law for elastic media. Let us now look at the normal component of the traction vector at $x_2 = 0$. We have

$$\begin{aligned} 0 = & -\lambda\kappa_p^2(1+R_{pp})\Phi_i e^{i\kappa_p x_1 \sin\theta_{ip}} - 2\mu\kappa_p^2\cos^2\theta_{ip}(1+R_{pp})\Phi_i e^{i\kappa_p x_1 \sin\theta_{ip}} \\ & + 2\mu\kappa_s^2 \sin\theta_{rs}\cos\theta_{rs} R_{ps}\Phi_i e^{i\kappa_s x_1 \sin\theta_{rs}} \end{aligned}$$

or

$$\kappa_p^2(\lambda + 2\mu\cos^2\theta_{ip})(1 + R_{pp}) = \mu\kappa_s^2 \sin(2\theta_{rs}) R_{ps}$$

or

$$\kappa_p^2\left[\frac{\lambda+2\mu}{\mu} - 2\sin^2\theta_{ip}\right](1+R_{pp}) = \kappa_s^2\sin(2\theta_{rs})R_{ps}\,.$$

We can get rid of λ and μ by using the relation

$$\frac{\lambda+2\mu}{\mu} = \frac{\rho c_p^2}{\rho c_s^2} = \frac{\kappa_s^2}{\kappa_p^2}$$

to get

$$\boxed{(\kappa_s^2 - 2\kappa_p^2\sin^2\theta_{ip})(1+R_{pp}) = \kappa_s^2\sin(2\theta_{rs})R_{ps}\,.} \tag{2.10}$$

We could solve equations (2.9) and (2.10) directly for the reflection coefficients R_{pp} and R_{ps}. However, it proves more useful to express these equations in terms of the phase velocities, a *ray parameter* (α) and two *impedances* Z which are defined as[§]

$$\alpha := \frac{\sin\theta_{ip}}{c_p} = \frac{\sin\theta_{rs}}{c_s} \quad \text{and} \quad Z_p := \frac{\rho c_p}{\cos\theta_{ip}}\,;\; Z_s := \frac{\rho c_s}{\cos\theta_{rs}}\,.$$

Using these definitions we get

$$\boxed{R_{pp} = \frac{4c_s^4\alpha^2\rho^2 - Z_p Z_s(1-2c_s^2\alpha^2)^2}{4c_s^4\alpha^2\rho^2 + Z_p Z_S(1-2c_s^2\alpha^2)^2}} \tag{2.11}$$

and

$$\boxed{R_{ps} = \frac{4c_s^2\alpha\rho Z_s(1-2c_s^2\alpha^2)}{4c_s^4\alpha^2\rho^2 + Z_p Z_S(1-2c_s^2\alpha^2)^2}\,.} \tag{2.12}$$

There is no reflected P-wave if $4c_s^4\alpha^2\rho^2 - Z_p Z_s(1-2c_s^2\alpha^2)^2 = 0$ and no reflected SV-wave if $c_s^2\alpha^2 = 1/2$.

[§] See Section 2.3.2 for a definition of impedance.

For mild steel with a Young's modulus (E) of 200 GPa, Poisson's ratio (ν) 0.3, and mass density 7850 kg/m^3, we have $\mu = E/[2(1+\nu)]$ =76.9 GPa, $\lambda = 2\mu\nu/(1-2\nu) =$ 115.4 GPa, $c_p = 5856$ m/s, and $c_s = 3130$ m/s. For diamond the corresponding values are 1220 GPa, 0.2, 3500 kg/m^3, 508 GPa, 339 GPa, 12051 m/s, and 19680 m/s, respectively. Plots of the amplitude and phase of the reflection coefficients for the two materials are shown in Figure 2.3.

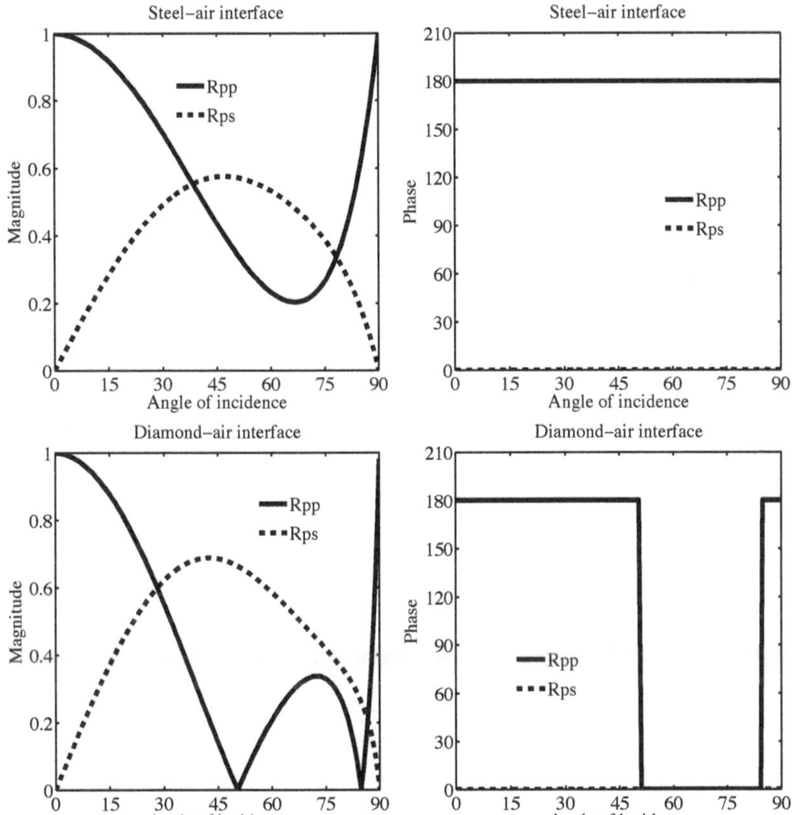

FIGURE 2.3
Reflection coefficients for the reflection of a P-wave at mild steel-air and diamond-air interfaces. The solid line is for R_{pp} and the dotted line is for R_{ps}. The plots on the left show the modulus of the complex reflection coefficient while the plots on the right show the phase. The top row is for steel and the bottom row is for diamond.

At normal incidence and grazing incidence there is no SV-wave. The reflection coefficient for the P-wave in mild steel is 180° out of phase with the incident wave while the SV-wave has the same phase. For diamond we see two angles of incidence at which there is no reflected P-wave and this is reflected in the phase plot.

2.2.3 Reflection of P-waves at the interface between two solids

Let us now consider the more complicated problem of a P-wave incident upon the interface between two isotropic and homogeneous solid materials as shown in Figure 2.4. In this case there is a transmitted P-wave and a transmitted SV-wave. Let the displacement potentials be

$$\begin{aligned}
\phi_i &= \Phi_i e^{i\kappa_{p1}(x_1 \sin\theta_i - x_2 \cos\theta_i)}; \qquad \phi_r = R_p \Phi_i e^{i\kappa_{p1}(x_1 \sin\theta_r + x_2 \cos\theta_r)} \\
\phi_t &= T_p \Phi_i e^{i\kappa_{p2}(x_1 \sin\theta_t - x_2 \cos\theta_t)} \\
\psi_r &= R_s \Phi_i e^{i\kappa_{s1}(x_1 \sin\theta_{rs} + x_2 \cos\theta_{rs})}; \quad \psi_t = T_s \Phi_i e^{i\kappa_{s2}(x_1 \sin\theta_{ts} - x_2 \cos\theta_{ts})}
\end{aligned}$$

where

$$\kappa_{p1} = \frac{\omega}{c_{p1}}; \quad \kappa_{p2} = \frac{\omega}{c_{p2}}; \quad \kappa_{s1} = \frac{\omega}{c_{s1}}; \quad \kappa_{s2} = \frac{\omega}{c_{s2}}$$

and c_{p1}, c_{s1} are the phase velocities in medium 1 and c_{p2}, c_{s2} are the phase velocities in medium 2.

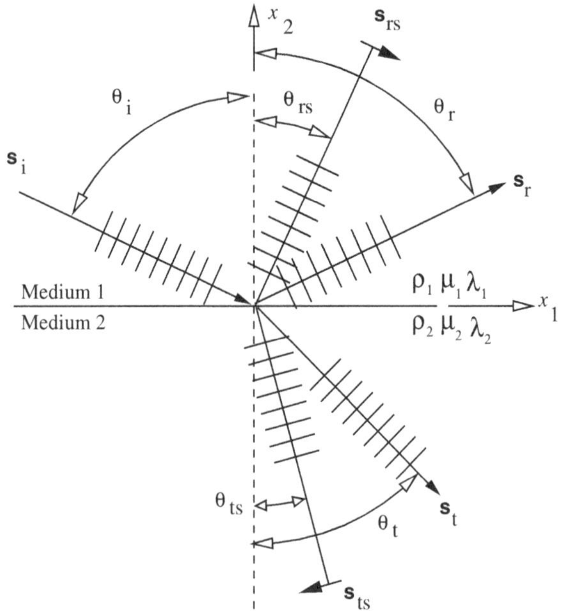

FIGURE 2.4
Reflection and transmission of a P-wave at an interface between two elastic solid media.

To find the four reflection and transmission coefficients we need the two traction continuity conditions $\sigma_{12}^{+} = \sigma_{12}^{-}$ and $\sigma_{22}^{+} = \sigma_{22}^{-}$, and two displacement continuity conditions $u_1^{+} = u_1^{-}$, $u_2^{+} = u_2^{-}$ where the $+$ sign indicates medium 1 and the $-$ sign indicates medium 2 at $x_2 = 0$. Following the same procedure as before, we can show that

$$\theta_i = \theta_r \; ; \quad \frac{\sin\theta_i}{c_{p1}} = \frac{\sin\theta_{rs}}{c_{s1}} = \frac{\sin\theta_t}{c_{p2}} = \frac{\sin\theta_{ts}}{c_{s2}} =: \alpha \; . \tag{2.13}$$

Using the traction conditions we arrive at a system of equations

$$\begin{aligned} &2\alpha c_{s1}^2 \rho_1^2 Z_{p2} Z_{s1}^2 Z_{s2}^2 \, R_p + c_{s1}^2 \rho_1 Z_{p1} Z_{p2} \left(\rho_1^2 - \alpha^2 Z_{s1}^2\right) Z_{s2}^2 \, R_s + 2\alpha c_{s2}^2 \rho_2^2 Z_{p1} Z_{s1}^2 Z_{s2}^2 \, T_p \\ &\quad - c_{s2}^2 \rho_2 Z_{p1} Z_{p2} Z_{s1}^2 \left(\rho_2^2 - \alpha^2 Z_{s2}^2\right) \, T_s = \alpha c_{s1}^2 \rho_1^2 Z_{p2} Z_{s1}^2 Z_{s2}^2 \end{aligned} \tag{2.14}$$

$$\begin{aligned} &\rho_1 \left[2\alpha^2 c_{s1}^2 Z_{p1}^2 - c_{p1}^2 \left(\rho_1^2 + \alpha^2 Z_{p1}^2\right)\right] Z_{p2}^2 Z_{s1} Z_{s2} \, R_p + 2\alpha c_{s1}^2 \rho_1^2 Z_{p1}^2 Z_{p2}^2 Z_{s2} \, R_s \\ &\quad + 2\alpha c_{s2}^2 \rho_2^2 Z_{p1}^2 Z_{p2}^2 Z_{s1} \, T_s + \rho_2 Z_{p1}^2 \left[c_{p2}^2 \rho_2^2 + \alpha^2 \left(c_{p2}^2 - 2c_{s2}^2\right) Z_{p2}^2\right] Z_{s1} Z_{s2} \, T_p \\ &\quad = -\rho_1 \left[2\alpha^2 c_{s1}^2 Z_{p1}^2 - c_{p1}^2 \left(\rho_1^2 + \alpha^2 Z_{p1}^2\right)\right] Z_{p2}^2 Z_{s1} Z_{s2} \end{aligned} \tag{2.15}$$

where we have used

$$Z_{p1} := \frac{\rho_1 c_{p1}}{\cos\theta_i} ; \; Z_{p2} := \frac{\rho_2 c_{p2}}{\cos\theta_t} ; \; Z_{s1} := \frac{\rho_1 c_{s1}}{\cos\theta_{rs}} ; \; Z_{s2} := \frac{\rho_2 c_{s2}}{\cos\theta_{ts}} .$$

From the displacement boundary conditions we have

$$\alpha Z_{s1} Z_{s2} \, R_p + \rho_1 Z_{s2} \, R_s - \alpha Z_{s1} Z_{s2} \, T_p + \rho_2 Z_{s1} \, T_s = -\alpha Z_{s1} Z_{s2} \tag{2.16}$$

$$\rho_1 Z_{p2} \, R_p - \alpha Z_{p1} Z_{p2} \, R_s + \rho_2 Z_{p1} \, T_p + \alpha Z_{p1} Z_{p2} \, T_s = \rho_1 Z_{p2} . \tag{2.17}$$

These four equations can be solved in closed form using any of a number of symbolic computation software packages. However, the resulting expressions for R_p, R_s, T_p, and T_s are quite complicated. The magnitude and phase of the coefficients for reflected and transmitted waves at a steel-diamond interface are shown in Figure 2.5. Note that these reflection coefficients do not represent the amplitudes of displacement. To compute the ratios of displacement amplitude we have to calculate $\boldsymbol{\nabla}\phi$ and $\boldsymbol{\nabla} \times \boldsymbol{\psi}$ and take ratios of the resulting quantities. Though we have idealized diamond as an isotropic material it is actually anisotropic. Exact solutions for reflection and transmission in general anisotropic media can be quite difficult to find because the shear and normal components of elastic waves in such media are often coupled.

2.2.4 Reflection of SH-waves at the interface between two solids

Another case that is of interest is when there is an SH-wave incident upon the interface. In that case no P- and SV-waves are excited at the interface since $u_1 = u_2 = 0$ and there is only a reflected SH-wave and a transmitted SH-wave. Since the displacement $u_3(x_1, x_2)$ satisfies a wave equation we can look for plane wave solutions directly in terms of the out-of-plane displacement, i.e.,

$$\begin{aligned} &u_{3i} = U_i e^{i\kappa_{s1}(x_1 \sin\theta_i - x_2 \cos\theta_i)} \; ; \quad u_{3r} = R U_i e^{i\kappa_{s1}(x_1 \sin\theta_r + x_2 \cos\theta_r)} \\ &u_{3t} = T U_i e^{i\kappa_{s2}(x_1 \sin\theta_t + x_2 \cos\theta_t)} \; . \end{aligned}$$

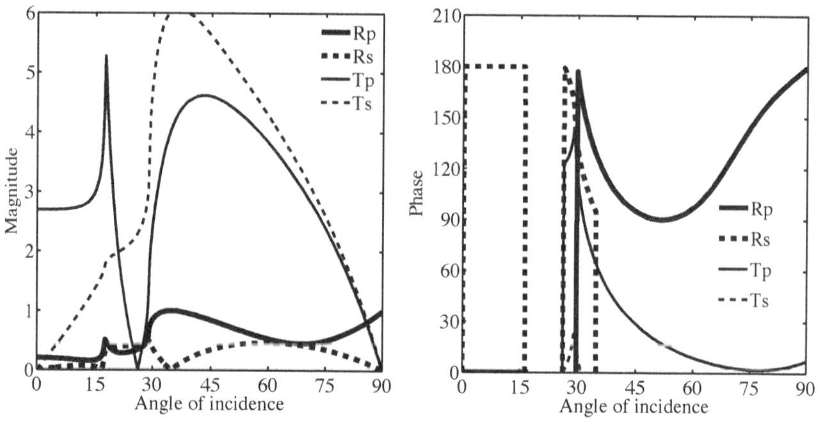

FIGURE 2.5
Reflection and transmission coefficients for the reflection of a P-wave at the interface of mild steel and diamond.

Following the same procedure as before, from the continuity of displacements at $x_2 = 0$ we get

$$\theta_i = \theta_t \quad \text{and} \quad \kappa_{s1} \sin\theta_i = \kappa_{s2} \sin\theta_t \quad \Longrightarrow \quad \frac{\sin\theta_i}{c_{s1}} = \frac{\sin\theta_t}{c_{s2}}$$

and

$$1 + R = T \ .$$

The only non-zero stress components in this case are σ_{13} and σ_{23}. Therefore, the continuity of tractions at the interface implies that $\sigma_{23}^{+} = \sigma_{23}^{-}$ which leads to

$$\mu_1(1 - R)\frac{\cos\theta_i}{c_{s1}} = \mu_2 T\frac{\cos\theta_t}{c_{s2}} \ .$$

Solving for R and T, we get

$$\boxed{R = \frac{\rho_1 c_{s1}\cos\theta_i - \rho_2 c_{s2}\cos\theta_t}{\rho_1 c_{s1}\cos\theta_i + \rho_2 c_{s2}\cos\theta_t}} \quad \text{and} \quad \boxed{T = \frac{2\rho_1 c_{s1}\cos\theta_i}{\rho_1 c_{s1}\cos\theta_i + \rho_2 c_{s2}\cos\theta_t}} \tag{2.18}$$

2.2.5 Negative refractive index and SH-waves

Let us define the refractive index for SH-waves as

$$n := \frac{c_0}{c_s} = \frac{c_0}{\sqrt{\mu/\rho}}$$

where c_0 is the shear wave speed in a reference medium. Then we can write the expressions for the reflection and transmission coefficients as

$$R = \frac{n_2\rho_1\cos\theta_i - n_1\rho_2\cos\theta_t}{n_2\rho_1\cos\theta_i + n_1\rho_2\cos\theta_t} \quad \text{and} \quad T = \frac{2n_2\rho_1\cos\theta_i}{n_2\rho_1\cos\theta_i + n_1\rho_2\cos\theta_t}. \tag{2.19}$$

It may be possible, at certain frequencies, for the effective dynamic density of a material to be negative (see Chapter 5 for examples). Let us consider a situation where the refractive index is also negative. In particular, let $n_2 = -n_1$ and $\rho_2 = -\rho_1$. Then,

$$n_1 = \frac{c_0}{\sqrt{\mu_1/\rho_1}} \implies \mu_1 = \frac{c_0^2}{n_1^2/\rho_1}$$

and

$$n_2 = \frac{c_0}{\sqrt{\mu_2/\rho_2}} \implies -n_1 = \frac{c_0}{\sqrt{\mu_2/-\rho_1}} \implies n_1 = \frac{ic_0}{\sqrt{\mu_2/\rho_1}} \implies \mu_2 = -\frac{c_0^2}{n_1^2/\rho_1}.$$

Therefore $\mu_2 = -\mu_1$ and we can get a negative index of refraction if we can create a material with negative density and negative shear modulus. Note that in this situation

$$\frac{\sin\theta_i}{\sin\theta_t} = \frac{c_{s1}}{c_{s2}} = \frac{n_2}{n_1} = -1$$

which implies that $\sin\theta_t = -\sin\theta_i$ or $\theta_t = -\theta_i$. We can also see that $R = 0$ and $T = 1$, i.e., there is no reflected SH-wave and the entire wave is transmitted though the interface.

2.3 Plane wave solutions in acoustics

Consider the acoustic wave equation (1.44) in terms of the scalar potential ϕ,

$$\nabla^2\phi = \frac{1}{c_0^2}\frac{\partial^2\phi}{\partial t^2}\,; \quad p = \rho_0\frac{\partial\phi}{\partial t}\,; \qquad \mathbf{v} = -\boldsymbol{\nabla}\phi\,.$$

Let us restrict ourselves to harmonic solutions of the form

$$\phi(\mathbf{x},t) = \widehat{\varphi}(\mathbf{x})\, e^{-i\omega t}\,; \quad p(\mathbf{x},t) = \widehat{p}(\mathbf{x})\, e^{-i\omega t}\,.$$

Then the governing equations for fixed frequency ω are

$$\nabla^2\widehat{\varphi} = -\frac{\omega^2}{c_0^2}\widehat{\varphi}\,; \quad \widehat{p} = -i\omega\rho_0\widehat{\varphi}\,; \quad \widehat{\mathbf{v}} = -\boldsymbol{\nabla}\widehat{\varphi}\,. \tag{2.20}$$

Since the particle velocities are harmonic, they must be derived from harmonic particle displacements of the form $\mathbf{u}(\mathbf{x},t) = \widehat{\mathbf{u}}(\mathbf{x}) \exp(-i\omega t)$. Therefore we have

$$\mathbf{v}(\mathbf{x},t) = \frac{\partial}{\partial t}[\mathbf{u}(\mathbf{x},t)] = -i\omega\widehat{\mathbf{u}}(\mathbf{x})\, e^{-i\omega t} = -\boldsymbol{\nabla}\widehat{\varphi}\, e^{-i\omega t}$$

which implies that

$$\widehat{\mathbf{u}}(\mathbf{x}) = \frac{1}{i\omega}\boldsymbol{\nabla}\widehat{\varphi}\,. \tag{2.21}$$

Consider plane wave solutions of the form

$$\widehat{\varphi}(\mathbf{x}) = \Phi_0\, e^{i\mathbf{k}\cdot\mathbf{x}}\,; \quad \widehat{p}(\mathbf{x}) = P_0\, e^{i\mathbf{k}\cdot\mathbf{x}}$$

where $\mathbf{k}$ is the wave vector (also called the wavenumber vector) discussed in Section 2.1. Substitution into equations (2.20) gives us, for a rectangular Cartesian coordinate system with $\mathbf{k} = k_i\mathbf{e}_i$,

$$-(k_1^2 + k_2^2 + k_3^2) = -\frac{\omega^2}{c_0^2}\,; \quad P_0 = -i\omega\rho_0\Phi_0\,.$$

The wave equation (first equation above) shows that the components of $\mathbf{k}$ are not all independent of ω while the second equation shows that the amplitude of the pressure is directly proportional to the amplitude of the scalar potential. We may also express the wave equation in the compact form

$$\mathbf{k}\cdot\mathbf{k} = (\|\mathbf{k}\|)^2 = \frac{\omega^2}{c_0^2}\,. \tag{2.22}$$

2.3.1 Reflection of an acoustic wave at a plane interface

Let us now look at the problem of a plane acoustic wave and its reflection and refraction at the interface between two media. Figure 2.6 shows a schematic of the situation. The normal to the wavefront is in the plane of the diagram. The wave is incident from a medium with density ρ_1 and phase speed c_1 at an angle θ_i to the normal to the interface. The medium on the other side of the interface has a density ρ_2 and phase speed c_2. The reflected wave makes an angle θ_r with the normal while the transmitted wave makes an angle θ_t to the normal. Let the incident, reflected, and transmitted wave vectors be $\mathbf{k}_i$, $\mathbf{k}_r$, and $\mathbf{k}_t$, respectively.

Let us choose an orthonormal basis $(\mathbf{e}_1, \mathbf{e}_2, \mathbf{e}_3)$ such that the $\mathbf{e}_1$ vector is in the x_1-direction and the $\mathbf{e}_2$ vector is in the x_2-direction. Then $\mathbf{x} = x_i\mathbf{e}_i$ and the wave vectors $\mathbf{k}_i$, $\mathbf{k}_r$, and $\mathbf{k}_t$ may be expressed in this basis as

$$\mathbf{k}_i = \|\mathbf{k}_i\|\sin\theta_i\,\mathbf{e}_1 - \|\mathbf{k}_i\|\cos\theta_i\,\mathbf{e}_2\,; \quad \mathbf{k}_r = \|\mathbf{k}_r\|\sin\theta_r\,\mathbf{e}_1 + \|\mathbf{k}_r\|\cos\theta_r\,\mathbf{e}_2$$
$$\mathbf{k}_t = \|\mathbf{k}_t\|\sin\theta_t\,\mathbf{e}_1 - \|\mathbf{k}_t\|\cos\theta_t\,\mathbf{e}_2\,.$$

Therefore the scalar potential for the incident plane wave can, after dropping the hats for convenience, be written as

$$\phi_i(\mathbf{x}) = \Phi_i\, e^{i\|\mathbf{k}_i\|\,(x_1\sin\theta_i - x_2\cos\theta_i)}\,.$$

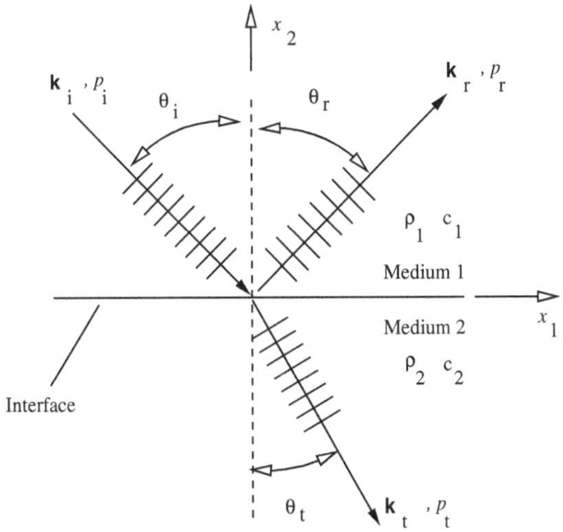

FIGURE 2.6
Reflection and refraction of an acoustic wave at the interface between two materials.

Similarly, the reflected and transmitted plane waves can be expressed as

$$\phi_r(\mathbf{x}) = \Phi_r\, e^{i\|\mathbf{k}_r\|(x_1 \sin\theta_r + x_2 \cos\theta_r)} \;; \quad \phi_t(\mathbf{x}) = \Phi_t\, e^{i\|\mathbf{k}_t\|(x_1 \sin\theta_t - x_2 \cos\theta_t)} \;.$$

Recall that the wave number, $\|\mathbf{k}\|$, is related to the phase speed (c) in a medium by $\|\mathbf{k}\| = \omega/c$. Hence $\|\mathbf{k}_i\| = \|\mathbf{k}_r\|$. Also, let $\Phi_r = R\Phi_i$ and $\Phi_t = T\Phi_i$ where R is the *reflection coefficient* and T is the *transmission coefficient*. Then the total potential in the medium of incidence, which is a superposition of the incident and reflected potentials, is given by

$$\phi_1(\mathbf{x}) = \phi_i(\mathbf{x}) + \phi_r(\mathbf{x}) = \Phi_i \left[e^{i\|\mathbf{k}_i\|(x_1 \sin\theta_i - x_2 \cos\theta_i)} + R\, e^{i\|\mathbf{k}_i\|(x_1 \sin\theta_r + x_2 \cos\theta_r)} \right] \;.$$

The pressure at the interface must be continuous, i.e., $\rho_1\phi_1 = \rho_2\phi_t$. Therefore,

$$\rho_1 \left[e^{i\|\mathbf{k}_i\|(x_1 \sin\theta_i - x_2 \cos\theta_i)} + R\, e^{i\|\mathbf{k}_i\|(x_1 \sin\theta_r + x_2 \cos\theta_r)} \right] = \rho_2\, T\, e^{i\|\mathbf{k}_t\|(x_1 \sin\theta_t - x_2 \cos\theta_t)} \;.$$

Since $x_2 = 0$ at the interface, the above equation reduces to

$$\frac{\rho_1}{\rho_2 T} \left[e^{i(\|\mathbf{k}_i\| \sin\theta_i - \|\mathbf{k}_t\| \sin\theta_t) x_1} + R\, e^{i(\|\mathbf{k}_i\| \sin\theta_r - \|\mathbf{k}_t\| \sin\theta_t) x_1} \right] = 1 \;.$$

For this relation to be valid for all values of x_1 we must have

$$\|\mathbf{k}_i\| \sin\theta_i - \|\mathbf{k}_t\| \sin\theta_t = 0 \;; \quad \|\mathbf{k}_i\| \sin\theta_r - \|\mathbf{k}_t\| \sin\theta_t = 0$$

which implies that

$$1 + R = \xi\, T \quad \text{where} \quad \xi := \frac{\rho_2}{\rho_1} \,. \tag{2.23}$$

These constraints also indicate that the law of reflection and Snell's law hold, i.e.,

$$\theta_i = \theta_r \ ; \quad \frac{\|\mathbf{k}_i\|}{\sin\theta_t} = \frac{\|\mathbf{k}_t\|}{\sin\theta_i} \quad \Leftrightarrow \quad \frac{\sin\theta_i}{c_1} = \frac{\sin\theta_t}{c_2} \,.$$

Another boundary condition at the interface of the two materials is that the normal velocity $v_2 = -\phi_{,2}$ is continuous. Now,

$$\phi_{1,2} = -i\,\|\mathbf{k}_i\|\,\cos\theta_i\,\Phi_i\left[e^{i\|\mathbf{k}_i\|\,(x_1\sin\theta_i - x_2\cos\theta_i)} - R\,e^{i\|\mathbf{k}_i\|\,(x_1\sin\theta_i + x_2\cos\theta_i)}\right]$$

$$\phi_{t,2} = -i\,\|\mathbf{k}_t\|\,\cos\theta_t\Phi_t\,e^{i\|\mathbf{k}_t\|\,(x_1\sin\theta_t - x_2\cos\theta_t)} \,.$$

Hence for waves in the incident medium, at $x_2 = 0$,

$$\|\mathbf{k}_i\|\,\cos\theta_i(1-R)e^{i\|\mathbf{k}_i\|x_1\sin\theta_i} = \|\mathbf{k}_t\|\,\cos\theta_t\,T\,e^{i\|\mathbf{k}_t\|x_1\sin\theta_t}$$

or

$$\frac{\omega}{c_1}\cos\theta_i(1-R)e^{i\|\mathbf{k}_i\|x_1\sin\theta_i} = \frac{\omega}{c_2}\cos\theta_t\,T\,e^{i\|\mathbf{k}_t\|x_1\sin\theta_t}$$

Using Snell's law we get

$$\frac{\cos\theta_i}{c_1}(1-R) = \frac{\cos\theta_t}{c_2}\,T = \left(\sqrt{\frac{c_1^2 - c_2^2\sin^2\theta_i}{c_1^2c_2^2}}\right)\,T$$

or

$$1 - R = \eta\,T \quad \text{where} \quad \eta := \sqrt{\frac{c_1^2 - c_2^2\sin^2\theta_i}{c_2^2\cos^2\theta_i}} = \sqrt{\frac{n^2 - \sin^2\theta_i}{\cos^2\theta_i}} \tag{2.24}$$

where the *refractive index* has been defined as $n := c_1/c_2$. The reflection and transmission coefficients can be found by solving equations (2.23) and (2.24) to get

$$R = \frac{\xi - \eta}{\xi + \eta} \,; \quad T = \frac{2}{\xi + \eta} \,. \tag{2.25}$$

If $\theta_i = \pi/2$ we have $R = -1$ and $T = 0$. This implies that grazing incidence is not allowed because a value of $R = 1$ means that the reflected wave will cancel out the incident wave. Another interesting situation is where $R = 0$, i.e., $\xi = \eta$. In that case

$$\xi\cos\theta_i - \sqrt{n^2 - \sin^2\theta_i} = 0 \quad \Longrightarrow \quad \sin\theta_i = \sqrt{\frac{\xi^2 - n^2}{\xi^2 - 1}} \,.$$

This angle of incidence is called *Brewster's angle* (θ_B) and at this angle the entire wave is transmitted across the interface without any reflection. Note that the above condition suggests that there are some constraints on the relative values of ξ and n.

To observe the similarity between the above equations and those for an SH-wave (2.19), we define $n_1 = c_0/c_1$ and $n_2 = c_0/c_2$, where c_0 is a reference sound velocity. Then, using $\xi = \rho_2/\rho_1$ and $\eta = n_2 \cos\theta_t / n_1 \cos\theta_i$ in equation (2.25), we have

$$R = \frac{n_1\rho_2\cos\theta_i - n_2\rho_1\cos\theta_t}{n_1\rho_2\cos\theta_i + n_2\rho_1\cos\theta_t} \quad \text{and} \quad T = \frac{2n_1\rho_2\cos\theta_i}{n_1\rho_2\cos\theta_i + n_2\rho_1\cos\theta_t} .$$

The similarity between this equation and that for an SH-wave in a solid suggests that if $n_2 = -n_1$ and $\rho_2 = -\rho_1$ then the bulk modulus $\kappa_2 = -\kappa_1$ because $c^2 = \kappa/\rho$.

2.3.2 Acoustic impedance

The *characteristic acoustic impedance* of a medium is defined as the ratio of the pressure to the acoustic wave speed in the medium

$$\widetilde{Z}_{\text{char}} = \frac{-i\omega\rho_0\Phi\, e^{i\mathbf{k}\cdot\mathbf{x}}}{-ik\Phi\, e^{i\mathbf{k}\cdot\mathbf{x}}} = \frac{\omega\rho_0}{k} = \rho_0 c_0 . \tag{2.26}$$

The *acoustic impedance* at an interface is the ratio of the pressure to the normal particle velocity at the interface,

$$Z_S = \frac{p}{v_n} .$$

The normal component of the velocity v_n is chosen such that it points into the medium characterized by Z_S. Recall that, for time harmonic waves, $p = -i\omega\rho\phi$. Therefore, for a homogeneous medium,

$$\mathbf{v} = -\boldsymbol{\nabla}\phi = -\frac{i}{\omega\rho}\boldsymbol{\nabla} p .$$

For plane harmonic waves with $\mathbf{v} = \widehat{\mathbf{v}}\exp(i\mathbf{k}\cdot\mathbf{x})$ and $p = \widehat{p}\exp(i\mathbf{k}\cdot\mathbf{x})$ we have

$$\boldsymbol{\nabla} p = i\mathbf{k}\widehat{p}\exp(i\mathbf{k}\cdot\mathbf{x}) \quad \Longrightarrow \quad \widehat{\mathbf{v}} = \frac{\mathbf{k}}{\omega\rho}\widehat{p} .$$

If the angle of incidence is θ, the normal component of the velocity is given by

$$v_n = \frac{k\cos\theta}{\omega\rho}\widehat{p} \quad \text{where} \quad k = \|\mathbf{k}\| .$$

Therefore, using $k = \omega/c$, we have

$$Z = \frac{\widehat{p}}{v_n} = \frac{\omega\rho}{k\cos\theta} = \frac{\rho c}{\cos\theta} . \tag{2.27}$$

Figure 2.7 shows a schematic of the situation and the directions of the normal components of the velocities in two media on opposite sides of an interface.

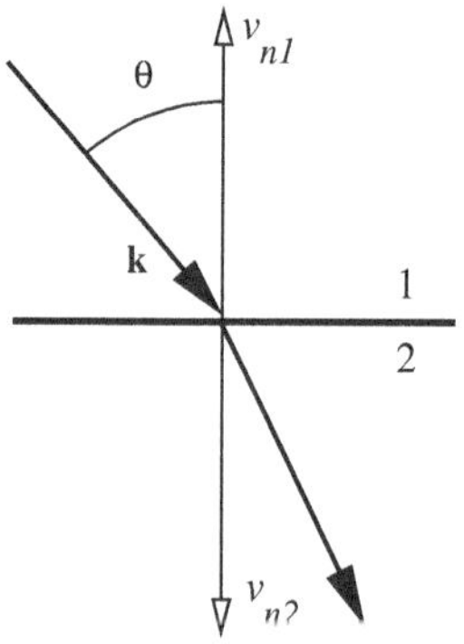

FIGURE 2.7
Normal components of velocities on the two sides of an interface.

If we define

$$Z_1 := \frac{\rho_1 c_1}{\cos\theta_i} \,; \quad Z_2 := \frac{\rho_2 c_2}{\cos\theta_t}$$

we have

$$1 - R = \left(\frac{\rho_2 Z_1}{\rho_1 Z_2}\right) T \quad \Longrightarrow \quad \eta = \xi\,\frac{Z_1}{Z_2} \quad \Longrightarrow \quad R = \frac{Z_2 - Z_1}{Z_2 + Z_1}\,.$$

Similarly,

$$T = \frac{2Z_2}{Z_2 + Z_1}\,.$$

At normal incidence $Z_1 = \rho_1 c_1$ and $Z_2 = \rho_2 c_2$. Notice that if the impedances match, i.e., $Z_1 = Z_2$, then there is no reflection ($R = 0$) and $T = \rho_1/\rho_2$. At the interface $p_1 = p_2$ and $v_{n1} = -v_{n2}$. Hence, $p_1/v_{n1} = Z_1 = -p_2/v_{n2}$ is also an acceptable interface continuity condition.

2.3.3 Reflection at a liquid-solid interface

It is often of interest to determine the reflection and transmission coefficients of an acoustic pressure wave incident from a fluid medium onto a solid medium. In that case there is an additional SV-wave in medium 2 in Figure 2.6 because the solid medium is able to support shear deformations.

Let the waves in medium 1 be described by *velocity* potentials

$$\phi_i = \Phi_i e^{i\kappa_1(x_1\sin\theta_i - x_2\cos\theta_i)} \;; \quad \phi_r = R\Phi_i e^{i\kappa_1(x_1\sin\theta_i + x_2\cos\theta_i)}$$

where $\kappa_1 = \|\mathbf{k}_i\|$. The corresponding pressures are $p_i = -i\omega\rho_1\phi_i$, $p_r = -i\omega\rho_1\phi_i$ and the displacements are $\mathbf{u}_i = 1/(i\omega)\boldsymbol{\nabla}\phi_i$, $\mathbf{u}_r = 1/(i\omega)\boldsymbol{\nabla}\phi_r$.

In our discussion on elastodynamics we have used *displacement* potentials where $\mathbf{u} = \boldsymbol{\nabla}\phi + \boldsymbol{\nabla} \times \boldsymbol{\psi}$. To remain consistent with the formulation we have used for acoustics, we would like the transmission coefficients in the solid to be also expressed in terms of *velocity* potentials. For time harmonic fields we have

$$\mathbf{u}(\mathbf{x},t) = \widehat{\mathbf{u}}(\mathbf{x})\,e^{-i\omega t}\ ;\ \phi(\mathbf{x},t) = \widehat{\varphi}(\mathbf{x})\,e^{-i\omega t}\ ;\ \boldsymbol{\psi}(\mathbf{x},t) = \widehat{\boldsymbol{\psi}}(\mathbf{x})\,e^{-i\omega t}\ .$$

Therefore,

$$\mathbf{v}(\mathbf{x},t) = \dot{\mathbf{u}} = -i\omega\widehat{\mathbf{u}}(\mathbf{x})\,e^{-i\omega t} = -i\omega\boldsymbol{\nabla}\widehat{\varphi}e^{-i\omega t} - i\omega\boldsymbol{\nabla}\times\widehat{\boldsymbol{\psi}}e^{-i\omega t}$$

or

$$\mathbf{v}(\mathbf{x},t) = -\boldsymbol{\nabla}\widetilde{\varphi} - \boldsymbol{\nabla}\times\widetilde{\boldsymbol{\psi}} \quad \text{where} \quad \widetilde{\varphi} := i\omega\phi\ ;\ \widetilde{\boldsymbol{\psi}} := i\omega\boldsymbol{\psi}\ .$$

Let the *velocity* potentials in medium 2 be

$$\widetilde{\varphi}_t = T_p\Phi_i e^{i\kappa_{p2}(x_1\sin\theta_t - x_2\cos\theta_t)}\ ;\ \widetilde{\psi}_t = T_s\Phi_i e^{i\kappa_{s2}(x_1\sin\theta_{ts} - x_2\cos\theta_{ts})}\ .$$

The displacements that derive from these velocity potentials are

$$\mathbf{u}_t = [\widetilde{\varphi}_{t,1}\mathbf{e}_1 + \widetilde{\varphi}_{t,2}\mathbf{e}_2]/(i\omega) \quad \text{and} \quad \mathbf{u}_{ts} = [\widetilde{\psi}_{t,2}\mathbf{e}_1 - \widetilde{\psi}_{t,1}\mathbf{e}_2]/(i\omega)\ .$$

Therefore the relevant stresses in medium 2 are, using equations (2.6) and (2.7),

$$\begin{aligned}\sigma_{12} &= \mu_2[2\widetilde{\varphi}_{t,12} + (\widetilde{\psi}_{t,22} - \widetilde{\psi}_{t,11})]/(i\omega)\\ \sigma_{22} &= [\lambda(\widetilde{\varphi}_{t,11} + \widetilde{\varphi}_{t,22}) + 2\mu_2(\widetilde{\varphi}_{t,22} - \widetilde{\psi}_{t,12})]/(i\omega)\ .\end{aligned}$$

The displacement boundary condition at the interface ($x_2 = 0$) is $u_{t2} + u_{ts2} = u_{i2} + u_{r2}$. We are assuming that the fluid is inviscid and slips in the tangential direction at the interface and hence there is no displacement continuity in the x_1-direction. The continuity of tractions requires that $\sigma_{12} = 0$ and $\sigma_{22} + (p_i + p_r) = 0$ at the interface. Using these conditions we get three equations for R, T_p, and T_s:

$$\frac{\cos\theta_i}{c_1}(1-R) - \left(\frac{\cos\theta_t}{c_{p2}}\right)\left(\frac{1}{\cos 2\theta_{ts}}\right)T_p = 0 \tag{2.28}$$

$$\frac{\rho_1}{\rho_2 c_{s2}}(1+R) - \left[\left(\frac{\cos\theta_t}{c_{p2}}\right)\left(\frac{2\sin\theta_{ts}\sin 2\theta_{ts}}{\cos 2\theta_{ts}}\right) + \frac{\cos 2\theta_{ts}}{c_{s2}}\right]T_p = 0 \tag{2.29}$$

and

$$T_s = \left(\frac{\cos\theta_t}{\sin\theta_t}\right)\left(\frac{2\sin^2\theta_{ts}}{\cos 2\theta_{ts}}\right)T_p \tag{2.30}$$

where the phase speed in medium 1 is c_1 and the P- and SV-wave phase speeds in medium 2 are c_{p2} and c_{s2}, respectively, and $\alpha = \sin\theta_i/c_1 = \sin\theta_t/c_{p2} = \sin\theta_{ts}/c_{s2}$. Explicit expressions for R, T_p, and T_s and a detailed analysis of the reflection coefficient can be found in Brekhovskikh (1960). The magnitude and phase of the reflection and transmission coefficients at a water-steel interface are shown in Figure 2.8 where the water has a density of 1000 kg/m^3 and a bulk modulus of 2.2 GPa. The discontinuous changes in slope at $\sim 14.5°$ and $\sim 28°$ occur because at these angles α is equal to $1/c_{p2}$ and $1/c_{s2}$, respectively.

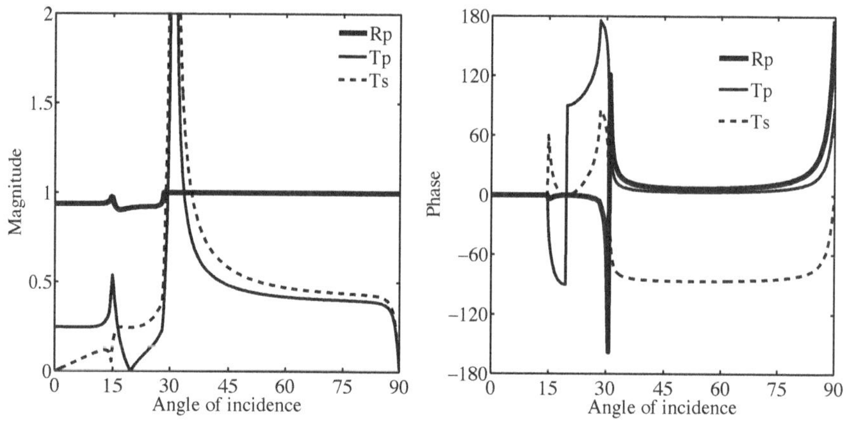

FIGURE 2.8
Reflection and transmission coefficients for the reflection of an acoustic wave at the interface of water and mild steel.

2.3.4 Negative refractive index

We have seen on p. 65 that a plane acoustic wave at the interface of two fluid media behaves like an SH-wave. Hence the discussion from Section 2.2.5 can be applied directly to that case. Let us now consider the case of a plane acoustic wave at the fluid-solid interface that we discussed in the previous section.

Let us define the refractive indices of the two media as

$$n_1 := c_0/c_1 = c_0/\sqrt{\kappa_1/\rho_1}$$

and

$$n_2 := c_0/c_{p2} = c_0/\sqrt{(\lambda_2 + 2\mu_2)/\rho_2} \quad n_3 := c_0/c_{s2} = c_0/\sqrt{\mu_2/\rho_2}\,.$$

Equations (2.28) and (2.29) can then be expressed in the form

$$n_1 \cos\theta_i \cos(2\theta_{ts})(1-R) - n_2 \cos\theta_t T_p = 0$$

$$n_3 \cos(2\theta_{ts})\frac{\rho_1}{\rho_2}(1+R) - \left[2n_2 \cos\theta_t \sin\theta_{ts} \sin 2\theta_{ts} + n_3 \cos^2(2\theta_{ts})\right] T_p = 0.$$

Let us look at the situation where $n_2 = -n_1$ and $\rho_2 = -\rho_1$ which implies that $\lambda_2 + 2\mu_2 = -\kappa_1$ and $\theta_t = -\theta_i$. From these results we see that $n_1/n_3 = \pm\sqrt{1/2 + \lambda_2/\kappa_1}$. If $\kappa_1 = 2\lambda_2$ we have $n_1/n_3 = \pm 1$. There is no solution when $n_1 = n_3$ and therefore we must have $n_1 = -n_3 \implies \theta_i = -\theta_{ts}$ in which case $R = 0$, $T_p = -\cos 2\theta_i$ and $T_s = \sin 2\theta_i$. The transmitted wave is out of phase with the incident up to an angle of incident of 45°.

Let us now consider the propagation of acoustic waves across the interface shown in Figure 2.9. If the pressure is the same at points at equal distances from the interface

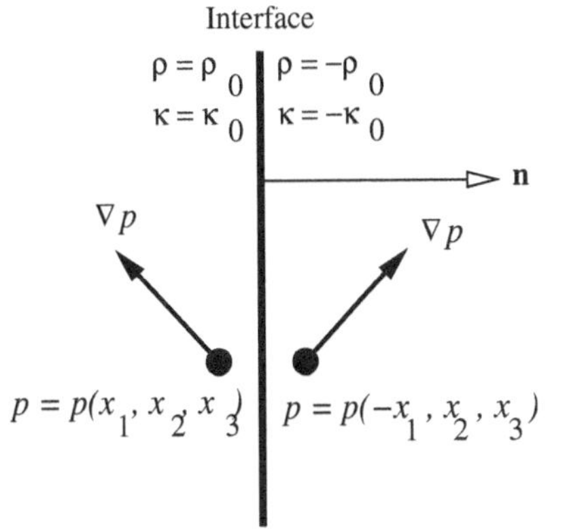

FIGURE 2.9
An interface separating a material with $\rho, \kappa > 0$ from one with $\rho, \kappa < 0$.

as shown, the vectors $\boldsymbol{\nabla} p$ are reflections of each other. If $\mathbf{n}$ is the normal to the interface, the quantity $\mathbf{n} \cdot \boldsymbol{\nabla} p$ changes sign across the interface. Since ρ also changes sign, $\mathbf{n} \cdot \mathbf{u}$ is continuous across the interface. Hence, the boundary condition across the interface is physical if the density on one side of the interface is negative.

2.4 Plane wave solutions in electrodynamics

Recall Maxwell's equations (1.89) for isotropic media at a fixed frequency ω shown below for convenience.

$$\boldsymbol{\nabla} \times \mathbf{E} = i\omega\,\mu\,\mathbf{H}\,; \quad \boldsymbol{\nabla} \times \mathbf{H} = -i\omega\,\varepsilon\,\mathbf{E}\,; \quad \boldsymbol{\nabla} \cdot \mathbf{B} = 0\,; \quad \boldsymbol{\nabla} \cdot \mathbf{D} = 0\,. \tag{2.31}$$

Let us look for plane wave solutions for the electric field of the form¶

$$\mathbf{E}(\mathbf{x}) = \mathbf{E}_0\, e^{i\,\mathbf{k}\cdot\mathbf{x}} \tag{2.32}$$

where $\|\mathbf{k}\| = 1/(2\pi\lambda)$ and λ is the wavelength. Then, using the first of equations (2.31) we have

$$\mathbf{H}(\mathbf{x}) = \mathbf{H}_0\, e^{i\,(\mathbf{k}\cdot\mathbf{x})}\,; \quad \mathbf{H}_0 := 1/(\omega\mu)\,\mathbf{k} \times \mathbf{E}_0\,. \tag{2.33}$$

On p. 36 we have seen that, for isotropic media, $\boldsymbol{\nabla} \cdot \mathbf{E} = 0$ which implies that $\mathbf{k} \cdot \mathbf{E}_0 = 0$, i.e., the electric field vector is normal to the wave vector. Similarly, since $\boldsymbol{\nabla} \cdot \mathbf{H} = 0$, we have $\mathbf{k} \cdot \mathbf{H}_0 = 0$.

¶Our discussion of electromagnetic plane waves is based on the descriptions given in Lorrain et al. (1988) and Brekhovskikh (1960).

Plugging equation (2.32) into the first of equations (1.91) we get

$$
\begin{aligned}
\nabla^2\mathbf{E} + \kappa^2\ \mathbf{E}(\mathbf{x}) &= \frac{\partial}{\partial x_m}\left[\frac{\partial}{\partial x_m}\left(E_{0n}\ e^{ik_lx_l}\right)\right]\mathbf{e}_n + \kappa^2\ \mathbf{E}_0\ e^{i\mathbf{k}\cdot\mathbf{x}} \\
&= \frac{\partial}{\partial x_m}\left[ik_l\frac{\partial x_l}{\partial x_m}E_{0n}e^{ik_lx_l}\right]\mathbf{e}_n + \kappa^2\ \mathbf{E}_0\ e^{i\mathbf{k}\cdot\mathbf{x}} \\
&= i^2k_mE_{0n}\ k_l\ \frac{\partial x_l}{\partial x_m}e^{ik_lx_l}\mathbf{e}_n + \kappa^2\ \mathbf{E}_0\ e^{i\mathbf{k}\cdot\mathbf{x}} \\
&= -k_mk_mE_{0n}\mathbf{e}_ne^{ik_lx_l} + \kappa^2\ \mathbf{E}_0\ e^{i\mathbf{k}\cdot\mathbf{x}} \\
&= -(\mathbf{k}\cdot\mathbf{k})\ \mathbf{E}_0\ e^{i\mathbf{k}\cdot\mathbf{x}} + \kappa^2\ \mathbf{E}_0\ e^{i\mathbf{k}\cdot\mathbf{x}} = \mathbf{0}
\end{aligned}
$$

where we have used rectangular Cartesian components for simplicity. If the above relation is to hold for all $\mathbf{x}$ we require that

$$
\mathbf{k}\cdot\mathbf{k} = \kappa^2 \qquad \Longrightarrow \qquad \|\mathbf{k}\| = \frac{\omega}{c}\ \text{for real}\ \mathbf{k}\,. \tag{2.34}
$$

Similarly, plugging the solution (2.33) into the second of equations (1.91) also leads to $\mathbf{k}\cdot\mathbf{k} = \kappa^2$ which is essentially the relation between the wavenumber and the phase speed at a given frequency.

2.4.1 Reflection of a plane wave at an interface

Figure 2.10 shows an electromagnetic wave that is incident upon an interface separating two media. Let us consider plane wave solutions to Maxwell's equations (2.31) at fixed frequency. Then we can express the waves shown in Figure 2.10 in the form

$$
\mathbf{E}_i = \mathbf{E}_{0i}\ e^{i(\mathbf{k}_i\cdot\mathbf{x})}\ ;\quad \mathbf{E}_r = \mathbf{E}_{0r}\ e^{i(\mathbf{k}_r\cdot\mathbf{x})}\ ;\quad \mathbf{E}_t = \mathbf{E}_{0t}\ e^{i(\mathbf{k}_t\cdot\mathbf{x})}
$$

where $\mathbf{k}_i, \mathbf{k}_r, \mathbf{k}_t$ are the wave vectors corresponding to the incident, reflected, and transmitted waves, respectively.

Continuity requires that the oscillations at the interface must have the same phase ($\theta = \mathbf{k}\cdot\mathbf{x}$), i.e., $\mathbf{k}_i\cdot\mathbf{x} = \mathbf{k}_r\cdot\mathbf{x} = \mathbf{k}_t\cdot\mathbf{x}$ for all $\mathbf{x}$ on the interface. Since $\mathbf{x} = x_1\mathbf{e}_1$ at the interface, the tangential components of $\mathbf{k}_i, \mathbf{k}_r, \mathbf{k}_t$ must be equal along the interface. Hence $\|\mathbf{k}_i\|\ \sin\theta_i = \|\mathbf{k}_r\|\ \sin\theta_r = \|\mathbf{k}_t\|\ \sin\theta_t$. Now,

$$
\kappa_i := \|\mathbf{k}_i\| = \frac{\omega}{c_1}\ ;\quad \kappa_r := \|\mathbf{k}_r\| = \frac{\omega}{c_1}\ ;\quad \kappa_t := \|\mathbf{k}_t\| = \frac{\omega}{c_2} \tag{2.35}
$$

where c_1 and c_2 are the phase velocities in medium 1 and medium 2, respectively. Hence we have

$$
\frac{\omega}{c_1}\sin\theta_i = \frac{\omega}{c_1}\sin\theta_r = \frac{\omega}{c_2}\sin\theta_t\ .
$$

This implies that

$$
\theta_i = \theta_r \qquad \text{and} \qquad \frac{\sin\theta_i}{\sin\theta_t} = \frac{c_1}{c_2}\,.
$$

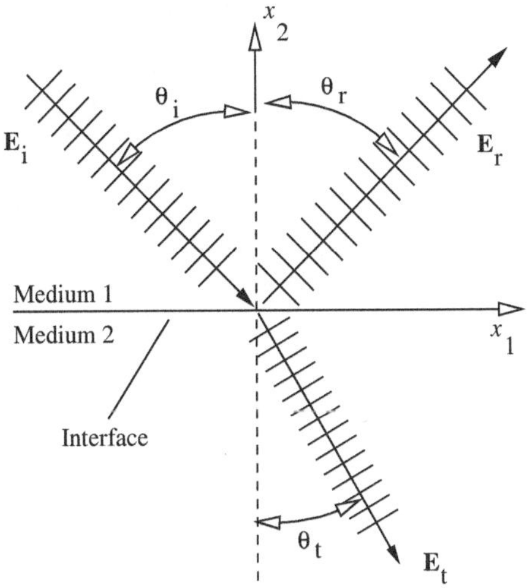

FIGURE 2.10
Reflection of an electromagnetic wave at an interface.

The refractive index is defined as

$$n := \frac{c_0}{c} \tag{2.36}$$

where c_0 is the phase velocity in vacuum. Therefore, we get Snell's law for electromagnetic waves,

$$\frac{\sin\theta_i}{\sin\theta_t} = \frac{n_2}{n_1}. \tag{2.37}$$

2.4.2 Wave polarized parallel to the plane of incidence

For an electromagnetic wave, the wave vector of $\mathbf{k}$ is not necessarily parallel or perpendicular to the fields $\mathbf{E}$ and $\mathbf{H}$. However, for plane waves we have seen that $\mathbf{k}\cdot\mathbf{E} = 0$ and $\mathbf{k}\cdot\mathbf{H} = 0$. Therefore plane waves have transverse polarization.

Consider an incident plane wave such that the electric field vector $\mathbf{E}_i(\mathbf{x},t)$ lies on the plane of incidence as shown in Figure 2.11. Let us choose an orthonormal basis $(\mathbf{e}_1, \mathbf{e}_2, \mathbf{e}_3)$ such that $\mathbf{e}_1$ is on the planar interface, $\mathbf{e}_2$ is normal to the plane of incidence, and $\mathbf{e}_3$ is normal to the interface.

The wave vectors $\mathbf{k}_i$, $\mathbf{k}_r$, and $\mathbf{k}_t$ may be expressed in this basis as

$$\mathbf{k}_i = \kappa_1 \sin\theta_i\, \mathbf{e}_1 - \kappa_1 \cos\theta_i\, \mathbf{e}_3\,; \quad \mathbf{k}_r = \kappa_1 \sin\theta_r\, \mathbf{e}_1 + \kappa_1 \cos\theta_r\, \mathbf{e}_3$$
$$\mathbf{k}_t = \kappa_2 \sin\theta_t\, \mathbf{e}_1 - \kappa_2 \cos\theta_t\, \mathbf{e}_3$$

where the notation introduced in equation (2.35) has been used. At fixed frequency, the incident, reflected, and transmitted plane waves can be expressed as

$$\mathbf{E}_i(\mathbf{x}) = \mathbf{E}_{0i}\, e^{i\mathbf{k}_i\cdot\mathbf{x}}\,; \quad \mathbf{E}_r(\mathbf{x}) = \mathbf{E}_{0r}\, e^{i\mathbf{k}_r\cdot\mathbf{x}}\,; \quad \mathbf{E}_t(\mathbf{x}) = \mathbf{E}_{0t}\, e^{i\mathbf{k}_t\cdot\mathbf{x}}\,.$$

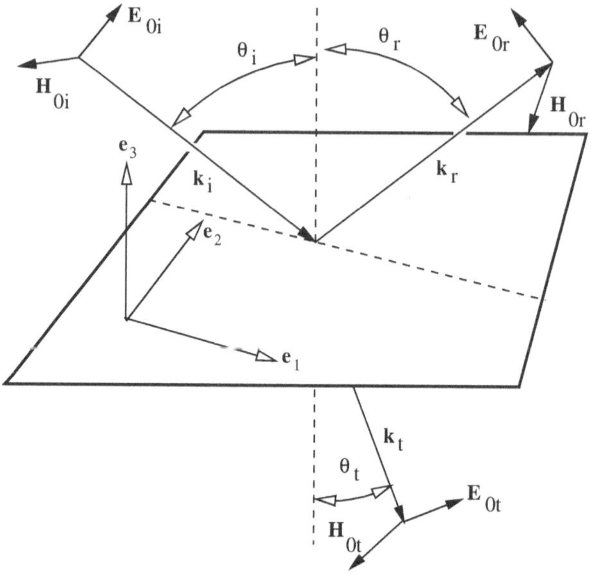

FIGURE 2.11
Plane wave polarized with the $\mathbf{E}_i$-vector parallel to the plane of incidence.

Since $\mathbf{k} \cdot \mathbf{E} = 0$, we can write these in terms of the basis $(\mathbf{e}_1, \mathbf{e}_2, \mathbf{e}_3)$ as

$$\begin{aligned} \mathbf{E}_i(\mathbf{x}) &= [E_i \cos\theta_i \, \mathbf{e}_1 + E_i \sin\theta_i \, \mathbf{e}_3] \, e^{i\kappa_1(x_1 \sin\theta_i - x_3 \cos\theta_i)} \\ \mathbf{E}_r(\mathbf{x}) &= [-E_r \cos\theta_r \, \mathbf{e}_1 + E_r \sin\theta_r \, \mathbf{e}_3] \, e^{i\kappa_1(x_1 \sin\theta_r + x_3 \cos\theta_r)} \\ \mathbf{E}_t(\mathbf{x}) &= [E_t \cos\theta_t \, \mathbf{e}_1 + E_t \sin\theta_t \, \mathbf{e}_3] \, e^{i\kappa_2(x_1 \sin\theta_t - x_3 \cos\theta_t)} \end{aligned} \tag{2.38}$$

where $E_i := \|\mathbf{E}_{0i}\|$, $E_r := \|\mathbf{E}_{0r}\|$, and $E_t := \|\mathbf{E}_{0t}\|$. Since the $\mathbf{E}$ vectors lie in the $(\mathbf{e}_1, \mathbf{e}_3)$-plane and since $\mathbf{H}_0 = 1/(\omega\mu)\, \mathbf{k} \times \mathbf{E}_0$, the magnetic field vector must lie in the $\mathbf{e}_2$-direction. In index notation we have $\mathbf{H}_0 = (\omega\mu)^{-1}\, e_{ijk} k_j E_{0k} \mathbf{e}_i$. Therefore, the magnetic field can be written as [‖]

$$\begin{aligned} &\mathbf{H}_i(\mathbf{x}) = [-H_i \mathbf{e}_2] e^{i\kappa_1(x_1 \sin\theta_i - x_3 \cos\theta_i)} \; ; \; \mathbf{H}_r(\mathbf{x}) = [-H_r \mathbf{e}_2] e^{i\kappa_1(x_1 \sin\theta_r + x_3 \cos\theta_r)} \\ &\mathbf{H}_t(\mathbf{x}) = [-H_t \mathbf{e}_2] e^{i\kappa_2(x_1 \sin\theta_t - x_3 \cos\theta_t)} \end{aligned}$$

where $H_i := \|\mathbf{H}_{0i}\| = \kappa_1 E_i/(\omega\mu_1)$, $H_r := \|\mathbf{H}_{0r}\| = \kappa_1 E_r/(\omega\mu_1)$, and $H_t := \|\mathbf{H}_{0t}\| = \kappa_2 E_t/(\omega\mu_2)$.

At the interface, continuity requires that the tangential components of the vectors $\mathbf{E}$ and $\mathbf{H}$ are continuous. Clearly, from the above equations, the $\mathbf{H}_0$ vectors are

[‖] These expressions indicate that the magnetic field component H_2 for this situation depends only on x_1 and x_3, i.e., $H_2 \equiv H_2(x_1, x_3)$ and $H_1 = H_3 = 0$. Recall from Section 1.4.6 that transverse magnetic (TM) fields have this property.

tangential to the interface. Also, at the interface $x_3 = 0$ and x_1 is arbitrary. Hence, continuity of the components of **H** at the interface can be achieved if $H_i + H_r = H_t$. In terms of the electric field, we then have

$$\frac{\kappa_1}{\omega\mu_1} E_i + \frac{\kappa_1}{\omega\mu_1} E_r = \frac{\kappa_2}{\omega\mu_2} E_t \; .$$

Recall that the refractive index is given by $n = c_0/c = c_0\ \kappa/\omega$. Therefore, we can write the above equation as

$$\boxed{\frac{n_1}{\mu_1}(E_i + E_r) = \frac{n_2}{\mu_2} E_t \; .} \tag{2.39}$$

The tangential components of the **E** vectors at the interface are given by $\mathbf{E} \times \mathbf{e}_3$, i.e.,

$$\begin{aligned}
\mathbf{E}_i \times \mathbf{e}_3 &= [-E_i \cos\theta_i\ \mathbf{e}_2]\, e^{i\kappa_1(x_1 \sin\theta_i - x_3\cos\theta_i)} \\
\mathbf{E}_r \times \mathbf{e}_3 &= [E_r \cos\theta_r\ \mathbf{e}_2]\, e^{i\kappa_1(x_1 \sin\theta_r + x_3\cos\theta_r)} \\
\mathbf{E}_t \times \mathbf{e}_3 &= [-E_t \cos\theta_t\ \mathbf{e}_2]\, e^{i\kappa_2(x_1 \sin\theta_t - x_3\cos\theta_t)} \; .
\end{aligned} \tag{2.40}$$

Using the arbitrariness of x_1 and from the continuity of the tangential components of the **E** vectors at the interface ($x_3 = 0$), we have

$$E_i \cos\theta_i - E_r \cos\theta_r = E_t \cos\theta_t \; .$$

Since $\theta_i = \theta_r$, we have

$$\boxed{(E_i - E_r)\cos\theta_i = E_t \cos\theta_t \; .} \tag{2.41}$$

From equations (2.39) and (2.41), we get two more relations:

$$\boxed{R_{\text{TM}} := \frac{E_r}{E_i} = \frac{\dfrac{n_2}{\mu_2}\cos\theta_i - \dfrac{n_1}{\mu_1}\cos\theta_t}{\dfrac{n_2}{\mu_2}\cos\theta_i + \dfrac{n_1}{\mu_1}\cos\theta_t}} \quad \text{and} \quad \boxed{T_{\text{TM}} := \frac{E_t}{E_i} = \frac{2\,\dfrac{n_1}{\mu_1}\cos\theta_i}{\dfrac{n_2}{\mu_2}\cos\theta_i + \dfrac{n_1}{\mu_1}\cos\theta_t}} \; . \tag{2.42}$$

Equations (2.39), (2.41), (2.42), are the *Fresnel equations* for TM-polarized electromagnetic waves. If we define $\mu_{r1} := \mu_1/\mu_0$ and $\mu_{r2} := \mu_2/\mu_0$ where μ_0 is the permeability of vacuum, we can write equations (2.42) as

$$R_{\text{TM}} = \frac{\dfrac{n_2}{\mu_{r2}}\cos\theta_i - \dfrac{n_1}{\mu_{r1}}\cos\theta_t}{\dfrac{n_2}{\mu_{r2}}\cos\theta_i + \dfrac{n_1}{\mu_{r1}}\cos\theta_t} \; ; \quad T_{\text{TM}} = \frac{2\,\dfrac{n_1}{\mu_{r1}}\cos\theta_i}{\dfrac{n_2}{\mu_{r2}}\cos\theta_i + \dfrac{n_1}{\mu_{r1}}\cos\theta_t} \; . \tag{2.43}$$

Note that there is *no transmitted wave* ($E_t = 0$) if $\theta_i = \pi/2$ and *no reflected wave* ($E_r = 0$) if

$$\frac{n_2}{\mu_{r2}} \cos\theta_i = \frac{n_1}{\mu_{r1}} \cos\theta_t \,. \tag{2.44}$$

For non-magnetic materials we have $\mu_{r1} = \mu_{r2} = 1$. Hence,

$$\frac{n_2}{n_1} = \frac{\cos\theta_t}{\cos\theta_i} \,. \tag{2.45}$$

Combining equation (2.45) with Snell's law (2.37), we get

$$\cos\theta_t \; \sin\theta_t - \cos\theta_i \; \sin\theta_i = 0 \quad \Longrightarrow \quad \cos(\theta_t + \theta_i) \; \sin(\theta_t - \theta_i) = 0 \,.$$

If $n_1 \neq n_2$ we have $\sin(\theta_t - \theta_i) \neq 0$. Hence,

$$\cos(\theta_t + \theta_i) = 0 \quad \Longrightarrow \quad \boxed{\theta_t + \theta_i = \frac{\pi}{2}} \,. \tag{2.46}$$

This is the condition that defines *Brewster's angle* ($\theta_i = \theta_B$) (see Figure 2.12). Plugging equation (2.46) into equation (2.37), we get

$$\frac{\sin\theta_B}{\sin(\pi/2 - \theta_B)} = \tan\theta_B = \frac{n_2}{n_1} \,.$$

This relation can be used to solve for Brewster's angle for various media. At Brewster's angle, we have $E_r/E_i = 0$. Hence the sign of the reflection coefficient E_r/E_i changes at Brewster's angle.

Brewster's angle is responsible for certain phenomena observed in rainbows. For an air-water interface with $n_1 = 1$ and $n_2 = 1.33$, we get a value of $\theta_B = 53^\circ$. This angle is very close to the angle that a rainbow ray makes with the normal to the surface of a raindrop at the point where it is reflected inside the drop. Hence only certain polarizations get reflected at this point and other polarizations are transmitted. At angles greater than the Brewster angle, the radiation is 90° out of phase with light reflected at less than the Brewster angle.

So far we have defined the reflection and transmission coefficients in terms of electric field amplitudes. We could alternatively use ratios of magnetic field amplitudes. If we define

$$R^H_{\text{TM}} := \frac{H_r}{H_i} \quad \text{and} \quad T^H_{\text{TM}} := \frac{H_t}{H_i}$$

and apply the continuity conditions at the interface, we get the relations

$$1 + R^H_{\text{TM}} = T^H_{\text{TM}} \quad \text{and} \quad (1 - R^H_{\text{TM}})\frac{\mu_1}{k_1}\cos\theta_i = T^H_{\text{TM}}\frac{\mu_2}{k_2}\cos\theta_t \,.$$

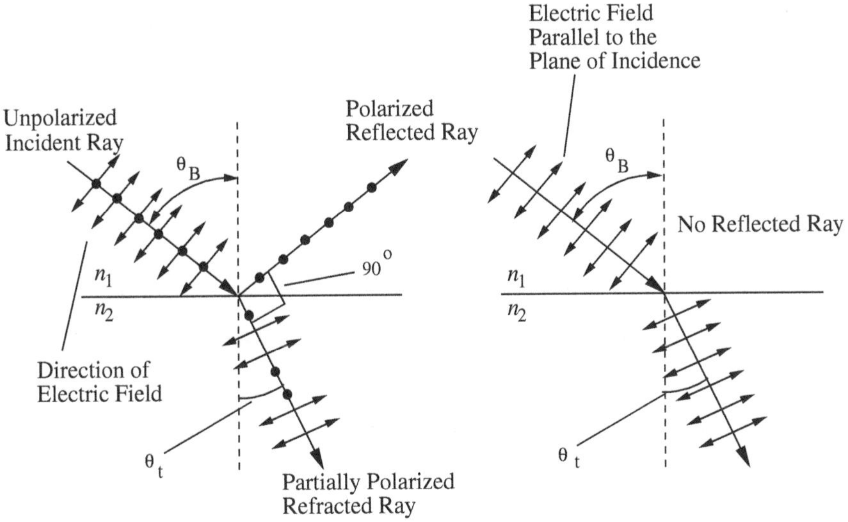

FIGURE 2.12
Brewster's angle and the polarization of light.

Using $k_1 = n_1\omega/c_0$ and $k_2 = n_2\omega/c_0$ and solving for the reflection and transmission coefficients gives

$$R^H_{\mathrm{TM}} = \frac{\dfrac{n_2}{\mu_2}\cos\theta_i - \dfrac{n_1}{\mu_1}\cos\theta_t}{\dfrac{n_2}{\mu_2}\cos\theta_i + \dfrac{n_1}{\mu_1}\cos\theta_t} \quad \text{and} \quad T^H_{\mathrm{TM}} = \frac{2\dfrac{n_2}{\mu_2}\cos\theta_i}{\dfrac{n_2}{\mu_2}\cos\theta_i + \dfrac{n_1}{\mu_1}\cos\theta_t}. \tag{2.47}$$

Comparison with equation (2.42) shows that the reflection coefficients are identical but the transmission coefficients differ.

2.4.3 Wave polarized perpendicular to the plane of incidence

For a plane wave with the $\mathbf{E}_i$ vector polarized perpendicular to the plane of incidence, we have**

$$\mathbf{E}_i(\mathbf{x}) = [E_i\ \mathbf{e}_2]e^{i\kappa_1(x_1\sin\theta_i - x_3\cos\theta_i)}\ ;\ \mathbf{E}_r(\mathbf{x}) = [E_r\ \mathbf{e}_2]e^{i\kappa_1(x_1\sin\theta_i + x_3\cos\theta_i)}$$
$$\mathbf{E}_t(\mathbf{x}) = [E_t\ \mathbf{e}_2]e^{i\kappa_2(x_1\sin\theta_t - x_3\cos\theta_t)}\ .$$

**These expressions indicate that the electric field component E_2 for this situation depends only on x_1 and x_3, i.e., $E_2 \equiv E_2(x_1, x_3)$ and $E_1 = E_3 = 0$. From Section 1.4.6 we know that transverse electric (TE) fields have this property.

Recall that $\mathbf{H}_0 = 1/(\omega\mu)\mathbf{k}\times\mathbf{E}_0$ and note that in this case $\mathbf{E}_0 = E_{m0}\mathbf{e}_m = E_{20}\mathbf{e}_2$. Therefore, $\mathbf{k}\times\mathbf{E}_0 = -k_3E_{20}\mathbf{e}_1 + k_1E_{20}\mathbf{e}_3$ and we have

$$\mathbf{H}_i(\mathbf{x}) = \frac{\kappa_1 E_i}{\omega\mu_1}(\cos\theta_i\ \mathbf{e}_1 + \sin\theta_i\ \mathbf{e}_3)e^{i\kappa_1(x_1\sin\theta_i - x_3\cos\theta_i)}$$

$$\mathbf{H}_r(\mathbf{x}) = \frac{\kappa_1 E_r}{\omega\mu_1}(-\cos\theta_i\ \mathbf{e}_1 + \sin\theta_i\ \mathbf{e}_3)e^{i\kappa_1(x_1\sin\theta_i + x_3\cos\theta_i)}$$

$$\mathbf{H}_t(\mathbf{x}) = \frac{\kappa_2 E_t}{\omega\mu_2}(\cos\theta_t\ \mathbf{e}_1 + \sin\theta_t\ \mathbf{e}_3)e^{i\kappa_2(x_1\sin\theta_t - x_3\cos\theta_t)}\ .$$

Continuity of tangential components of $\mathbf{E}$ at the interface requires that

$$E_i + E_r = E_t\ . \tag{2.48}$$

From the requirement of continuity of the tangential components of $\mathbf{H}$ at the interface and using the arbitrariness of x_1, we get

$$\frac{\kappa_1}{\omega\mu_1}(E_i - E_r)\ \cos\theta_i = \frac{\kappa_2}{\omega\mu_1}E_t\ \cos\theta_t\ . \tag{2.49}$$

Using the relation $\kappa = n\ \omega/c_0$, we get

$$\boxed{\frac{n_1}{\mu_1}(E_i - E_r)\ \cos\theta_i = \frac{n_2}{\mu_2}E_t\ \cos\theta_t\ .} \tag{2.50}$$

From equations (2.48) and (2.50), we get

$$\boxed{R_{\text{TE}} := \frac{E_r}{E_i} = \frac{\dfrac{n_1}{\mu_1}\cos\theta_i - \dfrac{n_2}{\mu_2}\cos\theta_t}{\dfrac{n_1}{\mu_1}\cos\theta_i + \dfrac{n_2}{\mu_2}\cos\theta_t}} \quad \text{and} \quad \boxed{T_{\text{TE}} := \frac{E_t}{E_i} = \frac{2\,\dfrac{n_1}{\mu_1}\cos\theta_i}{\dfrac{n_1}{\mu_1}\cos\theta_i + \dfrac{n_2}{\mu_2}\cos\theta_t}}\ . \tag{2.51}$$

Equations (2.48), (2.50), (2.51), are the Fresnel equations a wave polarized with the $\mathbf{E}_i$ vector perpendicular to the plane of incidence. We may also write the last two equations as

$$R_{\text{TE}} = \frac{\dfrac{n_1}{\mu_{r1}}\cos\theta_i - \dfrac{n_2}{\mu_{r2}}\cos\theta_t}{\dfrac{n_1}{\mu_{r1}}\cos\theta_i + \dfrac{n_2}{\mu_{r2}}\cos\theta_t} \quad \text{and} \quad T_{\text{TE}} = \frac{2\,\dfrac{n_1}{\mu_{r1}}\cos\theta_i}{\dfrac{n_1}{\mu_{r1}}\cos\theta_i + \dfrac{n_2}{\mu_{r2}}\cos\theta_t}\ . \tag{2.52}$$

From the above equations and Snell's law we can show that there is no reflected wave only if

$$\tan\theta_i = \tan\theta_t\ \frac{\mu_2}{\mu_1}\ .$$

This is only possible if there is no interface. Therefore, in the presence of an interface, there is always a reflected wave for this situation.

2.4.4 Electromagnetic impedance

The *characteristic electromagnetic impedance* of a medium at a fixed frequency is defined as

$$Z_0 := \frac{\|\mathbf{E}\|}{\|\mathbf{H}\|}$$

where $\|\mathbf{E}\|$ and $\|\mathbf{H}\|$ are the amplitudes of the electric and magnetic fields. For a plane wave with wave vector $\mathbf{k} = \kappa \mathbf{e}_1$, the relation $\mathbf{k} \cdot \mathbf{E} = 0$ implies that $\mathbf{E} = E_2\,\mathbf{e}_2 + E_3\,\mathbf{e}_3$. Therefore, $\mathbf{k} \times \mathbf{E} = -\kappa(E_3\,\mathbf{e}_2 - E_2\,\mathbf{e}_3)$ which implies that $\mathbf{H} = 1/(\omega\mu)\mathbf{k} \times \mathbf{E} = -\kappa/(\omega\mu)(E_3\,\mathbf{e}_2 - E_2\,\mathbf{e}_3)$. Using $\kappa = \omega/c = \omega\sqrt{\varepsilon\mu}$ and $n = c_0/c$, we have

$$Z_0 = \frac{\omega\mu}{\kappa} = \frac{\mu c_0}{n} = \sqrt{\frac{\mu}{\varepsilon}}. \tag{2.53}$$

For problems involving plane interfaces, the *normal impedance* is defined as the ratio of the tangential components of the electric and magnetic fields, i.e.,

$$Z := \frac{E_{\text{tang}}}{H_{\text{tang}}}.$$

Figure 2.13 shows a schematic of the two situations that can be superposed to find the solution for an electric field (E) with an arbitrary orientation. For the situation

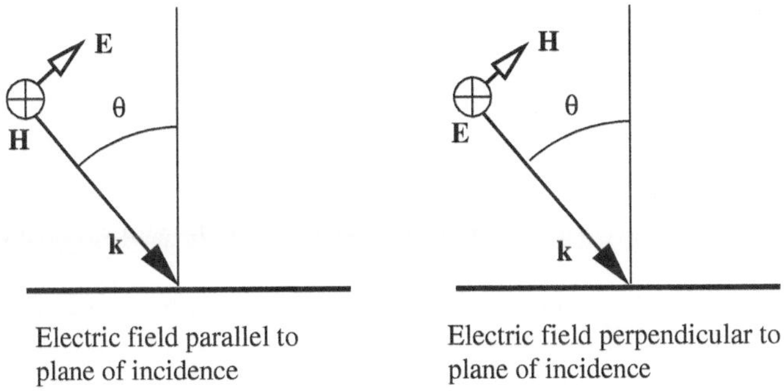

FIGURE 2.13
Schematic of electric fields parallel and perpendicular to the plane of incidence. The components of the **E** and **H** fields are scaled by factors of $\cos\theta$, depending on the polarization, to determine the tangential components.

where the **E** field is parallel to the plane of incidence and the **H** field is parallel to the interface we have $E_{\text{tang}} = E_i \cos\theta_i$ and $H_{\text{tang}} = H_i$. Hence,

$$Z_i = \frac{E_i}{H_i}\cos\theta_i = Z_0\cos\theta_i = c_0\cos\theta_i\frac{\mu_i}{n_i} = \cos\theta_i\sqrt{\frac{\mu_i}{\varepsilon_i}}. \tag{2.54}$$

For the situation where $\mathbf{E}$ is polarized perpendicular to the plane of incidence and parallel to the interface, we have $E_{\text{tang}} = E_i$ and $H_{\text{tang}} = H_i \cos\theta_i$ and the impdendance is

$$Z_i = \frac{E_i}{H_i \cos\theta_i} = \frac{Z_0}{\cos\theta_i} = c_0 \frac{\mu_i}{n_i} \frac{1}{\cos\theta_i} = \frac{1}{\cos\theta_i}\sqrt{\frac{\mu_i}{\varepsilon_i}}. \tag{2.55}$$

Applying the definition (2.54) to equation (2.47) leads to

$$R^H_{\text{TM}} = \frac{Z_1 - Z_2}{Z_1 + Z_2} \quad \text{and} \quad T^H_{\text{TM}} = \frac{2Z_1}{Z_1 + Z_2}.$$

Similarly, applying (2.55) to equation (2.51), we have

$$R_{\text{TE}} = \frac{Z_2 - Z_1}{Z_2 + Z_1} \quad \text{and} \quad T_{\text{TE}} = \frac{2Z_2}{Z_2 + Z_1}.$$

Observe the similarity between these results and those for acoustics in Section 2.3.2 (p. 65). Also, note that there is no reflection when $Z_1 = Z_2$. This is the condition needed for *impedance matched* media.

2.4.5 Negative refractive index

Recall Fresnel's equations (2.43) on p. 73 for the reflection and transmission coefficients for TM-polarized electromagnetic waves. Note that if $n_2 = -n_1$ and $\mu_2 = -\mu_1$, since $n = c_0/c = c_0/\sqrt{\varepsilon\mu}$ we must have $\varepsilon_2 = -\varepsilon_1$. Then, by Snell's law

$$\frac{\sin\theta_t}{\sin\theta_i} = -1 \quad \Longrightarrow \quad \theta_t = -\theta_i\,. \tag{2.56}$$

Hence, $R_{\text{TM}} = 0$ and $T_{\text{TM}} = 1$. So the radiation is transmitted at the angle $\theta_t = -\theta_i$ and none is reflected. In fact, an interface separating media with $\varepsilon_2 = -\varepsilon_1$ and $\mu_2 = -\mu_1$ behaves like a mirror. Consider the interface in Figure 2.14. For the situation shown in the figure, on the medium on the top of the interface, let $\mathbf{E}_t$ and $\mathbf{H}_t$ solve Maxwell's equations with $\mu = -\mu_1$ and $\varepsilon = -\varepsilon_1$, i.e.,

$$\boldsymbol{\nabla} \times \mathbf{E}_t + i\omega\,\mu_1\,\mathbf{H}_t = \mathbf{0}\,; \quad \boldsymbol{\nabla} \times \mathbf{H}_t - i\omega\,\varepsilon_1\,\mathbf{E}_t = \mathbf{0}\,.$$

Let the solutions on the top of the interface be of the form

$$\begin{aligned}
\mathbf{E}_t(\mathbf{x}) &= E_1(\mathbf{x})\mathbf{e}_1 + E_2(\mathbf{x})\mathbf{e}_2 + E_3(\mathbf{x})\mathbf{e}_3 \\
\mathbf{H}_t(\mathbf{x}) &= H_1(\mathbf{x})\mathbf{e}_1 + H_2(\mathbf{x})\mathbf{e}_2 + H_3(\mathbf{x})\mathbf{e}_3\,.
\end{aligned}$$

Let

$$\begin{aligned}
\mathbf{F}_t(\mathbf{x}) &:= \boldsymbol{\nabla} \times \mathbf{E}_t = (E_{3,2} - E_{2,3})\mathbf{e}_1 + (E_{1,3} - E_{3,1})\mathbf{e}_2 + (E_{2,1} - E_{1,2})\mathbf{e}_3 \\
&=: F_1(\mathbf{x})\mathbf{e}_1 + F_2(\mathbf{x})\mathbf{e}_2 + F_3(\mathbf{x})\mathbf{e}_3 \\
\mathbf{I}_t(\mathbf{x}) &:= \boldsymbol{\nabla} \times \mathbf{H}_t = (H_{3,2} - H_{2,3})\mathbf{e}_1 + (H_{1,3} - H_{3,1})\mathbf{e}_2 + (H_{2,1} - H_{1,2})\mathbf{e}_3 \\
&=: I_1(\mathbf{x})\mathbf{e}_1 + I_2(\mathbf{x})\mathbf{e}_2 + I_3(\mathbf{x})\mathbf{e}_3\,.
\end{aligned}$$

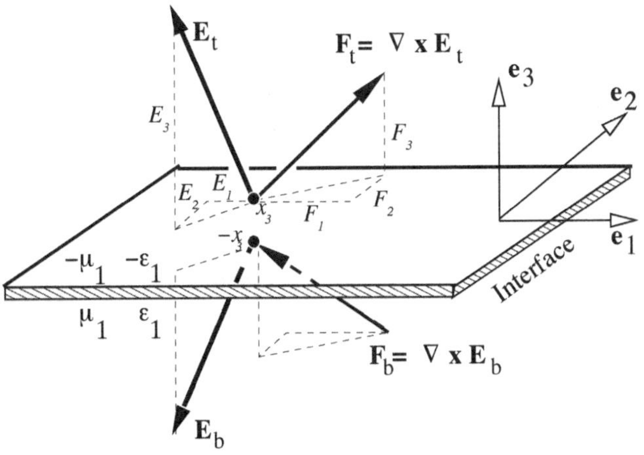

FIGURE 2.14
Reflection at an interface due to negative ε and μ.

In the medium at the bottom of the interface, let $\mathbf{E}_b$ and $\mathbf{H}_b$ solve Maxwell's equations with $\mu = -\mu_2$ and $\varepsilon = -\varepsilon_2$, i.e.,

$$\nabla \times \mathbf{E}_b + i\omega\,\mu_2\,\mathbf{H}_b = \mathbf{0}\,;\;\; \nabla \times \mathbf{H}_b - i\omega\,\varepsilon_2\,\mathbf{E}_b = \mathbf{0}.$$

Let us assume that the interface acts as a mirror such that the region at the bottom of the interface has reflected fields. Then, from Figure 2.14, we see that

$$\begin{aligned}
\mathbf{E}_b(\mathbf{x}) &= E_1(x_1,x_2,-x_3)\mathbf{e}_1 + E_2(x_1,x_2,-x_3)\mathbf{e}_2 - E_3(x_1,x_2,-x_3)\mathbf{e}_3 \\
&=: \widehat{E}_1(\mathbf{x})\mathbf{e}_1 + \widehat{E}_2(\mathbf{x})\mathbf{e}_2 + \widehat{E}_3(\mathbf{x})\mathbf{e}_3 \\
\mathbf{H}_b(\mathbf{x}) &= H_1(x_1,x_2,-x_3)\mathbf{e}_1 + H_2(x_1,x_2,-x_3)\mathbf{e}_2 - H_3(x_1,x_2,-x_3)\mathbf{e}_3 \\
&=: \widehat{H}_1(\mathbf{x})\mathbf{e}_1 + \widehat{H}_2(\mathbf{x})\mathbf{e}_2 + \widehat{H}_3(\mathbf{x})\mathbf{e}_3\,.
\end{aligned}$$

Therefore, at the bottom of the interface, we have

$$\begin{aligned}
\nabla \times \mathbf{E}_b &= (\widehat{E}_{3,2} - \widehat{E}_{2,3})\mathbf{e}_1 + (\widehat{E}_{1,3} - \widehat{E}_{3,1})\mathbf{e}_2 + (\widehat{E}_{2,1} - \widehat{E}_{1,2})\mathbf{e}_3 \\
&= (-E_{3,2} + E_{2,3})\mathbf{e}_1 + (-E_{1,3} + E_{3,1})\mathbf{e}_2 + (E_{2,1} - E_{1,2})\mathbf{e}_3 \\
&= -F_1\mathbf{e}_1 - F_2\mathbf{e}_2 + F_3\mathbf{e}_3 \\
\nabla \times \mathbf{H}_b &= -I_1\mathbf{e}_1 - I_2\mathbf{e}_2 + I_3\mathbf{e}_3\,.
\end{aligned}$$

For continuity of the fields at the interface ($x_3 = 0$) we can show that we must have

$$\mu_2 = -\mu_1 \qquad \text{and} \qquad \varepsilon_2 = -\varepsilon_1\,.$$

This implies that negative permeability and permittivity have the unusual property of reflecting the fields.

2.5 Wave propagation through a slab

Let us now consider wave propagation through the simplest multilayered system, the case where there are three layers. Usually, the two outer layers are the material while the slab in between is made of a different material. For simplicity, we assume that the planar dimensions of the layers is large enough that end effects are not important. Acoustic and electromagnetic waves can be treated in a similar manner and a detailed discussion of both can be found in Brekhovskikh (1960). An excellent treatment of electromagnetic waves in layered media can be found in Chew (1995).

2.5.1 Acoustic waves in a slab

Consider plane wave propagation through a slab wedged between two materials as shown in Figure 2.15. There are multiple reflections at the interface between medium 2 and medium 1. We will try to determine the reflection and transmission coefficients for the slab by looking at the behavior at a fixed value of x_1. Let us examine only the case of acoustic wave propagation.

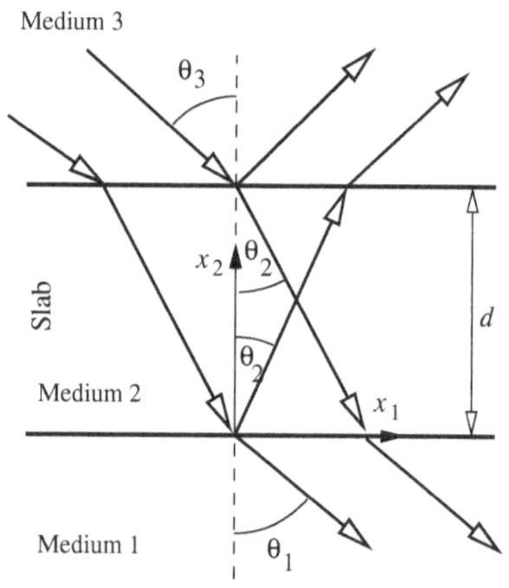

FIGURE 2.15
Ray diagram of wave propagation through a slab.

We assume that all three media are homogeneous and isotropic. The slab extends to infinity in the x_1 and x_3 directions and has a thickness d. The mass densities of the three media are ρ_1, ρ_2, ρ_3 and their phase velocities are c_1, c_2, c_3. For plane wave propagation at fixed frequency, the wave vectors in the three media have magnitudes

$k_1 = \omega/c_1$, $k_2 = \omega/c_2$, and $k_3 = \omega/c_3$. The velocity potential in medium 2 is

$$\phi_2 = A_2\, e^{ik_2(x_1 \sin\theta_2 - x_2 \cos\theta_2)} + B_2\, e^{ik_2(x_1 \sin\theta_2 + x_2 \cos\theta_2)} .$$

The pressure in the medium is given by $p = -i\omega\rho_2\phi_2$, i.e.,

$$p_2 = -i\omega\rho_2\, e^{ik_2 x_1 \sin\theta_2} \left[A_2\, e^{-ik_2 x_2 \cos\theta_2} + B_2\, e^{ik_2 x_2 \cos\theta_2}\right] .$$

The normal velocity in the medium is given by $v_n = -\phi_{2,2}$, i.e.,

$$v_{n2} = -ik_2 \cos\theta_2\, e^{ik_2 x_1 \sin\theta_2} \left[-A_2\, e^{-ik_2 x_2 \cos\theta_2} + B_2\, e^{ik_2 x_2 \cos\theta_2}\right] .$$

At the interface between medium 1 and medium 2, $x_2 = 0$, and the acoustic impedance is

$$\left.\frac{p_2}{v_{n2}}\right|_{x_2=0} = -\frac{\omega\rho_2}{k_2 \cos\theta_2} \frac{A_2 + B_2}{A_2 - B_2} . = -Z_2 \frac{A_2 + B_2}{A_2 - B_2}$$

where Z_2 is the normal acoustic impedance of medium 2. At the interface between medium 2 and medium 3, $x_2 = d$, and

$$\left.\frac{p_2}{v_{n2}}\right|_{x_2=d} = -Z_2 \frac{A_2\, e^{-ik_2 d \cos\theta_2} + B_2 e^{ik_2 d \cos\theta_2}}{A_2 e^{-ik_2 d \cos\theta_2} - B_2 e^{ik_2 d \cos\theta_2}} .$$

Continuity of pressure and the normal component of the velocity requires that (see Section 2.3.2, p. 65)

$$\left.\frac{p_2}{v_{n2}}\right|_{x_2=0} = -Z_1 = -\frac{\rho_1 c_1}{\cos\theta_1} \quad \text{and} \quad \left.\frac{p_2}{v_{n2}}\right|_{x_2=d} = -Z_{\text{in}}$$

where Z_{in} is the *input impedance* of medium 2. From the first condition we find that

$$\frac{B_2}{A_2} = \frac{Z_1 - Z_2}{Z_1 + Z_2} .$$

Substituting the above equation into the continuity condition for $x_2 = d$ gives

$$Z_{\text{in}} = Z_2 \frac{Z_2 \sin\alpha + iZ_1 \cos\alpha}{Z_1 \sin\alpha + iZ_2 \cos\alpha} \qquad \text{where} \quad \alpha := k_2 d \cos\theta_2 .$$

For the incident and reflected waves in medium 3, following our usual procedure gives

$$p_3 = -i\omega\rho_3\, e^{ik_3 x_1 \sin\theta_3} \left[A_3\, e^{-ik_3(x_2-d)\cos\theta_3} + B_3\, e^{ik_3(x_2-d)\cos\theta_3}\right]$$

$$v_{n3} = -ik_3 \cos\theta_3\, e^{ik_3 x_1 \sin\theta_3} \left[-A_3\, e^{-ik_3(x_2-d)\cos\theta_3} + B_3\, e^{ik_3(x_2-d)\cos\theta_3}\right] .$$

Therefore, at $x_2 = d$,

$$\left.\frac{p_3}{v_{n3}}\right|_{x_2=d} = -Z_3\,\frac{A_3+B_3}{A_3-B_3} = -Z_{\text{in}}\,.$$

Reorganizing gives us the reflection coefficient for the slab,

$$\boxed{R = \frac{B_3}{A_3} = \frac{Z_{\text{in}}-Z_3}{Z_{\text{in}}+Z_3} = \frac{(Z_2^2 - Z_1Z_3)\sin\alpha + iZ_2(Z_1-Z_3)\cos\alpha}{(Z_2^2 + Z_1Z_3)\sin\alpha + iZ_2(Z_1+Z_3)\cos\alpha}\,.} \tag{2.57}$$

For the transmitted wave below the interface between medium 2 and medium 1, we have

$$\begin{aligned} p_1 &= -i\omega\rho_1\, e^{ik_1x_1\sin\theta_1}\left[A_1\, e^{-ik_1x_2\cos\theta_1}\right] \\ v_{n1} &= -ik_1\cos\theta_1\, e^{ik_1x_1\sin\theta_1}\left[-A_1\, e^{-ik_1x_2\cos\theta_1}\right]. \end{aligned}$$

The continuity condition $p_1 = p_2$ at $x_2 = 0$ gives

$$\rho_1\, e^{ik_1x_1\sin\theta_1} A_1 = \rho_2\, e^{ik_2x_1\sin\theta_2}(A_2+B_2).$$

For this condition to hold for all x_1 we must have

$$\rho_1 A_1 = \rho_2(A_2+B_2)\,.$$

Similarly, for continuity of the pressure at $x_2 = d$ and $\forall x_1$, we require

$$\rho_2\left[A_2\, e^{-i\alpha} + B_2\, e^{i\alpha}\right] = \rho_3(A_3+B_3) = \rho_3(1+R)A_3$$

where $\alpha = k_2 d\cos\theta_2$. Expressing A_1 and A_3 in terms of A_2 and dividing gives us the transmission coefficient

$$\boxed{T = \frac{A_1}{A_3} = \left(\frac{\rho_3}{\rho_1}\right)\left[\frac{2Z_1Z_2}{Z_2(Z_1+Z_3)\cos\alpha - i(Z_2^2+Z_1Z_3)\sin\alpha}\right].} \tag{2.58}$$

For the situation where $Z_1 = Z_3$ and $\rho_1 = \rho_3$, and there is normal incidence, we have

$$R = \frac{(Z_2^2 - Z_1^2)\sin(k_2d)}{(Z_2^2+Z_1^2)\sin(k_2d) + 2iZ_1Z_2\cos(k_2d)}$$

and

$$T = \frac{2iZ_1Z_2}{(Z_2^2+Z_1^2)\sin(k_2d) + 2iZ_1Z_2\cos(k_2d)}$$

where $Z_1 = \rho_1c_1$, $Z_2 = \rho_2c_2$, and $k_2 = \omega/c_2$. This is the transmission coefficient measured by an impedance tube. Clearly, the computation of T becomes quite complicated as we add more layers to a structure. We will discuss waves in layered media in greater detail in Chapter 8.

2.5.2 Impedance tube measurements

In practice, it is not convenient to measure the transmission and reflection coefficients of a particular slab for each of a large range of frequencies. Instead, a broadband range of frequencies is generated at one go and the Fourier components are separated using analog or digital means. In the context of acoustics, such a measurement is carried out using an impedance tube, a schematic of which is shown in Figure 2.16.

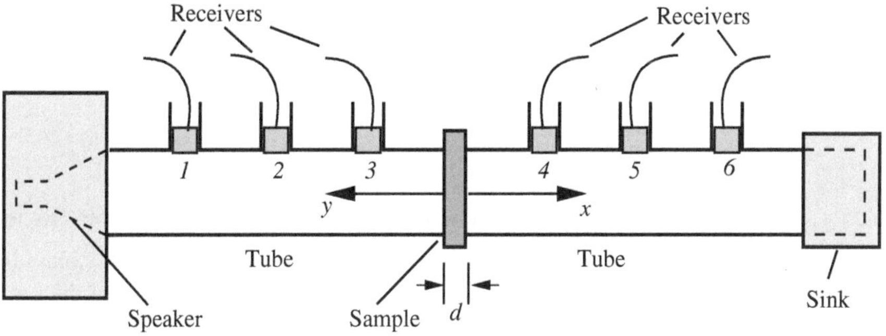

FIGURE 2.16
Schematic of an impedance tube experiment.

The diameter of the tube and the location of the speaker are such that the waves generated may be assumed to be plane. The microphone receivers 1, 2, and 3 record the incident and reflected waves from the sample while receivers 4, 5, and 6 record the transmitted waves. Any waves reflected from the end of the impedance tube are assumed to have relatively small amplitudes and are usually ignored in the analysis (this is also called the anechoic chamber assumption). If we choose not to ignore these reflections, for a particular frequency ω, the pressures at the receivers can be expressed in plane wave form as (see Hall et al. (2010)) for an example of this approach

$$p_1 = A_i e^{-iky_1} + B_r e^{iky_1} \; ; \quad p_2 = A_i e^{-iky_2} + B_r e^{iky_2} \; ; \quad p_3 = A_i e^{-iky_3} + B_r e^{iky_3}$$
$$p_4 = A_t e^{ikx_4} + B_t e^{-ikx_4} \; ; \quad p_5 = A_t e^{ikx_5} + B_t e^{-ikx_5} \; ; \quad p_6 = A_t e^{ikx_6} + B_t e^{-ikx_6}$$

where $k = 2\pi/\lambda = \omega/c$, x_i and y_j are the locations of receivers i, j relative to the surface of the specimen, A_i is the amplitude of the incident wave, B_r is the amplitude of the reflected wave in the medium of incidence, A_t is the amplitude of the transmitted wave, and B_t is the amplitude of reflected waves on the transmission side of the sample. In matrix form,

$$\begin{bmatrix} p_1 \\ p_2 \\ p_3 \\ p_4 \\ p_5 \\ p_6 \end{bmatrix} = \begin{bmatrix} e^{-iky_1} & e^{iky_1} & 0 & 0 \\ e^{-iky_2} & e^{iky_2} & 0 & 0 \\ e^{-iky_3} & e^{iky_3} & 0 & 0 \\ 0 & 0 & e^{ikx_4} & e^{-ikx_4} \\ 0 & 0 & e^{ikx_5} & e^{-ikx_5} \\ 0 & 0 & e^{ikx_6} & e^{-ikx_6} \end{bmatrix} \begin{bmatrix} A_i \\ B_r \\ A_t \\ B_t \end{bmatrix} \quad \text{or} \quad \underline{\underline{\mathsf{p}}} = \underline{\underline{\mathsf{K}}}\,\underline{\underline{\mathsf{a}}}.$$

Since we are trying to determine four constants A_i, B_r, A_t, and B_t, we need only four receivers. However, consider the situation where there are only four receivers, 1, 2, 4, 5. In that case,

$$\det(\underline{\underline{\mathsf{K}}}) = 4\sin[k(x_4 - x_5)]\sin[k(y_1 - y_2)]\,.$$

If $x_4 - x_5 = \lambda$ or $y_1 - y_2 = \lambda$, the matrix $\underline{\underline{\mathsf{K}}}$ is singular because $\det(\underline{\underline{\mathsf{K}}}) = 0$. Six receivers help reduce this possibility. But the matrix is no longer square and cannot be inverted directly. Instead, we solve a least-squares minimization problem of the form

$$\min_{\mathsf{a}} \left\| \underline{\underline{\mathsf{p}}} - \underline{\underline{\mathsf{K}}}\,\underline{\underline{\mathsf{a}}} \right\|^2 .$$

The solution of this problem is

$$\underline{\underline{\mathsf{a}}}_{\text{opt}} = (\underline{\underline{\mathsf{K}}}^T \underline{\underline{\mathsf{K}}})^{-1} \underline{\underline{\mathsf{K}}}^T \underline{\underline{\mathsf{p}}}\,.$$

If the matrix $\underline{\underline{\mathsf{K}}}^T \underline{\underline{\mathsf{K}}}$ is also singular, then we can use the Moore-Penrose pseudo-inverse to get an optimal (but not unique) solution.

To compute the transmission coefficient from the quantities A_i, B_r, A_t, B_t, we use the observation that the wave B_t is transmitted through the sample and provides a contribution to B_r, i.e., $B_r = RA_i + TB_t$, where R is the reflection coefficient and T is the transmission coefficient of the slab (assumed to be the same for incidence on either side of the sample). Similarly, the wave B_t is reflected at the surface of the sample and adds a contribution to A_t to give $A_t = TA_i + RB_t$. Solving these two equations gives us

$$R = \frac{A_t B_t - A_i B_r}{A_i^2 - B_t^2} \quad \text{and} \quad T = \frac{A_i A_t - B_r B_t}{A_i^2 - B_t^2}\,.$$

The transmission loss is usually converted to a sound reduction index (units of dB) using the relation

$$\text{TL (dB)} = 10\log_{10}(1/T^2)\,.$$

An example of the transmission loss through a locally resonant sonic material is shown in Figure 2.17. The plot is from impedance tube measurements on a specimen containing two angled steel beams with steel tip mass resonators. The two resonators were glued on medium-density fiberboard (MDF) wooden circles and then arranged in series along the axis of the tube, with a spring between them. A rubber ring was used to provide the elastic coupling. Note that the two beams were positioned at 180 degrees relative to each other.

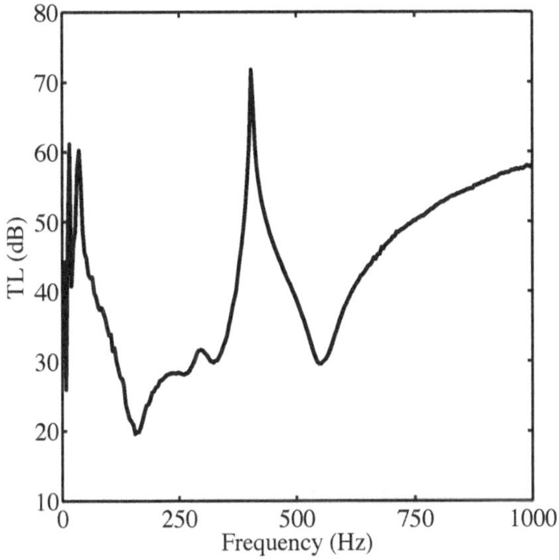

FIGURE 2.17

The sound reduction index as a function of frequency for a locally resonant sonic material containing beam resonators. The low frequency resonances are due to a rubber ring support that is used to clamp the specimen in the impedance tube. The peak sound reduction at 400 Hz is due to local resonances in the the structure. This experiment was performed at Industrial Research Limited, Auckland, New Zealand, by Andrew Hall.

2.5.3 Electromagnetic waves in a slab

Let us now examine the problem of reflection and transmission of electromagnetic waves in a three layer medium (see Figure 2.18). Our goal is to find the effective reflection and transmission coefficients in this medium. Once we know these coefficients, we can choose the materials in the layers to achieve a desired reflectivity or transmissivity.

In dealing with multiple layers it is more convenient to use the notation (x,y,z) instead of (x_1,x_2,x_3) for the coordinates. If the permittivity and magnetic permeability are scalars and locally isotropic in each layer we can write $\varepsilon \equiv \varepsilon(z)$ and $\mu \equiv \mu(z)$.

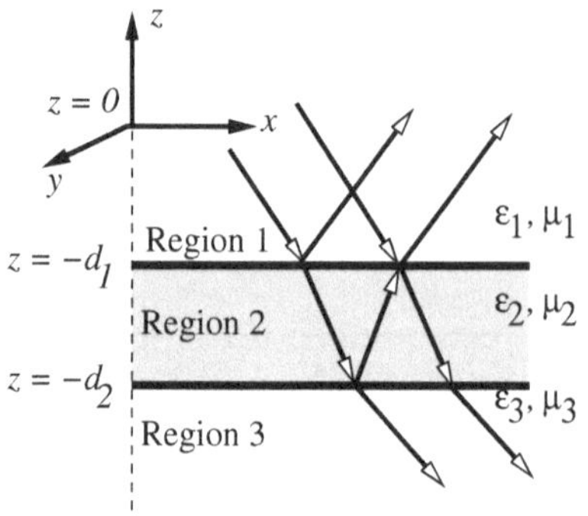

FIGURE 2.18
Reflection and transmission in a three layer medium (slab).

The wave equation for TE waves (1.93) (p. 38) can then be written as

$$\frac{\partial^2 E_y}{\partial x^2} + \left[\mu(z)\frac{\partial}{\partial z}\left(\frac{1}{\mu(z)}\frac{\partial}{\partial z}\right) + \omega^2\,\varepsilon(z)\,\mu(z)\right] E_y = 0\,. \tag{2.59}$$

Solutions to the above equation have the form

$$E_y(x,z) = \widetilde{E}_y(z)\,e^{\pm ik_x x}\,. \tag{2.60}$$

If we plug (2.60) into (2.59) we get an ordinary differential equation for $\widetilde{E}_y(z)$:

$$\left[\mu(z)\frac{d}{dz}\left(\frac{1}{\mu(z)}\frac{d\widetilde{E}_y}{dz}\right)\right] + \left[\omega^2\,\varepsilon(z)\,\mu(z) - k_x^2\right]\widetilde{E}_y = 0\,. \tag{2.61}$$

If we change the variable z appropriately, we can simplify (2.61) so that *within each layer* we have an equation of the form

$$\frac{d^2\widetilde{E}_y}{dz^2} + k_z^2\widetilde{E}_y = 0 \qquad \text{where} \qquad k_z^2 := \omega^2\,\varepsilon(z)\,\mu(z) - k_x^2\,. \tag{2.62}$$

Here k_z can be interpreted as the component of the wave vector $\mathbf{k}$ in the z-direction. Solutions of equation (2.62) have the form

$$\widetilde{E}_y(z) = E_0\,e^{\pm ik_z z} \tag{2.63}$$

where E_0 is the amplitude of the wave. Equation (2.63) clearly has the form of plane waves propagating in the $\pm z$-directions. Note that we would have come to the same conclusion had we started with the full TE wave equation which is valid for all frequencies, except that in that case

$$\widetilde{E}_y(z,t) = E_0\,e^{i(k_z z - \omega t)}\,.$$

We are now ready, once again, to compute the reflection and transmission coefficients at an interface between two layers using this expression for $\widetilde{E}$. Let us drop the tildes for convenience. The **E**-field in region 1 (see Figure 2.18) can be expressed as

$$E_{y1}(z) = E_0 \left[e^{-ik_{z1}z} + R\, e^{ik_{z1}z} \right] .$$

The first term of the left-hand side represents the incoming wave while the second term represents the reflected wave (hence the difference in the sign of z). The quantity R is a reflection coefficient. Similarly, in region 2,

$$E_{y2}(z) = E_0\, T\, e^{-ik_{z2}z}$$

where T is a transmission coefficient. Continuity of the tangential components of the **E** and **H** at the interface $1-2$ implies that

$$E_{y1} = E_{y2} \quad \text{and} \quad \mu_1^{-1} \frac{dE_{y1}}{dz} = \mu_2^{-1} \frac{dE_{y2}}{dz} .$$

Let us choose the coordinate system such that $z = 0$ at the interface. Proceeding as we have done a number of times in this chapter, we find that for incidence from layer 1 upon layer 2 we have

$$1 + R = T \quad \text{and} \quad \mu_1^{-1} k_{z1} (1 - R) = \mu_2^{-1} k_{z2} T \tag{2.64}$$

where R is the reflection coefficient and T is the transmission coefficient. Solving for R and T gives us

$$\boxed{R = \frac{\mu_2 k_{z1} - \mu_1 k_{z2}}{\mu_2 k_{z1} + \mu_1 k_{z2}} \quad \text{and} \quad T = \frac{2\mu_2 k_{z1}}{\mu_2 k_{z1} + \mu_1 k_{z2}}.} \tag{2.65}$$

These are the Fresnel equations that we had derived earlier. From equation (2.62) we know that $k_z^2 = \omega^2 \mu\varepsilon - k_x^2$. Therefore, if $\omega^2 \mu_2 \varepsilon_2 < k_{x2}^2$, then k_{z2} is purely imaginary. If k_{z1} is real, then the first of equations (2.65) implies that the numerator and the denominator are complex conjugates. This means that $|R| = 1$, i.e., there is perfect reflection at the interface. If such a situation exists, the wave in region 2 is called *evanescent.*

For multilayered systems it is more convenient to choose the origin of the coordinate system such that $z = -d_1$ at interface $1-2$ and $z = -d_2$ at interface $2-3$. Then, using a change of coordinates $z \leftarrow z + d_1$ in region 1, we have

$$\begin{aligned} E_{y1}(z) &= E_0 \left[e^{-ik_{z1}(z+d_1)} + \widetilde{R}_{12}\, e^{ik_{z1}(z+d_1)} \right] \\ &= E_0\, e^{-ik_{z1}d_1} \left[e^{-ik_{z1}z} + \widetilde{R}_{12}\, e^{ik_{z1}z}\, e^{2ik_{z1}d_1} \right] \end{aligned}$$

where $\widetilde{R}_{12}$ is the apparent reflection coefficient at the interface between regions 1 and 2 due to the slab. If we define $A_1 := E_0 \exp(-ik_{z1}d_1)$ we have

$$E_{y1}(z) = A_1 \left[e^{-ik_{z1}z} + \widetilde{R}_{12}\, e^{ik_{z1}z + 2ik_{z1}d_1} \right] .$$

Similarly, for region 2, we have

$$E_{y2}(z) = A_2 \left[e^{-ik_{z2}z} + \widetilde{R}_{23}\, e^{ik_{z2}z+2ik_{z2}d_2}\right]$$

where $\widetilde{R}_{23}$ is the apparent reflection coefficient for a downgoing wave at the interface between regions 2 and 3.

But the wave is transmitted in region 3 and there are no further reflections. Therefore, we have

$$E_{y3}(z) = A_3\, e^{-i\,k_{z3}\,z}$$

and

$$\widetilde{R}_{23} = R_{23} = \frac{\mu_3\, k_{z2} - \mu_2\, k_{z3}}{\mu_3\, k_{z2} + \mu_2\, k_{z3}}.$$

At this stage we don't know what $\widetilde{R}_{12}$ is. To find this quantity, note that the downgoing wave in region 2 equals the sum of the transmission of the downgoing wave in region 1 and a reflection of the upgoing wave in region 2 (see Figure 2.18). Hence, at the top interface $z = -d_1$,

$$A_2\, e^{ik_{z2}d_1} = T_{12}A_1\, e^{ik_{z1}d_1} + R_{21}A_2R_{23}\, e^{-ik_{z2}d_1+2ik_{z2}d_2} \tag{2.66}$$

where T_{12} is the transmission coefficient between regions 1 and 2 and $R_{21} = -R_{12}$ is the reflection coefficient of waves from region 2 incident upon region 1. Also, the upgoing wave in region 1 is the sum of the reflection of the incoming wave in region 1 and the transmission at interface 2-1 of the reflected wave at interface 2-3. Hence, at $z = -d_1$ we have

$$A_1\widetilde{R}_{12}\, e^{-ik_{z1}d_1+2ik_{z1}d_1} = R_{12}A_1\, e^{ik_{z1}d_1} + T_{21}A_2\, R_{23}\, e^{-ik_{z2}d_1+2ik_{z2}d_2}. \tag{2.67}$$

Eliminating A_2 from (2.66) gives

$$A_2 = \frac{T_{12}\, A_1\, e^{i(k_{z1}-k_{z2})d_1}}{1 - R_{21}R_{23}\, e^{2ik_{z2}(d_2-d_1)}}. \tag{2.68}$$

Plugging (2.68) into (2.67), we get

$$\boxed{\widetilde{R}_{12} = R_{12} + \frac{T_{12}T_{21}R_{23}\, e^{2ik_{z2}(d_2-d_1)}}{1 - R_{21}R_{23}\, e^{2ik_{z2}(d_2-d_1)}}} \tag{2.69}$$

which gives us an expression for the generalized reflection coefficient $\widetilde{R}_{12}$. We can use a similar approach to find the transmission coefficient for the wave going from layer 1 to layer 3:

$$\boxed{T_{13} = \frac{T_{12}T_{23}\, e^{ik_{z2}(d_2-d_1)}}{1 - R_{21}R_{23}\, e^{2ik_{z2}(d_2-d_1)}}} \tag{2.70}$$

We have considered only two internal reflections so far. How about further reflections? It turns out that equation (2.69) can be interpreted to include all possible internal reflections. To see this, let us assume that

$$0 \leq R_{21} R_{23}\, e^{2ik_{z2}\,(d_2-d_1)} < 1\,.$$

Then we can expand (2.69) in series form to get

$$\boxed{\widetilde{R}_{12} = R_{12} + T_{12}T_{21}R_{23}\, e^{2i\theta}\left[1 + R_{21}R_{23}\, e^{2i\theta} + R_{21}^2 R_{23}^2\, e^{4i\theta} + \ldots\right]} \tag{2.71}$$

where $\theta := k_{z2}(d_2 - d_1)$. This equation can be interpreted as shown in Figure 2.19. However, sometimes the series may not converge and it is preferable to use (2.69) for computations.

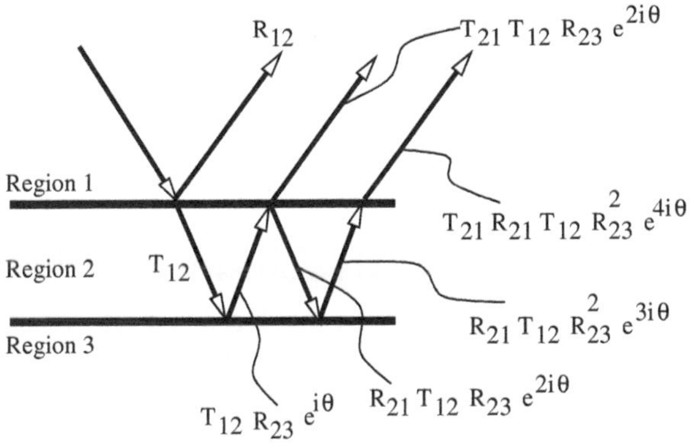

FIGURE 2.19
Interpretation of the series expansion of the generalized reflection coefficient $\widetilde{R}_{12}$.

The equivalent expression for the transmission coefficient is

$$\boxed{T_{13} = T_{12}T_{23}\, e^{i\theta}\left[1 + R_{21}R_{23}\, e^{2i\theta} + R_{21}^2 R_{23}^2\, e^{4i\theta} + \ldots\right].} \tag{2.72}$$

A similar expression can be used in acoustic impedance tube calculations (see, for example, Ho et al. (2005)).

Exercises

Problem 2.1 Show that for time harmonic plane waves, $\mathbf{u}(\mathbf{x},t) = \widehat{\mathbf{u}}\exp[i\omega(\mathbf{s}\cdot\mathbf{x} - t)]$, the governing equations for the dynamics of an isotropic and homogeneous linear elastic body decouple into the equations

$$\left(\rho - \frac{\mu}{c^2}\right)(\widehat{\mathbf{u}} \times \mathbf{s}) = \mathbf{0} \quad \text{and} \quad \left(\rho - \frac{\lambda + 2\mu}{c^2}\right)(\widehat{\mathbf{u}} \cdot \mathbf{s}) = 0 \,.$$

Problem 2.2 Derive the reflection coefficient relations given in equations (2.11) and (2.12) starting from equations (2.9) and (2.10).

Problem 2.3 Verify the reflection coefficient relations given in equations (2.14)–(2.17) using a symbolic computation tool if needed. Solve these equations numerically and plot the magnitude and phase of the reflection and transmission coefficients as a function of the angle of incidence for the interface between two materials with $\rho_1 = 820$, $c_{p1} = 1320$, $c_{s1} = 1.0e-4$, $\rho_2 = 1000$, $c_{p2} = 1500$, $c_{s2} = 1.0e-4$.

Problem 2.4 Consider a plane acoustic wave propagating from kerosene into water at room temperature. Kerosene has a density of 820 kg/m^3 and a sound speed of 1320 m/s while water has a density of 1000 kg/m^3 and a sound speed of 1500 m/s. Plot the magnitude and phase of the reflection coefficient (R) as a function of the angle of incidence. Is there any angle at which the entire energy of the wave is transmitted through the interface? At what angle does total internal reflection occur (i.e., the transmission coefficient becomes zero)? What happens as the angle of incidence is increased beyond the angle at which total internal reflection first occurs? How do your results compare with those from Problem 3.3?

Now consider the case where the materials absorb a small fraction of the energy of the acoustic wave. In that case we can add a damping factor (α) to the refractive index n, i.e, $n \to n(1 + i\alpha)$. Plot the magnitude and phase and a function of incidence angle for $\alpha = 0.01$. Is there total internal reflection in this situation?

Problem 2.5 We have defined the refractive index for acoustic waves propagating from a medium with phase velocity c_1 into a medium with phase velocity c_2 as $n = c_1/c_2$. If we choose a reference medium, e.g., air with a sound speed of c_0, we can have an alternative definition of the refractive indexes n_1 and n_2 of the two media given by $n_1 = c_0/c_1$ and $n_2 = c_0/c_2$ in which case $n = n_2/n_1$. We have mentioned earlier that waves cannot propagate in the medium if the phase velocity is imaginary. How then can waves propagate in a medium with a complex refractive index?

Problem 2.6 Maxwell's equations for an isotropic material at fixed frequency may be expressed as (1.89)

$$\boldsymbol{\nabla} \times \mathbf{E} = i\omega\mu\mathbf{H} \,; \;\; \boldsymbol{\nabla} \times \mathbf{H} = -i\omega\varepsilon\mathbf{E} \,; \;\; \boldsymbol{\nabla} \cdot \mathbf{H} = 0 \,; \;\; \boldsymbol{\nabla} \cdot \mathbf{E} = 0 \,.$$

Show that for a plane wave electric field $\mathbf{E}(\mathbf{x}) = \mathbf{E}_0 \exp(i\mathbf{k} \cdot \mathbf{x})$ the wave vector is perpendicular to the fields, i.e., $\mathbf{k} \cdot \mathbf{E}_0 = 0$. Then show that this implies that the magnetic field is also a plane wave of the form $\mathbf{H}(\mathbf{x}) = \mathbf{H}_0 \exp(i\mathbf{k} \cdot \mathbf{x})$ where

$$\mathbf{H}_0 = -\frac{1}{\omega\mu}(\mathbf{k} \times \mathbf{E}_0) \quad \text{and} \quad \mathbf{k} \cdot \mathbf{H}_0 = 0 \,.$$

Recall also that for fixed frequency $\nabla^2\mathbf{H} + (\omega^2/c^2)\,\mathbf{H} = \mathbf{0}$. Show that the above equation implies that for a plane wave

$$\mathbf{k} \cdot \mathbf{k} = (\|\mathbf{k}\|)^2 = \frac{\omega^2}{c^2} \,.$$

Problem 2.7 Express equations (2.51) in terms of electromagnetic impedances and then calculate the reflection and transmission coefficients for a medium that is impedance matched with a silicone rubber dielectric material.

Problem 2.8 Consider a slab of material in an impedance tube. The bulk modulus and density of air on both sides of the slab are 1.42×10^5 Pa and 1.20 kg/m^3, respectively. The Young's modulus (E), Poisson's ratio (ν), and density (ρ) of aluminum are 70 GPa, 0.33, and 2700 kg/m^3, respectively. Assume that the phase velocity in aluminum can be obtained from the relation $c = \sqrt{\kappa/\rho}$ where $\kappa = E/(3(1-2\nu))$ is the bulk modulus.

1. The transmission loss due to the slab is calculated using the relation

$$\text{TL (dB)} = 10\log_{10}\left(\frac{1}{T^2}\right)$$

   where T is the transmission coefficient. Plot the transmission loss for a 10 cm thick aluminum slab. Compare the transmission loss due to the solid slab with that for a similar slab made of aluminum foam with an aluminum volume fraction (f) of 10%. Assume that the effective foam density is given by $\rho_{\text{eff}} = f\rho_1 + (1-f)\rho_2$ and that the effective foam Young's modulus is given by $E_{\text{eff}} = E(\rho_{\text{eff}}/\rho)^2$. The Poisson's ratio of the foam is 0.33. What does the imaginary part of the transmission coefficient indicate? What is the effect of slab density on the transmission loss?

2. Next plot the transmission losses for aluminum and aluminum slabs for a fixed frequency as a function of slab thickness. Assume a frequency of 100 Hz and keep in mind that ω has units of radians/s and not cycles/s. Such a plot is called a mass law plot in acoustics. What would the mass law effect be if ρ_{eff} were a function of frequency and the system had a resonance frequency of 100 Hz?

Problem 2.9 For the situation in Figure 2.18 show that the transmission coefficient for the slab can be expressed as

$$T_{13} = \frac{T_{12}T_{23}\, e^{ik_{z2}(d_2-d_1)}}{1 - R_{21}R_{23}\, e^{2ik_{z2}(d_2-d_1)}}.$$

Also verify that the series expansion of the above equation is

$$T_{13} = T_{12}T_{23}\, e^{i\theta} + T_{12}T_{23}R_{21}R_{23}\, e^{3i\theta} + T_{12}T_{23}R_{21}^2R_{23}^2\, e^{5i\theta} + \ldots.$$

3

Sources and Scattering

> The separation of the solution into an unmarred incident wave and a scattered wave—a useful dichotomy for long wavelengths, where there is no shadow—does not describe the short wave solution, where there is a sharp-edged shadow as well as an outgoing wave, reflected from the sphere's surface.
>
> PHILIP M. MORSE, *In at the beginnings: A physicist's life*, 1977.

So far we have explored the reflection and transmission of plane waves when they are incident on plane surfaces. But if we look at waves in a region close to a point source of energy we are likely to find that the waves are spherical and that the amplitude of the waves decreases as we move further away from the source. If the source is a line source the waves are likely to be cylindrical. On the other hand, if the region of interest is far from the source region the waves may be well represented by plane waves of constant amplitude but the interface might be curved. As we will discover in this chapter, both these situations can be represented by superpositions of plane waves. For curved interfaces we will find closed form solutions for single scattering of acoustic, elastodynamic, and electrodynamic waves incident upon spherical objects. These solutions can then be used as the starting point for multiple scattering calculations or to verify the accuracy of numerical techniques needed for more complicated geometries. We will also take a brief look at multiple scattering and explore multiple scattering from an array of cylinders.

Single and multiple scattering solutions have been used widely to determine the effective response of metamaterials and photonic/phononic crystals, and to find the effectiveness of coated cylinders and spheres as cloaks. We will not discuss solutions for spheres and cylinders with anisotropic coatings because general solutions can be quite difficult to find without using numerical tools.

3.1 Plane wave expansion of sources

It is often useful to expand sources in terms of plane waves so that the results discussed in the previous chapter may be used directly on the basis of superposition. In this section we look at the expansion of point sources in terms of plane waves for a homogeneous medium. We start with the wave equation at fixed frequency ω and in the presence of a point source at the origin,

$$\nabla^2\varphi(\mathbf{x}) + \frac{\omega^2}{c^2}\varphi(\mathbf{x}) = -\delta(\mathbf{x}) \tag{3.1}$$

where $\delta(\mathbf{x})$ is the Dirac delta function and c is the phase speed. Note that, with the addition of boundary conditions, φ is the Green's function for the differential operator $\mathcal{L} := \nabla^2 + \omega^2/c^2$. Let us define $k := \omega/c$. Then, in rectangular Cartesian coordinates, the above equation has the form

$$\left[\frac{\partial^2}{\partial x_1^2} + \frac{\partial^2}{\partial x_2^2} + \frac{\partial^2}{\partial x_3^2} + k^2\right] \varphi(\mathbf{x}) = -\delta(x_1)\,\delta(x_2)\,\delta(x_3)\,. \tag{3.2}$$

We can also use the expressions for the Laplacian in curvilinear coordinates in Section 1.2.3 to show that in cylindrical polar coordinates (with $x_1 = r\cos\theta$, $x_2 = r\sin\theta$ and $x_3 = z$) the wave equation has the form

$$\frac{1}{r}\left[\frac{\partial}{\partial r}\left(r\frac{\partial\varphi}{\partial r}\right)\right] + \frac{1}{r^2}\frac{\partial^2\varphi}{\partial\theta^2} + \frac{\partial^2\varphi}{\partial z^2} + k^2\varphi(r,\theta,z) = -\delta(r)\delta(\theta)\delta(z)\,. \tag{3.3}$$

Similarly, for spherical coordinates with $x_1 = r\cos\theta\sin\phi$, $x_2 = r\sin\theta\sin\phi$, and $x_3 = r\cos\phi$, we have

$$\begin{aligned}&\frac{1}{r^2}\left[\frac{\partial}{\partial r}\left(r^2\frac{\partial\varphi}{\partial r}\right)\right] + \frac{1}{r^2\sin\phi}\left[\frac{1}{\sin\phi}\frac{\partial^2\varphi}{\partial\theta^2} + \frac{\partial}{\partial\phi}\left(\sin\phi\frac{\partial\varphi}{\partial\phi}\right)\right]\\ &\quad + k^2\varphi(r,\theta,\phi) = -\delta(r)\delta(\theta)\delta(\phi)\,.\end{aligned} \tag{3.4}$$

3.1.1 Two-dimensional scalar wave equation

Let us first look at the two-dimensional scalar wave equation which can be used for acoustics, TE- and TM-waves, antiplane shear elasticity (SH-waves), etc. In the presence of a point source, it is convenient to express the problem in polar coordinates (cylindrical coordinates without z-dependence) because the solution is symmetric about the origin and therefore there is no θ-dependence.

Expressed in two-dimensional cylindrical coordinates, equation (3.1) then becomes

$$\left[\frac{d^2}{dr^2} + \frac{1}{r}\frac{d}{dr} + k^2\right] \varphi(r) = -\delta(r)\delta(\theta)\,. \tag{3.5}$$

So far we have looked only at homogeneous plane wave solutions where k has been real. Let us now consider the inhomogeneous case where we assume that k has a small positive imaginary part (it is a slightly lossy material), i.e.,

$$k = k' + ik'' \qquad \text{where} \quad k'' > 0. \tag{3.6}$$

The radial part of equation (3.5) can be expressed in self-adjoint form as

$$\frac{d}{d\rho}\left(\rho\frac{d\varphi}{d\rho}\right) + \rho\varphi = -\delta(\rho) \quad \text{where} \quad \rho = kr\,.$$

The homogeneous part of this equation is Bessel's equation* and has the solutions

$$\varphi_1(\rho) = J_0(\rho)\ ;\ \ \varphi_2(\rho) = H_0^{(1)}(\rho)\ .$$

Here $J_n(x)$ is a Bessel function of the first kind and $H_n^{(1)}(x)$ is a Hankel function of the first kind defined as $H_0^{(1)}(x) := J_0(x) + i\,Y_0(x)$ where $Y_n(x)$ is a Bessel function of the second kind. The first of these solutions has been chosen because J_0 is finite at $r = 0$. Also, the waves generated by a point source become plane waves of the form e^{ikr} as $r \longrightarrow \infty$. The second solution has been chosen because $H_0^{(1)}(kr) \sim \exp(ikr)$ as $r \longrightarrow \infty$.

For the solution of the inhomogeneous equation we follow the Green's function approach and recall that (see Arfken and Weber (2005), section 10.5, for details) if the Green's function $g(\rho,\rho_0)$ solves the equation

$$\frac{d}{d\rho}\left[\rho\frac{dg(\rho,\rho_0)}{d\rho}\right] + \rho g(\rho,\rho_0) = -\delta(\rho-\rho_0)$$

then

$$g(\rho,\rho_0) = -\frac{1}{C}\begin{cases}\varphi_1(\rho)\,\varphi_2(\rho_0) & \text{for} \quad \rho < \rho_0 \\ \varphi_1(\rho_0)\,\varphi_2(\rho) & \text{for} \quad \rho_0 < \rho\end{cases}$$

where C is a constant. The function $g(\rho,\rho_0)$ is finite at $\rho = 0$ and $\rho = \rho_0$, and has the asymptotic behavior of φ at $\rho \longrightarrow \infty$. The Wronskian is (see NIST (2010), section 10.5)

$$W = \varphi_1\varphi_2' - \varphi_2\varphi_1' = J_0(\rho)\,[H_0^{(1)}(\rho)]' - [J_0(\rho)]'\,H_0^{(1)}(\rho) = \frac{2i}{\pi\rho}\ .$$

We can also show that the Wronskian is given by $W = C/\rho$ (see Arfken and Weber (2005), section 9.6). Hence, $C = 2i/\pi$ and the solutions are

$$g(\rho,\rho_0) = -\frac{\pi}{2i}\begin{cases}J_0(\rho)\,H_0^{(1)}(\rho_0) & \text{for} \quad \rho < \rho_0 \\ J_0(\rho_0)\,H_0^{(1)}(\rho) & \text{for} \quad \rho_0 < \rho.\end{cases}$$

Each of these solutions is for the radial direction. To find the complete solution we have to consider the $\delta(\theta)$ term in equation (3.5). If $G(\rho,\rho_0)$ is the Green's function for the full equation we can show that†

$$G(\rho,\rho_0) = \left(\frac{1}{2\pi}\right)\left(-\frac{\pi}{2i}\right)J_0(\rho_<)\,H_0^{(1)}(\rho_>)$$

*Bessel's equation is the ordinary differential equation of the form

$$\frac{d^2J_n}{dx^2} + \frac{1}{x}\frac{dJ_n}{d\rho} + \left[1 - \frac{n^2}{x^2}\right]J_n = 0\ .$$

†See the discussion starting with equation (3.8) on solutions in rectangular coordinates using Fourier transforms for the approach that is used to find the $1/2\pi$ factor.

where the subscripts $<$ and $>$ and are a compact way of indicating the two cases. For our problem $\rho >= \rho_0 = 0$, i.e, $J_0(\rho_0) = J_0(0) = 1$, and only the second solution is of interest. Therefore we have

$$\boxed{\varphi(r) = \frac{i}{4} H_0^{(1)}(kr)\,.} \tag{3.7}$$

Now let us look at (3.1) in two-dimensional rectangular Cartesian coordinates. The wave equation may then be written in the form

$$\frac{\partial^2 \varphi}{\partial x^2} + \frac{\partial^2 \varphi}{\partial y^2} + k^2 = -\delta(x)\,\delta(y)\,. \tag{3.8}$$

We can solve (3.8) using Fourier transforms. Let us assume that the function $\varphi(x,y)$ has an inverse Fourier transform,

$$\varphi(x,y) = \frac{1}{2\pi}\int_{-\infty}^{\infty} \widehat{\varphi}(k_x,y)\, e^{ik_x x}\, \mathrm{d}k_x\,. \tag{3.9}$$

Also note that the Fourier representation of the delta function is

$$\delta(x) = \frac{1}{2\pi}\int_{-\infty}^{\infty} e^{ik_x x}\, \mathrm{d}k_x\,. \tag{3.10}$$

Plugging (3.9) and (3.10) into (3.1) gives

$$\frac{1}{2\pi}\int_{-\infty}^{\infty}\left[\frac{\partial^2}{\partial x^2} + \frac{\partial^2}{\partial y^2} + k^2\right]\left[\widehat{\varphi}(k_x,y)\, e^{ik_x x}\right] \mathrm{d}k_x = -\frac{1}{2\pi}\int_{-\infty}^{\infty} \delta(y)\, e^{ik_x x}\, \mathrm{d}k_x$$

or

$$\int_{-\infty}^{\infty}\left[-k_x^2 + \frac{\partial^2}{\partial y^2} + k^2\right] \widehat{\varphi}(k_x,y)\, e^{ik_x x}\, \mathrm{d}k_x = -\int_{-\infty}^{\infty} \delta(y)\, e^{ik_x x}\, \mathrm{d}k_x\,.$$

Since the above equation holds for all values of x, the Fourier components on both sides must be identical, i.e.,

$$\left[\frac{\partial^2}{\partial y^2} + k^2 - k_x^2\right] \widehat{\varphi}(k_x,y) = -\delta(y)\,.$$

Defining

$$k_y^2 := k^2 - k_x^2$$

we get

$$\left[\frac{\partial^2}{\partial y^2} + k_y^2\right] \widehat{\varphi}(k_x,y) = -\delta(y)\,. \tag{3.11}$$

Note that now the equation has a source only at $y = 0$. Away from the source (i.e., $y \neq 0$), the right-hand side of (3.11) is zero, and the solution corresponds to the homogeneous part of the equation. Therefore,

$$\widehat{\varphi}(k_x,y) = A_1\, e^{ik_y y} + A_2\, e^{-ik_y y} \qquad \text{for} \quad y \neq 0\,. \tag{3.12}$$

This solution must be matched with the singularity at $y = 0$ which can be achieved by requiring that the solution have discontinuous second derivatives at $y = 0$ (see Stakgold (2000) for details). If we consider only waves that are damping away from the source rather that those growing exponentially, we have[‡]

$$\widehat{\varphi}(k_x, y) = \frac{i}{2k_y}\, e^{ik_y|y|} \,. \tag{3.13}$$

Plugging (3.13) into (3.9) gives

$$\boxed{\varphi(x,y) = \frac{i}{4\pi}\int_{-\infty}^{\infty} \frac{1}{k_y}\, e^{ik_x x + ik_y|y|}\, \mathrm{d}k_x \,.} \tag{3.14}$$

Equation (3.14) is a plane wave solution for the wave equation with a point source and is the two-dimensional version of the *Weyl integral*. Note that the denominator in (3.14) contains $k_y = \sqrt{k^2 - k_x^2}$. Hence, when $k_x = \pm k$ the solution blows up and there are branch points at these locations as shown in Figure 3.1. In a lossless medium, $k'' \to 0$ and these points appear as poles on the real k_x-axis. The integral in equation (3.14) can be computed using the residue theorem. The region between the two poles is where waves are allowed to propagate in the y-direction while the region outside the poles is where these waves are evanescent in the y-direction. *So the point source has been converted into a sum of propagating plane waves and some evanescent terms.*

If we now compare the solutions (3.7) and (3.14), we get a definition for the Hankel function,

$$H_0^{(1)}(kr) = \frac{1}{\pi}\int_{-\infty}^{\infty} \frac{1}{k_y}\, e^{ik_x x + ik_y|y|}\, \mathrm{d}k_x \,.$$

Also, differentiating (3.8) with respect to y and x, we get

$$\left[\frac{\partial^2}{\partial x^2} + \frac{\partial^2}{\partial y^2} + k^2\right] \frac{\partial\varphi(x,y)}{\partial y} = -\delta(x)\, \frac{\partial\delta(y)}{\partial y} \tag{3.15}$$

$$\left[\frac{\partial^2}{\partial x^2} + \frac{\partial^2}{\partial y^2} + k^2\right] \frac{\partial\varphi(x,y)}{\partial x} = -\delta(y)\, \frac{\partial\delta(x)}{\partial x}. \tag{3.16}$$

The products $\delta(x)\,\frac{\partial\delta(y)}{\partial y}$ and $\delta(y)\,\frac{\partial\delta(x)}{\partial x}$ correspond to *dipole sources* in the y- and x-directions, respectively. Taking derivatives of equation (3.14), we have

$$\boxed{\begin{aligned} \varphi_y(x,y) &:= \frac{\partial\varphi(x,y)}{\partial y} = -\frac{\operatorname{sgn}(y)}{4\pi}\int_{-\infty}^{\infty} e^{i\,k_x\,x + i\,k_y\,|y|}\, \mathrm{d}k_x \\ \varphi_x(x,y) &:= \frac{\partial\varphi(x,y)}{\partial x} = -\frac{1}{4\pi}\int_{-\infty}^{\infty} \frac{k_x}{k_y}\, e^{i\,k_x\,x + i\,k_y\,|y|}\, \mathrm{d}k_x \,. \end{aligned}} \tag{3.17}$$

These are the plane wave expansions of dipoles in the y- and x-directions respectively. Taking higher derivatives gives results for quadrupoles and other multipoles.

[‡]The formula in equation (3.13) can be derived using the Green's function procedure discussed earlier.

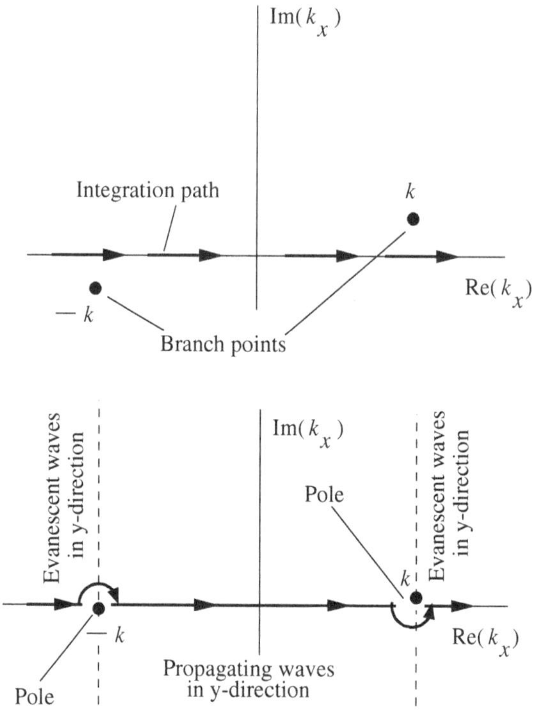

FIGURE 3.1
Poles and integration path for plane wave solutions corresponding to a point source.

3.1.2 Three-dimensional scalar wave equation

We have determined that a two-dimensional point source could be expanded into plane waves. We may think of such a point source as a line source in three dimensions. We can similarly try to expand true three-dimensional point sources in terms of plane waves. To do that, let us start with a three-dimensional scalar wave equation in spherical coordinates (3.4) and consider the particular case where there is not variation in the θ and ϕ directions. As before, assume that k has a small positive imaginary part, i.e., $k = k' + ik''$. In that case the wave equation takes the form

$$\frac{1}{r^2}\left[\frac{\partial}{\partial r}\left(r^2\frac{\partial\varphi}{\partial r}\right)\right] + k^2\varphi(r,\theta,\phi) = -\delta(r)\delta(\theta)\delta(\phi) .$$

The solution of the above equation can be obtained using the same procedure as before and the solution is (see Arfken and Weber (2005), p. 598)

$$\boxed{\varphi(r) = \frac{1}{4\pi r}e^{ikr} .} \tag{3.18}$$

If we express equation (3.1) in three-dimensional rectangular Cartesian coordinates we get

$$\left[\frac{\partial^2}{\partial x^2}+\frac{\partial^2}{\partial y^2}+\frac{\partial^2}{\partial z^2}+k^2\right]\,\varphi(x,y,z)=-\delta(x)\,\delta(y)\,\delta(z)\,. \tag{3.19}$$

Let us try to solve (3.19) using Fourier transforms. To do that, we assume that a Fourier transform of $\varphi(x,y,z)$ exists and the inverse Fourier transform has the form

$$\varphi(x,y,z)=\frac{1}{8\pi^3}\int_{-\infty}^{\infty}\widehat{\varphi}(\mathbf{k})\,e^{i\mathbf{k}\cdot\mathbf{x}}\,\mathrm{d}\mathbf{k} \tag{3.20}$$

where $\mathbf{k}=k_x\mathbf{e}_1+k_y\mathbf{e}_2+k_z\mathbf{e}_3$, $\mathbf{k}\cdot\mathbf{x}=k_xx+k_yy+k_zz$, and $\mathrm{d}\mathbf{k}:=\mathrm{d}k_x\,\mathrm{d}k_y\,\mathrm{d}k_z$. Plugging (3.20) into (3.19) and using the Fourier expansion of the three-dimensional delta function,

$$\delta(x)\,\delta(y)\,\delta(z)=\frac{1}{8\pi^3}\int_{-\infty}^{\infty}e^{i,\mathbf{k}\cdot\mathbf{x}}\,\mathrm{d}\mathbf{k}$$

gives (for x,y,z not all zero)

$$\frac{1}{8\pi^3}\int_{-\infty}^{\infty}\left[-k_x^2-k_y^2-k_z^2+k^2\right]\,\widehat{\varphi}(\mathbf{k})\,e^{i\mathbf{k}\cdot\mathbf{x}}\,\mathrm{d}\mathbf{k}=-\frac{1}{8\pi^3}\int_{-\infty}^{\infty}e^{i\mathbf{k}\cdot\mathbf{x}}\,\mathrm{d}\mathbf{k}\,.$$

Since the above equation holds for all values of x, the Fourier components must agree, i.e., $[-k_x^2-k_y^2-k_z^2+k^2]\,\widehat{\varphi}(\mathbf{k})=-1$. Therefore,

$$\widehat{\varphi}(\mathbf{k})=-\frac{1}{k^2-\mathbf{k}\cdot\mathbf{k}}\,. \tag{3.21}$$

Plugging (3.21) into (3.20) gives

$$\boxed{\varphi(x,y,z)=-\frac{1}{8\pi^3}\int_{-\infty}^{\infty}\frac{1}{k^2-\mathbf{k}\cdot\mathbf{k}}\,e^{i\mathbf{k}\cdot\mathbf{x}}\,\mathrm{d}\mathbf{k}\,.} \tag{3.22}$$

We can write the above in expanded form as

$$\varphi(x,y,z)=-\frac{1}{8\pi^3}\int_{-\infty}^{\infty}\int_{-\infty}^{\infty}\int_{-\infty}^{\infty}\frac{1}{k^2-k_x^2-k_y^2-k_z^2}\,e^{i\,(k_xx+k_yy+k_zz)}\,\mathrm{d}k_x\mathrm{d}k_y\mathrm{d}k_z\,.$$

Let us consider the integral over k_z first. The poles are at $k^2-\mathbf{k}\cdot\mathbf{k}=0$ which implies that $k_z=\pm\sqrt{k^2-k_x^2-k_y^2}$. If $k_z=k_z'+ik_z''$ we have $\exp(ik_zz)=\exp(ik_z'z)\exp(-k_z''z)$. Therefore, for $z>0$ the integral is exponentially decreasing when $k_z''=\mathrm{Im}(k_z)\to\infty$. We can take advantage of this feature and split the integral over k_z into the sum of an integral along the real k_z line and an integral over an arc of a circle of radius infinity in the complex k_z-plane and take care of the poles by adding to the integral the sum of the residues at each of the poles.[§] See Figure 3.2 for a sketch of the situation.

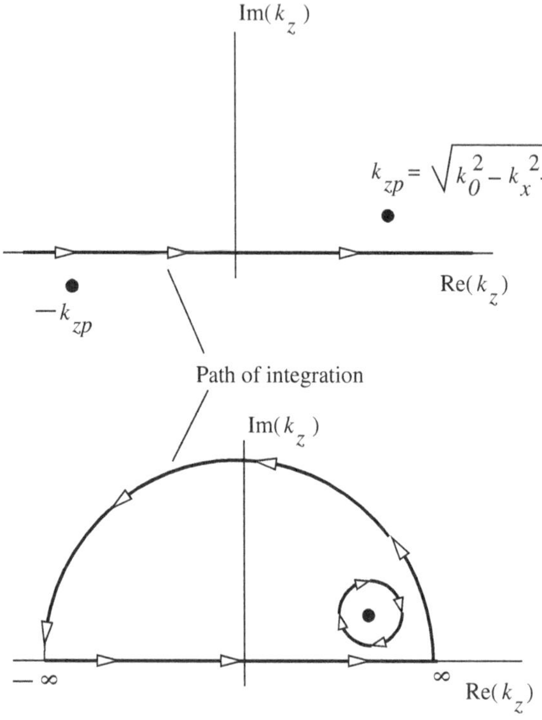

FIGURE 3.2
Poles and integration path for integration over k_z for $z > 0$.

Using the residue theorem we can show that

$$\varphi(x,y,z) = \frac{i}{8\pi^2} \int_{-\infty}^{\infty} \int_{-\infty}^{\infty} \frac{1}{k_{zp}} e^{ik_x x + ik_y y + ik_{zp} z} \, dk_x \, dk_y$$

where k_{zp} is the value of k_z at the poles, i.e., $k_{zp} := \pm\sqrt{k^2 - k_x^2 - k_y^2}$. When $z < 0$,

[§]Recall that if $g(z) = f(z)/(z - z_0)$, i.e., $g(z)$ has a pole at z_0, and if $f(z)$ is analytic and non-singular at z_0, then the residue at z_0 is $f(z_0)$. The residue theorem states that

$$\oint g(z) \, dz = 2\pi i \sum \text{residues}$$

where the contour integral is taken over a piecewise smooth, positively oriented, simple closed curve that does not pass through any of the poles of $g(z)$. The sum of the residues is taken over all poles that lie inside the integration contour.

one takes the semicircular contour C in the lower half plane and picks up the residue at $-k_{zp}$. The result for all z can therefore be written as

$$\boxed{\varphi(x,y,z) = \frac{i}{8\pi^2}\int_{-\infty}^{\infty}\int_{-\infty}^{\infty}\frac{1}{k_{zp}}\,e^{ik_x x+ik_y y+ik_{zp}|z|}\,\mathrm{d}k_x\,\mathrm{d}k_y\,.} \tag{3.23}$$

The integral is over propagating waves parallel to the xy-plane and evanescent plane waves in the z-direction when k_{zp} is imaginary, i.e., $k_x^2+k_y^2>k^2$. Note that we could have chosen to carry out the integration over k_x in which case we would have had propagating waves parallel to the yz-plane and evanescent waves in the x-direction. Such solutions are physically identical because we can choose our coordinate system arbitrarily.

Comparing equations (3.23) and (3.18), we get the *Weyl identity* (Weyl, 1919) for the solution of the wave equation in spherical coordinates

$$\boxed{\frac{1}{r}\,e^{ikr} = \frac{i}{2\pi}\int_{-\infty}^{\infty}\int_{-\infty}^{\infty}\frac{1}{k_z}\,e^{ik_x x+ik_y y+ik_z|z|}\,\mathrm{d}k_x\,\mathrm{d}k_y} \tag{3.24}$$

where $k_z = \sqrt{k^2-k_x^2-k_y^2}$. It is possible to establish a relation of this form because the solutions of the wave equation with homogeneous boundary conditions form a vector space and we can choose certain basis vectors (also called eigenfunctions) to span this space. The Weyl integral uses plane waves as a basis to express the solution for a point source boundary condition.

We have seen while dealing with plane waves in the previous chapter that it is convenient to work with the magnitude of the wave vector and the angle it makes with the two-dimensional Cartesian coordinate axes. In a similar manner, we can express (3.24) in terms of the angle ϕ between $\mathbf{k}$ and the z-axis ($\mathbf{e}_3$), and the angle θ between $\mathbf{k}$ and the x-axis ($\mathbf{e}_1$). In that case the wave vector is

$$\mathbf{k} = k_x\mathbf{e}_1 + k_y\mathbf{e}_2 + k_z\mathbf{e}_3 = \|\mathbf{k}\|\,(\sin\phi\cos\theta\,\mathbf{e}_1 + \sin\phi\sin\theta\,\mathbf{e}_2 + \cos\phi\,\mathbf{e}_3)\,.$$

Then the determinant of the Jacobian matrix of the transformation is

$$\det(J) = \det\begin{bmatrix}\dfrac{\partial k_x}{\partial\phi} & \dfrac{\partial k_x}{\partial\theta}\\ \dfrac{\partial k_y}{\partial\phi} & \dfrac{\partial k_y}{\partial\theta}\end{bmatrix} = k^2\sin\phi\cos\phi$$

where we have used $\|\mathbf{k}\| = k$. Therefore,

$$\mathrm{d}k_x\,\mathrm{d}k_y = k^2\sin\phi\cos\phi\,\mathrm{d}\theta\,\mathrm{d}\phi \qquad\Longrightarrow\qquad \frac{\mathrm{d}k_x\,\mathrm{d}k_y}{k_z} = k\sin\phi\,\mathrm{d}\theta\,\mathrm{d}\phi\,.$$

Because $k_z = \sqrt{k^2-k_x^2-k_y^2}$, k_z varies from k to $i\infty$ when k_x, k_y vary from 0 to $\pm\infty$. Since $\cos\phi = k_z/k$, the limits of ϕ are 0 to $\pi/2-\infty$ and the path of integration in the complex plane has to be chosen accordingly. The limits for θ go from 0 to 2π. The Weyl integral can then be written as

$$\boxed{\frac{1}{r}e^{ikr} = \frac{ik}{2\pi}\int_0^{\pi/2-i\infty}\int_0^{2\pi} e^{ik(\cos\theta\sin\phi x+\sin\theta\sin\phi y+\cos\phi\,|z|)}\,\sin\phi\,d\theta d\phi\,.} \tag{3.25}$$

The waves corresponding to complex values of ϕ are inhomogeneous evanescent waves. These waves are propagated with short wavelengths along the xy-plane and with an exponentially decaying amplitude in the z-direction.

Evaluation of these integrals for situations involving interfaces can be quite involved and techniques such as the method of steepest descents have been used, usually in situations where the distance from the source to the interface is much larger than the wavelength. For examples see Brekhovskikh (1960), Aki and Richards (1980), and Jackson (1999).

3.1.3 Expansion of electromagnetic dipole sources

So far we have dealt with the generic wave equation at fixed frequency. We can use the same techniques for more general situations in elastodynamics and acoustics involving dipole and higher-order sources. Acoustic dipoles in isotropic media are equivalent to elastodynamic P-wave dipoles which can be interpreted at the volume expansion and contraction of two small spheres in close proximity to each other. In elastodynamics we can also have S-wave dipoles which can be thought of as harmonic shear motions along a dislocation. Details of elastodynamic dipoles and their plane wave expansions can be found in Aki and Richards (1980) and Brekhovskikh (1960). Acoustic dipoles are discussed in detail in Skudrzyk (1971) and Morse and Ingard (1986).

Let us instead focus on Maxwell's equations at fixed frequency in the presence of sources and examine the form of the solutions when the source is a current dipole. Consider sources of charge and current that are harmonic, i.e.,

$$\rho(\mathbf{r},t) = \widehat{\rho}(\mathbf{r})\,e^{-i\omega t} \quad \text{and} \quad \mathbf{J}(\mathbf{r},t) = \widehat{\mathbf{J}}(\mathbf{r})\,e^{-i\omega t}\,.$$

In the presence of these sources we have to return back to the full Maxwell's equations given in (1.69) and (1.70). At a fixed frequency ω, and for isotropic media, these equations have the form

$$\boldsymbol{\nabla}\times\widehat{\mathbf{E}} = i\omega\widehat{\mathbf{B}}\;;\;\; \boldsymbol{\nabla}\times\widehat{\mathbf{H}} = -i\omega\widehat{\mathbf{D}}+\widehat{\mathbf{J}}\;;\;\; \boldsymbol{\nabla}\cdot\widehat{\mathbf{B}} = 0\;;\;\; \boldsymbol{\nabla}\cdot\widehat{\mathbf{D}} = \widehat{\rho}\;;\;\; \widehat{\mathbf{B}} = \widehat{\mu}\widehat{\mathbf{H}}\;;\;\; \widehat{\mathbf{D}} = \widehat{\varepsilon}\widehat{\mathbf{E}}$$

and the continuity of charge requires that $\widehat{\rho} = \boldsymbol{\nabla}\cdot\widehat{\mathbf{J}}/(i\omega)$. We can combine the above equations together to get

$$\boldsymbol{\nabla}\times\boldsymbol{\nabla}\times\mathbf{E}(\mathbf{r}) - k^2\,\mathbf{E}(\mathbf{r}) = i\omega\mu\,\mathbf{J}(\mathbf{r}) \tag{3.26}$$

where $k^2 = \omega^2\varepsilon\mu$ and the hats have been dropped for simplicity. Using the identity

$$\boldsymbol{\nabla}\times\boldsymbol{\nabla}\times\mathbf{E} = \boldsymbol{\nabla}(\boldsymbol{\nabla}\cdot\mathbf{E}) - \nabla^2\mathbf{E}$$

we get

$$\nabla(\nabla\cdot\mathbf{E}(\mathbf{r})) - \nabla^2\mathbf{E}(\mathbf{r}) - k^2\,\mathbf{E}(\mathbf{r}) = i\omega\mu\,\mathbf{J}(\mathbf{r})\,. \tag{3.27}$$

Now, for an isotropic homogeneous medium the constitutive relations and the continuity of charge lead to

$$\nabla\cdot\mathbf{E} = \frac{1}{i\omega\varepsilon}\nabla\cdot\mathbf{J}\,.$$

Plugging this into (3.27) we get

$$\frac{1}{i\omega\varepsilon}\nabla(\nabla\cdot\mathbf{J}(\mathbf{r})) - \nabla^2\mathbf{E}(\mathbf{r}) - k^2\,\mathbf{E}(\mathbf{r}) = i\omega\mu\,\mathbf{J}(\mathbf{r})\,.$$

Since $k^2 = \omega^2\mu\varepsilon$ we can write the above equation as

$$-\frac{i\omega\mu}{k^2}\nabla(\nabla\cdot\mathbf{J}(\mathbf{r})) - \nabla^2\mathbf{E}(\mathbf{r}) - k^2\,\mathbf{E}(\mathbf{r}) = i\omega\mu\,\mathbf{J}(\mathbf{r})$$

or

$$\boxed{\left[\nabla^2 + k^2\right]\,\mathbf{E}(\mathbf{r}) = -i\omega\mu\left[\frac{1}{k^2}\nabla(\nabla\cdot\mathbf{J}(\mathbf{r})) + \mathbf{J}(\mathbf{r})\right]\,.} \tag{3.28}$$

This equation has the form of the scalar wave equation

$$\left[\nabla^2 + k^2\right]\,\varphi(\mathbf{r}) = s(\mathbf{r})\,. \tag{3.29}$$

The only difference is that (3.28) consists of three scalar wave equations and the source term is given by

$$\mathbf{s}(\mathbf{r}) := -i\omega\mu\left[\frac{1}{k^2}\nabla(\nabla\cdot\mathbf{J}(\mathbf{r})) + \mathbf{J}(\mathbf{r})\right]\,. \tag{3.30}$$

We have already seen the three-dimensional Green's function for equation (3.29) in the presence of a source at the origin in equation (3.18). When the source is not at the origin but at the location $\mathbf{r}_0$, the Green's function becomes

$$G(\mathbf{r},\mathbf{r}_0) = \frac{e^{ik\|\mathbf{r}-\mathbf{r}_0\|}}{4\pi\|\mathbf{r}-\mathbf{r}_0\|}\,.$$

Therefore, using the Green's function method, we can find the solution of the scalar wave equation (3.29) (see Arfken and Weber (2005) Chapters 9 and 10 and Chew (1995) pp. 24–28 for details) as

$$\varphi(\mathbf{r}) = -\frac{1}{4\pi}\int\frac{e^{ik\|\mathbf{r}-\mathbf{r}'\|}}{\|\mathbf{r}-\mathbf{r}'\|}s(\mathbf{r}')\,\mathrm{d}r'\,. \tag{3.31}$$

In an analogous manner we can find the solution of (3.28), and we get

$$\boxed{\mathbf{E}(\mathbf{r}) = \frac{i\omega\mu}{4\pi}\int\frac{e^{ik|\mathbf{r}-\mathbf{r}'|}}{|\mathbf{r}-\mathbf{r}'|}\left[\frac{1}{k^2}\nabla(\nabla\cdot\mathbf{J}(\mathbf{r}')) + \mathbf{J}(\mathbf{r}')\right]\,\mathrm{d}\mathbf{r}'\,.} \tag{3.32}$$

For *electric dipole fields*, if one has a point current source directed in the $\mathbf{n}$-direction ($\mathbf{n}$ is a unit vector), then the current density is given by

$$\mathbf{J}(\mathbf{r},\mathbf{r}') = I\ell\,\delta(\mathbf{r}-\mathbf{r}')\mathbf{n}(\mathbf{r}') \tag{3.33}$$

where I is the current, ℓ is the length of the moment arm, and $I\ell$ is the current dipole moment, i.e., as $\ell \to 0$ and $I \to \infty$, $I\ell$ remains constant, and δ is the Dirac delta function. Note that the direction $\mathbf{n}$ is usually assumed fixed. If the origin is taken at the point $\mathbf{r}'$, we get

$$\mathbf{J}(\mathbf{r}) = I\ell\,\delta(\mathbf{r})\mathbf{n} \qquad \text{with} \quad \delta(\mathbf{r}) = \delta(x)\delta(y)\delta(z)\,. \tag{3.34}$$

Plugging (3.34) into (3.32) gives

$$\mathbf{E}(\mathbf{r}) = \frac{i\omega\mu I\ell}{4\pi}\int \frac{e^{ikr}}{r}\left[\frac{1}{k^2}\nabla[\nabla\cdot\{\mathbf{n}\delta(\mathbf{r}')\}] + \mathbf{n}\delta(\mathbf{r}')\right]d\mathbf{r}'$$

or

$$\boxed{\mathbf{E}(\mathbf{r}) = \frac{i\omega\mu I\ell}{4\pi}\left[\frac{1}{k^2}\nabla(\nabla\cdot\mathbf{n}) + \mathbf{n}\right]\frac{e^{ikr}}{r}\,.} \tag{3.35}$$

Also, from $\nabla\times\mathbf{E} = i\omega\mu\mathbf{H}$ and using the identity $\nabla\times\nabla\mathbf{E} = 0$, the magnetic field is given by

$$\boxed{\mathbf{H}(r) = \frac{I\ell}{4\pi}\nabla\times\left(\mathbf{n}\frac{e^{ikr}}{r}\right)\,.} \tag{3.36}$$

Substituting the Weyl identity (3.24) into these expressions gives formulae for $\mathbf{E}$ and $\mathbf{H}$ in terms of plane waves. Equivalent expressions in terms of the electrostatic dipole moment can be found in Jackson (1999) (p. 410). The plane wave expansion of the electric field (3.35) can therefore be expressed as (see Pendry (2000) for an example)

$$\mathbf{E}(\mathbf{r}) = \int_{-\infty}^{\infty}\int_{-\infty}^{\infty}\mathbf{E}_\mu(k_x,k_y)\,e^{ik_xx+ik_yy+ik_z|z|}\,dk_x\,dk_y \tag{3.37}$$

where we make explicit the fact that k_z is depends on k_x and k_y, and

$$\mathbf{E}_\mu(k_x,k_y) = -\frac{\omega\mu I\ell}{8\pi^2 k_z}\left[\frac{1}{k^2}\nabla(\nabla\cdot\mathbf{n}) + \mathbf{n}\right]\,.$$

Note that in the above expression k is constant and not integrated.

3.2 Single scattering from spheres

We have looked at the expansion of spherical waves into plane waves. Let us now investigate the situation of the expansion of plane waves into concentric spherical waves. Such an expansion is useful when dealing with scattering problems involving spherical objects.

We start with the observation that the Legendre polynomials[¶] $P_n(\cos\theta)$ are orthogonal and form a complete set. This implies that a function can be expressed as a linear combination of these polynomials and we can assume that the resulting series converges in the interval $[-1,1]$ (see, for example, Arfken and Weber (2005) p. 757). Then we can write a plane wave of unit amplitude as

$$e^{ikx_3} = e^{ikr\cos\phi} = \sum_{n=0}^{\infty} a_n P_n(\cos\phi) . \tag{3.38}$$

We can find the coefficients a_n using the orthogonality of the Legendre polynomials. Multiplying the above equation with $P_m(\cos\phi)$, and integrating from -1 to 1, leads to

$$a_n = \frac{2n+1}{2}\left[\int_{-1}^{1} e^{ikr\cos\phi} P_n(\cos\phi)\,\mathrm{d}(\cos\phi)\right] .$$

It can be shown that the above integral evaluates to $2i^n j_n(kr)$ (see, for example, NIST (2010) eq. 10.54.2) where $j_n(x)$ is a spherical Bessel function. Therefore, we have

$$a_n = (2n+1)i^n j_n(kr) \qquad \text{where} \quad j_n(kr) = \sqrt{\frac{\pi}{2kr}} J_{n+1/2}(kr)$$

and where $J_n(kr)$ is the Bessel's function of the first kind. The expansion of the plane wave in equation (3.38) can therefore be written in terms of spherical waves as

$$\boxed{e^{ikx_3} = e^{ikr\cos\phi} = \sum_{n=0}^{\infty} i^n (2n+1) j_n(kr) P_n(\cos\phi) .} \tag{3.39}$$

The above representation is appropriate for incoming waves and could have been derived starting with the spherical Bessel functions as a basis set.

3.2.1 Scattering of acoustic waves from a liquid sphere

Consider a plane harmonic acoustic wave incident from a fluid medium (bulk modulus κ_1 and density ρ_1) onto a fluid sphere (bulk modulus κ_2 and density ρ_2) as shown in Figure 3.3. This situation can represent an acoustic lens.[‖]

[¶]Legendre polynomials solve Legendre's equation which can be expressed as

$$\frac{1}{\sin\theta}\frac{d}{d\theta}\left[\sin\theta\frac{dP_n(\cos\theta)}{d\theta}\right] + n(n+1)P_n(\cos\theta) = 0 .$$

[‖]Our discussion of acoustic waves incident upon a fluid sphere is based on Skudrzyk (1971).

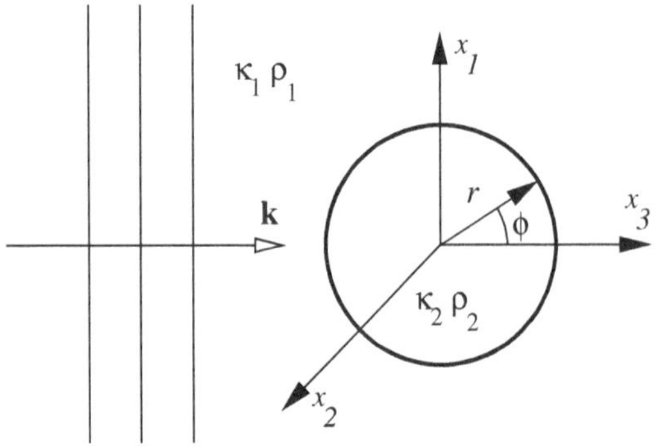

FIGURE 3.3
Plane acoustic wave incident on a liquid sphere with $\mathbf{k} = k_1\mathbf{e}_1 = (\omega/c_1)\mathbf{e}_1$.

Let the plane incident wave be of unit amplitude and let it be derived from the potential φ_i. Using equation (3.39) we can write this potential as

$$\varphi_i = e^{ik_1 r\cos\phi} = \sum_{n=0}^{\infty} i^n(2n+1)j_n(k_1 r)P_n(\cos\phi)$$

where $k_1 = \omega/c_1 = \omega/\sqrt{\kappa_1/\rho_1}$. The scattered waves and the waves that are refracted and reflected inside the sphere must satisfy the Helmholtz equation (1.52)

$$\nabla^2\varphi + k^2\varphi = 0\,.$$

We can show that solutions to the above equation in spherical coordinates, that take the symmetry of the problem into account, have the form

$$\varphi = \sum_{n=0}^{\infty} C_n z_n(kr)P_n(\cos\phi)$$

where C_n are unknown constants and z_n are spherical Bessel or Hankel functions of various kinds.

Then, for the outgoing scattered waves from the sphere to have the right behavior at large distances we can express them in terms of spherical Hankel functions, i.e.,

$$\varphi_s = \sum_{n=0}^{\infty} A_n h_n(k_1 r)P_n(\cos\phi) \qquad \text{where} \quad h_n(z) := h_n^{(1)}(z) = \sqrt{\frac{\pi}{2z}}H_{n+1/2}^{(1)}(z)$$

and A_n are unknown coefficients. $H_n^{(1)}(z)$ is the Hankel function of the first kind. The waves refracted and reflected inside the sphere have to be bounded at the origin and are assumed to have the form

$$\varphi_q = \sum_{n=0}^{\infty} B_n j_n(k_2 r)P_n(\cos\phi)$$

where B_n are unknown coefficients, j_n is the spherical Bessel function, and $k_2 = \omega/c_2 = \omega/\sqrt{\kappa_2/\rho_2}$. Without loss of generality, we can choose the coefficients A_n and B_n such that $A_n = i^n(2n+1)a_n$ and $B_n = i^n(2n+1)b_n$ where a_n and b_n are scaled coefficients. A ray representation of the situation is shown in Figure 3.4.

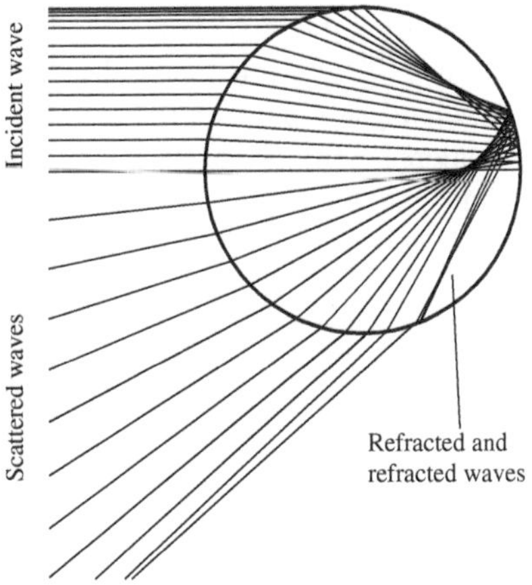

FIGURE 3.4

Ray representation of the incident, scattered, and internal waves for a spherical fluid acoustic lens. Of the scattered waves only the backscattered portion is shown.

The boundary conditions at the surface of the sphere are continuity of pressure and the radial component of velocity. Recall from (2.20) that the pressure is given by $p = -i\omega\rho\varphi$. Also, the gradient of a scalar in spherical coordinates (r, θ, ϕ) is

$$\nabla\varphi = \frac{\partial\varphi}{\partial r}\mathbf{e}_r + \frac{1}{r\sin\phi}\frac{\partial\varphi}{\partial\theta}\mathbf{e}_\theta + \frac{1}{r}\frac{\partial\varphi}{\partial\phi}\mathbf{e}_\phi \,.$$

Hence the radial component of the velocity is $v = -\partial\varphi/\partial r$. The continuity conditions then imply that at $r = a$ (a is the radius of the sphere)

$$\begin{aligned} p_i + p_s = p_q \quad &\Longrightarrow \quad \rho_1(\varphi_i + \varphi_s) = \rho_2\varphi_q \\ v_i + v_s = v_q \quad &\Longrightarrow \quad \frac{\partial}{\partial r}(\varphi_i + \varphi_s) = \frac{\partial\varphi_q}{\partial r} \,. \end{aligned}$$

Equating components for each value of n we have

$$\begin{aligned} \rho_1\left[j_n(k_1a) + a_n h_n(k_1a)\right] &= \rho_2 b_n j_n(k_2a) \\ k_1\left[j_n{}'(k_1a) + a_n h_n{}'(k_1a)\right] &= k_2 b_n j_n{}'(k_2a) \,. \end{aligned}$$

Define $\xi := \rho_2/\rho_1$ and $\eta := k_2/k_1 = c_1/c_2$. Solving the two equations for a_n and b_n

gives

$$a_n = -\frac{\eta j_n(k_1 a) j_n'(k_2 a) - \xi j_n'(k_1 a) j_n(k_2 a)}{\eta h_n(k_1 a) j_n'(k_2 a) - \xi h_n'(k_1 a) j_n(k_2 a)} \quad \text{and}$$

$$b_n = -\frac{i/(k_1 a)^2}{\eta h_n(k_1 a) j_n'(k_2 a) - \xi h_n'(k_1 a) j_n(k_2 a)}.$$

where we have used the identity $h_n(k_1a)j_n'(k_1a) - h_n'(k_1a)j_n(k_1a) = -i/(k_1a)^2$. Note that if the sphere is made of the same material as the medium of incidence, i.e., if $\rho_1 = \rho_2$ and $k_1 = k_2$, then $\xi = 1$ and $\eta = 1$. Therefore we have $a_n = 0$, which means that no waves are scattered from the sphere.

Substituting the coefficients into the expressions for the scattered pressure and the pressure inside the sphere, we get

$$\begin{aligned} p_s &= i\omega\rho_1 \sum_{n=0}^{\infty} \frac{i^n(2n+1)h_n(k_1 r)P_n(\cos\phi)[\eta j_n(k_1 a) j_n'(k_2 a) - \xi j_n'(k_1 a) j_n(k_2 a)]}{\eta h_n(k_1 a) j_n'(k_2 a) - \xi h_n'(k_1 a) j_n(k_2 a)} \\ p_q &= -\frac{\omega\rho_2}{k_1^2 a^2} \sum_{n=0}^{\infty} \frac{i^n(2n+1) j_n(k_2 r) P_n(\cos\phi)}{\eta h_n(k_1 a) j_n'(k_2 a) - \xi h_n'(k_1 a) j_n(k_2 a)}. \end{aligned} \tag{3.40}$$

Similarly, the radial components of the velocities are given by

$$\begin{aligned} v_s &= k_1 \sum_{n=0}^{\infty} \frac{i^n(2n+1)h_n'(k_1 r)P_n(\cos\phi)[\eta j_n(k_1 a) j_n'(k_2 a) - \xi j_n'(k_1 a) j_n(k_2 a)]}{\eta h_n(k_1 a) j_n'(k_2 a) - \xi h_n'(k_1 a) j_n(k_2 a)} \\ v_q &= \frac{ik_2}{k_1^2 a^2} \sum_{n=0}^{\infty} \frac{i^n(2n+1) j_n'(k_2 r) P_n(\cos\phi)}{\eta h_n(k_1 a) j_n'(k_2 a) - \xi h_n'(k_1 a) j_n(k_2 a)}. \end{aligned} \tag{3.41}$$

The magnitude of the normalized radial pressure distribution (p_q/p_i) along $\phi = 0$ from the center of the sphere and the magnitude of the normalized angular pressure distribution at $r/a = 0.56$ are shown in Figure 3.5. The peak pressure shows the point at which the wave is focused inside the sphere. Note that dispersion will be observed when the refractive index is a function of frequency.

3.2.2 Acoustic scattering cross-section

Several measures are used to quantify the amount of scattering due to an object. A widely used measure is the *scattering cross-section* which can be defined at the ratio of the total energy scattered per unit time to the amount of energy delivered by the incident wave per unit time per unit area. The scattering cross-section is therefore an area (m^2 in SI units) that measures the relative intensity of the scattered wave with respect to the incident wave. The scattered energy is measured by summing the acoustic power flow across the surface of a sphere of radius b centered on the object and outside its surface. For a plane incident wave, the incident power is calculated over a unit area normal to the direction of propagation of the wave.

Let us now determine the scattering cross-section for the case where an acoustic plane wave is incident on a sphere. Consider a sphere of radius b centered on the

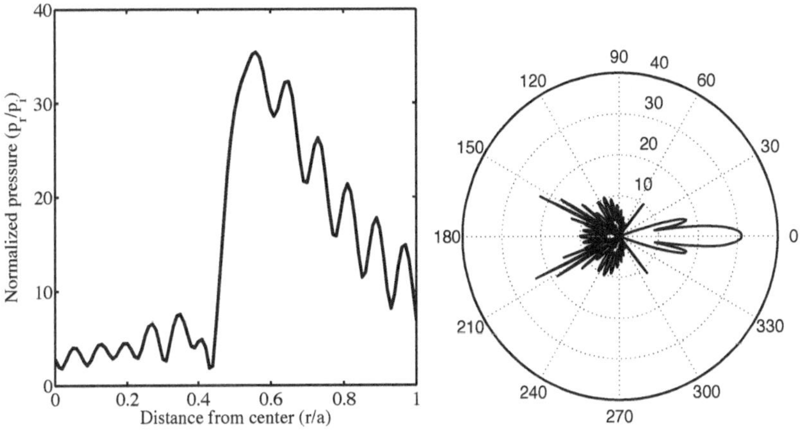

FIGURE 3.5
Normalized radial (left) and angular (right) pressure distribution inside the sphere. The values used in these plots are $n_2 = c_1/c_2 = 2.25$, $\rho_2/\rho_1 = n_2^2$, and $a/\lambda = 3.05$ where the wavelength $\lambda = 2\pi c_1/\omega$, $c_1 = 1550$m/s, and ω is the frequency. The polar plot is for $r = 0.56$ and shows $20\log_{10}(p_q/p_i)$ as a function of ϕ.

sphere of radius a where $b > a$. The time rate at which energy is carried away by the scattered wave is given by

$$\mathcal{P}_s|_{r=b} = \int_{\Gamma_b} p_s v_s \,\mathrm{d}\Gamma_b = 2\pi b^2 \int_0^{\pi} p_s v_s \sin\phi \,\mathrm{d}\phi$$

where Γ_b is the surface of the sphere of radius b, p_s is the pressure in the scattered wave, and v_s is the radial velocity of the scattered wave. Since only the magnitudes of p_s and v_s contribute to the energy, we express the above integral as

$$\mathcal{P}_s|_{r=b} = 2\pi b^2 \int_0^{\pi} \tfrac{1}{2}(p_s\overline{v_s} + \overline{p_s}v_s)\sin\phi\,\mathrm{d}\phi = \pi b^2 \int_{-1}^{1} (p_s\overline{v_s} + \overline{p_s}v_s)\,\mathrm{d}(\cos\phi)$$

where $\overline{z}$ is the complex conjugate of z. Let us integrate each of the terms inside the integral after plugging in the expression for ϕ_s. Then we have, using $\zeta = \cos\phi$,

$$\begin{aligned}\int_{-1}^{1} p_s\overline{v_s}\mathrm{d}\zeta &= \int_{-1}^{1}\left[-i\omega\rho_1 \sum_{n=0}^{\infty} A_n h_n(k_1 b)P_n(\zeta)\right]\left[-k_1 \sum_{m=0}^{\infty} \overline{A_m}\,\overline{h_m}'(k_1 b)P_m(\zeta)\right]\mathrm{d}\zeta \\ &= i\omega\rho_1 k_1 \sum_{n=0}^{\infty}\sum_{m=0}^{\infty} A_n\overline{A_m}h_n(k_1 b)\overline{h_m}'(k_1 b)\int_{-1}^{1} P_n(\zeta)P_m(\zeta)\mathrm{d}\zeta\end{aligned}$$

where we have assumed real k_1 and ρ_1 for simplicity. From the orthogonality of the Legendre polynomials we know that

$$\int_{-1}^{1} P_n(\zeta)P_m(\zeta)\mathrm{d}\zeta = \frac{2}{2n+1}\delta_{mn}\,.$$

Therefore,

$$\int_{-1}^{1} p_s\overline{v_s}\,\mathrm{d}\zeta = 2i\omega\rho_1 k_1 \sum_{n=0}^{\infty} \frac{1}{2n+1} A_n\overline{A_n}\, h_n(k_1 b)\,\overline{h_n}'(k_1 b)\,.$$

Similarly, we can show that

$$\int_{-1}^{1} \overline{p_s} v_s\,\mathrm{d}\zeta = -2i\omega\rho_1 k_1 \sum_{n=0}^{\infty} \frac{1}{2n+1} \overline{A_n} A_n\, \overline{h_n}(k_1 b)\, h_n'(k_1 b)\,.$$

Adding the terms and using $h_n(z)\overline{h_n}'(z) - \overline{h_n}(z)h_n'(z) = -2i/z^2$ gives

$$\mathcal{P}_s\big|_{r=b} = \frac{4\pi\omega\rho_1}{k_1} \sum_{n=0}^{\infty} \frac{\|A_n\|^2}{2n+1}$$

where $\|A_n\|^2 = A_n\overline{A_n}$ is the square of the magnitude of A_n. For the incident wave, the energy flux per unit area is

$$\mathcal{P}_i = \tfrac{1}{2}(p_i\overline{v_i} + \overline{p_i}v_i) = i\omega\rho_1 k_1$$

since $\varphi_i = \exp(ik_1x_1)$. Therefore the scattering cross-section is

$$\boxed{\gamma_{cs} = \frac{4\pi}{k_1^2} \sum_{n=0}^{\infty} \frac{\|A_n\|^2}{2n+1}\,.} \tag{3.42}$$

Scattering cross-sections for fluid spheres with positive and negative materials are shown in Figure 3.6. Notice that the scattering cross-section appears to settle to a constant value for wavelengths that are large compared to the radius of the sphere.

3.2.3 Scattering by an elastic sphere in an elastic medium

Let us now consider an isotropic elastic medium with elastic moduli λ_1 and μ_1 and density ρ_1. Let there be a embedded sphere of a different material with parameters λ_2, μ_2, and ρ_2 that is perfectly bonded to the host medium. Recall that in the general case there are P-waves and two types of S-waves that can be propagated in the two media. In the particular case when the inhomogeneity is a sphere, we can only have SV-waves but no SH-waves because of symmetry. This problem has been solved using several techniques. We follow the approach taken by Ying and Truell (1956) and Brill and Gaunaurd (1987). Alternative approaches can be found in Willis (1980) and Korneev and Johnson (1993).

Recall that the displacements in an isotropic linear elastic medium can be derived from scalar and vector potentials as

$$\mathbf{u} = \boldsymbol{\nabla}\varphi + \boldsymbol{\nabla}\times\boldsymbol{\psi}\,.$$

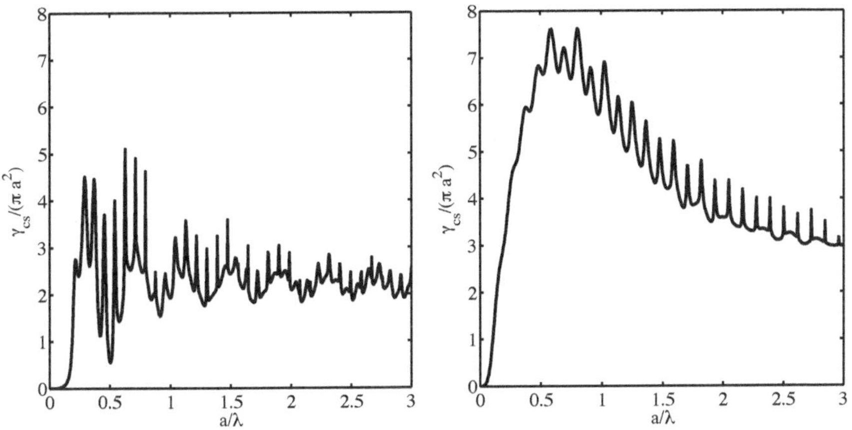

FIGURE 3.6
Normalized scattering cross-sections as a function of sphere radius. The incident medium has a sound speed of 1550 m/s and a density of 1000 kg/m^3. The plot on the left is for a sphere with refractive index $n_2 = 2.25$ with respect to the incident medium and a density of 5060 kg/m^3. The figure on the right is for a sphere made of a negative material with $n_2 = -1$ and $\rho_2 = -1000$ kg/m^3. The radius of the sphere is a and λ is the wavelength of the incident plane wave.

Due to the spherical symmetry of the problem there is no θ dependence, i.e., $\varphi = \varphi(r,\phi)$ and $\boldsymbol{\psi} = \boldsymbol{\psi}(r,\phi)$. The spherical symmetry of the scattered waves and the waves inside the sphere can be used to represent the vector potential $\boldsymbol{\psi}$ in terms of a Debye scalar potential $\Pi(r,\phi,t)$ such that $\boldsymbol{\psi} = \boldsymbol{\nabla} \times (\Pi r \mathbf{e}_r)$. The potential is chosen such that

$$\nabla^2 \Pi + k_s^2 \Pi = 0 \quad \Longrightarrow \quad \boldsymbol{\nabla} \cdot (\boldsymbol{\nabla}\boldsymbol{\psi})^T + k_s^2 \boldsymbol{\psi} = \mathbf{0}$$

where $k_s^2 = \omega^2/c_s^2$. Then**

$$\boldsymbol{\nabla}\varphi = \frac{\partial \varphi}{\partial r}\mathbf{e}_r + \frac{1}{r}\frac{\partial \varphi}{\partial \phi}\mathbf{e}_\phi \; ; \quad \boldsymbol{\psi} = -\frac{\partial \Pi}{\partial r}\mathbf{e}_\theta \; ; \quad \boldsymbol{\nabla} \cdot \boldsymbol{\psi} = 0 \quad \text{and}$$

$$\boldsymbol{\nabla} \times \boldsymbol{\psi} = -\frac{1}{r\sin\phi}\left[\cos\phi \frac{\partial \Pi}{\partial \phi} + \sin\phi \frac{\partial^2 \Pi}{\partial \phi^2}\right]\mathbf{e}_r + \frac{1}{r}\left[\frac{\partial \Pi}{\partial \phi} + r\frac{\partial^2 \Pi}{\partial r \partial \phi}\right]\mathbf{e}_\phi$$

**In components with respect to a spherical coordinate basis $(\mathbf{e}_r, \mathbf{e}_\theta, \mathbf{e}_\phi)$ we have

$$\boldsymbol{\nabla}\varphi = \frac{\partial \varphi}{\partial r}\mathbf{e}_r + \frac{1}{r\sin\phi}\frac{\partial \varphi}{\partial \theta}\mathbf{e}_\theta + \frac{1}{r}\frac{\partial \varphi}{\partial \phi}\mathbf{e}_\phi$$

$$\boldsymbol{\nabla} \times \mathbf{v} = \frac{1}{r\sin\phi}\left[\frac{\partial}{\partial \phi}(\sin\phi\, v_\theta) - \frac{\partial v_\phi}{\partial \theta}\right]\mathbf{e}_r + \frac{1}{r}\left[\frac{\partial (r v_\phi)}{\partial r} - \frac{\partial v_r}{\partial \phi}\right]\mathbf{e}_\theta + \frac{1}{r}\left[\frac{1}{\sin\phi}\frac{\partial v_r}{\partial \theta} - \frac{\partial (r v_\theta)}{\partial r}\right]\mathbf{e}_\phi .$$

We can use the Helmholtz equation for Π to eliminate derivatives with respect to ϕ in the equation for the radial displacement. The displacement components can then be expressed in the simplified form

$$u_r = \frac{\partial\varphi}{\partial r} + \frac{\partial^2}{\partial r^2}(\Pi r) + k_s^2 \Pi r \, ; \quad u_\phi = \frac{1}{r}\frac{\partial}{\partial\phi}\left[\varphi + \frac{\partial}{\partial r}(\Pi r)\right] . \tag{3.43}$$

Recall that the stress in an isotropic and homogeneous elastic medium is given by

$$\boldsymbol{\sigma} = \lambda(\boldsymbol{\nabla}\cdot\mathbf{u})\boldsymbol{1} + \mu[\boldsymbol{\nabla}\mathbf{u} + (\boldsymbol{\nabla}\mathbf{u})^T] .$$

At the surface of the sphere we require the continuity of normal and shear tractions (and also radial and tangential displacements). Therefore the stress components of interest are σ_{rr}, $\sigma_{r\phi}$, and $\sigma_{r\theta}$ which we find to be

$$\begin{aligned}
\sigma_{rr} &= \frac{\lambda}{r^2}\left[\cot\phi\frac{\partial\varphi}{\partial\phi} + \frac{\partial^2\varphi}{\partial\phi^2} + 2r\frac{\partial\varphi}{\partial r} + r^2\frac{\partial^2\varphi}{\partial r^2}\right] \\
&\quad + \frac{\mu}{r^2}\left[2\left\{\cot\phi\frac{\partial\Pi}{\partial\phi} + \frac{\partial^2\Pi}{\partial\phi^2} - r\left(\cot\phi\frac{\partial^2\Pi}{\partial r\partial\phi} + \frac{\partial^3\Pi}{\partial r\partial\phi^2}\right)\right\} + 2r^2\frac{\partial^2\varphi}{\partial r^2}\right] \\
\sigma_{r\phi} &= \frac{\mu}{r^2}\left[-\frac{\partial\varphi}{\partial\phi} + \cot^2\phi\frac{\partial\Pi}{\partial\phi} - \cot\phi\frac{\partial^2\Pi}{\partial\phi^2} - \frac{\partial^3\Pi}{\partial\phi^3} + r\left(2\frac{\partial^2\varphi}{\partial r\partial\phi} + \frac{\partial^2\Pi}{\partial r\partial\phi} + r\frac{\partial^3\Pi}{\partial r^2\partial\phi}\right)\right] \\
\sigma_{r\theta} &= 0 .
\end{aligned}$$

We can use the Helmholtz equation for φ, i.e., $\nabla^2\varphi + k_p^2\varphi = 0$, the Helmholtz equation for Π, and $\boldsymbol{\nabla}\cdot(\boldsymbol{\nabla}\times\boldsymbol{\psi}) = 0$ to simplify the above expressions into the form

$$\begin{aligned}
\sigma_{rr} &= -\lambda k_p^2\varphi + 2\mu\left[\frac{\partial^2}{\partial r^2}\left(\varphi + \frac{\partial}{\partial r}(\Pi r)\right) + k_s^2\frac{\partial}{\partial r}(\Pi r)\right] \\
\sigma_{r\phi} &= \frac{2\mu}{r}\frac{\partial}{\partial r}\left[\frac{\partial}{\partial\phi}\left(\varphi + \frac{\partial}{\partial r}(\Pi r)\right)\right] - \frac{\mu}{r^2}\frac{\partial}{\partial\phi}\left(\varphi + \frac{\partial}{\partial r}(\Pi r)\right) + \mu k_s^2\frac{\partial\Pi}{\partial\phi} .
\end{aligned} \tag{3.44}$$

Let us consider the case when the incident wave is a plane harmonic P-wave. Then the displacement $\mathbf{u}^i = \boldsymbol{\nabla}\varphi_i$ can be derived from the potential

$$\varphi_i = e^{ik_{1p}r\cos\phi} = \sum_{n=0}^{\infty} i^n(2n+1)\,j_n(k_{1p}r)\,P_n(\cos\phi)$$

where $k_{1p} = \omega/c_{1p} = \omega/\sqrt{(\lambda_1 + 2\mu_1)/\rho_1}$ and $\Pi_i = 0$. Using the ideas developed for the acoustic wave problem, we realize that the scattered waves have the form

$$\varphi_s = \sum_{n=0}^{\infty} i^n(2n+1)A_n h_n(k_{1p}r)\,P_n(\cos\phi)$$

$$\Pi_s = \sum_{n=0}^{\infty} i^n(2n+1)B_n h_n(k_{1s}r)\,P_n(\cos\phi)$$

where $k_{1s} = \omega/c_{1s} = \omega/\sqrt{\mu_1/\rho_1}$. Similarly, for the waves inside the sphere, we have

$$\varphi_q = \sum_{n=0}^{\infty} i^n (2n+1) C_n j_n(k_{2p} r) P_n(\cos\phi)$$

$$\Pi_q = \sum_{n=0}^{\infty} i^n (2n+1) D_n j_n(k_{2s} r) P_n(\cos\phi)$$

where $k_{2p} = \omega/c_{2p} = \omega/\sqrt{(\lambda_2 + 2\mu_2)/\rho_2}$ and $k_{2s} = \omega/c_{2s} = \omega/\sqrt{\mu_2/\rho_2}$. Then

$$\lambda_1 = \omega^2 \rho_1 \left(\frac{1}{k_{1p}^2} - \frac{2}{k_{1s}^2} \right) ; \; \lambda_2 = \omega^2 \rho_2 \left(\frac{1}{k_{2p}^2} - \frac{2}{k_{2s}^2} \right) ; \; \mu_1 = \frac{\omega^2 \rho_1}{k_{1s}^2} ; \; \mu_2 = \frac{\omega^2 \rho_2}{k_{2s}^2} .$$

Now we can apply the continuity conditions at the surface of the sphere ($r = a$) which are

$$u_r^i + u_r^s = u_r^q \; ; \; u_\phi^i + u_\phi^s = u_\phi^q \; ; \; \sigma_{rr}^i + \sigma_{rr}^s = \sigma_{rr}^q \; ; \; \sigma_{r\phi}^i + \sigma_{r\phi}^s = \sigma_{r\phi}^q .$$

Using these conditions we get a system of four equations in A_n, B_n, C_n, and D_n which can be solved for these coefficients. Detailed expressions can be found in Ying and Truell (1956) and Brill and Gaunaurd (1987).

The scattering cross-section can be found using the procedure that we used in the acoustic case. For a time harmonic wave, the rate of transport of energy across a sphere of radius $b > a$ is given by

$$\mathcal{P} = \pi i \omega b^2 \int_0^\pi [\sigma_{rr}^s \bar{u}_r^s + \sigma_{r\theta}^s \bar{u}_\theta^s + \sigma_{r\phi}^s \bar{u}_\phi^s - \bar{\sigma}_{rr}^s u_r^s + \bar{\sigma}_{rr}^s u_r^s + \bar{\sigma}_{rr}^s u_r^s]_{r=b} \sin\phi \, d\phi .$$

We can then show that the scattering cross-section has the form

$$\gamma_{cs} = 4\pi \sum_{n=0}^{\infty} (2n+1) \left[\|A_n\|^2 + n(n+1) \frac{k_{s1}}{k_{p1}} \|B_n\|^2 \right] .$$

3.2.4 Scattering of electromagnetic radiation from a sphere

In this section we will consider the scattering of a plane electromagnetic wave incident on a sphere of refractive index n. The solution of this problem was first given by Gustav Mie (1908) and it is called the *Mie solution.* We present a sketch of the solution here. Detailed derivations can be found in Kerker (1969) and Ishimaru (1978).

Consider the sphere shown in Figure 3.7. We set up our coordinate system such that the origin is at the center of the sphere. The sphere has a magnetic permeability of μ and a permittivity ε. The medium outside the sphere has a permittivity ε_0 and a permeability μ_0. The electric field is oriented parallel to the x_1-axis and the x_2-axis points out of the plane of the paper.

Let us now consider the situation where the material inside the sphere is non-magnetic, i.e., it has the magnetic permeability of vacuum. If the medium outside the sphere is also non-magnetic, we may write

$$\mu = \mu_0 \; ; \; \varepsilon = \varepsilon_r \varepsilon_0 = n^2 \varepsilon_0$$

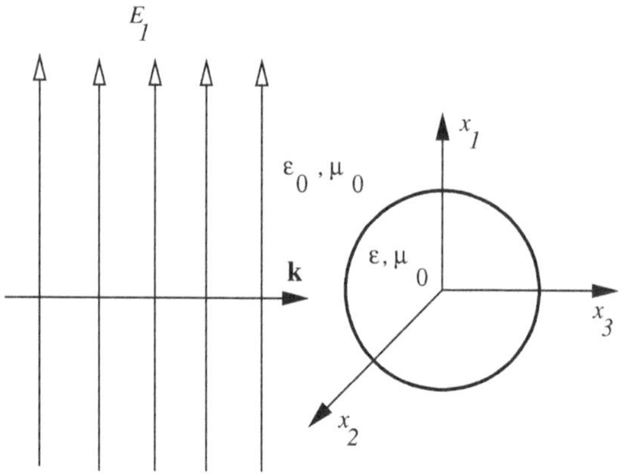

FIGURE 3.7
Scattering of a plane electromagnetic wave from a sphere.

where ε_r is the relative permittivity of the material inside the sphere. The incident plane wave is given by

$$\mathbf{E} = e^{ikx_3}\,\mathbf{e}_1 = e^{ikr\cos\phi}\,\mathbf{e}_1 \tag{3.45}$$

where $\mathbf{e}_1$ is the unit vector in the x_1-direction.

Recall that Maxwell's equations can be expressed in terms of Hertz vector potentials†† as

$$\begin{aligned} \mathbf{E} &= \boldsymbol{\nabla}\times\boldsymbol{\nabla}\times\boldsymbol{\Pi}_e - \mu\,\boldsymbol{\nabla}\times\frac{\partial\boldsymbol{\Pi}_m}{\partial t} \\ \mathbf{H} &= \boldsymbol{\nabla}\times\boldsymbol{\nabla}\times\boldsymbol{\Pi}_m + \varepsilon\,\boldsymbol{\nabla}\times\frac{\partial\boldsymbol{\Pi}_e}{\partial t}\,. \end{aligned} \tag{3.46}$$

For spherically symmetric time harmonic problems, an important class of Hertz vector potentials are those of the form

$$\boldsymbol{\Pi}_e = ur\,\mathbf{e}_r = u\mathbf{r} \quad \text{and} \quad \boldsymbol{\Pi}_m = vr\,\mathbf{e}_r = v\mathbf{r} \tag{3.47}$$

where $\mathbf{e}_r$ is the radial unit vector from the origin in a spherical coordinate system and r is the radial coordinate. The functions $u(r,\theta,\phi)$ and $v(r,\theta,\phi)$ are scalar potentials, called *Debye potentials*, which satisfy the homogeneous wave equations

$$(\nabla^2+k^2)u = 0 \qquad \text{and} \qquad (\nabla^2+k^2)v = 0\,. \tag{3.48}$$

One important result is that every electromagnetic field defined in a source-free region between two concentric spheres can be represented there by two Debye potentials (Wilcox, 1957). Let the time harmonic fields be given by

$$\mathbf{E} = \widehat{\mathbf{E}}\,e^{-i\omega t}\ ;\ \mathbf{H} = \widehat{\mathbf{H}}\,e^{-i\omega t}\ ;\ u = \hat{u}\,e^{-i\omega t}\ ;\ v = \hat{v}\,e^{-i\omega t}\,. \tag{3.49}$$

††Hertz vector potentials have been discussed on p. 30. See equation (1.75).

Plugging these into (3.46), using the Debye potentials, and dropping the hats gives the Maxwell equations at fixed frequency:

$$\begin{aligned} \mathbf{E} &= \nabla \times \nabla \times (u\mathbf{r}) + i\omega\mu\, \nabla \times (v\mathbf{r}) \\ \mathbf{H} &= \nabla \times \nabla \times (v\mathbf{r}) - i\omega\varepsilon\, \nabla \times (u\mathbf{r})\,. \end{aligned} \tag{3.50}$$

The electric and magnetic fields have components in all three coordinate directions. Following the procedure that we have used for acoustic and elastic problems, we will split the potentials u and v outside the sphere into incident and scattered fields. Thus, $u = u_i + u_s$ and $v = v_i + v_s$, where the subscript i indicates an incident field and the subscript s indicates a scattered field. Inside the sphere, the potentials are denoted by $u = u_r$ and $v = v_r$, where the subscript r indicates a refracted + reflected field. We require that these potentials satisfy wave equations of the form given in (3.48), i.e.,

$$\begin{aligned} (\nabla^2 + k^2)u_i &= 0 \qquad \text{and} \qquad (\nabla^2 + k^2)v_i = 0 \\ (\nabla^2 + k^2)u_s &= 0 \qquad \text{and} \qquad (\nabla^2 + k^2)v_s = 0 \\ (\nabla^2 + k^2n^2)u_r &= 0 \qquad \text{and} \qquad (\nabla^2 + k^2n^2)v_r = 0\,. \end{aligned}$$

Since each of these satisfies a scalar wave equation, we can express each potential in terms of spherical harmonics. In particular, the Debye potentials associated with the incident field given in equation (3.45) can be expressed using spherical harmonics as (see Ishimaru (1978) for details)

$$\begin{aligned} ru_i &= \frac{1}{k^2}\sum_{n=1}^{\infty} \frac{i^{n-1}(2n+1)}{n(n+1)}\, \psi_n(kr)\, P_n^1(\cos\phi)\, \cos\theta \\ rv_i &= \frac{1}{\eta k^2}\sum_{l=1}^{\infty} \frac{i^{n-1}(2n+1)}{n(n+1)}\, \psi_n(kr)\, P_n^1(\cos\phi)\, \sin\theta \end{aligned}$$

where $\eta = \sqrt{\mu_0/\varepsilon_0}$, $\psi_n(kr) := kr\,j_n(kr)$ where $j_n(kr)$ is a spherical Bessel function of the first kind, and P_n^1 is an associated Legendre polynomial.[‡‡] The functions $\psi_n(kr)$ are chosen such that they are regular at the origin. The scattered fields have a similar expansion

$$\begin{aligned} ru_s &= \frac{-1}{k^2}\sum_{n=1}^{\infty} \frac{i^{n-1}(2n+1)}{n(n+1)}\, A_n\, \zeta_n(kr)\, P_n^1(\cos\phi)\, \cos\theta \\ rv_s &= \frac{-1}{\eta k^2}\sum_{n=1}^{\infty} \frac{i^{n-1}(2n+1)}{n(n+1)}\, B_n\, \zeta_n(kr)\, P_n^1(\cos\phi)\, \sin\theta \end{aligned}$$

[‡‡]The associated Legendre polynomials solve

$$\frac{d}{dx}\left[(1-x^2)\,\frac{dP_n^m}{dx}\right] + \left[l(l+1) - \frac{m^2}{1-x^2}\right] P_n^m = 0\,.$$

where $\zeta_n(kr) = krh_n(kr)$ and $h_n(kr) = h_n^{(1)}(kr)$ is the spherical Hankel function of the first kind which solves Bessel's equation but decays at infinity. Inside the sphere, the expansion of the fields takes the form

$$ru_r = \frac{1}{k^2 n^2} \sum_{n=1}^{\infty} \frac{i^{n-1}(2n+1)}{n(n+1)} C_n \, \psi_n(knr) \, P_n^1(\cos\phi) \, \cos\theta$$

$$rv_r = \frac{1}{\eta k^2 n^2} \sum_{n=1}^{\infty} \frac{i^{n-1}(2n+1)}{n(n+1)} D_n \, \psi_n(knr) \, P_n^1(\cos\phi) \, \sin\theta \, .$$

We require continuity of the tangential components of the fields at the surface of the sphere. In spherical coordinates, the components of the fields between in a source-free region between two concentric spheres are given by

$$E_r = \left(\frac{\partial^2}{\partial r^2} + k^2\right)(ru) \, ; \qquad H_r = \left(\frac{\partial^2}{\partial r^2} + k^2\right)(rv)$$

$$E_\theta = \frac{1}{r\sin\phi} \frac{\partial^2}{\partial r \partial\theta}(ru) - ik\sqrt{\mu/\varepsilon}\frac{\partial v}{\partial\phi} \, ; \qquad H_\theta = \frac{1}{r\,\sin\phi} \frac{\partial^2}{\partial r \partial\theta}(rv) + ik\sqrt{\mu/\varepsilon}\frac{\partial u}{\partial\phi}$$

$$E_\phi = \frac{1}{r}\frac{\partial^2}{\partial r \partial\phi}(ru) + \frac{ik\sqrt{\mu/\varepsilon}}{\sin\phi}\frac{\partial v}{\partial\theta} \, ; \qquad H_\phi = \frac{1}{r}\frac{\partial^2}{\partial r \partial\phi}(rv) - \frac{ik\sqrt{\mu/\varepsilon}}{\sin\phi}\frac{\partial u}{\partial\theta} \, .$$

It can be shown that for $E_\theta, E_\phi, H_\theta, H_\phi$ (tangential components of **E** and **H**) to be continuous across the surface of the sphere at $r = a$, it is sufficient that $n^2 u$, $\partial/\partial r(ru)$ and $\partial/\partial r(rv)$ be continuous. Applying these conditions, we get

$$\boxed{\begin{aligned} A_n &= \frac{\psi_n(\alpha)\,\psi_n'(\beta) - n\,\psi_n(\beta)\,\psi_n'(\alpha)}{\zeta_n(\alpha)\,\psi_n'(\beta) - n\,\psi_n(\beta)\,\zeta_n'(\alpha)} \\ B_n &= \frac{n\,\psi_n(\alpha)\,\psi_n'(\beta) - \psi_n(\beta)\,\psi_n'(\alpha)}{n\,\zeta_n(\alpha)\,\psi_n'(\beta) - \psi_n(\beta)\,\zeta_n'(\alpha)} \end{aligned}} \tag{3.51}$$

where $\alpha := ka$ and $\beta := kna$. The scattered fields E_θ, E_ϕ far from the sphere are given by

$$E_\theta = -\frac{i\,e^{ikr}}{kr} S_1(\phi)\,\sin\theta \quad \text{and} \quad E_\phi = \frac{i\,e^{ikr}}{kr} S_1(\phi)\,\cos\theta \tag{3.52}$$

where

$$S_1(\phi) = \sum_{n=1}^{\infty} \frac{2n+1}{n(n+1)} [A_n\,\pi_n(\cos\phi) + B_n\,\tau_n(\cos\phi)]$$

$$S_2(\phi) = \sum_{n=1}^{\infty} \frac{2n+1}{n(n+1)} [A_n\,\tau_n(\cos\phi) + B_n\,\pi_n(\cos\phi)]$$

and

$$\pi_n(\cos\phi) := \frac{P_n^1(\cos\phi)}{\sin\phi} \, ; \quad \tau_n(\cos\phi) := \frac{d}{d\phi} P_n^1(\cos\phi) \, . \tag{3.53}$$

Note that the tangential components of **E** fall off as $1/r$ while the radial component falls off as $1/r^2$. The Mie solution was used to explain many features of rainbows that could not be explained by an earlier theory called the Airy theory.

3.3 Multiple scattering

So far we have considered scattering of waves from one obstacle. This situation is called single or independent scattering. This approximation is excellent when the spacing between individual scatterers is large compared to their size and the wavelength of the incident waves. Mie theory, which is based on single scattering, explains most features of rainbows (see Adam (2002) for an excellent review on ray and wave theories of rainbows). However, scattering of light by clouds and that of sound by dense distributions of bubbles is not well described by single scattering. This lacuna is addressed by multiple (or dependent) scattering. The main feature of multiple scattering is that the effect on the phase of waves due to adjacent scatterers is taken into account.*

3.3.1 The self-consistent method

The "self-consistency" assumption is that we can replace the effect due to individual scatterers in a region with a homogeneous *effective field* caused by the interactions between all those scatterers. We can then calculate the effect of a single scatterer on the effective field to get the multiple scattering solution. The self-consistent method has been used in numerous effective medium theories and a detailed description in that context can be found in Milton (2002). In multiple scattering, this approach is also called the Foldy-Lax self-consistent method. Variations of the technique include the Korringa-Kohn-Rostoker (KKR) approximation and the coherent potential approximation (CPA).

The total field (ϕ) is expressed in terms of the incident (ϕ^{inc}) and scattered (ϕ^{sc}) fields as

$$\phi = \phi^{\text{inc}} + \sum_{j=1}^{N} \phi_j^{\text{sc}}$$

where N is the number of scatterers. The effective field experienced by the m-th scatterer is given by

$$\phi_m^{\text{eff}} = \phi - \phi_m^{\text{sc}} = \phi^{\text{inc}} + \sum_{\substack{j=1 \\ j\neq m}}^{N} \phi_j^{\text{sc}}\,.$$

If ϕ_j is the field incident on the j-th scatterer, we can calculate the scattered field if we know the scattering coefficient T_j for that body. For linear problems, we can

*Multiple scattering has a vast literature. Some important contributions to the literature include the early work by Rayleigh (Rayleigh, 1892) and Kasterin, Foldy (Foldy, 1945) and Lax (Lax, 1951). Our brief review of a simple self-consistent approach to multiple scattering is based on Martin (2006). Details of the Green's function-based multiple scattering method pioneered by Foldy can also be found in Martin (2006). Several applications and physical insights on multiple scattering can be found in Sheng (2006).

superpose these scattered fields to get

$$\phi_m^{\text{eff}} = \phi^{\text{inc}} + \sum_{\substack{j=1 \\ j \neq m}}^{N} T_j \phi_j .$$

In principle, we have solved the multiple scattering problem because we can now calculate ϕ_m^{sc} due to the effective field using the techniques discussed in the previous sections. However, as we can guess, it is not obvious how we can find T_j. Much of the literature on multiple scattering is concerned with finding T_j.

A special version of this approach was applied by Rayleigh (Rayleigh, 1892) to a regular array of cylindrical scatterers with circular cross-sections.* Consider the infinite array of spheres shown in Figure 3.8. The spheres are arranged so that their centers are in the $x_1 - x_2$ plane. A plane wave is incident on the array from the x_3-direction. We want to find the scattering coefficient T_j. As usual, we assume that the field satisfies the Helmholtz equation at a fixed frequency ω.

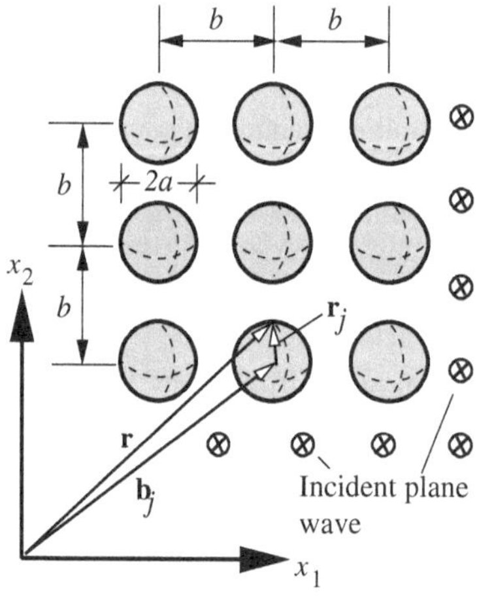

FIGURE 3.8
Multiple scattering from an array of spheres.

Let the incident field be of the form $\phi = e^{ikx_3}$ where k is the wavenumber. Outside the j-th sphere, the total field is†

$$\phi(\mathbf{r}) = \sum_n \left[A_j^n \psi^n(\mathbf{r}_j) + B_j^n \chi^n(\mathbf{r}_j) \right]$$

*This method is sometimes called the Rayleigh multipole method.

†Note that in this relation A^n is a short form of A_m^n where the subscript m indicates the m-th term of the series expansion of ψ^n. A similar interpretation should be applied for B^n.

where ψ^n are the solutions that are regular for all $\mathbf{r}_j$, and χ^n are the solutions that are singular at $\mathbf{r}_j = \mathbf{0}$ and satisfy the appropriate radiation conditions for large r_j (i.e., they correspond to outgoing waves). In the previous sections we have seen examples of ψ and χ which satisfy the Helmholtz equation in spherical coordinates. The solutions are often called multipoles and involve series expansions which have been suppressed here for compactness.

The periodic nature of the geometry makes the problem easier to solve because the neighborhood of each sphere, if we sit on the sphere, looks identical to that of the other spheres. That implies

$$A^n_j = A^n \qquad \text{and} \qquad B^n_j = B^n\,.$$

For simplicity let us assume that $\partial\phi/\partial r = 0$ at $r = a$ for each sphere. This particular boundary condition corresponds to the hard-sphere approximation in acoustics or the no-slip boundary condition in fluid mechanics. Then we have

$$0 = \sum_n \left[A^n \psi^n_{,r}(a) + B^n \chi^n_{,r}(a)\right]$$

where we have taken some liberties with the notation to reduce complication. Since we are using superposition, each sphere can be considered independently and we can write

$$B^n = T^n A^n \qquad \text{where} \qquad T^n := \frac{\psi^n_{,r}(a)}{\chi^n_{,r}(a)}\,. \tag{3.54}$$

We will get other expressions for T^n if we use other boundary conditions. Also, near sphere j, the *effective* incident field must be equal to the sum of the actual incident field and the scattered field from all spheres other than j. Therefore,

$$\sum_n A^n \psi^n(\mathbf{r}_j) = e^{ikx_3} + \sum_n B_n \sum_{\substack{p \\ p\neq j}} \chi^n(\mathbf{r}_p) = e^{ikx_3} + \sum_n T^n A^n \sum_{\substack{p \\ p\neq j}} \chi^n(\mathbf{r}_p)\,.$$

To solve for A^n, it is convenient and often possible to express the incident plane wave and the scattered waves in terms of ψ^n using eigenfunction expansions of the form

$$e^{ikx_3} = \sum_n C^n \psi^n(\mathbf{r}_j) \qquad \text{and} \qquad \chi^n(\mathbf{r}_p) = \sum_q S^{nq}(\mathbf{b}_j - \mathbf{b}_p)\psi^q(\mathbf{r}_j)$$

where $\mathbf{b}_j$ is the center of the j-th sphere. Then we have

$$\sum_n A^n \psi^n(\mathbf{r}_j) = \sum_n C^n \psi^n(\mathbf{r}_j) + \sum_n T^n A^n \sum_{\substack{p \\ p\neq j}} \sum_q S^{nq}(\mathbf{b}_j - \mathbf{b}_p)\psi^q(\mathbf{r}_j)\,.$$

Equating terms gives us the equation for A^n,

$$A^n = C^n + \sum_q T^q A^q \sum_{\substack{p \\ p\neq j}} S^{qn}(\mathbf{b}_j - \mathbf{b}_p)\,. \tag{3.55}$$

Several situations where this general approach can be used can be found in Martin (2006).

3.3.2 Acoustic scattering from multiple circular cylinders

Let us start with the situation where there is one circular cylinder of radius a located at the origin of the coordinate system. We assume that the cylinder can be approximated by a two-dimensional circle. The incident plane wave, a fixed frequency, has the form

$$\phi^{\text{inc}} = Ae^{ik(x_1\cos\theta_i - x_2\sin\theta_i)}$$

where θ_i is the angle of incidence and $k = \|\mathbf{k}\|$ is the wavenumber. If we introduce a polar coordinate system at the center of the circle such that $x_1 = r\cos\theta$ and $x_2 = r\sin\theta$ we can express the above as

$$\phi^{\text{inc}} = Ae^{ikr(\cos\theta\cos\theta_i - \sin\theta\sin\theta_i)} = Ae^{ikr\cos(\theta+\theta_i)} =: Ae^{ikr\cos\alpha}\,.$$

Figure 3.9 shows a schematic of the situation.

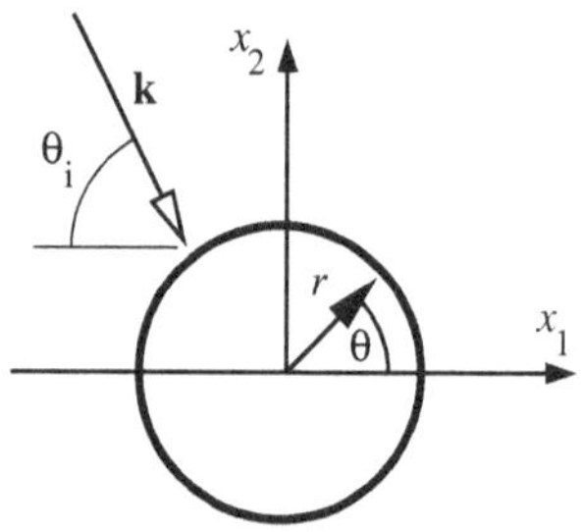

FIGURE 3.9
Scattering from a single circular cylinder.

Using the generating function for the Bessel function of the first kind (J_n) (see NIST (2010)), we can expand the plane wave into cylindrical waves,

$$\phi^{\text{inc}} = AJ_0(kr) + \sum_{m=1}^{\infty} 2i^m AJ_m(kr)\cos(m\alpha)\,.$$

As we have seen before, for reasons relating to the behavior of the scattered waves at ∞, the scattered wave has the form

$$\phi^{\text{sc}} = B_0H_0(kr) + \sum_{m=1}^{\infty} 2i^m B_mH_m(kr)\cos(m\alpha)$$

where H_m are the cylindrical Hankel functions of the first kind. Let us assume that no waves are transmitted through the sphere which implies that the normal velocity is zero on the surface of the cylinder, i.e., $\mathbf{v}\cdot\mathbf{e}_r = -\boldsymbol{\nabla}\phi\cdot\mathbf{e}_r = \partial\phi/\partial r = 0$ where $\mathbf{e}_r$ is the radial basis vector. Applying the boundary condition at $r = a$ gives

$$[AJ_0'(ka) + B_0H_0'(ka)] + \sum_{m=1}^{\infty} 2i^m[AJ_m'(ka) + B_mH_m'(ka)]\cos(m\alpha) = 0\,.$$

For this expression to be valid for all α, we must have

$$AJ'_m(ka) + B_m H'_m(ka) = 0 \qquad m = 0, 1, \ldots \infty$$

or

$$B_m = -AJ'_m(ka)/H'_m(ka)\,.$$

We can then calculate the scattered field and other related quantities in the usual manner. For multiple scattering problems it is more convenient to express the incident and scattered waves in the form

$$\phi^{\rm inc} = \sum_{m=-\infty}^{\infty} A_m \psi_m(\mathbf{r}) \qquad \text{and} \qquad \phi^{\rm sc} = \sum_{m=-\infty}^{\infty} B_m \chi_m(\mathbf{r})$$

where $\mathbf{r} = r\mathbf{e}_r$, $A_m \leftarrow Ai^m e^{im\theta_i}$, $B_m \leftarrow B_m i^m e^{im\theta_i}$,

$$\psi_m(\mathbf{r}) = J_m(kr)\, e^{im\theta} \qquad \text{and} \qquad \chi_m(\mathbf{r}) = H_m(kr)\, e^{im\theta}\,.$$

In this case, the boundary condition at $r = a$ leads to

$$A_m J'_m(ka) + B_m H'_m(ka) = 0 \qquad m = 0, \pm 1, \pm 2, \ldots \infty\,.$$

We can then express B_m in terms of A_m in the form

$$B_m = \left[\frac{J'_m(ka)}{H'_m(ka)}\right] A_n = \sum_{n=-\infty}^{\infty} T_{mn} A_n \quad \text{where} \quad T_{mn} := \left[\frac{J'_m(ka)}{H'_m(ka)}\right] \delta_{mn}\,.$$

In matrix form,

$$\underline{\underline{\mathsf{B}}} = \underline{\underline{\mathsf{T}}}\,\underline{\underline{\mathsf{A}}}\,.$$

The matrix $\underline{\underline{\mathsf{T}}}$ is called the *T-matrix* and is diagonal in this case. As $r \to \infty$, the scattered field tends to (see NIST (2010) for asymptotic expansions of Bessel functions)

$$\phi^{\rm sc}(\mathbf{r}) \sim \sqrt{\tfrac{2}{\pi}}\, e^{-\pi i/4} \frac{e^{ikr}}{\sqrt{kr}} f(\theta)$$

where

$$f(\theta) = \sum_{m,n=-\infty}^{\infty} (-1)^m e^{im\theta} T_{mn} A_n\,.$$

The quantity $f(\theta)$ is called the *far-field pattern*.

Let us now consider the situation where there are two circular cylinders with radii a_1 and a_2. Let the centers of the cylinders be at $\mathbf{r} = \mathbf{b}_1$ and $\mathbf{r} = \mathbf{b}_2$. Define local two-dimensional polar coordinates (r_1, θ_1) and (r_2, θ_2) with origins at the centers of the two cross-sections. Then the location of any point in the two-dimensional plane with respect to the global origin is given by $\mathbf{r} = \mathbf{b}_1 + \mathbf{r}_1 = \mathbf{b}_2 + \mathbf{r}_2$. See Figure 3.8 to get a feel for the notation.

Let the incident field in the neighborhood of cylinder j be

$$\phi_j^{\text{inc}} = \sum_{m=-\infty}^{\infty} A_j^m \psi^m(\mathbf{r}_j).$$

The scattered field due to cylinder j can be expressed as

$$\phi_j^{\text{sc}} = \sum_{m=-\infty}^{\infty} B_j^m \chi^m(\mathbf{r}_j).$$

Therefore the total scattered field outside either cylinder is

$$\phi^{\text{sc}} = \sum_{m=-\infty}^{\infty} B_1^m \chi^m(\mathbf{r}_1) + \sum_{m=-\infty}^{\infty} B_2^m \chi^m(\mathbf{r}_2).$$

Even though the solutions $\chi^m(\mathbf{r}_j)$ are singular at the center of the j-th cylinder, they are regular at the center of the other cylinder and we can write

$$\chi^m(\mathbf{r}_2) = \sum_{n=-\infty}^{\infty} S^{mn}(\mathbf{b})\psi^n(\mathbf{r}_1) \quad \text{for } r_1 < \|\mathbf{b}\|$$

where $\mathbf{b} = \mathbf{r}_2 - \mathbf{r}_1$. The explicit form of the above relation for cylindrical problems is called *Graf's addition theorem* (see Martin (2006), p. 39) which states that

$$H_m(kr_2)e^{im\theta_2} = \sum_{n=-\infty}^{\infty} H_{m-n}(kb)e^{i(m-n)\beta} J_n(kr_1)e^{in\theta_1}$$

where $b = \|\mathbf{b}\|$ and $\beta = \arccos(\mathbf{b}\cdot\mathbf{e}_1/\|\mathbf{b}\|)$. We can write the total field near cylinder $j = 1 (r_1 < b)$ as

$$\begin{aligned}\phi = \phi^{\text{inc}} + \phi^{\text{sc}} &= \sum_{m=-\infty}^{\infty} A_1^m \psi^m(\mathbf{r}_1) + \sum_{m=-\infty}^{\infty} B_1^m \chi^m(\mathbf{r}_1) + \sum_{m=-\infty}^{\infty} B_2^m \chi^m(\mathbf{r}_2) \\ &= \sum_{m=-\infty}^{\infty} \left[B_1^m \chi^m(\mathbf{r}_1) + \left(A_1^m + \sum_{n=-\infty}^{\infty} S^{nm}(\mathbf{b}) B_2^n \right) \psi^m(\mathbf{r}_1) \right].\end{aligned}$$

Applying the boundary condition $\partial\phi/\partial r_1 = 0$ at $r_1 = a_1$ gives

$$H'_m(ka_1)B_1^m + J'_m(ka_1)\sum_{n=-\infty}^{\infty} S^{nm}(\mathbf{b})B_2^n = -J'_m(ka_1)A_1^m \qquad m = 0, \pm 1, \pm 2, \ldots, \pm\infty.$$

Repeating the same process for the other cylinder leads to

$$H'_m(ka_2)B_2^m + J'_m(ka_2)\sum_{n=-\infty}^{\infty} S^{nm}(-\mathbf{b})B_1^n = -J'_m(ka_2)A_2^m \qquad m = 0, \pm 1, \pm 2, \ldots, \pm\infty.$$

This system of equation can be solved for the coefficients B^m if A^m are known. For N cylinders, we can use the same approach to get a system of equations

$$H'_m(ka_p)B_p^m + J'_m(ka_p)\sum_{\substack{q=1\\q\neq p}}^{N}\sum_{n=-\infty}^{\infty} S^{nm}(\mathbf{b}_p - \mathbf{b}_q)B_q^n = -J'_m(ka_p)A_p^m. \tag{3.56}$$

The solution of this system of equations is usually completed numerically by truncating the series. Multiple scattering approaches have been used quite widely in the calculation of the effective behavior of arrays (or random distributions) of coated spheres and cylinders; representative examples include Liu et al. (2005) and Torrent and Sánchez-Dehesa (2008b). Detailed treatments of multiple scattering can be found in Sheng (2006) and Martin (2006).

Exercises

Problem 3.1 The acoustic potential satisfies Helmholtz equation

$$\nabla^2\varphi + k^2\varphi = 0\,.$$

Show that for the problem depicted in Figure 3.3 the solutions of this equation for outgoing waves can be expressed in the form

$$\varphi = \sum_{n=0}^{\infty} A_n h_n(kr) P_n(\cos\theta)$$

and those for incoming waves can be expressed in the form

$$\varphi = \sum_{n=0}^{\infty} B_n j_n(kr) P_n(\cos\theta)\,.$$

Problem 3.2 A classic problem in scattering is that of acoustic waves from an infinitely long fluid cylinder of circular cross-section. Show that, for harmonic waves, the solutions of the governing equation $\nabla^2 p + k^2 p = 0$ have the form

$$p = \sum_{n=0}^{\infty} A_n \cos(n\theta) \begin{bmatrix} J_n(kr) \\ H_n^{(1)}(kr) \end{bmatrix}$$

where $J_n(z)$ is a Bessel function of the first kind and $H_n^{(1)}$ is a Hankel function of the first kind. Then follow the standard procedure of matching boundary conditions at the surface of the cylinder to show for an incident plane wave of unit amplitude $p^i = \exp(ik_1 x)$ and a scattered wave given by

$$p^s = \sum_{n=0}^{\infty} B_n \cos(n\theta) H_n^{(1)}(k_1 r)$$

that coefficient B_n has the form

$$B_n = -\frac{e_n i^n}{1 + iC_n} \quad \text{with} \quad C_n := \frac{\dfrac{J_n'(k_2a)Y_n(k_1a)}{J_n(k_2a)J_n'(k_1a)} - \xi\eta\dfrac{Y_n'(k_1a)}{J_n'(k_1a)}}{\dfrac{J_n'(k_2a)J_n(k_1a)}{J_n(k_2a)J_n'(k_1a)} - \xi\eta}$$

where $Y_n(z)$ is a Bessel function of the second kind, $\xi = \rho_2/\rho_1$, $\eta = c_2/c_1$, and $e_n = 1$ if $n = 0$ and $e_n = 2$ otherwise.

Problem 3.3 If a is the radius of the sphere and the wavelength of the incident plane wave is much larger than a we have the condition $ka \ll 1$. Find the approximate solutions for the acoustic lens problem for this case. What are the asymptotic solutions when $ka \gg 1$? (Hint: Use formulas from NIST (2010).)

Problem 3.4 From equation (3.40) identify the conditions under which there can be resonance within the system. Plot the scattered pressure as a function of the frequency when the sphere in Figure 3.3 is made of air. The external medium is water.

Problem 3.5 Consider a plane harmonic acoustic wave incident from a medium with bulk modulus κ_1 and density ρ_1 upon a sphere with bulk modulus κ_2 and density ρ_2. Let the radius of the sphere be a. The sphere is coated with a fluid layer of density $\rho_3 = \rho_2$, radius b, and a bulk modulus given by

$$\frac{b^3}{\kappa_3} = \frac{a^3}{\kappa_1} + \frac{b^3 - a^3}{\kappa_2}.$$

Find the scattering cross-section of the coated sphere.

Problem 3.6 A vector displacement potential for shear waves can be expressed in terms of scalar Debye potentials (Π, χ) as

$$\boldsymbol{\psi} = \boldsymbol{\nabla} \times (\Pi r \mathbf{e}_r) + \boldsymbol{\nabla} \times (\boldsymbol{\nabla} \times (\chi r \mathbf{e}_r)).$$

Express the displacement and stress components for an elastic material in terms of these potentials. What are the forms of the potential that can be used to represent an incident plane SV-wave and a plane SH-wave? What are the appropriate forms of the expansions for the scattered fields?

Problem 3.7 Consider three identical infinitely long rigid circular cylinders arranged in an equilateral triangle. Assume that a plane wave with wavenumber k is incident on the system along the line joining two of the cylinders and that $ka = 2$ where a is the radius of the cylinder. Assume that the cylinders are separated by a distance $b = \pi a$. Calculate the effective potential field (ϕ) around a cylinder at a distance $r = 0.5b$ as a function of angle.

4

Electrodynamic Metamaterials

Accordingly we see that we must look for substances with $\varepsilon < 0$ and $\mu < 0$ primarily among gyrotropic media.

V. G. VESELAGO, The electrodynamics of substances with simultaneously negative values of ε and μ, 1964.

In Section 1.4.5 we introduced the possibility of materials with a negative electromagnetic refractive index. The non-reflecting, yet mirror-like effect produced by such materials when plane harmonic waves are incident on them has been discussed in Section 2.4.5. Recall that a negative refractive index material requires that the permittivity (ε) and the magnetic permeability (μ) be simultaneously negative. Electrodynamic metamaterials have the potential of providing a means to achieve such unusual properties.

A universal consensus of the definition of metamaterials does not seem to exist yet. Some authors define metamaterials as those whose properties depend strongly on the geometry of the microstructure but *appear* not to depend on the properties of the constituents. This definition is not accurate because the effective properties of metamaterials do depend on the properties of the constituents, and indeed they must. Another definition is that metamaterials are those materials whose properties do not reflect everyday experience such as negative refractive indexes or negative Poisson's ratios. However, it is generally acknowledged the unusual properties expected of metamaterials usually require some specific resonances in the structure.

In this chapter we will discuss ways in which electrodynamic metamaterials can be realized and some possible uses for these materials. The literature on the subject is already quite vast and is expanding at a rapid rate and detailed expositions on recent developments can be found in Ramakrishna and Grzegorczyk (2009), Marqués et al. (2008), and Sarychev and Shalaev (2007). Our discussion of electrodynamic metamaterials in this chapter covers only some elementary aspects.*

*The material in this chapter is based primarily on Prof. G. W. Milton's lecture notes and Pendry (2000), Smith et al. (2004), Ramakrishna (2005), Milton et al. (2005), and Milton and Nicorovici (2006).

4.1 A model for the permittivity of a medium

Consider a simple model of an electron bound to an atom by a harmonic force† as shown in Figure 4.1. Let the system be under the influence of a slowly varying electric field. The applied field distorts charge distributions and produces an induced dipole moment in each molecule of the medium which is being represented by the bound charge.

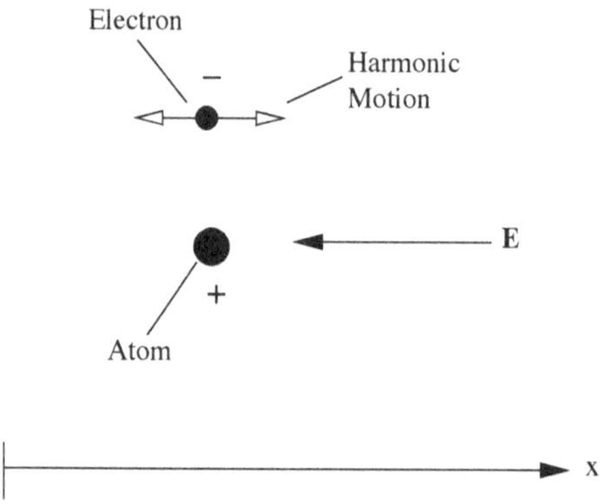

FIGURE 4.1
An electron bound by a harmonic force to an atom.

Let m be the mass of the electron and let $-e$ be its charge. Let the charge be bound under the action of a restoring force $\mathbf{F} = \mathbf{F}_0 \exp(-i\omega_0 t)$ where ω_0 is the frequency of the harmonic force binding the electron to the atom. If $\mathbf{x}$ is the position of the electron, its motion must have the form $\mathbf{x} = \mathbf{x}_0 \exp(-i\omega_0 t)$ if linearity applies. From Newton's second law we have $\mathbf{F} = m\ddot{\mathbf{x}}$. Therefore, for a given frequency ω_0, we have $\mathbf{F}_0 = -m\omega_0^2 \mathbf{x}_0$.

Let the electric field be $\mathbf{E}(\mathbf{x},t)$ and assume that it varies slowly in space over the distance that the electron moves, i.e., $\mathbf{E}(\mathbf{x},t) \approx \mathbf{E}(t)$. Under the action of the electric field the electron is displaced from its equilibrium position by a distance $\mathbf{x}$ given by the equation of motion

$$m\left(\frac{d^2\mathbf{x}}{dt^2} + \gamma\frac{d\mathbf{x}}{dt} + \omega_0^2\,\mathbf{x}\right) = -e\,\mathbf{E}(t) \tag{4.1}$$

†A detailed description can in found in Jackson (1999), Sections 4.6 and 7.5.

where γ is a damping coefficient due to interactions with obstacles. Let

$$\mathbf{x}(t) = \widehat{\mathbf{x}}\, e^{-i\omega t} \quad \text{and} \quad \mathbf{E}(t) = \widehat{\mathbf{E}}\, e^{-i\omega t}\,. \tag{4.2}$$

Plugging equations (4.2) into equation (4.1) and solving for $\widehat{\mathbf{x}}$, we get

$$\widehat{\mathbf{x}} = -\frac{e}{m}\left(\omega_0^2 - \omega^2 - i\omega\gamma\right)^{-1}\widehat{\mathbf{E}}\,.$$

Let $q(\mathbf{x},t)$ be the charge density of the electron-atom system, i.e.,

$$q(\mathbf{x},t) = -\delta[\mathbf{x}(t) - \mathbf{x}_0]\, e + \delta(\mathbf{x}_0 - \mathbf{x}_0)\, e$$

where $\delta(\mathbf{x})$ is the Dirac delta function and $\mathbf{x}_0$ is the position of the atom. The first term is due to the electron while the second term is due to the positively charged atom. Then, the dipole moment contributed by the single electron-atom system is given by

$$\mathbf{p}(t) = \int q(x,t)\, \mathbf{x}(t)\, \mathrm{d}x = -e\left[\int \delta[\mathbf{x}(t) - \mathbf{x}_0]\, \mathbf{x}(t)\, \mathrm{d}x + \cancelto{0}{\int \delta(0)\, \mathbf{x}(t)\, \mathrm{d}x}\right]$$

or $\mathbf{p}(t) = -e\,\mathbf{x}(t)$. Therefore,

$$\widehat{\mathbf{p}} = -e\,\widehat{\mathbf{x}} = \frac{e^2}{m}\left(\omega_0^2 - \omega^2 - i\omega\gamma\right)^{-1}\widehat{\mathbf{E}}\,.$$

Suppose now that there are N_j electrons per unit volume with binding frequencies ω_j and damping constants γ_j and. Then we can write the polarization as

$$\widehat{\mathbf{P}} = \left[\sum_j \frac{a_j}{\omega_j^2 - \omega^2 - i\omega\gamma_j}\right]\widehat{\mathbf{E}} \quad \text{where} \quad a_j := \frac{e^2 N_j}{m}\,. \tag{4.3}$$

Recall that the electric displacement is related to the electric field and the polarization by

$$\widehat{\mathbf{D}} = \varepsilon_0\,\widehat{\mathbf{E}} + \widehat{\mathbf{P}} = \varepsilon\,\widehat{\mathbf{E}} \tag{4.4}$$

where ε_0 is the permittivity of vacuum. Therefore, from equations (4.3) and (4.4) we get an expression for the frequency dependence of the permittivity of the medium

$$\varepsilon(\omega) = \varepsilon_0 + \sum_j \frac{a_j}{\omega_j^2 - \omega^2 - i\omega\gamma_j}\,.$$

A plot of the real and imaginary parts of the permittivity as a function of frequency is shown in Figure 4.2. We can see that close to some resonance frequencies it is possible for the permittivity to be negative. Recall from equation (1.105) that there is dissipation when the imaginary part of the permittivity is greater than zero. The figure suggests that dissipative behavior is pronounced in the region close to the resonance frequencies. For the special case where $\omega_j = \omega_0$ and $\gamma_j = \gamma$, we have

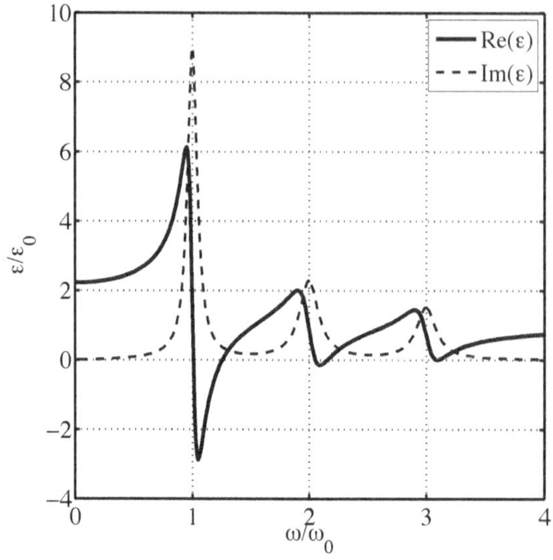

FIGURE 4.2
Schematic plot of the dielectric constant of a dissipative medium with three resonance frequencies. The abscissa is normalized by the first resonance frequency.

$$\boxed{\varepsilon(\omega) = \varepsilon_0 + \sum_j \frac{a_j}{\omega_0^2 - \omega^2 - i\omega\gamma}.} \tag{4.5}$$

Comparing equation (4.5) with (5.12) (p. 155) we observe that they have the same form which implies that the effective permittivity is analogous to the effective mass. This also implies that the electric field is analogous to the velocity. Similarly, the polarization is analogous to the momentum.

If $\omega_0 \neq 0$ and in the limit $\omega \to 0$, we recover the expression for the dielectric constant in the static limit. On the other hand, if there are free electrons in the medium then $\omega_0 = 0$ for these electrons and the permittivity is singular at $\omega = 0$. In that case we can separate out the contribution of the bound electrons and the free electrons and can write the expression for ε in the form

$$\varepsilon(\omega) = \varepsilon_0 + \varepsilon_b - \sum_j \frac{a_j}{\omega^2 + i\omega\gamma} = \bar{\varepsilon}_0 + \frac{i}{\omega}\sum_j \frac{a_j}{\gamma - i\omega}.$$

Recall the expression on the permittivity from equation (1.92),

$$\varepsilon(\omega) = \varepsilon_0 + \frac{i}{\omega}\sigma$$

where σ is the electrical conductivity. Comparing the above with the expression from our simple model, we find that

$$\sigma(\omega) = \sum_j \frac{a_j}{\gamma - i\omega} = \frac{e^2 N}{m(\gamma - i\omega)} \quad \text{where } N := \sum_j N_j \,. \tag{4.6}$$

This is the *Drude model* for electrical conductivity where N is the number of free electrons per unit volume in the medium. For high frequencies, $\omega \gg \omega_0$, equation (4.5) simplifies to

$$\varepsilon(\omega) \approx \varepsilon_0 - \sum_j \frac{a_j}{\omega^2} = \varepsilon_0 \left(1 - \frac{\omega_p^2}{\omega^2}\right) \qquad \text{where } \omega_p^2 := \frac{e^2 N}{m\varepsilon_0} \,.$$

The quantity ω_p is called the *plasma frequency* and depends only on the total number of electrons (N) in the medium. For most naturally occurring materials $\omega_p < \omega$. However, if $\omega_p > \omega$ the value of the permittivity (ε) is negative. If the medium has the magnetic permeability of vacuum (μ_0), the wave vector for a plane wave is given by

$$k^2 = \omega^2 \varepsilon\mu = \omega^2 \varepsilon_0 \mu_0 \left(1 - \omega_p^2/\omega^2\right) = \varepsilon_0 \mu_0 \left(\omega^2 - \omega_p^2\right) \,.$$

This *dispersion relation* indicates that for $\omega < \omega_p$ we have an imaginary k and only evanescent waves exist below the plasma frequency. For $\omega > \omega_p$, i.e., for negative ε, a propagating wave can exist. This wave is called a *plasmon.*

4.2 Negative permittivity materials

At optical frequencies, many charge neutral metal systems have a negative permittivity because of plasma-like behavior that can be described in terms of a plasma frequency. Can we design composite materials that exhibit a negative permittivity analogous to that observed above the plasma frequency in metals? This question was addressed by Pendry et al. (1998) who showed that arrays of thin wires could have a negative effective permittivity for frequencies significantly lower than the optical range, i.e., GHz rather than THz.

It is worth noting that a composite made of conductors and dielectric materials may have properties which are quite different from those of the constituents. Several exact relations and bounds on the effective dielectric properties of composites are discussed in Milton (2002). However, these relations typically do not consider the effect of local resonances that lead to the plasma-like behavior that is a characteristic of electrodynamic materials.

To calculate the effective permittivity of an infinite array of thin wires, let us assume that the wires are of radius a and separated by a distance b where $b \gg a$. A schematic of the situation is shown in Figure 4.3. An electric field $\mathbf{E}_0$ is applied parallel to the wires in the z-direction. The wavelength of the incident radiation is

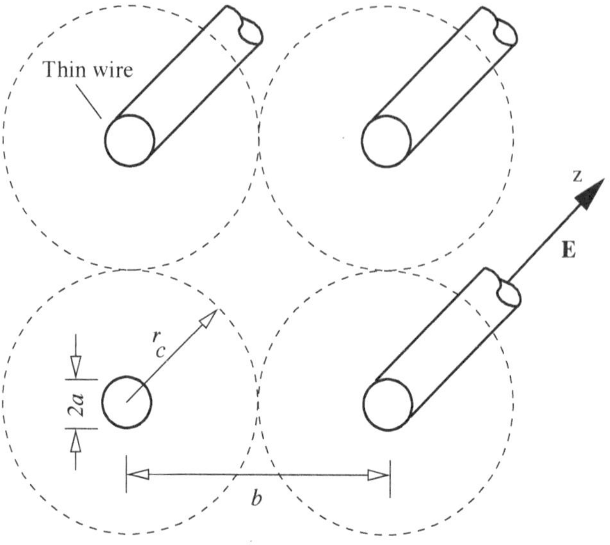

FIGURE 4.3
An array of thin wires arranged in a regular grid.

assumed to be much larger than b so that we can operate in the quasi-static limit. Recall Maxwell's equation for the magnetic field

$$\nabla \times \mathbf{H} = \frac{\partial \mathbf{D}}{\partial t} + \mathbf{J}_f =: \frac{\partial \widetilde{\mathbf{D}}}{\partial t} \,.$$

For the situation shown in the figure, the field $\mathbf{D}$ can be assumed to be uniform for long wavelengths while the free current density $\mathbf{J}_f$ is localized within each wire. If we consider a plasmon-like longitudinal wave, we can write the above relation in the form

$$\nabla \times \mathbf{H} = -\frac{I}{\pi r_c^2}\mathbf{e}_z$$

where I is the current through a single wire and r_c is the effective cross-section of the wire. If we express this equation in cylindrical coordinates and use the condition that the magnetic field is zero at $r = r_c$, we get

$$H_\theta = \begin{cases} \dfrac{I}{2\pi r}\left(1 - \dfrac{r^2}{r_c^2}\right) & \text{for } 0 < r \le r_c \\ 0 & \text{for } r > r_c \end{cases} \,.$$

The magnetic field can be derived from a vector potential $\mathbf{A}$ such that $\mathbf{H} = \mu^{-1}\,\nabla \times \mathbf{A}$. Once again expressing the equation in cylindrical coordinates and using $A_z = 0$ at

$r = r_c$, we have

$$A_z = \begin{cases} \dfrac{I\mu_0}{2\pi}\left[\dfrac{1}{2}\left(\dfrac{r^2}{r_c^2} - 1\right) + \ln\left(\dfrac{r_c}{r}\right)\right] & \text{for } 0 < r \leq r_c \\ 0 & \text{for } r > r_c \end{cases} .$$

To avoid magnetic induction between adjacent wires, the value of b can be chosen such that $b \geq 2r_c$. For the case where $r_c = b/2$ we can find the potential at the wire ($r \ll b$) to a good approximation using

$$A_z = \frac{I\mu_0}{2\pi}\left[\frac{1}{2}\left(\frac{4r^2}{b^2} - 1\right) + \ln\left(\frac{b}{2r}\right)\right] \approx \frac{I\mu_0}{2\pi}\ln\left(\frac{b}{r}\right) .$$

For our problem, $I = \pi a^2 Nve$ where N is electron density in the wires, v is the mean electron velocity, and e is the charge. Hence we have

$$\mathbf{A}(r) = \frac{\mu_0 a^2 Nve}{2}\ln\left(\frac{b}{r}\right)\mathbf{e}_z .$$

This approximation is reasonably good because of the symmetry of the lattice. If we assume that the electrons flow on the surface of the wire, the momentum per unit length of the wire is

$$\mathbf{p} = \pi a^2 Ne\mathbf{A}(a) = (\pi a^2 Nv)\frac{\mu_0 a^2 Ne^2}{2}\ln\left(\frac{b}{a}\right)\mathbf{e}_z =: (\pi a^2 Nv) m_{\text{eff}}\,\mathbf{e}_z$$

where m_{eff} is the effective mass of an electron in the wire. Assuming that the electrons are confined to the wires, the effective electron density of the system is

$$N_{\text{eff}} = \frac{\pi a^2}{b^2}N$$

where N is the electron density of the wires. Then the plasma frequency of the system is given by

$$\boxed{\omega_p^2 = \frac{e^2 N_{\text{eff}}}{\varepsilon_0 m_{\text{eff}}} = \frac{2\pi c^2}{b^2 \ln(b/a)}} \qquad (4.7)$$

where $c^2 = 1/(\varepsilon_0\mu_0)$. For aluminum wires with $N \sim 10^{29}$ m^{-3}, $a = 1$ μm, and $b =$ 10 mm, the effective mass is $m_{\text{eff}} \sim 15m_p$, where m_p is the mass of a proton. The plasma frequency is $\omega_p \sim 2$ GHz. For frequencies larger than this value negative effective permittivities can be expected if we use the Drude model for the effective permittivity,

$$\varepsilon^{\text{eff}}(\omega) = \varepsilon_0\left(1 - \frac{\omega_p^2}{\omega^2}\right) .$$

If we consider dissipation due to finite conductivity in the thin wires we get the relation

$$\boxed{\varepsilon^{\text{eff}}(\omega) = \varepsilon_0\left(1 - \frac{\omega_p^2}{\omega^2 + i\Gamma\omega}\right)} \qquad \text{with} \qquad \Gamma := \frac{\varepsilon_0 b^2 \omega_p^2}{\pi a^2 \sigma} \tag{4.8}$$

where σ is the electrical conductivity of the material. Note that the above relation can be also be derived using AC circuit elements (inductance and capacitance).

An alternative approach to computing the effective permittivity of a cubic lattice of thin wires can be found in Sarychev et al. (2000). In this approach, the electric field outside the wires is approximated as the sum of a constant field $\mathbf{E}_0 = E_0\mathbf{e}_z$ and a dipole field. Then the electric field in a unit cell has the form

$$\mathbf{E}(r) = \begin{cases} \mathbf{E}_0 J_0(k_m r) & \text{for } r < a \\ \mathbf{E}_0\left[J_0(k_m a) - k_m a J_1(k_m a)\ln(r/a)\right] & \text{for } r > a \end{cases}$$

where a subscript m indicates the properties of the wire material, $k_m = k\sqrt{\varepsilon_m \mu_m}$, $k = \omega/c = 2\pi/\lambda$, and J_0, J_1 are Bessel functions of the first kind. The effective permittivity of the system is defined by the relation

$$\langle \mathbf{D} \rangle = \varepsilon^{\text{eff}}(\langle \mathbf{E} \rangle - \langle \mathbf{L} \rangle)$$

with

$$\langle \mathbf{D} \rangle = \frac{1}{b^3}\int \varepsilon(\mathbf{r})\mathbf{E}(\mathbf{r})d\mathbf{r} \;;\; \langle \mathbf{E} \rangle = \frac{1}{b^3}\int \mathbf{E}(\mathbf{r})d\mathbf{r} \;;\; \langle \mathbf{L} \rangle = \frac{1}{2\pi b^3}\int [\mathbf{r}\times(\nabla\times\mathbf{E})]d\mathbf{r}$$

and $\varepsilon(\mathbf{r}) = \varepsilon_m$ for $r < a$; $\varepsilon(\mathbf{r}) = \varepsilon_0$ for $r > a$. The effective permittivity is found to be

$$\varepsilon^{\text{eff}}(\omega) = \varepsilon_0\left[1 - \frac{f[b^2 J\omega^2 + 4\pi(J-1)c^2]}{\omega^2\pi a^2 J(1+L) - 4\pi c^2}\right]$$

where $c^2 = 1/(\varepsilon_0\mu_0)$, ε_m is the permittivity of the metal wires,

$$f := \frac{\pi a^2}{b^2} \;;\; J := \varepsilon_m F(a\omega\sqrt{\varepsilon_m}/c) \;;\; F(x) := \frac{2J_1(x)}{xJ_0(x)} \;;\; L := 2\ln\left(\frac{b}{\sqrt{2}a}\right) + \frac{\pi}{2} - 3.$$

A universally accepted homogenization procedure for metamaterials continues to be sought, though some progress has been made (see, for instance, Felbacq and Bouchitté (1997, 2005)). For an array of wires, Felbacq and Bouchitté use a homogenization procedure to derive a relation of the form

$$\varepsilon^{\text{eff}}(\omega) = \varepsilon_0\left[1 - \frac{2\pi f c^2}{\omega^2 + 2i\omega\pi f c^2 \varepsilon_0 \langle\sigma\rangle^{-1}}\right] \tag{4.9}$$

with $c^2 = 1/(\varepsilon_0\mu_0)$, $f\eta^2\lambda^2 = -1/\ln(2\pi a/b)$, $\eta = b/\lambda$ is a factor that is chosen such that the radio of b to the incident wavelength, λ, is small, $\langle\sigma\rangle$ is the average conductivity of the domain which is kept constant by scaling with the filling ratio of the unit cell, and σ is the conductivity of the wires. Though the responses of the three models are similar for perfect conductors, significant differences are observed under other situations.

From a practical point of view, materials consisting of arrays of long thin wires are tedious to construct as connectivity is usually required between each unit cell. An alternative design is to use resonant structures (see Schurig et al. (2006b) for an example).

4.3 Artificial magnetic metamaterials

The fact that artificial magnetic materials may be created from relatively non-magnetic materials was first briefly hinted at by Shelkunoff and Friis in their textbook on antenna design (Shelkunoff and Friis (1952), pp. 584–585). The idea was developed in more detail by Pendry and co-workers (Pendry et al., 1999). In that work, split-ring resonators were used to develop a magnetic material containing non-magnetic components. A schematic of the split ring resonator is shown in Figure 4.4.

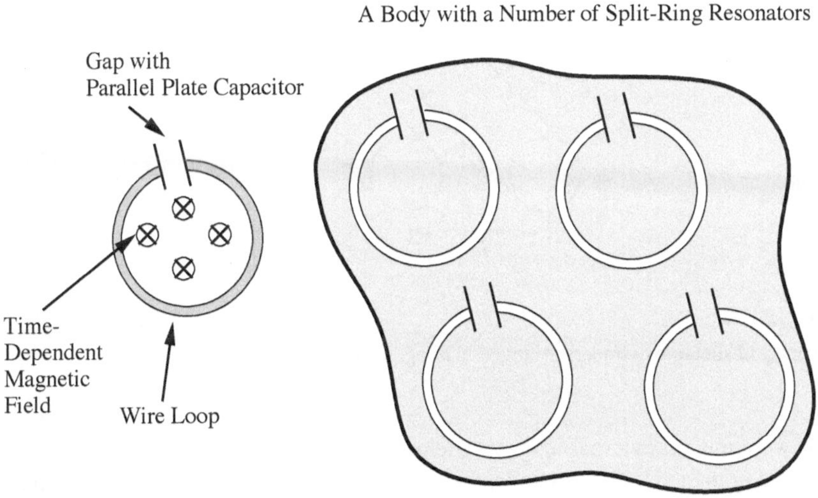

FIGURE 4.4
A split-ring resonator can be used to create artificial magnetic materials with possibly negative magnetic permeability.

If the magnetic field intensity **H** is time dependent and the magnetization vector **M** is zero, then

$$\frac{\partial \mathbf{H}}{\partial t} \neq 0 \quad \Longrightarrow \quad \frac{\partial \mathbf{B}}{\partial t} \neq 0 \quad \Longrightarrow \quad \boldsymbol{\nabla} \times \mathbf{E} \neq 0\,.$$

Therefore, there is a non-zero electric field around the loop which implies that there is a current in the split ring. Now if we place a parallel plate capacitor in the gap, charges build up in the capacitor and the current oscillates back and forth in the ring as the field **H** changes. The result is that the ring resonates and the net magnetic dipole moment **M** becomes non-zero. Hence we have an artificial magnetic structure composed of non-magnetic materials.

Let us examine a simpler structure that produces artificial magnetism. The structure consists of an array of cylinders with conducting surfaces. The cylinders are of radius a and are spaced a distance b from each other as shown in Figure 4.5. This structure was used by Pendry et al. (1999) to motivate the split-ring resonator (SRR) design that has dominated electromagnetic metamaterials.

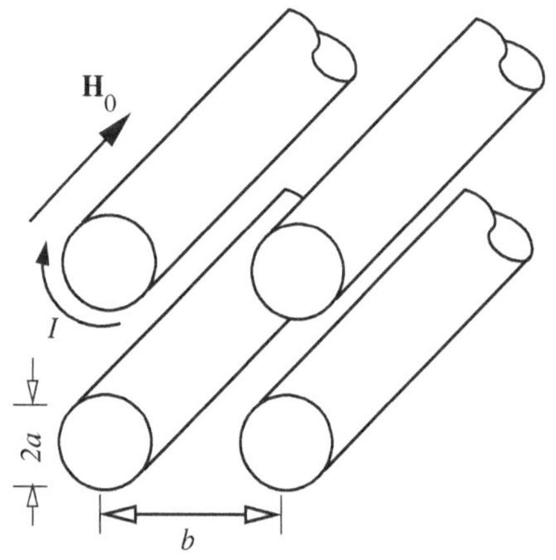

FIGURE 4.5
An array of metallic cylinders arranged in a regular grid.

If we apply an external magnetic field $\mathbf{H}_0 = H_0\mathbf{e}_z$ (at fixed frequency ω) where $\mathbf{e}_z$ is along the axis of the cylinders, a current I (per unit length) is produced on the surface of the cylinders. We can solve Maxwell's equation relating the magnetic field and the current density to find the magnetic field (H) inside the cylinders. Noting that the current sources include the surface current I and the depolarizing sources at the ends of the cylinders, we can write

$$H = H_0 + I - fI \quad \text{where} \quad f := \frac{\pi a^2}{b^2}\,.$$

Recall that the electromotive force (emf) is given by

$$\mathcal{E} = \oint \mathbf{E}\cdot d\mathbf{l} = -\frac{\partial}{\partial t}\int \mathbf{B}\cdot\mathbf{n}\,dA = -\frac{\partial}{\partial t}\int \mu\mathbf{H}\cdot\mathbf{n}\,dA\,.$$

Therefore we have, around the circumference of the cylinder,

$$\frac{I}{\sigma}(2\pi a) = i\omega\mu_0[H_0 + (1-f)I](\pi a^2)$$

where σ is the electrical conductivity of the cylinder surface per unit area. Solving for I gives

$$I = \frac{-\,i\omega\pi a^2\mu_0 H_0}{i\omega\pi a^2\mu_0(1-f) - 2\pi a/\sigma}\,. \tag{4.10}$$

Now that we have an expression for the current, we can proceed with finding the effective fields in the structure which are defined by

$$\langle\mathbf{B}\rangle = \boldsymbol{\mu}^{\text{eff}}\cdot\langle\mathbf{H}\rangle \qquad \text{and} \qquad \langle\mathbf{D}\rangle = \boldsymbol{\varepsilon}^{\text{eff}}\cdot\langle\mathbf{E}\rangle$$

where $\langle(\bullet)\rangle$ indicates an averaged quantity. The crucial point in homogenizing metamaterial structures is that the averages are taken not over the volume but over surfaces and line by using Maxwell's equation in integral form:

$$\oint \mathbf{E}\cdot d\mathbf{l} = -\frac{\partial}{\partial t}\int \mathbf{B}\cdot\mathbf{n}\,dA \qquad \text{and} \qquad \oint \mathbf{H}\cdot d\mathbf{l} = \frac{\partial}{\partial t}\int \mathbf{D}\cdot\mathbf{n}\,dA\,.$$

For a periodic structure in which the unit cell can be represented by a cube, Pendry et al. (1999) suggest that $\langle\mathbf{H}\rangle$ be computed by averaging the **H**-field along each of the three axes of the unit cell. If the three lattice vectors are $\mathbf{b}_i = b\mathbf{e}_i$, $i = 1,2,3$, we have

$$\langle\mathbf{H}\rangle = \frac{1}{b}\sum_{i=1}^{3}\mathbf{e}_i\oint_{\mathbf{r}=\mathbf{0}}^{\mathbf{r}=\mathbf{b}_i}\mathbf{H}\cdot d\mathbf{r}\,.$$

This approach works as long as the edges of the unit cell do not intersect any of the internal structures. On the other hand, the average **B**-field is computed by summing averages over three orthogonal faces. If the surface normals to the three faces are $\mathbf{n}_i = \mathbf{e}_i$ and the corresponding faces are denoted A_i, then

$$\langle\mathbf{B}\rangle = \frac{1}{b^2}\sum_{i=1}^{3}\int_{A_i}\mathbf{B}\cdot\mathbf{n}_i\,dA\,.$$

The j-th diagonal component of the effective magnetic permeability, in the $(\mathbf{e}_1,\mathbf{e}_2,\mathbf{e}_3)$ basis, can then be calculated from

$$\left[\boldsymbol{\mu}^{\text{eff}}\right]_{jj} = \frac{\langle\mathbf{B}\rangle_j}{\langle\mathbf{H}\rangle_j} \qquad (\text{no sum over } j)\,. \tag{4.11}$$

Using this approach, in the case of the array of cylinders, the average **B**-field is

$$\langle \mathbf{B} \rangle = \mu_0 H_0 \mathbf{e}_3$$

and the average **H**-field is

$$\langle \mathbf{H} \rangle = (H_0 - fI)\mathbf{e}_3 \,.$$

Using these average fields and equations (4.10) and (4.11), we get

$$\boxed{\mu^{\text{eff}} = \mu_0 \left[1 - \frac{\omega a \mu_0 f}{\omega a \mu_0 + 2i/\sigma} \right] \,.} \tag{4.12}$$

As $\sigma \to \infty$ we have $\mu^{\text{eff}} \to (1-f)\mu_0$. Also, we always have $0 \leq \mu^{\text{eff}} \leq \mu_0$. Note that the structure also has an effective permittivity which is such that the structure retains causality.

Let us now consider the case of the split-ring resonator shown in Figure 4.6. Each ring has width $2w$, height h, inner radius a, and a gap width g. We will assume that the magnetic field due to the currents in the ring can be approximated as equivalent to that in a continuous cylinder. The ring is place in a time-harmonic magnetic field $\mathbf{H}_0$ which is oriented in the z-direction (along the axis of the split rings or an equivalent cylinder). We will apply a AC circuit analogy to the problem and treat the ring as an inductance and resistance in series with a capacitance.

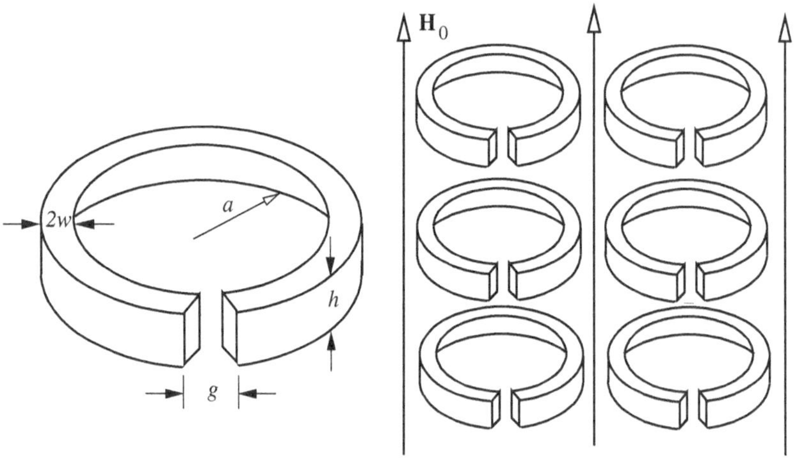

FIGURE 4.6
A split-ring resonator (SRR) and a composite made of an array of SRRs.

We can write the electromotive force along the rings as

$$\mathcal{E} = \oint \mathbf{E} \cdot d\mathbf{l} = \left(-i\omega L - \frac{1}{i\omega C} + R \right) I(2\pi a)$$

where L is the inductance of the rings, C is the capacitance of the gap (and the surface of rings), R is the resistance, and I is the total current. Various expressions for L, C, and R can be derived depending on the approximations made (see, for example, Merlin (2009), Delgado et al. (2009)). To keep things simple we will assume that the rings are thin and the gap is small, i.e., $a \gg w \gg g$ and we will approximate the cross-section of the ring as circular ($2wh \approx \pi w^2$). We will also assume, as we have done previously, that the rings are in vacuum.

If we ignore the gap and approximate the curved geometry of the ring as locally straight, we can set up a cylindrical coordinate system (ρ, θ, ζ) at the center of the cross-section with the ζ-axis oriented along the split ring. For an external magnetic field oriented along the ζ-axis, we have $\partial E_\zeta / \partial \zeta = 0$. Then the non-zero fields inside the ring are (Merlin, 2009)

$$E_\zeta(\rho) = i \frac{2I}{cw\sqrt{\varepsilon_m}} \frac{J_0(k\rho\sqrt{\varepsilon_m})}{J_1(kw\sqrt{\varepsilon_m})} \quad \text{and} \quad H_\theta(\rho) = \frac{2I}{cw} \frac{J_1(k\rho\sqrt{\varepsilon_m})}{J_1(kw\sqrt{\varepsilon_m})} \tag{4.13}$$

where I is the total current in the SRR, $c = \omega/k$ is the phase velocity in vacuum, k is the magnitude of the wave vector, ε_m is the relative permittivity of the material of the SRR, and J_0, J_1 are Bessel functions of the first kind. We can get the field in the gap using the above relations but will ignore that effect in the following. The resistance in the SRR can be calculated directly from the first of equations (4.13) using

$$R = \frac{2\pi a}{I} E_\zeta(a) = i \frac{4\pi a}{cw\sqrt{\varepsilon_m}} \frac{J_0(ka\sqrt{\varepsilon_m})}{J_1(kw\sqrt{\varepsilon_m})} \approx -\frac{4\pi a}{cw\sqrt{\varepsilon_m}}.$$

We can approximate the gap capacitance as (see Feynman et al. (1964), better estimates can be found in Delgado et al. (2009))

$$C = \frac{4w^2}{gc^2\mu_0}.$$

The inductance can be approximated from the second of equations (4.13) and the relation $LI^2 = (1/\mu) \int_\Omega \|\mathbf{B}\|^2 \mathrm{d}\Omega$, to give

$$L = 4\pi a \mu_0 \ln\left(\frac{2\pi a}{w}\right).$$

Higher-order effects can be incorporated by adding a *kinetic* impedance term due to the motion of electrons to the above expression. Once again we use the relation

$$\mathcal{E} = \oint \mathbf{E} \cdot d\mathbf{l} = i\omega \int \mu \mathbf{H} \cdot \mathbf{n}\, dA$$

to get

$$\left(-i\omega L - \frac{1}{i\omega C} + R\right) I(2\pi a) = i\omega\mu_0[H_0 - fI](\pi a^2)$$

where $f = \pi a^2/b^2$. We can solve this equation for the current I and follow the procedure that we used for the cylinders to find the effective magnetic permeability of the split-ring resonator array. The real part of the effective magnetic permeability for this structure is found to be negative for a small range of frequencies.

In the seminal paper by Pendry et al. (1999), μ^{eff} for a number of other structures has been estimated. The most popular of these has been the double split-ring resonator shown in Figure 4.7. Several variations of this design have been developed. For a square array of hollow cylinders of inner radius a, spaced b apart, and with

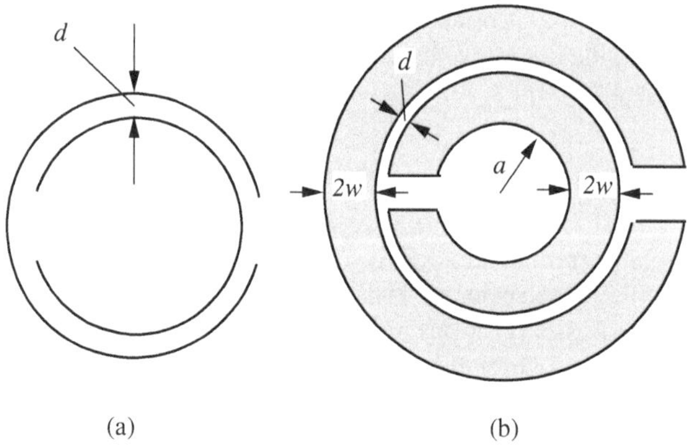

FIGURE 4.7
Double split ring resonators. a) Thin rings. b) Thick rings with small gap.

cross-sections in the shape of double-split rings (Figure 4.7(a)), the effective magnetic permeability is found to be (Pendry et al., 1999)

$$\boxed{\mu^{\text{eff}} = \mu_0\left[1 + \frac{f\omega^2}{\omega_0^2 - \omega^2 - i\omega\Gamma}\right]} \tag{4.14}$$

with

$$f := \frac{\pi a^2}{b^2}\,; \quad \Gamma := \frac{2}{a\mu_0\sigma}\,; \quad \omega_0^2 = \frac{1}{\pi\mu_0 C a^2}\,; \quad C = \frac{\pi a}{3dc^2\mu_0}\,.$$

where σ is the electrical conductivity of the rings per unit area, μ_0 is the permeability of vacuum, and C is the capacitance per unit area between the two split cylinders. Note that this expression has a form similar to equation (4.5) and at some frequencies we will get a magnetic permeability which has a negative real part. The value of ω at which $\mu^{\text{eff}} = 0$ has been called the magnetic plasma frequency (ω_{mp}). Many metamaterial designs have been shown to have the above form of effective magnetic permeability. Some of those designs can be seen in Smith et al. (2004) and Cai and Shalaev (2009).

The Pendry et al. (1999) approach to homogenization of resonant structures has a heuristic basis. Some progress has been made on the homogenization of these structures for special situations, see Kohn and Shipman (2008) and Bouchitté and Schweizer (2010). For instance, Bouchitté and Schweizer (2010) show rigorously that the split-ring resonator geometry has an effective magnetic permeability of the form shown in equation (4.14). Recent developments (Milton, 2010) show that hierarchical multiple rank laminate structures can be used to achieve artificial magnetic metamaterials.

4.4 Negative refraction and perfect lenses

We have seen in the previous sections that it is possible to manufacture materials with negative permittivity and negative magnetic permeability over a small range of frequencies. In Section 2.4.5 we have seen that negative index materials can be produced if both ε and μ are negative. The first such material was produced by Smith et al. (2000) by using a combination of thin wires and double split rings. Since then several designs have been explored with the primary aim of designing a slab lens that has a resolution which is not limited by the incident wavelength. Excellent reviews of progress in negative index materials and new designs can be found in Shalaev (2007), Valentine et al. (2008), and Cai and Shalaev (2009).

Let us now explore the holy grail of electromagnetic metamaterial research, the search for a perfect lens. We saw in Section 2.4.5 that an interface separating media with permittivities $\varepsilon_1 = -\varepsilon_2$ and magnetic permeabilities $\mu_1 = -\mu_2$ "behaves like a mirror" (see Figure 2.14). Milton (2010) showed that any pair of $\boldsymbol{\varepsilon}^{\text{eff}}$ and $\boldsymbol{\mu}^{\text{eff}}$, including tensors with negative real eigenvalues, can be realized at a fixed frequency by using a hierarchical lamination procedure. Therefore, materials with $\varepsilon = \mu = -1$ are not physically prohibited and can be manufactured in principle.

Let us examine how a slab of such a material immersed in another medium can act like a perfect lens. Let the permittivity and permeability of the surrounding medium be $\varepsilon = \mu = 1$ (normalized with respect to the values for free space). Let the normalized permittivity and permeability of the slab be $\varepsilon = \mu = -1$. Hence, the first interface between the medium and the slab acts as a mirror in that it reflects the electric field $\mathbf{E}$. The second interface also acts as a mirror and reflects the field $\mathbf{E}$ to the original orientation as depicted in Figure 4.8.

If the source is located at a distance d_0 from the first interface, and the slab has a

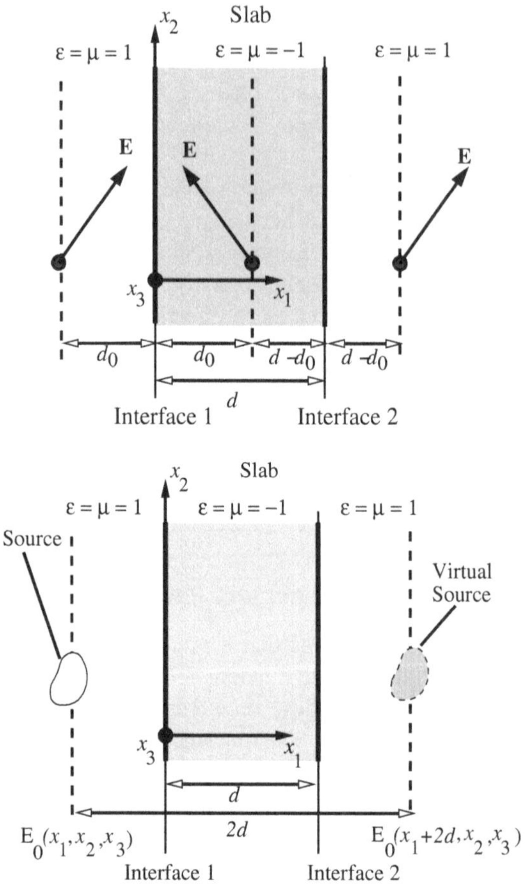

FIGURE 4.8
The effect of a slab of material with $\varepsilon = \mu = -1$ on the field **E**. The effect of a negative index slab is to make the source appear to be at the right of the slab.

thickness d, we have

$$\mathbf{E}(x_1 + 2d, x_2, x_3) = \mathbf{E}(x_1, x_2, x_3) \qquad \text{for} \quad -d < x_1 < 0\,. \tag{4.15}$$

Therefore, the effect of the slab is just a translation. The same is true for the **H**, **D**, and **B** fields.

Let $\mathbf{E}_0$ be the field which solves the electromagnetic problem for a given source in the absence of a slab. Let us now insert a negative index slab in the field. The effect of the slab is that the fields appear to move to the right of the slab, i.e., to the right of the slab it appears as if all the fields have been moved a distance $2d$. In other words, it appears that the source has been moved a distance $2d$ to the right. This implies that the slab works as a “perfect lens” in the sense that the image to the right of the slab is not diffraction limited. This observation was first made by Pendry (2000) and was a surprising result because most lenses were though to be diffraction limited.

Consider for instance the ordinary lens shown in Figure 4.9(a). From geometric optics we expect the rays from the source to be focused at a point. However, if we consider the wave nature of electromagnetic radiation, several Fourier components of the wave are superimposed at the focal point and the maximum resolution of the image can never be greater than $\lambda/2$ where λ is the wavelength. On the other hand,

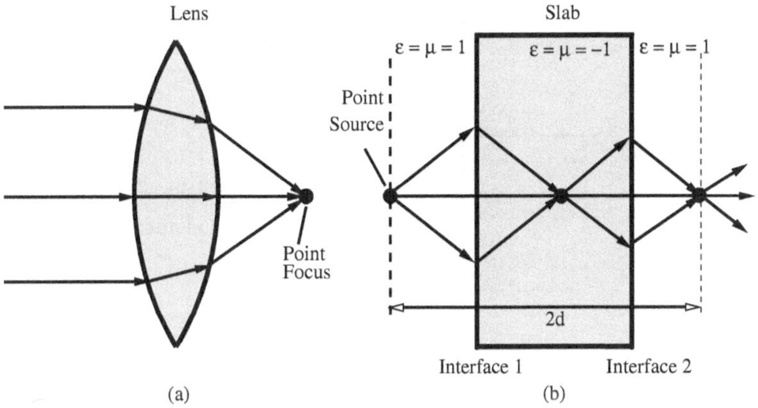

FIGURE 4.9

Focusing due to an ordinary lens versus a "perfect" lens. a) Ordinary lens. b) "Perfect" slab lens.

the lensing effect with a slab of negative index material is expected to lead to a point source being exactly represented at the focal point. This idea dates back to Veselago (1968). From Figure 4.9(b) we observe that there will appear to be sources inside the lens and at the focal point when a negative index slab is used as a lens. However, a point source leads to a singularity in the Maxwell equations and there should be no singularities where there are no physical point sources. This is a paradox.

The paradox can be resolved by observing that, in fact, a solution does not exist to the time harmonic equations if $\varepsilon = \mu = -1$ in the slab, However, if we let

$$\varepsilon = -1 + i\delta \qquad \text{and} \qquad \mu = -1 + i\delta$$

and let $\delta \to 0$, then we do have a solution. This is equivalent to assuming that there is some loss in the material due to the electrical conductivity of the material (see Section 1.4.8 for more on electromagnetic dissipation and its relation to electrical conductivity).

For $d_0 > d/2$ (see Figure 4.10), the fields blow up to infinity within a strip of width $2(d - d_0)$ starting from the focal point within the slab to the focal point outside the slab. They develop more and more oscillations (in space), i.e., at finer and finer length scales. In the remaining regions, the field converge to Pendry's solution. Therefore, the image looks like a point source only on one side of the lens if $\delta \neq 0$. However, in the limit that $\delta \to 0$, the image also looks like a point source (see Milton et al. (2005) for details).

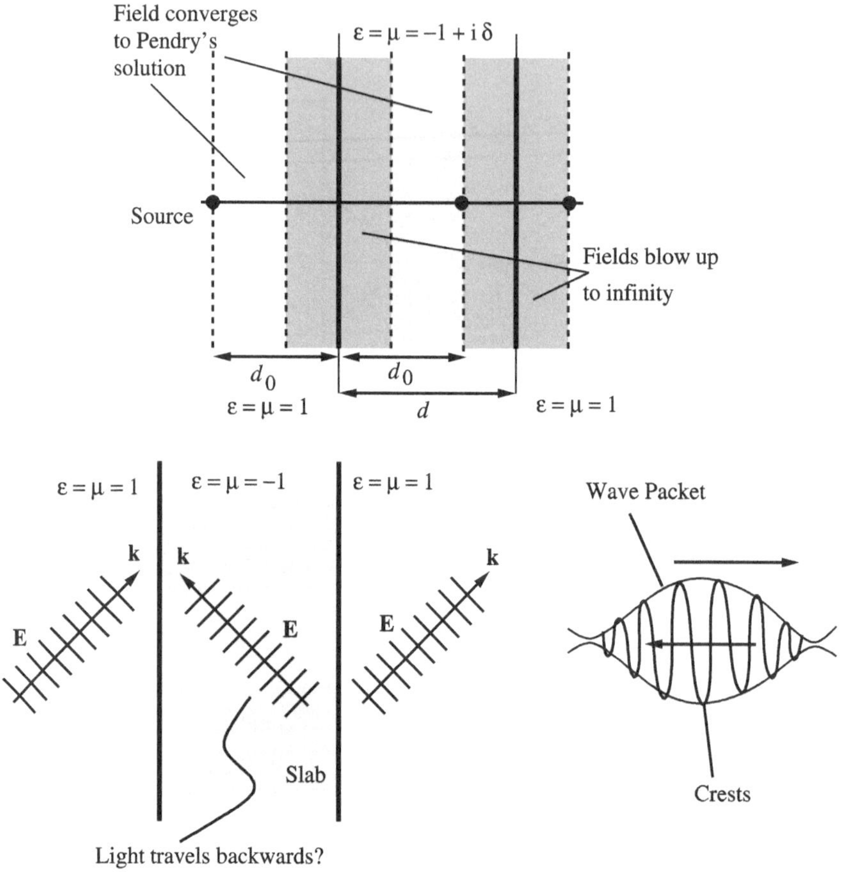

FIGURE 4.10
Behavior of fields around and inside a "perfect" lens and the direction of the wave vector and the wave packets.

If we look at the wave vectors of the electromagnetic waves, then from the reflected direction of the wave vector inside the lens it appears that light travels backwards inside a negative refractive index lens. But one has to remember that it is the wave crests that are traveling backwards and transport of energy is in the direction propagation of the EM waves (and therefore it is not useful to think of the direction of the wave vector as the direction of the Poynting vector).[‡] Note that the phase velocity is negative if the refractive index of the material is negative.

4.4.1 Pendry's argument

To get a feel for the insight that Pendry (2000) provided, let us examine the reflection and transmission coefficients for a negative index slab when a point dipole source is placed in front of it. Recall from equation (3.37) that we can expand a dipole source into plane waves using the Weyl identity to get an expression of the form

$$\mathbf{E}(\mathbf{r}) = \int_{-\infty}^{\infty}\int_{-\infty}^{\infty} \mathbf{E}_{\mu}(k_x,k_y)\, e^{ik_x x+ik_y y+ik_z|z|}\, \mathrm{d}k_x\, \mathrm{d}k_y$$

where $k_z^2 = \omega^2\varepsilon\mu - k_x^2 - k_y^2$. If the dipole source is placed in front of a lens, the effect of the lens is to apply a phase correction to each of the Fourier components inside the integral such that at some distance from the source we get an image of the dipole source (note that the image is not a source itself). When $k_x^2 + k_y^2 < \omega^2\varepsilon\mu$ we have $k_z^2 > 0$ and this situation corresponds to propagating waves in the z-direction. But when $k_x^2 + k_y^2 > \omega^2\varepsilon\mu$ we get $k_z^2 < 0$, i.e., k_z is imaginary with a positive imaginary part (needed for the integral to converge). This situation corresponds to the case where the waves decay exponentially with z and are evanescent. The requirement that the waves propagate in the direction of the lens limits the magnitude of $\mathbf{k}$ that is allowable. Therefore the maximum resolution of an image can never be greater than the wavelength λ under ordinary circumstances.

Pendry's argument was that this resolution limit did not apply to negative refractive index lenses. If the z-direction is along the axis of the lens, then from equation (2.65) (p. 87) for TE-waves, we know that the reflection and transmission coefficients at the interface between two layers are given by

$$R = \frac{\mu_2 k_{z1} - \mu_1 k_{z2}}{\mu_2 k_{z1} + \mu_1 k_{z2}} \quad \text{and} \quad T = \frac{2\mu_2 k_{z1}}{\mu_2 k_{z1} + \mu_1 k_{z2}}.$$

We also know from equations (2.69) and (2.70) that the effective reflection and transmission coefficients for the slab can be expressed as

$$\widetilde{R}_{12} = R_{12} + \frac{T_{12}T_{21}R_{23}\, e^{2ik_{z2}(d_2-d_1)}}{1-R_{21}R_{23}\, e^{2ik_{z2}(d_2-d_1)}} \quad \text{and} \quad T_{13} = \frac{T_{12}T_{23}\, e^{ik_{z2}(d_2-d_1)}}{1-R_{21}R_{23}\, e^{2ik_{z2}(d_2-d_1)}}. \tag{4.16}$$

For the case of a slab of relative magnetic permeability μ_r with vacuum on both sides we have

$$R_{12} = \frac{\mu_r k_{z1} - k_{z2}}{\mu_r k_{z1} + k_{z2}} \quad \text{and} \quad R_{21} = R_{23} = \frac{k_{z2} - \mu_r k_{z1}}{k_{z2} + \mu_r k_{z1}}. \tag{4.17}$$

[‡]The Poynting vector is $\mathbf{S} = \mathbf{E}\times\mathbf{H}$ and represents the direction of energy flow. In negative index slabs, the wave vector and the Pointing vector point in opposite directions. This does not mean that the flow of energy is not in the direction of the Poynting vector, just that the phase velocities are in another direction.

Also,

$$T_{12} = \frac{2\mu_r k_{z1}}{\mu_r k_{z1} + k_{z2}} \quad \text{and} \quad T_{21} = T_{23} = \frac{2k_{z2}}{k_{z2} + \mu_r k_{z1}}. \tag{4.18}$$

At this point it is worth recalling that the above equations were derived with $k_z^2 = \omega^2\mu\varepsilon - k_x^2$. Let us examine the case where the incident waves are evanescent, i.e., $k_x^2 > \omega^2\mu\varepsilon$. Then, $k_z = i\sqrt{k_x^2 - \omega^2\mu\varepsilon}$ and we have

$$k_{z1} = i\sqrt{k_x^2 - \omega^2\mu_0\varepsilon_0} \quad \text{and} \quad k_{z2} = i\sqrt{k_x^2 - \omega^2\mu_r\varepsilon_r\mu_0\varepsilon_0}\,. \tag{4.19}$$

Substituting (4.17) and (4.18) into (4.16), using (4.19), and taking the limit $\mu_r \to -1$ and $\varepsilon_r \to -1$, we get

$$\lim_{\substack{\mu_r\to-1\\ \varepsilon_r\to-1}} \widetilde{R}_{12} = 0 \quad \text{and} \quad \lim_{\substack{\mu_r\to-1\\ \varepsilon_r\to-1}} T_{13} = e^{-ik_{z2}(d_2-d_1)} = e^{-ik_{z1}(d_2-d_1)}\,.$$

Since k_{z1} is imaginary we find that T_{13} now increases with increasing values $k_x^2 + k_y^2$, i.e., instead of exponential decay we now observe an *amplification* of the wave. This discovery led Pendry to conclude that both propagating and evanescent parts can contribute to the image of a source and there is no limit on the allowable value of $\|\mathbf{k}\|$. Hence perfect imaging of an object is possible with a negative index slab. Note also that if $\varepsilon_r = \mu_r = -1$ there is no reflection and the slab is perfectly impedance matched with the vacuum.

The idea of negative refraction is now well established. High-quality sub-diffraction lensing has been achieved in situations where the magnetic field can be assumed to be decoupled from the electric field by utilizing naturally occurring negative permittivity materials such as silver (Fang et al., 2005). Perfect lensing using man-made metamaterials has been hampered by the large losses associated with resonance and the limited range of frequencies over which such an effect can be achieved (Smith et al., 2003, Zharov et al., 2005) (see Zhang and Liu (2008) for some recent developments in this field). Attention has moved to active boosting of evanescent fields to reduce loss and to transformation-based methods for sub-wavelength imaging because of their potential for broadband applications with low loss. Some of the ideas behind transformation-based imaging are discussed in Chapter 6.

Exercises

Problem 4.1 Show that the effective permittivity of an array of thin wires with finite conductivity can be expressed in the form

$$\varepsilon(\omega) = \varepsilon_0\left(1 - \frac{\omega_p^2}{\omega^2 + i\Gamma\omega}\right) \qquad \text{with} \qquad \Gamma := \frac{\varepsilon_0 b^2\omega_p^2}{\pi a^2\sigma}.$$

Problem 4.2 Derive the effective permittivity relation in equation (4.9) using the approach in Felbacq and Bouchitté (2005).

Problem 4.3 Derive the effective magnetic permeability for the single split-ring resonator array in Figure 4.6 using the approach used for the array of cylinders.

Problem 4.4 Show that

$$\mu^{\text{eff}} = \mu_0 \left[1 + \frac{f\omega^2}{\omega_0^2 - \omega^2 - i\omega\Gamma} \right]$$

for the double split cylinder geometry using the assumptions in Pendry et al. (1999).

Problem 4.5 Show that for evanescent TM-waves incident on a slab lens in vacuum with axis along the z-direction, the effective transmission coefficient is

$$\lim_{\substack{\mu_r \to -1 \\ \varepsilon_r \to -1}} T = e^{-ik_{zi}d}$$

where d is the thickness of the slab and k_{zi} is the the z-component of the wave vector in the medium of incidence.

Problem 4.6 Show that evanescent pressure waves in an acoustic medium can also be amplified by a negative refractive index acoustic slab lens.

5

Acoustic and Elastodynamic Metamaterials

> Clearly, there is a need for some generalized continuum elastodynamic equations which govern the response of bodies with or without voids.
>
> GRAEME W. MILTON AND JOHN R. WILLIS, On modifications of Newton's second law and linear continuum elastodynamics, 2007.

In most studies of composite behavior, the effective mass density of a composite is assumed to be given by the volume fraction weighted average of the densities of the constituents. For instance, for a composite with n constituents, the effective mass density is defined as

$$\rho_{\text{eff}} = \langle \rho \rangle = \frac{1}{V}\sum_{i=1}^{n} m_i = \sum_{i=1}^{n} f_i \, \rho_i \, ; \qquad f_i := \frac{V_i}{V}, \; \rho_i := \frac{m_i}{V_i}$$

where V is the total volume occupied by the composite, V_i are the volumes occupied by the constituent phases, and m_i are the masses of the constituents. The unstated assumption in this definition is the effective density is unaffected by the relative motion of the components of the composite.

When low-frequency relative motion is considered in two-phase composites with a fluid matrix, we find that Ament's relation (Ament, 1953, Geertsma and Smit, 1961, Martin et al., 2010) holds, i.e.,

$$\rho_{\text{eff}} = \rho_m \left(\frac{1 - fQ}{1 + 2fQ} \right) \quad \text{where} \quad Q := \frac{\rho_m - \rho_i}{\rho_m + 2\rho_i}$$

where ρ_m is the mass density of the fluid matrix, ρ_i is the mass density of the inclusions (or scatterers), and f is the volume fraction of the scatterers. Similar relations have been found by Waterman and Truell (1961), Kuster and Toksöz (1974), and Berryman (1980). The frequency is implicit in the above relations. Clearly, the dynamic effective density can be different from the static density.

We may also observe a dynamic mass density effect in a poroelastic material where the fluid can move relative to the solid. Consider the example of a porous rock containing some water (see Figure 5.1). Both the rock grains and the water are connected. In this case, the water will move with a different frequency than the rock and the density of the composite will be dependent on the frequency. However, even when the relative motion of the constituents of the composite is considered, it is often found that the effective density is just the average density (Martin et al., 2010).

We get a frequency-dependent density if all the constituents do not move in lock step. In fact, lock step motion almost never occurs in ordinary materials because there are thermal vibrations at the micro-scale. At a molecular level, we can imagine a crystal with tungsten or lead atoms attached by single bonds to the lattice (see Figure 5.1). Presumably, the resonant frequency of such molecules is very high so we will see the frequency dependence of the mass only at very high frequencies.

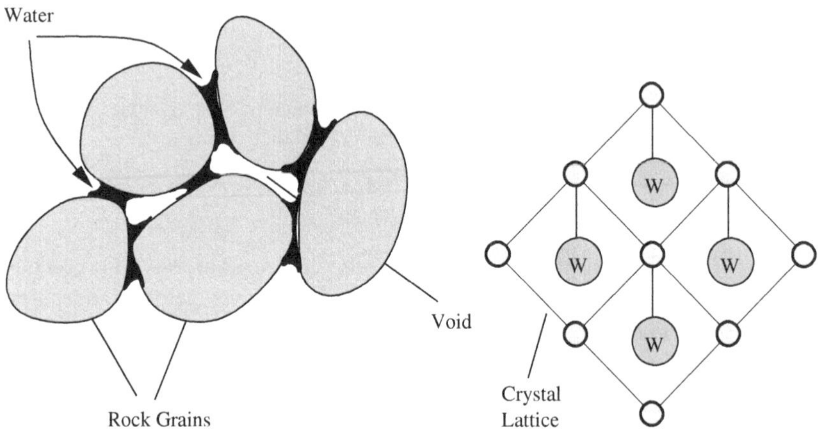

FIGURE 5.1
Frequency dependence of mass can be caused by the relative motion of the components of a structure. Left: A porous rock containing water and voids. Right: A crystal lattice containing tungsten atoms.

There are many other macroscopic situations in which lock step motion does not occur. Sheng et al. (2003) have shown in experiments that designed composite materials can indeed have frequency-dependent masses. An example of such a material is shown in Figure 5.2. The material is composed of a number of small coated balls in a matrix. The outer matrix is made of epoxy, the balls are steel, and the coating is a layer of silicone rubber. At certain frequencies, the steel balls can move in the direction opposite to the applied force, thus creating a composite with a negative effective mass. Such materials are often called *locally resonant sonic materials* (LRSM) (Liu et al., 2000).

In this chapter we will explore the following questions about mass density.

- Can there be situations where the relative motion of the constituents of a composite leads to a frequency-dependent effective dynamic density that is different from the static density?
- Are there situations where the effective mass can be negative and/or complex?
- Are there situations where the effective mass density is an anisotropic, tensor quantity?

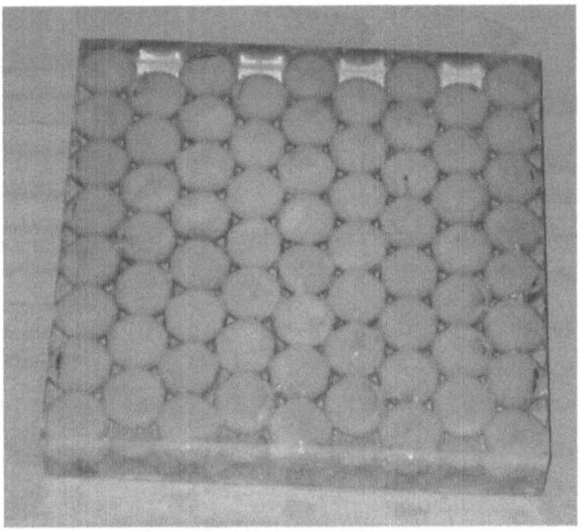

FIGURE 5.2
A composite material with frequency-dependent mass. The composite consists of steel balls coated with a layer of silicone in an epoxy matrix.

We will also show that an apparently rigid body can have a frequency-dependent mass moment of inertia if it contains hidden gyroscopic elements.

A similar set of questions arises in the theory of elastic composites. Elastic stability is guaranteed if the elastic bulk and shear moduli are positive. After the discovery that stable isotropic materials with negative Poisson's ratios can be designed (Lakes, 1993), an important question in elasticity has been

Are there stable elastic materials with negative moduli?

Also, the Willis equations for the elastodynamics of composites Willis (1997) suggest that the stress in certain composite materials is a function not only of the strain but also of the velocity. A question that is of interest in this regard is

Can we design materials that are described by the Willis equations?

We will discuss the Willis equations and their extension to materials with voids and show that structures can indeed be designed to exhibit Willis-like behavior.

Another important class of unusual composites, called extremal composites, will also be examined in this chapter. We will focus on pentamode extremal materials because of their potential use as cloaking materials and also because these models may be used to describe certain granular materials.

5.1 Dynamic mass density

Let us start by exploring the difference between "dynamic" mass density and the "static" mass density.* The static mass density is defined as the mass per unit volume. In contrast, the dynamic density is defined as the inertial mass density that appears in Newton's law $f = ma$ where f is the force, m is the inertial mass, and a is the acceleration.

Is the dynamic mass density the same as the static mass density? Interest in this question was re-ignited after the discovery of an elastic material with local resonances by Sheng and coworkers (Sheng et al., 2003, Liu et al., 2005). A mathematical analysis of the problem was provided by Avila et al. (2005). Several models of composites with differing dynamic and static mass densities have been explored since 2003. A number of these were proposed by Milton and Willis (2007). We discuss some of these models in this section. We then explore a model where the mass density is complex. The model is then extended so that we get a composite for which the dynamic mass density is anisotropic.

5.1.1 A simple model with negative mass

Consider a rigid bar with n voids each of width d as shown in Figure 5.3.† Each cavity contains a spherical ball of mass m and radius r. The ball is attached to the walls of the cavity by springs with spring constant K. A force $F(t)$, where t is the time, is applied on the left side of the rigid bar. Our aim is to find the response of the bar as a function of time.

Let us consider a balance of linear momentum in the composite system. We assume that all quantities depend harmonically on time and that a one-dimensional approximation of the problem is adequate. The spring attached to the left wall of the cavity exerts a force $f_1(t)$ on the wall while the spring attached to the right wall exerts a force of $f_2(t)$ on the wall. Then the internal and external forces acting on a unit cell of the bar are given by

$$F(t) = \mathrm{Re}(\widehat{F}\, e^{-i\omega t})\ ;\ \ f_1(t) = \mathrm{Re}(\widehat{f}_1\, e^{-i\omega t})\ ;\ \ f_2(t) = \mathrm{Re}(\widehat{f}_2\, e^{-i\omega t})$$

where the amplitudes $\widehat{F}$, $\widehat{f}_1$, and $\widehat{f}_2$ are generally complex. Let the time-dependent position of the left side of each cavity be given by

$$X(t) = X_0 + U_0(t) = X_0 + \mathrm{Re}(\widehat{U}_0\, e^{-i\omega t})$$

*Recent work on the subject has focused on the possibility of a "negative" dynamic mass density. Note that, for some applications in acoustics, it is often the difference between the static and the dynamic mass that is more important than the possibility of a negative dynamic mass.

†Our discussion is based on Milton and Willis (2007). A similar model was proposed earlier by Maysenhölder (2003). The spring constant may be complex valued to allow for materials with viscous damping.

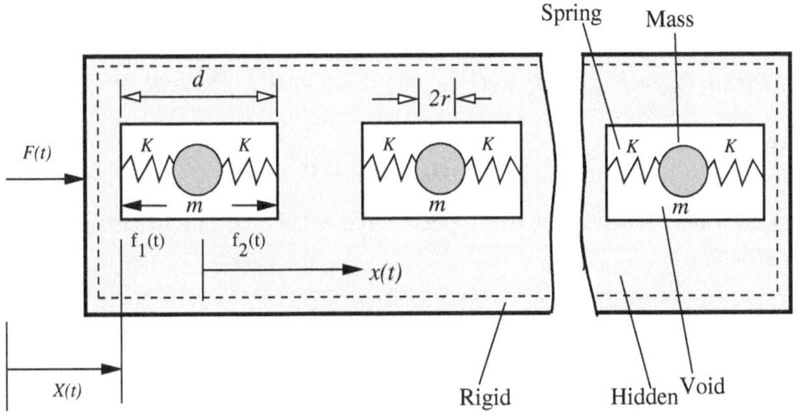

FIGURE 5.3
A rigid bar containing n voids. Each void contains a spherical ball that is attached to the bar by springs.

where X_0 is the initial position and $\widehat{U}_0$ is the complex valued displacement of the bar. Then the velocity of the rigid bar is

$$V_0(t) = \frac{dX}{dt} = \text{Re}(-i\omega\, \widehat{U}_0\, e^{-i\omega t}) =: \text{Re}(\widehat{V}_0\, e^{-i\omega t})\,.$$

Assume that the rigid bar has mass M_0. Therefore, the linear momentum of the rigid bar is

$$P_{\text{bar}}(t) = M_0\, V_0(t) = M_0\, \text{Re}(\widehat{V}_0\, e^{-i\omega t})\,.$$

If $U(t)$ is the relative displacement of the ball, the position of the ball is given by

$$x(t) = X(t) + \frac{d}{2} + U(t) = X_0 + \frac{d}{2} + [U_0(t) + U(t)] =: X_0 + \frac{d}{2} + u(t)\,.$$

For harmonic motions, we can write

$$x(t) = X_0 + \frac{d}{2} + \text{Re}(\widehat{u}\, e^{-i\omega t}) \tag{5.1}$$

where $\widehat{u}$ is the complex valued displacement of each ball. Therefore, the velocity of each ball is

$$v(t) = \frac{dx}{dt} = \text{Re}(-i\omega\, \widehat{u}\, e^{-i\omega t}) =: \text{Re}(\widehat{v}\, e^{-i\omega t})\,.$$

If there are n balls, the total linear momentum of the balls is

$$P_{\text{ball}}(t) = n\, m\, \frac{dx}{dt} = n\, m\, \text{Re}(\widehat{v}\, e^{-i\omega t})\,.$$

Then the total linear momentum of the system is

$$P(t) = P_{\text{bar}} + P_{\text{ball}} = M_0 \,\text{Re}(\widehat{V}_0\, e^{-i\omega t}) + n\, m\, \text{Re}(\widehat{v}\, e^{-i\omega t}) =: \text{Re}(\widehat{P}\, e^{-i\omega t})\,.$$

Therefore,

$$\widehat{P} = M_0\, \widehat{V}_0 + n\, m\, \widehat{v}\,. \tag{5.2}$$

From Newton's second law, the applied force equals the rate of change of linear momentum,

$$F(t) = \frac{d}{dt}[P(t)]\,.$$

Hence, for harmonic forces, we have

$$\widehat{F} = -i\omega\, \widehat{P} = -i\omega(M_0 \widehat{V}_0 + nm\widehat{v})\,. \tag{5.3}$$

Note that $\widehat{v}$ is unobservable since it is in the hidden part of the bar and we need to relate $\widehat{P}$ directly to the observable velocity $\widehat{V}_0$.

Let us now consider the free-body diagram of each spring inside a cavity as shown in Figure 5.4.

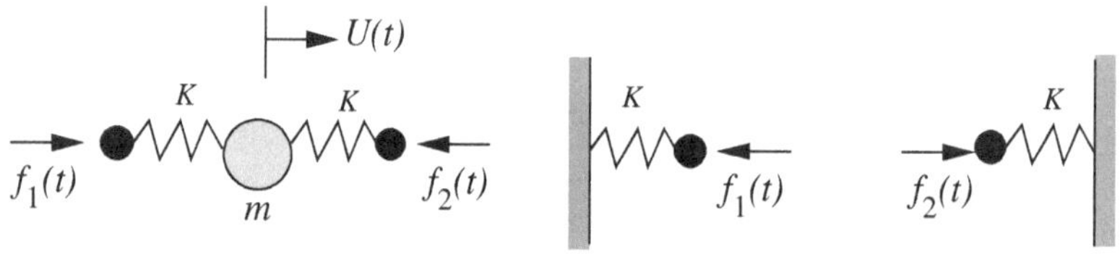

FIGURE 5.4
Free-body diagrams for the spring-mass system and for each spring.

Hooke's law for each spring implies that (note that $u(t)$ is positive in the positive x-direction)

$$-f_1(t) = K\, u(t) = f_2(t)$$

where K is the complex spring constant. Recall (from equation 5.1) that the displacement of the spring is given by

$$U(t) = \text{Re}(\widehat{u}\, e^{-i\omega t}) - \text{Re}(\widehat{U}_0\, e^{-i\omega t})\,.$$

Using the assumed harmonic forms of $f_1(t)$, $f_2(t)$ and $U(t)$, we then have

$$\text{Re}(\widehat{f}_1\, e^{-i\omega t}) = -\text{Re}(\widehat{f}_2\, e^{-i\omega t}) = K\left[\text{Re}(\widehat{U}_0\, e^{-i\omega t}) - \text{Re}(\widehat{u}\, e^{-i\omega t})\right]$$

or

$$\widehat{f}_1 = -\widehat{f}_2 = K\,(\widehat{U}_0 - \widehat{u})\,. \tag{5.4}$$

Next, considering the free body diagram of the spring-mass system (see Figure 5.4 and equation (5.1)), the balance of linear momentum for the spring-mass system implies that

$$f_1(t) - f_2(t) = m\,\frac{d^2x}{dt^2} = m\,\mathrm{Re}(-\omega^2\,\widehat{u}\,e^{-i\omega t})\,.$$

Therefore, substituting in the harmonic forms of $f_2(t)$ and $f_1(t)$, we get

$$\widehat{f_1} - \widehat{f_2} = -m\,\omega^2\,\widehat{u}\,. \tag{5.5}$$

From equations (5.4) and (5.5), we have

$$\widehat{f_1} - \widehat{f_2} = -m\,\omega^2\,\widehat{u} = 2\,K\,(\widehat{U}_0 - \widehat{u})$$

or

$$\widehat{u} = \frac{2\,K}{2\,K - m\,\omega^2}\,\widehat{U}_0\,.$$

Now $\widehat{V}_0 = -i\omega\widehat{U}_0$ and $\widehat{v} = -i\omega\widehat{u}$. Hence,

$$\widehat{v} = \frac{2\,K}{2\,K - m\,\omega^2}\,\widehat{V}_0\,. \tag{5.6}$$

Plugging equation (5.6) into equation (5.2), we get

$$\widehat{P} = \left(M_0 + \frac{2\,K\,n\,m}{2\,K - m\,\omega^2}\right)\widehat{V}_0 =: M\,\widehat{V}_0\,.$$

M is the *effective dynamic mass* of the model and is given by

$$\boxed{M(\omega) = M_0 + \frac{2Knm}{2K - m\,\omega^2} = M_0\left[1 - \frac{\alpha\omega_r^2}{\omega^2 - \omega_r^2}\right]} \tag{5.7}$$

where ω_r is the resonance frequency,

$$\omega_r := \sqrt{\frac{2\,K}{m}} \quad \text{and} \quad \alpha := \frac{nm}{M_0}.$$

The effective mass depends on the frequency ω *and is different from the static mass.* At $\omega = 0$, the effective mass is equal to the rest mass $M_0 + nm$. Close the resonance frequency, the effective mass can take *high positive values or negative values*. This observation is important in the design of acoustic metamaterials where the high absolute value of the effective mass can be exploited. We can think of such materials being considerably heavier close to resonance and therefore having a higher transmission loss than materials that have a larger static density. A normalized plot of the effective mass versus the frequency in shown in Figure 5.5.

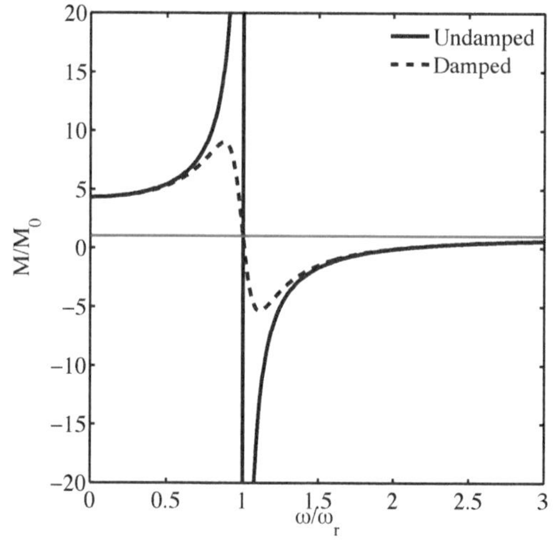

FIGURE 5.5
The effective mass of the bar as a function of the frequency. ω_r is the resonance frequency.

5.1.2 Complex mass density

In the previous section we have considered springs with a real spring constant. Let us now consider the situation where there is a small amount of dissipation in the system, i.e., the springs have an imaginary part. To achieve springs with an imaginary part to the spring constant, let us use the one-dimensional Maxwell model depicted in Figure 5.6. A complex spring constant can be obtained if we replace the springs in Figure 5.3 with an elastic spring (with real spring constant k) in series with a dashpot (with a real viscosity η).

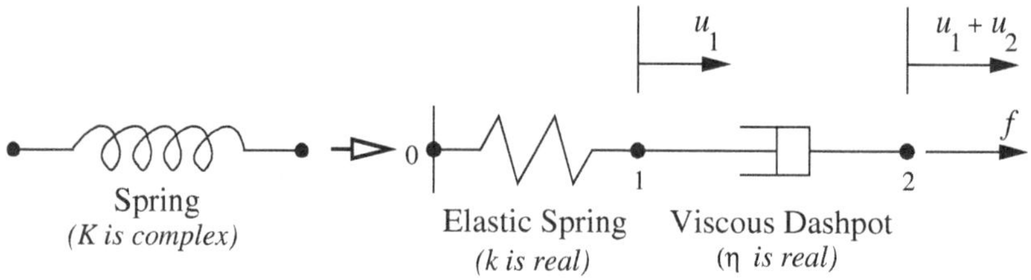

FIGURE 5.6
The Maxwell model for a spring with complex spring constant.

Let u_1 be the displacement of the elastic spring and let u_2 be that of the dashpot under the action of a force f. For the elastic spring, we have

$$u_1(t) = \frac{f(t)}{k}. \tag{5.8}$$

For the dashpot,

$$\frac{du_2}{dt} = \frac{f(t)}{\eta}. \tag{5.9}$$

Once again, assuming that $f(t)$, $u_1(t)$, and $u_2(t)$ can be expressed as harmonic functions of t, we have

$$u_1(t) = \widehat{u}_1\, e^{-i\omega t}\ ;\ u_2(t) = \widehat{u}_2\, e^{-i\omega t}\ ;\ f(t) = \widehat{f}\, e^{-i\omega t}\,. \tag{5.10}$$

Plugging equations (5.10) into equations (5.8) and (5.9), we get

$$\widehat{u}_1 = \frac{\widehat{f}}{k} \qquad \text{and} \qquad -\,i\omega\,\widehat{u}_2 = \frac{\widehat{f}}{\eta}. \tag{5.11}$$

Recall that the displacement $u(t)$ of the sphere of mass m inside the cavity is related to the applied force $f(t)$ by the relation

$$u(t) = \frac{f(t)}{K} \qquad \Longrightarrow \qquad \widehat{u} = \frac{\widehat{f}}{K}.$$

Since $\widehat{u} = \widehat{u}_1 + \widehat{u}_2$, we have

$$\frac{\widehat{f}}{K} = \frac{\widehat{f}}{k} - \frac{1}{i\omega}\frac{\widehat{f}}{\eta}$$

or

$$\frac{1}{K} = \frac{1}{k} + \frac{i}{\omega\eta}.$$

So, if the springs behave like Maxwell elements, the effective mass of the system is

$$M(\omega) = M_0 + \frac{2\,n\,m}{2 - \dfrac{1}{K}\,m\,\omega^2} = M_0 + \frac{2\,n\,m}{2 - \dfrac{1}{k}\,m\,\omega^2 - \dfrac{i}{\omega\eta}\,m\,\omega^2}$$

or

$$\boxed{M(\omega) = M_0 + \frac{2\,n\,k}{\dfrac{2\,k}{m} - \omega^2 - \dfrac{i\omega\,k}{\eta}} = M_0\left[1 - \frac{\alpha\omega_r^2}{\omega^2 - \omega_r^2 + i\gamma\omega}\right]} \tag{5.12}$$

where $\omega_r^2 := 2k/m$, $\gamma := k/\eta$, and $\alpha := nm/M_0$. This model is remarkably similar to a simple model for the frequency dependent permittivity $\varepsilon(\omega)$ that has inspired the design of negative permittivity and negative magnetic permittivity materials (see Section 4.1 for details). Clearly, the effective mass of the system can be a complex quantity if dissipation is introduced into the system via a dashpot.

If the damping constant k/η is small compared to $2k/m$, then $M(\omega)$ is approximately real for reasonably low values of ω. This is true for a spring with a small amount of damping. If, in addition, $\omega^2 < 2k/m$, then the effective mass is positive. This is usually true for low frequencies. However, if $\omega^2 > 2k/m$, the effective mass becomes negative. We can therefore choose the ratio of the spring stiffness to the mass of the ball in such a way that the effective mass becomes negative even at low frequencies.

A plot of the magnitude and phase of the complex effective mass is shown in Figure 5.7. The effect of the resonance is that the absolute dynamic mass of the bar appears to be much large than the static mass and there is a change of phase. However, for a range of frequencies larger than the resonance frequency, the dynamic mass is actually smaller than the static mass. The amount of phase change, and hence the region where the dynamic mass appears to be negative, depends quite strongly on the amount of damping in the spring.

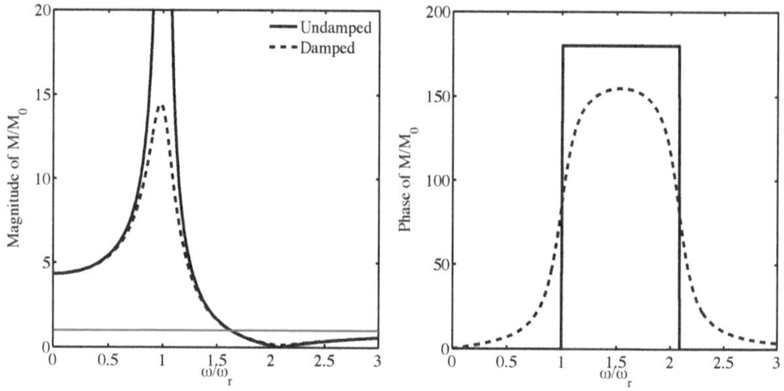

FIGURE 5.7

Magnitude and phase of the complex effective mass of the bar as a function of frequency; ω_r is the resonance frequency for when the springs are not dissipative.

A Lagrangian-based approach

Spring-mass systems can be quite a convenient way to analyze the expected response of acoustic and elastodynamic systems. The method we have used to calculate the effective mass in the previous section can be quite cumbersome if multiple springs and masses are involved. Instead, the dynamic response of the bar to a harmonic forcing can be calculated in a straightforward way by starting with the Lagrangian of the motion. For the hidden spring-mass system in Figure 5.3 the Lagrangian is

$$\mathcal{L} = \frac{1}{2}\, k_1\, (u_1 - U)^2 + \frac{1}{2}\, k_2\, (U - u_1)^2 - \frac{1}{2}\, m_1\, \dot{u}_1^2$$

where u_1 is the displacement of the mass m_1, U is the displacement of the bar, and k_1 and k_2 are the spring stiffnesses of the two springs in the cavity. The Euler-Lagrange equations of the system are given by‡

$$\frac{\partial \mathcal{L}}{\partial u_i} - \frac{d}{dt}\left(\frac{\partial \mathcal{L}}{\partial \ddot{u}_i}\right) = 0\,.$$

For a cavity with a single mass,

$$k_1\,(u_1 - U) - k_2\,(U - u_1) + m_1\,\ddot{u}_1 = 0\,.$$

Let us assume that the displacements vary harmonically due to the applied harmonic force, i.e.,

$$u_i = \text{Re}[\hat{u}_i\,\exp(-i\,\omega\,t)]\,.$$

This assumption leads, after dropping the hats for simplicity, to the equation

$$u_1 = \frac{k_1 + k_2}{(k_1 + k_2) - \omega^2\,m_1}\,U\,.$$

The momentum of a representative element of the bar in harmonic space is given by

$$P = M_0\,U + m_1\,u_1 = \left(M_0 + \frac{k_1 + k_2}{\omega_r^2 - \omega^2}\right)\,U =: M_{\text{eff}}\,U \tag{5.13}$$

where M_0 is the mass of the bar excluding the ball and the resonance frequency is $\omega_r = (k_1 + k_2)/m_1$. Note that this expression is identical to that in equation (5.7). The effective mass is negative for higher frequencies and tends to M_0 as the ratio $k_i/\omega \to 0$.

To compare our model with results from impedance tube experiments, we use the effective mass to calculate the displacement of the bar as a function of the applied force. In Fourier space, the displacement of the bar (u) and its velocity (v) are given by

$$u = -\frac{f}{M_{\text{eff}}\,\omega^2} \quad \text{and} \quad v = -i\omega u\,.$$

The impedance of the system is given by

$$Z = \frac{f}{v} = -i\,M_{\text{eff}}\,\omega\,. \tag{5.14}$$

‡The Lagrangian used above assumes that the system is conservative, i.e., there is not dissipation. Dissipation can be included after deriving the equation of motion by adding a damping term to the equation. Alternatively, one can add a Rayleigh dissipation term to the Lagrangian and use modified Euler-Lagrange equations,

$$\mathcal{D} = -\frac{\lambda}{2}\left[\frac{d}{dt}(\|u_i - u_j\|)\right]^2 \;;\; \frac{\partial \mathcal{L}}{\partial u_i} - \frac{d}{dt}\left(\frac{\partial \mathcal{L}}{\partial \dot{u}_i}\right) = -\frac{\partial \mathcal{D}}{\partial \dot{u}_i}\,.$$

Alternatively we could start with an appropriate Hamiltonian. See Ostoja-Starzewski (2002) for a review of lattice models in elasticity and for examples of Lagrangians and Hamiltonians.

We can then calculate the transmission loss in the system by using the relation

$$R = 20 \log_{10} \left| 1 + \frac{Z}{2\, \rho_0\, c_0} \right| . \tag{5.15}$$

The single mass and two springs give us a flavor of what to expect when there are multiple masses and springs in series. Let us now consider the situation where there are n masses connected in series with $n+1$ springs as shown in Figure 5.8.

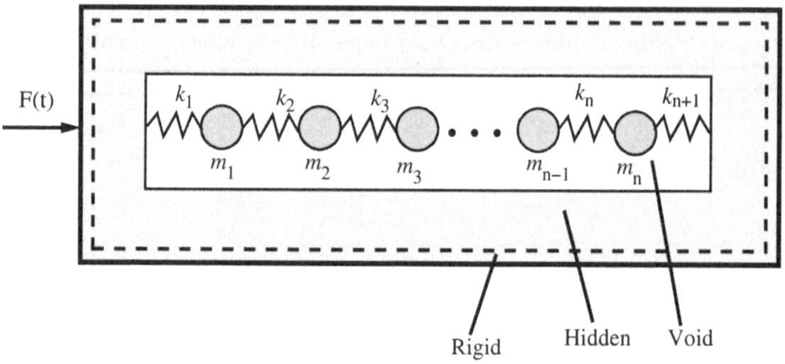

FIGURE 5.8
A rigid bar containing cavities with a number of springs and masses.

In this case the Lagrangian of the system is given by

$$\mathcal{L} = \tfrac{1}{2}\, k_1\, (u_1 - U)^2 + \tfrac{1}{2}\, k_2\, (u_2 - u_1)^2 + \tfrac{1}{2}\, k_3\, (u_3 - u_2)^2 + \cdots + \tfrac{1}{2}\, k_n\, (u_n - u_{n-1})^2 + \tfrac{1}{2}\, k_{n+1}\, (U - u_n)^2 - \tfrac{1}{2}\, m_1\, \dot{u}_1^2 - \tfrac{1}{2}\, m_2\, \dot{u}_2^2 - \ldots \tfrac{1}{2}\, m_n\, \dot{u}_n^2$$

The Euler-Lagrange equations of the system, in matrix form, are given by

$$\begin{bmatrix} (k_1+k_2) - m_1\omega^2 & -k_2 & 0 & \ldots & 0 \\ -k_2 & (k_2+k_3) - m_2\omega^2 & -k_3 & \ldots & 0 \\ & & \ddots & & \\ 0 & \ldots & 0 & -k_n & (k_n + k_{n+1}) - m_n\omega^2 \end{bmatrix} \begin{bmatrix} u_1 \\ u_2 \\ \vdots \\ u_n \end{bmatrix} = \begin{bmatrix} k_1\, U \\ 0 \\ \vdots \\ k_{n+1}\, U \end{bmatrix}.$$

Note that this system of equations may be written as

$$(\mathbf{K} - \omega^2 \mathbf{M})\mathbf{u} = \mathbf{f} \tag{5.16}$$

where $\mathbf{K}$ is a tridiagonal matrix and $\mathbf{M}$ is a diagonal matrix. The system of equations can be inverted to find $\mathbf{u}$. Following the approach used for the single mass system, we can write the total momentum as

$$P = M_0 U + \sum_{i=1}^{n} m_i u_i = \left(M_0 + \sum_{i=1}^{n} m_i \frac{u_i}{U} \right) U =: M_{\text{eff}}\, U \tag{5.17}$$

We can calculate the impedance of the system and the transmission loss using equations (5.14) and (5.15).

Let us now replace the rigid material of the bar with an elastic material. In that case, one possible spring-mass model for the bar is of the form shown in Figure 5.9.

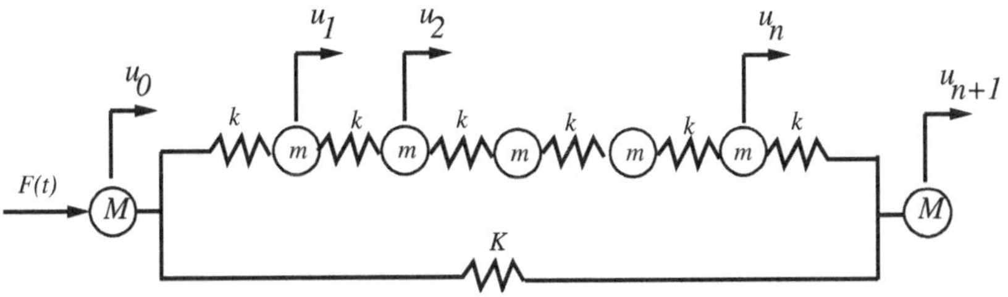

FIGURE 5.9
Representation of an elastic bar containing cavities with a number of springs and masses.

The Lagrangian of the system is

$$\begin{aligned}\mathcal{L} = &\tfrac{1}{2}\, k_1\, (u_1 - u_0)^2 + \tfrac{1}{2}\, k_2\, (u_2 - u_1)^2 + \tfrac{1}{2}\, k_3\, (u_3 - u_2)^2 + \cdots + \tfrac{1}{2}\, k_n\, (u_n - u_{n-1})^2 \\ &+ \tfrac{1}{2}\, k_{n+1}\, (u_{n+1} - u_n)^2 - \tfrac{1}{2}\, m_1\, \dot{u}_1^2 - \tfrac{1}{2}\, m_2\, \dot{u}_2^2 - \cdots - \tfrac{1}{2}\, m_n\, \dot{u}_n^2 \\ &+ \tfrac{1}{2}\, K_0\, (u_{n+1} - u_0)^2 - \tfrac{1}{2}\, \tfrac{M_0}{2}\, \dot{u}_0^2 - \tfrac{1}{2}\, \tfrac{M_0}{2}\, \dot{u}_{n+1}^2 - F_1\, u_0 - F_2 u_{n+1} + \frac{K_0}{2} u_0^2 + \frac{K_0}{2} u_{n+1}^2\end{aligned}$$

The Euler-Lagrange equations of the system are given by

$$\begin{aligned} K_0 u_0 - k_1\, (u_1 - u_0) - K_0\, (u_{n+1} - u_0) - \tfrac{M_0}{2}\, \omega^2\, u_0 &= F_1 \\ k_1\, (u_1 - u_0) - k_2\, (u_2 - u_1) - m_1\, \omega^2\, u_1 &= 0 \\ k_2\, (u_2 - u_1) - k_3\, (u_3 - u_2) - m_2\, \omega^2\, u_2 &= 0 \\ &\vdots \\ k_n\, (u_n - u_{n-1}) - k_{n+1}\, (u_{n+1} - u_n) - m_n\, \omega^2\, u_n &= 0 \\ K_0 u_{n+1} + k_{n+1}\, (u_{n+1} - u_n) + K_0\, (u_{n+1} - u_0) - \tfrac{M_0}{2}\, \omega^2\, u_{n+1} &= F_2 \end{aligned}$$

In matrix form, we get an equation equivalent to (5.16),

$$(\mathbf{K} - \omega^2\, \mathbf{M})\, \mathbf{u} = \mathbf{f}$$

where

$$\mathbf{K} = \begin{bmatrix} (2K_0+k_1) & 0 & 0 & 0 & \dots & 0 & 0 & -K_0 \\ -k_1 & (k_1+k_2) & -k_2 & 0 & \dots & 0 & 0 & 0 \\ 0 & -k_2 & (k_2+k_3) & -k_3 & \dots & 0 & 0 & 0 \\ & & \ddots & & & & & \\ 0 & 0 & 0 & 0 & \dots & -k_n & (k_n+k_{n+1}) & -k_{n+1} \\ -K_0 & 0 & 0 & 0 & \dots & 0 & -k_{n+1} & (2K_0+k_{n+1}) \end{bmatrix}$$

and

$$\mathbf{M} = \begin{bmatrix} \frac{M_0}{2} & 0 & \dots & 0 & & \\ 0 & m_1 & 0 & \dots & 0 & \\ 0 & 0 & m_2 & 0 & \dots & 0 \\ & & \ddots & & & \\ 0 & \dots & \dots & 0 & m_n & 0 \\ 0 & \dots & \dots & 0 & 0 & \frac{M_0}{2} \end{bmatrix} ; \quad \mathbf{u} = \begin{bmatrix} u_0 \\ u_1 \\ u_2 \\ \vdots \\ u_n \\ u_{n+1} \end{bmatrix} ; \quad \mathbf{f} = \begin{bmatrix} F_1 \\ 0 \\ 0 \\ \vdots \\ 0 \\ F_2 \end{bmatrix} .$$

The systems of equations can be inverted directly to obtain $\mathbf{u}$ for a given $\mathbf{F}$. Alternatively, we can calculate the eigenvalues and eigenvectors of the system

$$(\mathbf{K} - \omega^2\, \mathbf{M})\, \mathbf{u} = \mathbf{0}$$

and determine the solution using modal superposition. Let λ_j be the eigenvalues and $\mathbf{n}_j$ be corresponding eigenvectors. Then

$$(\mathbf{K} - \lambda_j\, \mathbf{M})\, \mathbf{n}_j = \mathbf{0} \quad \Longrightarrow \quad \mathbf{n}_j^T \mathbf{K} \mathbf{n}_j = \lambda_j\, \mathbf{n}_j^T \mathbf{M} \mathbf{n}_j$$

where $j = 1 \dots n+2$. Normalizing the mass so that $\mathbf{n}_j^T \mathbf{M} \mathbf{n}_j = 1$ gives

$$\mathbf{n}_j^T \mathbf{K} \mathbf{n}_j = \lambda_j .$$

Since the eigenvectors form a basis for the system, the solution vector $\mathbf{u}$ can be expressed as

$$\mathbf{u} = \sum_{j=1}^{n+2} u_j\, \mathbf{n}_j .$$

Plugging this back into the original system of equations gives

$$\mathbf{K} \sum u_j \mathbf{n}_j - \omega^2\, \mathbf{M} \sum u_j \mathbf{n}_j = \mathbf{F}.$$

Premultiplying by $\mathbf{n}_j^T$ leads to a system of $n+2$ equations

$$u_j\, (\mathbf{n}_j^T \mathbf{K}\, \mathbf{n}_j - \omega^2\, \mathbf{n}_j^T \mathbf{M}\, \mathbf{n}_j) = \mathbf{n}_j^T \mathbf{F}$$

or

$$u_j\, (\lambda_j - \omega^2) = \mathbf{n}_j^T \mathbf{F}.$$

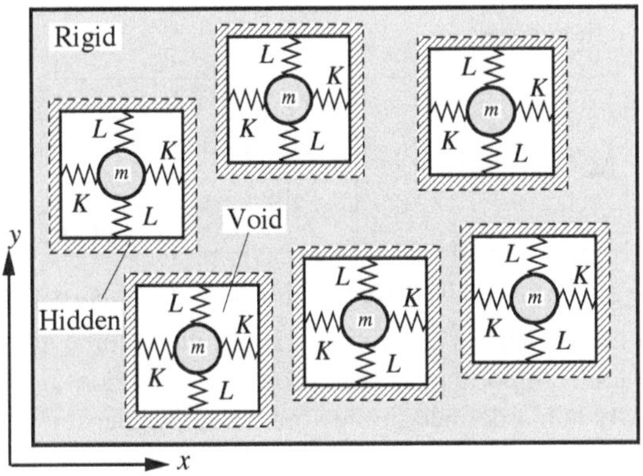

FIGURE 5.10
Schematic of a material with an anisotropic mass density. In this case the springs in each cavity are parallel to each other.

Solving for u_j gives

$$u_j = \frac{\mathbf{n}_j^T \mathbf{F}}{\lambda_j - \omega^2}.$$

We can calculate $\mathbf{u}$ given u_j and $\omega, \lambda_j, \mathbf{n}_j$.[§] Given an applied force f at the '0' location, we can calculate the displacement u_{n+1}. The velocity of the bar at that point is then given by $v = -i\,\omega\,u_{n+1}$. The impedance, effective mass, and transmission loss of the system can then be calculated using the usual procedure.

5.1.3 Anisotropic mass

Another material behavior that is of considerable interest is the possibility of materials with anisotropic mass densities. Let us first examine a two-dimensional extension of the model in Figure 5.3 and then examine an older version of the idea.[¶]

Consider the rigid body containing cavities shown in Figure 5.10. Here K and L are complex spring constants in the x- and y-directions. From equation (5.7), the effective mass along the x-direction is given by

$$M_x(\omega) = M_0 + \frac{2\,K\,n\,m}{2\,K - m\,\omega^2} \tag{5.18}$$

while that along the y-direction is given by

$$M_y(\omega) = M_0 + \frac{2\,L\,n\,m}{2\,L - m\,\omega^2}. \tag{5.19}$$

[§]Note that standard eigenvalue solvers may not be accurate when complex spring stiffnesses are involved and a direct solve is often preferable for small systems.

[¶]The spring-mass models discussed in this section were first developed by Milton and Willis (2007).

In matrix form, we then have

$$\underline{\underline{\mathsf{M}}}(\omega) = M_0 \underline{\mathsf{I}} + n\, m \begin{bmatrix} \dfrac{2\,K}{2\,K - m\,\omega^2} & 0 \\ 0 & \dfrac{2\,L}{2\,L - m\,\omega^2} \end{bmatrix} \tag{5.20}$$

where $\underline{\mathsf{I}}$ is the 2×2 identity matrix. Hence, the effective mass is clearly *anisotropic*. Note that from a macroscopic perspective it is not the volume averaged velocity in the composite which is important. In fact, such a quantity does not even make sense because the velocity is not defined in the void phase. Rather it is the velocity of the matrix that is the relevant quantity in this model.

One can generalize the model one step further by having the springs be oriented at different angles to each other as shown in Figure 5.11. Let $\boldsymbol{R}$ be the rotation that is needed to orient each set of springs with the x- and y-axes. Also, let the springs in each cavity have different spring constants and let the masses in each cavity be different. In matrix form, if $\underline{\underline{\mathsf{R}}}_j$ is the rotation matrix for cavity j containing a mass

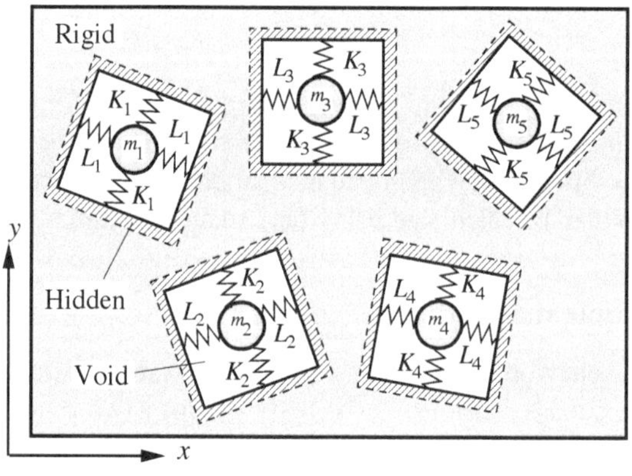

FIGURE 5.11

Schematic of a material in which the springs in each cavity are oriented at various angles to each other.

m_j, and if K_j and L_j are the complex spring constants for that cavity, the effective mass for a system with n cavities can be written as

$$\underline{\underline{\mathsf{M}}}(\omega) = M_0 \underline{\mathsf{I}} + \sum_{j=1}^{n} \underline{\underline{\mathsf{R}}}_j^T \begin{bmatrix} \dfrac{2\,K_j\, m_j}{2\,K_j - m_j\,\omega^2} & 0 \\ 0 & \dfrac{2\,L_j\, m_j}{2\,L_j - m_j\,\omega^2} \end{bmatrix} \underline{\underline{\mathsf{R}}}_j\,. \tag{5.21}$$

Observe that the eigenvalues of $\underline{\underline{\mathsf{M}}}(\omega)$ can depend on ω. The effective dynamic mass density in the presence of resonances and the dynamic homogenization of composites in general continues to be an area of active research. Some approaches that have been explored can be found in Torrent and Sánchez-Dehesa (2006), Mei et al. (2007), Torrent and Sánchez-Dehesa (2008b), Farhat et al. (2009), and Craster et al. (2010).

5.1.4 General expression for frequency-dependent mass

Consider a body containing deformable internal structures and a rigid matrix. Instead of applying velocities that vary harmonically in time, let us apply a time varying velocity $\mathbf{v}(t)$ to the body and observe the relation between the velocity and the momentum $\mathbf{p}(t)$. If we assume a linear response of the system there will be some linear constitutive relation of the form

$$\mathbf{p}(t) = \int_{-\infty}^{\infty} \boldsymbol{H}(t-\tau) \cdot \mathbf{v}(\tau)\, \mathrm{d}\tau\,. \tag{5.22}$$

The kernel $\boldsymbol{H}(t-\tau)$ is second-order tensor valued and may possibly be singular, i.e., delta functions are allowed. Also, since both $\mathbf{p}$ and $\mathbf{v}$ are physical and real, $\boldsymbol{H}$ must be real. Causality implies that $\boldsymbol{H}(s) = \boldsymbol{0}$ when $s = t - \tau < 0$ (or $t < \tau$) since the inertial force cannot depend on velocities in the future. Taking the Fourier transform of equation (5.22) and using the convolution theorem, we get

$$\widehat{\mathbf{p}}(\omega) = \boldsymbol{M}(\omega) \cdot \widehat{\mathbf{v}}(\omega) \tag{5.23}$$

where

$$\boldsymbol{M}(\omega) := \widehat{\boldsymbol{H}}(\omega) = \int_{-\infty}^{\infty} \boldsymbol{H}(s)\, e^{i\omega s}\, \mathrm{d}s\,. \tag{5.24}$$

The quantity $\boldsymbol{M}(\omega)$ can be shown to satisfy the Cauchy-Riemann analyticity equations only if $\mathrm{Im}(\omega) > 0$. This is a consequence of causality, $s \geq 0$, and the fact that the integral in equation (5.24) only converges in the upper half of the complex ω plane. Hence, $\boldsymbol{M}(\omega)$ is analytic in ω when $\mathrm{Im}(\omega) > 0$. The quantity $\boldsymbol{H}(s)$ is real. The complex conjugate of $\boldsymbol{M}(\omega)$ is given by

$$\overline{\boldsymbol{M}}(\omega) = \int_{-\infty}^{\infty} \boldsymbol{H}(s)\, e^{-i\overline{\omega} s}\, \mathrm{d}s = \boldsymbol{M}(-\overline{\omega}) \tag{5.25}$$

where $\overline{z}$ denotes the complex conjugate of a complex number z. Assume that, for large enough frequencies, the dynamic mass tends toward the static mass, i.e.,

$$\lim_{\omega \to \infty} \boldsymbol{M}(\omega) = M_0\, \boldsymbol{1}\,. \tag{5.26}$$

The real and imaginary parts of $\boldsymbol{M}$ are not independent and can be related through Kramers-Kronig relations. Equation (5.24) can be used to establish the Kramers-Kronig equations for the material. To do that, recall Cauchy's formula for a function f which is analytic on a domain that is enclosed in a piecewise smooth curve C:

$$f(z) = \frac{1}{2\pi i} \oint_C \frac{f(\zeta)}{\zeta - z}\, \mathrm{d}\zeta\,.$$

Since the function $\boldsymbol{M}(\omega)$ is analytic on the upper-half ω plane, for any point z in a closed contour C in the upper-half ω plane, we have

$$\boldsymbol{M}(z) - M_0\,\boldsymbol{1} = \frac{1}{2\pi i}\oint_C \frac{\boldsymbol{M}(\omega') - M_0\,\boldsymbol{1}}{\omega' - z}\,\mathrm{d}\omega' . \tag{5.27}$$

Let us now choose the contour C such that it consists of the real ω-axis and a great semicircle at infinity in the upper-half plane (see Figure 5.12).

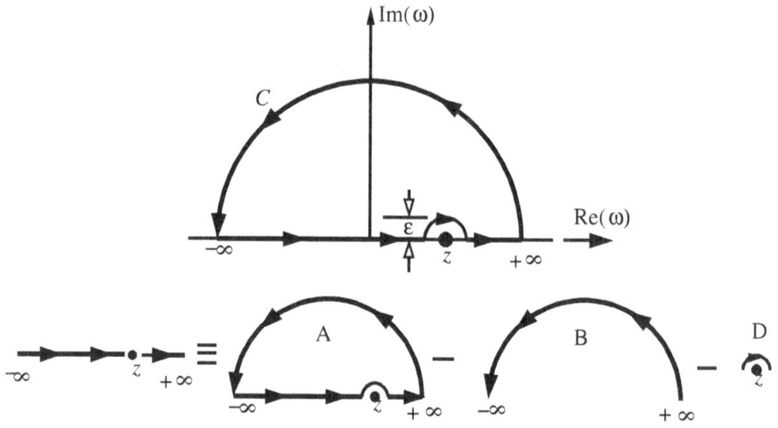

FIGURE 5.12
The closed curve C that is used to evaluate the integral in equation (5.27).

Also, from equation (5.26) we observe that

$$\boldsymbol{M}(\omega) - M_0\,\boldsymbol{1} = \boldsymbol{0} \qquad \text{as} \qquad \omega \to \infty .$$

Hence, there is no contribution to the integral in equation (5.27) due to the semicircular part of the contour and we just have to perform an integration only over the real line:

$$\boldsymbol{M}(z) = M_0\,\boldsymbol{1} + \frac{1}{2\pi i}\int_{-\infty}^{\infty} \frac{\boldsymbol{M}(\omega') - M_0\,\boldsymbol{1}}{\omega' - z}\,\mathrm{d}\omega' . \tag{5.28}$$

Now, let us consider the integral

$$\boldsymbol{I} = \int_{-\infty}^{\infty} \frac{\boldsymbol{M}(\omega') - M_0\,\boldsymbol{1}}{\omega' - z}\,\mathrm{d}\omega' . \tag{5.29}$$

From Figure 5.12 we observe that this integral can be expressed as the integral over the path A minus the sum of the integrals over the paths B and D. There is a pole at the point $\omega' = z$ (in the figure, the contour is shown as a semicircle of radius ε centered at the pole). In the limit $\varepsilon \to 0$, this integral may be interpreted to mean the Cauchy principal value. From the Cauchy-Goursat theorem, the integral over the closed path A is zero and we can write

$$0 = P\int_{-\infty}^{\infty} \frac{\boldsymbol{M}(\omega') - M_0\,\boldsymbol{1}}{\omega' - z}\,\mathrm{d}\omega' + \oint_B \frac{\boldsymbol{M}(\omega') - M_0\,\boldsymbol{1}}{\omega' - z}\,\mathrm{d}\omega' + \oint_D \frac{\boldsymbol{M}(\omega') - M_0\,\boldsymbol{1}}{\omega' - z}\,\mathrm{d}\omega'$$

where

$$P\int_{-\infty}^{\infty} \frac{\boldsymbol{M}(\omega') - M_0\,\boldsymbol{1}}{\omega' - z}\,\mathrm{d}\omega' = \int_{-\infty}^{z-\varepsilon} \frac{\boldsymbol{M}(\omega') - M_0\,\boldsymbol{1}}{\omega' - z}\,\mathrm{d}\omega' + \int_{z-\varepsilon}^{\infty} \frac{\boldsymbol{M}(\omega') - M_0\,\boldsymbol{1}}{\omega' - z}\,\mathrm{d}\omega' .$$

We have seen that since $\boldsymbol{M}(\omega) - M_0\boldsymbol{1} \to 0$ as $\omega \to \infty$, the integral over B is zero. The integral over the path D around the pole is obtained from the residue theorem (where the value is divided by two because the integral is over a semicircle), i.e.,

$$\oint_D \frac{\boldsymbol{M}(\omega') - M_0\,\boldsymbol{1}}{\omega' - z}\,\mathrm{d}\omega' = -\pi i\,[\boldsymbol{M}(z) - M_0\boldsymbol{1}] .$$

Therefore,

$$P\int_{-\infty}^{\infty} \frac{\boldsymbol{M}(\omega') - M_0\,\boldsymbol{1}}{\omega' - z}\,\mathrm{d}\omega' = \pi i\,[\boldsymbol{M}(z) - M_0\boldsymbol{1}]$$

or

$$\boldsymbol{M}(z) = M_0\,\boldsymbol{1} + \frac{1}{\pi i} P\int_{-\infty}^{\infty} \frac{\boldsymbol{M}(\omega') - M_0\,\boldsymbol{1}}{\omega' - z}\,\mathrm{d}\omega' .$$

Letting $z = \omega + i\varepsilon$ and taking the limit as $\varepsilon \to 0$, we get

$$\boxed{\boldsymbol{M}(\omega) = M_0\,\boldsymbol{1} + \frac{1}{\pi i} P\int_{-\infty}^{\infty} \frac{\boldsymbol{M}(\omega') - M_0\,\boldsymbol{1}}{\omega' - \omega}\,\mathrm{d}\omega' .} \tag{5.30}$$

Note that, in the above equation, both ω and ω' are real. Expanding equation (5.30) into real and imaginary parts and collecting terms, we get the first form of the Kramers-Kronig relations

$$\boxed{\begin{aligned} \mathrm{Re}[\boldsymbol{M}(\omega)] &= M_0\,\boldsymbol{1} + \frac{1}{\pi} P\int_{-\infty}^{\infty} \frac{\mathrm{Im}[\boldsymbol{M}(\omega')]}{\omega' - \omega}\,\mathrm{d}\omega' \\ \mathrm{Im}[\boldsymbol{M}(\omega)] &= -\frac{1}{\pi} P\int_{-\infty}^{\infty} \frac{\mathrm{Re}[\boldsymbol{M}(\omega')] - M_0\,\boldsymbol{1}}{\omega' - \omega}\,\mathrm{d}\omega' . \end{aligned}} \tag{5.31}$$

Therefore, the real part of the frequency dependent mass can be determined if we know the imaginary part and vice versa. We can also eliminate the negative frequencies from equations (5.31). Recall from equation (5.25) that $\overline{\boldsymbol{M}}(\omega) = \boldsymbol{M}(-\overline{\omega})$. Since, in equations (5.31), ω and ω' are real, we have $\overline{\boldsymbol{M}}(\omega) = \boldsymbol{M}(-\omega)$. This implies that

$$\mathrm{Re}[\boldsymbol{M}(\omega)] = \mathrm{Re}[\boldsymbol{M}(-\omega)] \qquad \text{and} \qquad \mathrm{Im}[\boldsymbol{M}(\omega)] = -\mathrm{Im}[\boldsymbol{M}(-\omega)] .$$

Consider the first of equations (5.31). We can write this relation as

$$\begin{aligned}
\mathrm{Re}[\boldsymbol{M}(\omega)] &= M_0\,\boldsymbol{1} + \frac{1}{\pi}\left\{P\int_{-\infty}^{0}\frac{\mathrm{Im}[\boldsymbol{M}(\omega')]}{\omega'-\omega}\,\mathrm{d}\omega' + P\int_{0}^{\infty}\frac{\mathrm{Im}[\boldsymbol{M}(\omega')]}{\omega'-\omega}\,\mathrm{d}\omega'\right\}\\
&= M_0\,\boldsymbol{1} + \frac{1}{\pi}\left\{P\int_{0}^{\infty}\frac{\mathrm{Im}[\boldsymbol{M}(-\omega')]}{-\omega'-\omega}\,\mathrm{d}\omega' + P\int_{0}^{\infty}\frac{\mathrm{Im}[\boldsymbol{M}(\omega')]}{\omega'-\omega}\,\mathrm{d}\omega'\right\}\\
&= M_0\,\boldsymbol{1} + \frac{1}{\pi}\left\{-P\int_{0}^{\infty}\frac{-\mathrm{Im}[\boldsymbol{M}(\omega')]}{\omega'+\omega}\,\mathrm{d}\omega' + P\int_{0}^{\infty}\frac{\mathrm{Im}[\boldsymbol{M}(\omega')]}{\omega'-\omega}\,\mathrm{d}\omega'\right\}\\
&= M_0\,\boldsymbol{1} + \frac{1}{\pi}P\int_{0}^{\infty}\mathrm{Im}[\boldsymbol{M}(\omega')]\left[\frac{1}{\omega'+\omega}+\frac{1}{\omega'-\omega}\right]\,\mathrm{d}\omega'\\
&= M_0\,\boldsymbol{1} + \frac{2}{\pi}P\int_{0}^{\infty}\frac{\omega'\,\mathrm{Im}[\boldsymbol{M}(\omega')]}{\omega'^2-\omega^2}\,\mathrm{d}\omega'\,.
\end{aligned}$$

Similarly, the second of equations (5.31) may be written as

$$\begin{aligned}
\mathrm{Im}[\boldsymbol{M}(\omega)] &= -\frac{1}{\pi}\left\{P\int_{-\infty}^{0}\frac{\mathrm{Re}[\boldsymbol{M}(\omega')]-M_0\,\boldsymbol{1}}{\omega'-\omega}\,\mathrm{d}\omega' + P\int_{0}^{\infty}\frac{\mathrm{Re}[\boldsymbol{M}(\omega')]-M_0\,\boldsymbol{1}}{\omega'-\omega}\,\mathrm{d}\omega'\right\}\\
&= -\frac{1}{\pi}\left\{-P\int_{0}^{\infty}\frac{\mathrm{Re}[\boldsymbol{M}(-\omega')]-M_0\,\boldsymbol{1}}{\omega'+\omega}\,\mathrm{d}\omega' + P\int_{0}^{\infty}\frac{\mathrm{Re}[\boldsymbol{M}(\omega')]-M_0\,\boldsymbol{1}}{\omega'-\omega}\,\mathrm{d}\omega'\right\}\\
&= -\frac{1}{\pi}\left\{P\int_{0}^{\infty}(\mathrm{Re}[\boldsymbol{M}(-\omega')]-M_0\,\boldsymbol{1})\left[-\frac{1}{\omega'+\omega}+\frac{1}{\omega'-\omega}\right]\,\mathrm{d}\omega'\right\}\\
&= -\frac{2\,\omega}{\pi}\left\{P\int_{0}^{\infty}\frac{\mathrm{Re}[\boldsymbol{M}(-\omega')]-M_0\,\boldsymbol{1}}{\omega'^2-\omega^2}\,\mathrm{d}\omega'\right\}\,.
\end{aligned}$$

Therefore, the alternative form of the Kramers-Kronig relations is

$$\boxed{\begin{aligned}
\mathrm{Re}[\boldsymbol{M}(\omega)] &= M_0\,\boldsymbol{1} + \frac{2}{\pi}P\int_{0}^{\infty}\frac{\omega'\,\mathrm{Im}[\boldsymbol{M}(\omega')]}{\omega'^2-\omega^2}\,\mathrm{d}\omega'\\
\mathrm{Im}[\boldsymbol{M}(\omega)] &= -\frac{2\,\omega}{\pi}\left\{P\int_{0}^{\infty}\frac{\mathrm{Re}[\boldsymbol{M}(-\omega')]-M_0\,\boldsymbol{1}}{\omega'^2-\omega^2}\,\mathrm{d}\omega'\right\}\,.
\end{aligned}} \qquad (5.32)$$

Dissipation and frequency-dependent mass

Consider harmonically varying force ($\mathbf{f}(t)$) and velocity ($\mathbf{v}(t)$) given by

$$\mathbf{f}(t) = \mathrm{Re}(\widehat{\mathbf{f}}\,e^{-i\omega t}) \quad \text{and} \quad \mathbf{v}(t) = \mathrm{Re}(\widehat{\mathbf{v}}\,e^{-i\omega t})\,.$$

Then,

$$\widehat{\mathbf{f}} = -i\omega\boldsymbol{M}(\omega)\cdot\widehat{\mathbf{v}}$$

which implies that

$$\mathrm{Re}(\widehat{\mathbf{f}}) = \omega\,[\mathrm{Re}(\boldsymbol{M})\cdot\mathrm{Im}(\widehat{\mathbf{v}}) + \mathrm{Im}(\boldsymbol{M})\cdot\mathrm{Re}(\widehat{\mathbf{v}})]$$
$$\mathrm{Im}(\widehat{\mathbf{f}}) = \omega\,[\mathrm{Im}(\boldsymbol{M})\cdot\mathrm{Im}(\widehat{\mathbf{v}}) - \mathrm{Re}(\boldsymbol{M})\cdot\mathrm{Re}(\widehat{\mathbf{v}})]\,.$$

The average rate of work done on the system in a cycle of oscillation will be

$$\begin{aligned} W &= \frac{\omega}{2\,\pi}\int_0^{2\pi/\omega} \mathbf{f}(t)\cdot\mathbf{v}(t)\,\mathrm{d}t \\ &= \frac{\mathrm{Re}(\widehat{\mathbf{f}})\cdot\mathrm{Re}(\widehat{\mathbf{v}}) + \mathrm{Im}(\widehat{\mathbf{f}})\cdot\mathrm{Im}(\widehat{\mathbf{v}})}{2} \\ &= \omega\,[\mathrm{Re}(\widehat{\mathbf{v}})\cdot\mathrm{Im}[\boldsymbol{M}(\omega)]\cdot\mathrm{Re}(\widehat{\mathbf{v}}) + \mathrm{Im}(\widehat{\mathbf{v}})\cdot\mathrm{Im}[\boldsymbol{M}(\omega)]\cdot\mathrm{Im}(\widehat{\mathbf{v}})]. \end{aligned}$$

This quadratic form will be non-negative for all choices of $\widehat{\mathbf{v}}$ if and only if $\mathrm{Im}(\boldsymbol{M}(\omega))$ is positive semidefinite for all real $\omega > 0$. Note that the quadratic form does not contain $\mathrm{Re}[\boldsymbol{M}(\omega)]$. Since the work done in a cycle should be zero in the absence of dissipation, this implies that the imaginary part of the mass is connected to the energy dissipation (for instance, into heat). Therefore, a physical restriction on the behavior of such materials is that

$$\boxed{\mathrm{Im}[\boldsymbol{M}(\omega)] \geq 0\,.} \tag{5.33}$$

5.2 Frequency-dependent moment of inertia

Another model which achieves a frequency-dependent inertia is the rotating ring shown in Figure 5.13. The model was developed by Milton and Willis (2007) and is related to gyrocontinua and micromorphic elasticity (see D'Eleuterio and Hughes (1984), Brocato and Capriz (2001), Grekova and Maugin (2005), and Erofeyev (2003)). The ring has mass m and is contained in a spherical cavity inside a rigid body. A point mass m_1 may be attached to the ring. We assume that $m_1 = 0$ in the following discussion. Let us also assume that the rotating top spins without friction on its axis with frequency ω_r.

Let the point O on the rigid body be subject to a harmonic linear velocity, $\mathbf{v}_\mathrm{o}$, and a harmonic angular velocity, $\boldsymbol{\Omega}_\mathrm{o}$, given by

$$\mathbf{v}_\mathrm{o}(t) = \varepsilon\mathrm{Re}(\widehat{\mathbf{v}}_\mathrm{o}\,e^{-i\omega t}) \quad \text{and} \quad \boldsymbol{\Omega}_\mathrm{o}(t) = \varepsilon\mathrm{Re}(\widehat{\boldsymbol{\Omega}}_\mathrm{o}\,e^{-i\omega t})$$

where ε is a small real parameter, i.e, the amplitude of oscillations of the system is small. The linear velocity is caused by an external force and the angular velocity by an external torque. Let C be the center of the top (which is also the center of the spherical cavity) and let P be a point in the rigid body. Then, the velocities of the two points are given by

$$\mathbf{v}_\mathrm{c}(t) = \mathbf{v}_\mathrm{o}(t) + \boldsymbol{\Omega}_\mathrm{o}(t)\times\mathbf{X}_\mathrm{c} \quad \text{and} \quad \mathbf{v}_\mathrm{p}(t) = \mathbf{v}_\mathrm{o}(t) + \boldsymbol{\Omega}_\mathrm{o}(t)\times\mathbf{X}_\mathrm{p} \tag{5.34}$$

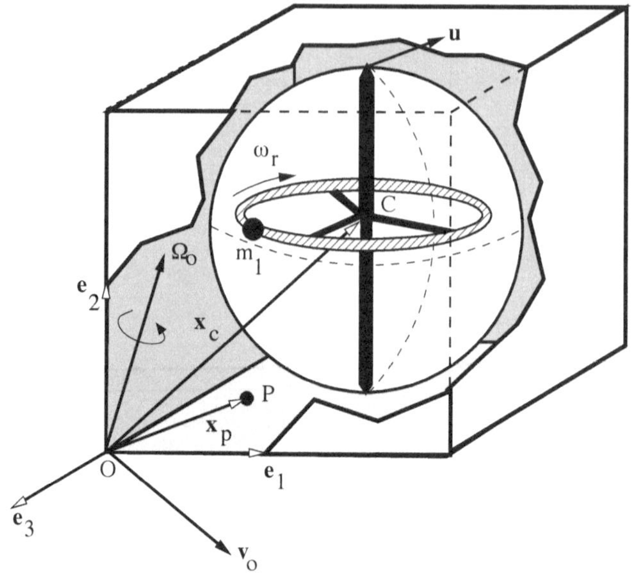

FIGURE 5.13
A spherical cavity inside a rigid body containing a rotating wheel.

where $\mathbf{X}_c$ is the initial position of point C and $\mathbf{X}_p$ is the initial position of point P. Therefore, ignoring other possible rigid body motions, the displacements of these points are

$$\begin{aligned} \mathbf{u}_c(t) &= \varepsilon\mathrm{Re}\left(\widehat{\mathbf{u}}_o e^{-i\omega t}\right) + \varepsilon\mathrm{Re}\left(\widehat{\boldsymbol{\theta}}_o e^{-i\omega t}\right) \times \mathbf{X}_c \\ \mathbf{u}_p(t) &= \varepsilon\mathrm{Re}\left(\widehat{\mathbf{u}}_o e^{-i\omega t}\right) + \varepsilon\mathrm{Re}\left(\widehat{\boldsymbol{\theta}}_o e^{-i\omega t}\right) \times \mathbf{X}_p . \end{aligned} \tag{5.35}$$

where

$$\widehat{\mathbf{u}}_o := i\widehat{\mathbf{v}}_o/\omega \quad \text{and} \quad \widehat{\boldsymbol{\theta}}_o := i\widehat{\boldsymbol{\Omega}}_o/\omega .$$

Then the positions of points C and P at time t are given by

$$\begin{aligned} \mathbf{x}_c(t) &= \mathbf{X}_c + \varepsilon\mathrm{Re}\left(\widehat{\mathbf{u}}_o e^{-i\omega t}\right) + \varepsilon\mathrm{Re}\left(\widehat{\boldsymbol{\theta}}_o\, e^{-i\omega t}\right) \times \mathbf{X}_c \\ \mathbf{x}_p(t) &= \mathbf{X}_p + \varepsilon\mathrm{Re}\left(\widehat{\mathbf{u}}_o e^{-i\omega t}\right) + \varepsilon\mathrm{Re}\left(\widehat{\boldsymbol{\theta}}_o\, e^{-i\omega t}\right) \times \mathbf{X}_p . \end{aligned} \tag{5.36}$$

Let us define $\mathbf{u}_o(t) := \varepsilon\mathrm{Re}(\widehat{\mathbf{u}}_o e^{-i\omega t})$ and $\boldsymbol{\theta}_o(t) := \varepsilon\mathrm{Re}(\widehat{\boldsymbol{\theta}}_o e^{-i\omega t})$. We can then write equations (5.36) as

$$\mathbf{x}_c(t) = \mathbf{X}_c + \mathbf{u}_o(t) + \boldsymbol{\theta}_o(t) \times \mathbf{X}_c \quad \text{and} \quad \mathbf{x}_p(t) = \mathbf{X}_p + \mathbf{u}_o(t) + \boldsymbol{\theta}_o(t) \times \mathbf{X}_p .$$

If we now move the origin to the point C, the displacements are

$$\mathbf{x}_c(t) = \mathbf{u}_o(t) \quad \text{and} \quad \mathbf{x}_p(t) = \mathbf{X}_p + \mathbf{u}_o(t) + \boldsymbol{\theta}_o(t) \times \mathbf{X}_p \tag{5.37}$$

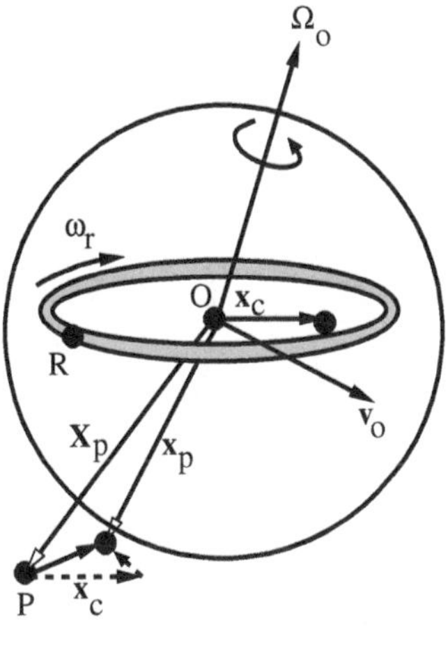

FIGURE 5.14
Displacements of points inside the rigid body relative to the center of the cavity.

and we have the situation shown in Figure 5.14.

Let $\mathbf{X}_r$ be the initial position of a point R on the ring. If the angular velocity of the ring is ω_r, in the absence of other motions, the position of the point is given by

$$\overline{\mathbf{x}}_r(t) = \boldsymbol{Q}(t) \cdot \mathbf{X}_r$$

where the rotation matrix $\boldsymbol{Q}(t)$ has components (with respect to a basis $(\mathbf{e}_1, \mathbf{e}_2)$ in the plane of the ring and $\mathbf{e}_3$ out-of-plane)

$$\boldsymbol{Q}(t) \equiv \underline{\underline{Q}} = \begin{bmatrix} \cos(\omega_r t) & \sin(\omega_r t) & 0 \\ -\sin(\omega_r t) & \cos(\omega_r t) & 0 \\ 0 & 0 & 1 \end{bmatrix} . \tag{5.38}$$

Equation (5.38) assumes that the axis of rotation of the ring is aligned with the $\mathbf{e}_3$-axis. Let us keep the basis $(\mathbf{e}_1, \mathbf{e}_2, \mathbf{e}_3)$ fixed and allow the axis of the ring to rotate around a fixed axis $\mathbf{b}$ (see Figure 5.15), i.e., the axis of the ring wobbles around the $\mathbf{b}$-axis.

Allowing only infinitesimal rigid body rotations around $\mathbf{b}$, the rotation of point R can be expressed as

$$\boldsymbol{R}(t) = \boldsymbol{Q}(t) + \boldsymbol{B}(t) \cdot \boldsymbol{Q}(t) = [\boldsymbol{1} + \boldsymbol{B}(t)] \cdot \boldsymbol{Q}(t) \tag{5.39}$$

where the $\boldsymbol{B}(t)$ is the skew-symmetric matrix corresponding to the axial vector $\mathbf{b}$, defined as $\boldsymbol{B} \cdot \mathbf{a} = \mathbf{b} \times \mathbf{a}$ for all vectors $\mathbf{a}$. Let the infinitesimal rotation vector $\mathbf{b}$ be subject to small perturbations (i.e., changes in the spin velocity) and let us assume that it has the harmonic form

$$\mathbf{b} = \varepsilon \operatorname{Re}(\widehat{\mathbf{b}} e^{-i\omega t}) \quad \Longrightarrow \quad \boldsymbol{B} = \varepsilon \operatorname{Re}(\widehat{\boldsymbol{B}} e^{-i\omega t})$$

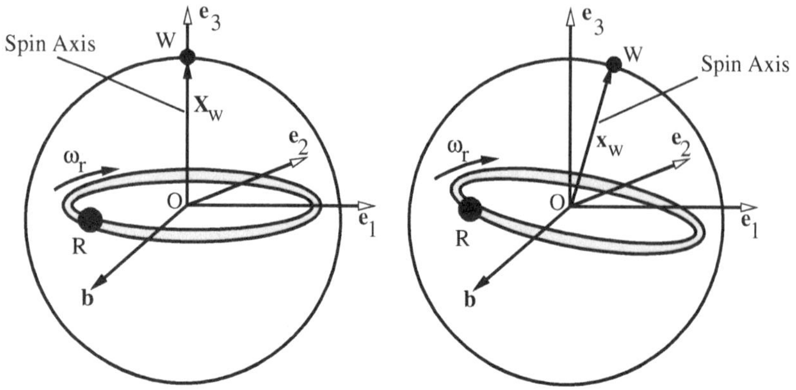

FIGURE 5.15
Rotation of the axis of the ring around an axis **b**.

where $\widehat{\mathbf{b}}$ and $\widehat{\boldsymbol{B}}$ are complex amplitudes. If the point of contact between the spindle and the cavity wall is frictionless, points on the ring experience only the translations of the body directly. Rigid body rotations are not transferred in the form of the small wobbles described above. Therefore, the position of point R is given by

$$\overline{\mathbf{x}}_r(t) = \mathbf{u}_o(t) + \boldsymbol{R}(t) \cdot \mathbf{X}_r \,. \tag{5.40}$$

Let W be the location of the top of the spindle and let its initial position be $\mathbf{X}_w$. Then, at time t, the position vector of W is (from the second of equations (5.37)

$$\mathbf{x}_w(t) = \mathbf{X}_w + \mathbf{u}_o(t) + \boldsymbol{\theta}_o(t) \times \mathbf{X}_w \,. \tag{5.41}$$

Looking at the position of the same point from the point of view of the ring-spindle structure inside the cavity, due to the rotation of the spindle about the vector **b**, the vector $\mathbf{X}_w$ takes the position

$$\overline{\mathbf{x}}_w(t) = \mathbf{u}_o(t) + \mathbf{X}_w + \boldsymbol{B}(t) \cdot \mathbf{X}_w = \mathbf{X}_w + \mathbf{u}_o(t) + \mathbf{b}(t) \times \mathbf{X}_w \,. \tag{5.42}$$

Since the motion of point W must match, comparing equations (5.41) and (5.42), we have

$$\boldsymbol{\theta}_o = \mathbf{b} \,.$$

If $\mathbf{X}_s$ is the initial position of the center of mass (S) of the rigid body (excluding the ring), the position of location S at time t is

$$\mathbf{x}_s(t) = \mathbf{X}_s + \mathbf{u}_o(t) + \boldsymbol{\theta}_o(t) \times \mathbf{X}_s \,. \tag{5.43}$$

Let us now calculate the linear momentum of the ring. Let the initial position of point R on the ring be parametrized by the angle φ such that

$$\mathbf{X}_r(\varphi) = r_0 \cos\varphi \, \mathbf{e}_1 + r_0 \sin\varphi \, \mathbf{e}_2$$

where r_0 is the radius of the ring. Then the momentum of an element of the ring of arc length $\mathrm{d}\varphi$ is

$$\mathrm{d}\mathbf{p}_{\mathrm{r}}(\varphi,t) = \left(\frac{m\,\mathrm{d}\varphi}{2\pi}\right)\frac{d}{dt}[\overline{\mathbf{x}}_{\mathrm{r}}(\varphi,t)]\,. \tag{5.44}$$

From equation (5.40) we have

$$\frac{d\overline{\mathbf{x}}_{\mathrm{r}}}{dt} = \frac{d\mathbf{u}_{\mathrm{o}}}{dt} + \frac{d\boldsymbol{R}}{dt}\cdot\mathbf{X}_{\mathrm{r}}(\varphi) = \varepsilon\,\mathrm{Re}(-i\omega\widehat{\mathbf{u}}_{\mathrm{o}}\,e^{-i\omega t}) + \frac{d\boldsymbol{R}}{dt}\cdot\mathbf{X}_{\mathrm{r}}(\varphi)\,.$$

Integrating equation (5.44) and using the above relation we have

$$\mathbf{p}_{\mathrm{r}}(t) = \int_0^{2\pi}\mathrm{d}\mathbf{p}_{\mathrm{r}}(\varphi,t) = \varepsilon m\,\mathrm{Re}(\widehat{\mathbf{v}}_{\mathrm{o}}\,e^{-i\omega t})\,. \tag{5.45}$$

The linear momentum of the rigid body (excluding the ring) can be calculated from the motion of its center of mass, S, i.e.,

$$\mathbf{p}_{\mathrm{s}}(t) = M_s\frac{d\mathbf{x}_{\mathrm{S}}}{dt} = M_s\left[\frac{d\mathbf{u}_{\mathrm{o}}}{dt} + \frac{d\boldsymbol{\theta}_{\mathrm{o}}}{dt}\times\mathbf{X}_{\mathrm{s}}\right]$$

where M_s is the mass of the rigid body. Using the definitions of $\mathbf{u}_{\mathrm{o}}$ and $\boldsymbol{\theta}_{\mathrm{o}}$, we have

$$\mathbf{p}_{\mathrm{s}}(t) = \varepsilon M_s\,\mathrm{Re}\left[(\widehat{\mathbf{v}}_{\mathrm{o}} + \widehat{\boldsymbol{\Omega}}_{\mathrm{o}}\times\mathbf{X}_{\mathrm{s}})\,e^{-i\omega t}\right]\,. \tag{5.46}$$

Therefore the total linear momentum of the system is

$$\mathbf{p}(t) = \mathbf{p}_{\mathrm{r}}(t) + \mathbf{p}_{\mathrm{s}}(t) = \varepsilon\,\mathrm{Re}\,(\widehat{\mathbf{p}}\,e^{-i\omega t}]$$

where

$$\boxed{\widehat{\mathbf{p}} := (M_s + m)\widehat{\mathbf{v}}_{\mathrm{o}} + M_s\,\widehat{\boldsymbol{\Omega}}_{\mathrm{o}}\times\mathbf{X}_{\mathrm{s}}\,.} \tag{5.47}$$

The angular momentum of the ring is given by

$$\mathbf{q}_{\mathrm{r}}(t) = \int_0^{2\pi}\overline{\mathbf{x}}_{\mathrm{r}}(\varphi,t)\times\mathrm{d}\mathbf{p}_{\mathrm{r}}(\varphi,t)\,.$$

If ignore terms of order ε^2, we can express this relation in the form

$$\begin{aligned}\mathbf{q}_{\mathrm{r}}(t) = {} & \frac{m\omega_r}{2\pi}\left[\int_0^{2\pi}[\boldsymbol{Q}\cdot\mathbf{X}_{\mathrm{r}}(\varphi)]\times[\widetilde{\boldsymbol{Q}}\cdot\mathbf{X}_{\mathrm{r}}(\varphi)]\mathrm{d}\varphi + \int_0^{2\pi}[\boldsymbol{Q}\cdot\mathbf{X}_{\mathrm{r}}(\varphi)]\times[\widetilde{\boldsymbol{A}}\cdot\mathbf{X}_{\mathrm{r}}(\varphi)]\mathrm{d}\varphi\right. \\ & \left. + \int_0^{2\pi}[\boldsymbol{A}\cdot\mathbf{X}_{\mathrm{r}}(\varphi)]\times[\widetilde{\boldsymbol{Q}}\cdot\mathbf{X}_{\mathrm{r}}(\varphi)]\mathrm{d}\varphi\right] \\ & + \frac{m}{2\pi}\int_0^{2\pi}[\boldsymbol{Q}\cdot\mathbf{X}_{\mathrm{r}}(\varphi)]\times[\mathrm{Re}(-i\omega\boldsymbol{A})\cdot\mathbf{X}_{\mathrm{r}}(\varphi)]\mathrm{d}\varphi\end{aligned}$$

where $\widetilde{\boldsymbol{Q}} := \boldsymbol{Q}(\omega_r t + \pi/2)$, $\boldsymbol{A} := \boldsymbol{B}\cdot\boldsymbol{Q}$ and $\widetilde{\boldsymbol{A}} := \boldsymbol{B}\cdot\widetilde{\boldsymbol{Q}}$. Also, since $Q_{i3} = \delta_{i3}$, $\widetilde{Q}_{i3} = 0$, and

$$\int_0^{2\pi} X_p(\varphi)X_q(\varphi)\mathrm{d}\varphi = r_0^2\pi(\delta_{pq} - \delta_{p3}\delta_{q3})$$

we can express the angular momentum of the ring in the form

$$\mathbf{q}_\mathrm{r}(t) = \frac{mr_0^2}{2}e_{ijk}\Big[\omega_r\big(Q_{jp}\widetilde{Q}_{kp} + B_{km}Q_{jp}\widetilde{Q}_{mp} + B_{jm}Q_{mp}\widetilde{Q}_{kp}\big) \\ - i\omega(B_{km}Q_{jp}Q_{mp} - \delta_{j3}B_{k3})\Big]\mathbf{e}_i\,.$$

Recall that the skew-symmetric matrix $\boldsymbol{B}$ corresponding to an axial vector $\mathbf{b}$ can be expressed as

$$\boldsymbol{B} \equiv B_{ij} \equiv \begin{bmatrix} 0 & b_3 & -b_2 \\ -b_3 & 0 & b_1 \\ b_2 & -b_1 & 0 \end{bmatrix}.$$

The angular momentum of the ring can then be written as

$$\mathbf{q}_\mathrm{r}(t) = mr_0^2\omega_r\begin{bmatrix} b_2 \\ -b_1 \\ -1 \end{bmatrix} + mr_0^2\begin{bmatrix} 3i\omega b_1/2 \\ 3i\omega b_2/2 \\ i\omega b_3 \end{bmatrix}.$$

Assuming that there is no friction between the spindle and the walls of the cavity inside the rigid body, b_3 must be equal to zero. Using the relation $\mathbf{b} = \boldsymbol{\theta}_\mathrm{o}$, we find that the angular momentum of the ring is given by

$$\mathbf{q}_\mathrm{r}(t) = \mathbf{q}_0 + \mathrm{Re}(\widehat{\mathbf{q}}_\mathrm{r}\, e^{-i\omega t})$$

where $\mathbf{q}_0 = -mr_0^2\omega_r\mathbf{e}_3$ and

$$\widehat{\mathbf{q}}_\mathrm{r} = \frac{\varepsilon mr_0^2}{2}\left[-\left(\frac{2i\omega_r\widehat{\Omega}_2}{\omega} + 3\widehat{\Omega}_1\right)\mathbf{e}_1 + \left(\frac{2i\omega_r\widehat{\Omega}_1}{\omega} - 3\widehat{\Omega}_2\right)\mathbf{e}_2\right] \tag{5.48}$$

where $\widehat{\boldsymbol{\Omega}}_\mathrm{o} = \widehat{\Omega}_1\mathbf{e}_1 + \widehat{\Omega}_2\mathbf{e}_2 + \widehat{\Omega}_3\mathbf{e}_3$. The angular momentum of the rigid body is given by

$$\mathbf{q}_\mathrm{s}(t) = M_s\mathbf{x}_\mathrm{s}(t)\times\frac{d\mathbf{x}_\mathrm{s}}{dt}.$$

If we substitute the expression for $\mathbf{x}_\mathrm{s}$ from equation (5.43), use the fact that $d\boldsymbol{\theta}_\mathrm{o}/dt = \boldsymbol{\Omega}_\mathrm{o}$, and ignore terms containing ε^2, we have

$$\mathbf{q}_\mathrm{s}(t) = M_s\left[\mathbf{X}_\mathrm{s}\times\mathbf{v}_\mathrm{o} + \mathbf{X}_\mathrm{s}\times(\boldsymbol{\Omega}_\mathrm{o}\times\mathbf{X}_\mathrm{s})\right].$$

If we express the second term above in the form $\boldsymbol{K}_0\cdot\boldsymbol{\Omega}_\mathrm{o}$, where $\boldsymbol{K}_0$ is the moment of inertia, we have

$$\mathbf{q}_\mathrm{s}(t) = \boldsymbol{K}_0\cdot\boldsymbol{\Omega}_\mathrm{o} + M_s\mathbf{X}_\mathrm{s}\times\mathbf{v}_\mathrm{o} = \varepsilon\,\mathrm{Re}\left[\left(\boldsymbol{K}_0\cdot\widehat{\boldsymbol{\Omega}}_\mathrm{o} + M_s\mathbf{X}_\mathrm{s}\times\widehat{\mathbf{v}}_\mathrm{o}\right)e^{-i\omega t}\right]$$

or

$$\mathbf{q}_s(t) = \mathrm{Re}(\widehat{\mathbf{q}}_s e^{-i\omega t})$$

where

$$\widehat{\mathbf{q}}_s = \boldsymbol{K}_0 \cdot \widehat{\boldsymbol{\Omega}}_o + M_s \mathbf{X}_s \times \widehat{\mathbf{v}}_o . \tag{5.49}$$

Therefore the total angular momentum of the system is

$$\mathbf{q} = \mathbf{q}_r + \mathbf{q}_s = \mathbf{q}_0 + \mathrm{Re}(\widehat{\mathbf{q}} e^{-i\omega t})$$

with

$$\widehat{\mathbf{q}} = \widehat{\mathbf{q}}_r + \widehat{\mathbf{q}}_s = M_s \mathbf{X}_s \times \widehat{\mathbf{v}}_o + \boldsymbol{K}(\omega) \cdot \widehat{\boldsymbol{\Omega}}_o \tag{5.50}$$

and

$$\boxed{\boldsymbol{K}(\omega) = \boldsymbol{K}_0 + \frac{\varepsilon m r_0^2}{2} \begin{bmatrix} -3 & -2i\omega_r/\omega & 0 \\ 2i\omega_r/\omega & -3 & 0 \\ 0 & 0 & 0 \end{bmatrix}} . \tag{5.51}$$

Thus we have an apparently rigid body with a frequency-dependent moment of inertia.

The stability of such structures and their nonlinear dynamics, particularly when the rotating ring has nonuniform mass could lead to even richer behavior. We can imagine, following D'Eleuterio and Hughes (1984), a long and thin elastic composite with a microstructure consisting of elements with frequency-dependent inertia. The dynamics of such a structure could be controlled by controlling the motion of the ring. For other applications of gyroelastic media see Peck and Cavender (2004).

5.3 Negative elastic moduli

Early work on negative elastic moduli in the context of extremely dissipative viscoelastic materials can be found in Lakes et al. (2001). More recent work on ultrasonic metamaterials can be found in Fang et al. (2006). The aim of much recent work on elastodynamic metamaterials has been to design materials with negative moduli and thus extend the range of material behaviors available to the designer.

5.3.1 A Helmholtz resonator model

Consider the array of Helmholtz resonators shown in Figure 5.16 (an example of such a resonator in everyday life is a soda bottle which resonates when you blow over the top of the neck.) The resonator can be thought of as a spring-mass system where the air inside the cavity acts as a spring and the water in the narrow neck acts as a mass. There is some frequency at which the spring-mass system resonates.

A model of the Helmholtz resonator is shown in Figure 5.17. For simplicity, we assume that each cavity has a square cross-section as do the piston arms. The

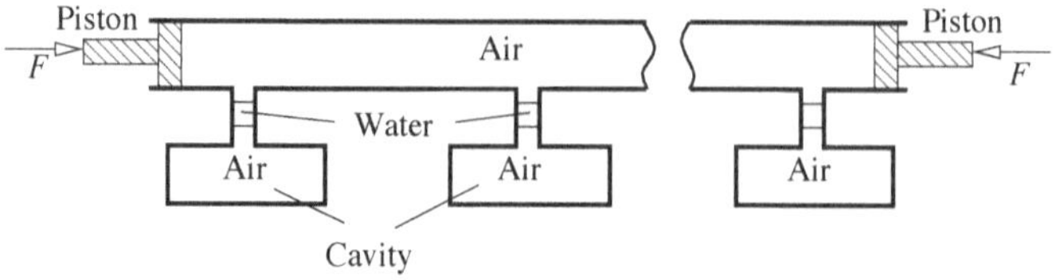

FIGURE 5.16
An array of Helmholtz resonators.

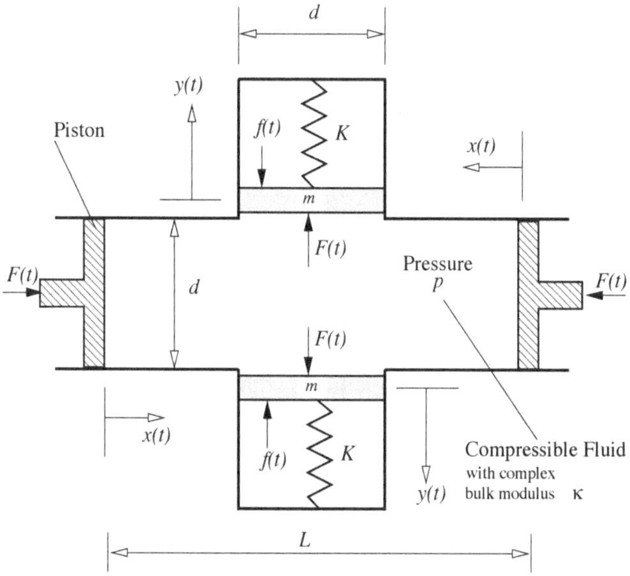

FIGURE 5.17
A mechanical realization of a Helmholtz resonator. One can devise similar models using electrical circuit theory (see, for instance, Morse and Ingard (1986)).

cross-sectional area is assumed to be d^2. The air in each cavity is modeled with a spring of complex spring constant K and the water in the neck is modeled as a rigid body of mass m. The piston is filled with a compressible fluid with complex bulk modulus κ.

Let the force applied on the pistons be $F(t)$. Then the pressure in the fluid is

$$p(t) = \frac{F(t)}{d^2}\,. \tag{5.52}$$

Since the fluid transmits the pressure to the mass and the area of cross-section of the cavity is d^2, the force applied by the fluid on the mass is also $F(t)$. Let $f(t)$ be the force that the spring applies on the mass. By symmetry, the same force is applied to both cavities shown in Figure 5.17. Let $x(t)$ denote the position of the piston and let $y(t)$ be the position of the mass.

Assume harmonic time dependence of the quantities $F(t)$, $f(t)$, $x(t)$, and $y(t)$,

$$\begin{aligned} F(t) &= \mathrm{Re}(\widehat{F}\, e^{-i\omega t})\,; \;\; f(t) = \mathrm{Re}(\widehat{f}\, e^{-i\omega t}) \\ x(t) &= \mathrm{Re}(\widehat{x} e^{-i\omega t})\,; \;\; y(t) = \mathrm{Re}(\widehat{y} e^{-i\omega t})\,. \end{aligned} \tag{5.53}$$

Also assume that $p = 0$ at $t = 0$ which leads to

$$p(t) = \mathrm{Re}(\widehat{p}\, e^{-i\omega t}) = \frac{1}{d^2}\mathrm{Re}(\widehat{F}\, e^{-i\omega t}) \quad \Longrightarrow \quad \widehat{p} = \frac{\widehat{F}}{d^2}\,. \tag{5.54}$$

Newton's law implies that

$$m\,\frac{d^2y}{dt^2} = F(t) - f(t)\,. \tag{5.55}$$

Substituting equations (5.53) into equation (5.55), we get

$$-m\omega^2 \widehat{y} = \widehat{F} - \widehat{f}\,. \tag{5.56}$$

From Hooke's law

$$f(t) = K\, y(t) \qquad \Longrightarrow \qquad \widehat{f} = K\,\widehat{y}\,. \tag{5.57}$$

Combining equations (5.56), (5.57), and (5.54) gives

$$\widehat{F} = (K - m\,\omega^2)\,\widehat{y} \quad \Longrightarrow \quad \widehat{p} = \frac{(K - m\,\omega^2)}{d^2}\,\widehat{y}\,. \tag{5.58}$$

The change in volume of the fluid due to the motions of the pistons and the masses is given by $\Delta V(t) = 2\, d^2\, [y(t) - x(t)]$. From the constitutive relation for the fluid,

$$p(t) = -\kappa\,\frac{\Delta V(t)}{V_0} = -\kappa\,\frac{2\, d^2\, [y(t) - x(t)]}{V_0} \tag{5.59}$$

where κ is the complex bulk modulus of the fluid and V_0 is the initial volume. Substituting equations (5.53) and (5.54) into equation (5.59), we get

$$\widehat{p} = \frac{2\,\kappa\, d^2\, (\widehat{x} - \widehat{y})}{V_0}\,. \tag{5.60}$$

Eliminating $\widehat{y}$ from equations (5.58) and (5.60) leads to

$$\widehat{p} = \frac{2\,\kappa\, d^2}{V_0}\left[\widehat{x} - \frac{d^2}{(K - m\,\omega^2)}\,\widehat{p}\right]\,.$$

Solving for $\widehat{p}$, we have

$$\widehat{p} = \frac{2\,\kappa\, d^2}{V_0}\left[1 + \frac{2\,\kappa\, d^4}{V_0(K - m\,\omega^2)}\right]^{-1}\,\widehat{x}\,.$$

Therefore,

$$\widehat{F} = d^2\, \widehat{p} = \frac{2\,\kappa\, d^4}{V_0}\left[1 + \frac{2\,\kappa\, d^4}{V_0(K - m\,\omega^2)}\right]^{-1} \widehat{x}\,. \tag{5.61}$$

By definition, the Young's modulus relates the stress to the strain. For our model this means that

$$\widehat{F}/d^2 = E\,\widehat{\Delta L}/L_0$$

where E is a complex Young's modulus, $\Delta L(t)$ is the change in length, and L_0 is the initial length of the region between the two pistons. Since $\Delta L(t) = 2x(t)$, we have

$$\widehat{F} = 2d^2 E\widehat{x}/L_0\,. \tag{5.62}$$

From equations (5.61) and (5.62) we can deduce an expression for the Young's modulus of the form

$$\boxed{E(\omega) = \frac{\kappa\, d^2\, L_0}{V_0}\left[1 + \frac{2\,\kappa\, d^4}{V_0(K - m\,\omega^2)}\right]^{-1}.} \tag{5.63}$$

If κ and K are real (purely elastic springs with no damping) and positive, then a plot of E as a function of ω has the form shown in Figure 5.18. *So it will appear that the system will have a negative Young's modulus for frequencies higher than* $\omega_r = \sqrt{K/m}$. Several other negative modulus materials can be envisaged using a similar analogy.

Equation (5.63) can also be written as

$$E(\omega) = \frac{E_0(\omega_r^2 - \omega^2)}{\omega_r^2 - \omega^2 + \beta}$$

where

$$E_0 := \frac{\kappa d^2 L_0}{V_0} \quad \text{and} \quad \beta := \frac{2E_0 d^2}{L_0 m} = \frac{2E_0 d^2}{L_0 K}\omega_r^2\,.$$

Alternatively, we can write the equation in terms of a characteristic time $\tau := \sqrt{m/K}$,

$$E(\omega) = \frac{E_0(1 - \omega^2\tau^2)}{1 + (\beta - \omega^2)\tau^2}\,.$$

Let us now consider a damped system where Newton's law (5.55) takes the form

$$m\left(\frac{d^2y}{dt^2} + \gamma\frac{dy}{dt}\right) = F(t) - f(t)$$

where γ is a damping factor. The Fourier transformed version of the above equation is

$$-m(\omega^2 + i\omega\gamma)\widehat{y} = \widehat{F} - \widehat{f}\,.$$

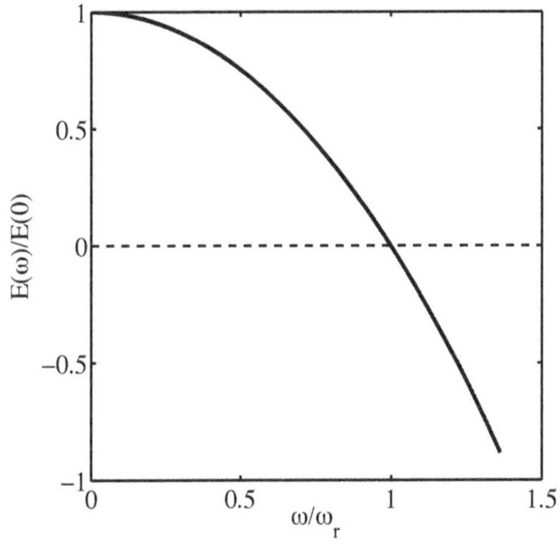

FIGURE 5.18
Effective Young's modulus of Helmholtz resonator as a function of frequency. The modulus has been normalized by the value of E at $\omega = 0$ while the frequency has been normalized by the resonance frequency ω_r. For this particular plot we have assumed $\kappa_{\text{fluid}} = 2.2$ GPa, $\rho_{\text{piston}} = 8100$ kg/m^3, $d_{\text{piston}} = 1$ cm, $h_{\text{piston}} = 5$ mm, $E_{\text{spring}} = 70$ GPa, $r_{\text{spring}} = 5$ mm, $L_{\text{spring}} = 1$ cm, $K = 0.001 E_{\text{spring}} r^2_{\text{spring}} / L_{\text{spring}}$, $L_0 = 5$ cm, and $V_0 = L_0 d^2$.

Comparing the above equation with (5.56), and after a small amount of algebra, we see that the effective Young's modulus in this case has the form

$$E(\omega) = E_0 \left[1 + \frac{\beta}{\omega_r^2 - \omega^2 - i\omega\gamma} \right]^{-1} = E_0 \left[1 - \frac{\alpha \omega_r^2}{\omega^2 - \omega_r^2 + i\omega\gamma} \right]^{-1} \tag{5.64}$$

where $\alpha := 2E_0 d^2/(L_0 K)$. For the situation where the cavity and the pipe contain the same fluid, the bulk modulus of the fluid (κ) and the spring constant (K) are related and α becomes a purely geometrical factor.

For a system containing a number of Helmholtz resonators along the length of the pipe, we can use the analogy of a number of springs in series. If the j-th Helmholtz resonator has a piston mass m_j, spring stiffness K_j, damping factor γ_j, and geometrical factors d_j, L_j, V_j, the effective Young's modulus of the system is

$$\boxed{E(\omega) = E_0 \sum_j \left[1 - \frac{\alpha_j \omega_j^2}{\omega^2 - \omega_j^2 + i\omega\gamma_j} \right]^{-1}} \tag{5.65}$$

where ω_j is the resonance frequency of the j-th resonator. This function is analytic over the entire complex ω-plane except for isolated poles in the $\text{Im}(\omega) < 0$ part of the complex plane. We can also show that $E(-\overline{\omega}) = \overline{E}(\omega)$, where $\overline{(\bullet)}$ indicates the complex conjugate. If we were to use a time-dependent form of Hooke's law and rewrite equation (5.57) as

$$f(t) = \int_{-\infty}^{\infty} K(t-t')\,y(t')\,\mathrm{d}t' \qquad \Longrightarrow \qquad \widehat{f}(\omega) = \widehat{K}(\omega)\,\widehat{y}(\omega)$$

then causality requires that $E(\omega)$ is analytic only if $\mathrm{Im}(\omega) \geq 0$ (see the discussion in Section 1.2.5, p. 10).

5.3.2 Negative moduli and viscoelasticity

Since linear viscoelasticity also deals with frequency dependent moduli, it is natural at this stage to ask whether negative moduli are allowed in viscoelastic materials. A generalized Maxwell model is commonly used in linear viscoelasticity (see Figure 5.19). Let us examine the relation between the applied force (F) and the displacement (u) for such a model. Recall that for a single Maxwell element (j), the

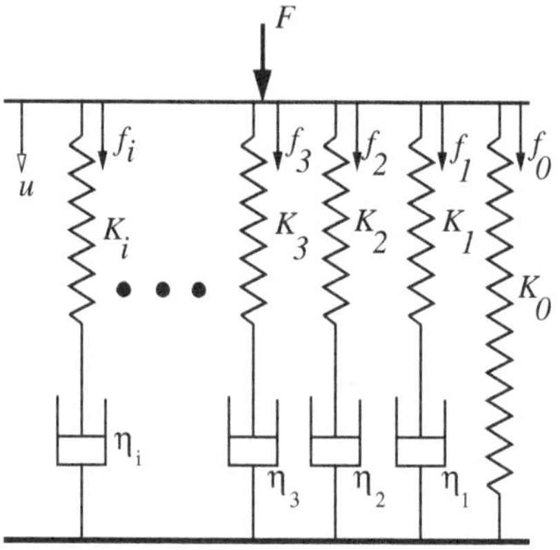

FIGURE 5.19

A generalized Maxwell model of viscoelasticity.

displacements in the spring and the dashpot are given by

$$K_j u_1(t) = f_j(t) \qquad \text{and} \qquad \eta_j \ddot{u}_2(t) = f_j(t)\,.$$

Fourier transforming these equations gives

$$\widehat{u}_1(\omega) = \widehat{f}_j(\omega)/K_j \qquad \text{and} \qquad -i\omega\,\widehat{u}_2(\omega) = \widehat{f}_j(\omega)/\eta_j\,.$$

The total displacement of the Maxwell element is

$$\widehat{u}_j(\omega) = \widehat{u}_1 + \widehat{u}_2 = [1/K_j + i/(\omega\,\eta_j)]\;\widehat{f}_j(\omega)\,.$$

From a balance of forces for the generalized Maxwell model we have (after dropping the hats for convenience),

$$F = f_0 + \sum f_j = \left[K_0 + \sum_{j=1}^{N} \left(\frac{1}{K_j} + \frac{i}{\omega\, \eta_j} \right)^{-1} \right] u =: K_{\text{eff}}\, u\ .$$

If we define the *relaxation time* as $\tau_j := \eta_j / K_j$, the *effective spring constant* of the model can be expressed as

$$K_{\text{eff}} = K_0 + \sum_{j=1}^{N} \left(\frac{1}{K_j} + \frac{i}{\omega\, \eta_j} \right)^{-1} = K_0 + \sum_{j=1}^{N} \frac{K_j}{1 + i/(\omega \tau_j)}.$$

A continuum version can be derived from the discrete Maxwell model by taking the limit as $N \to \infty$ (see, for instance, Christensen (2003)). Then, the frequency-dependent Young's modulus (E) can be expressed as

$$E_{\text{eff}}(\omega) = E_0 + \int_0^\infty \frac{H(\tau)}{1 + i/(\omega\tau)}\, d\tau = E_0 + \int_0^\infty \frac{H(\tau)\, [1 - i/(\omega\tau)]}{1 + 1/(\omega^2\tau^2)}\, d\tau$$

where $H(\tau) \geq 0$ is the *relaxation spectrum*, and $E_0 \geq 0$. If we discretize the spectrum we get

$$\boxed{E_{\text{eff}}(\omega) = E_0 + \sum_j \frac{H_j\, [1 - i/(\omega\tau_j)]}{1 + 1/(\omega^2\tau_j^2)}} \tag{5.66}$$

where $H_j \geq 0$, $\tau_j > 0$, and $E_0 \geq 0$. This implies that $\text{Re}(E_{\text{eff}}(\omega)) \geq 0$ and $\text{Im}(E_{\text{eff}}(\omega)) \leq 0$ for all ω. Therefore, *linear viscoelastic models cannot represent negative elastic moduli and can fail badly when used to model materials that are better represented by Helmholtz resonator models.* In fact, for viscoelastic material models, we find that $E(\omega)$ is analytic in the entire complex plane except for isolated poles (at intervals of $\omega\tau$) in the negative imaginary axis located at $\omega = -i/\tau_j$.

On the other hand, for the frequency-dependent model that we have discussed in the previous section, the effective modulus is generally analytic only in the upper-half ω-plane (see Figure 5.20). Also, for such materials, $\overline{E}(\omega) = E(-\overline{\omega})$, where $\overline{(\bullet)}$ indicates the complex conjugate. Note that we did not consider the mass when we derived the modulus of the Maxwell model. The relation between viscoelastic models of the Maxwell type and general frequency-dependent materials continues to be an open question.

Fading memory and frequency dependent mass

A justification of the Maxwell model can be provided by considering the behavior of viscoelastic materials.‖ Consider an experiment where a bar of viscoelastic material

‖The model that is discussed here is based on the description given in Christensen (2003). The inconsistency between the fading memory description of materials and general frequency-dependent materials was pointed out by G. W. Milton.

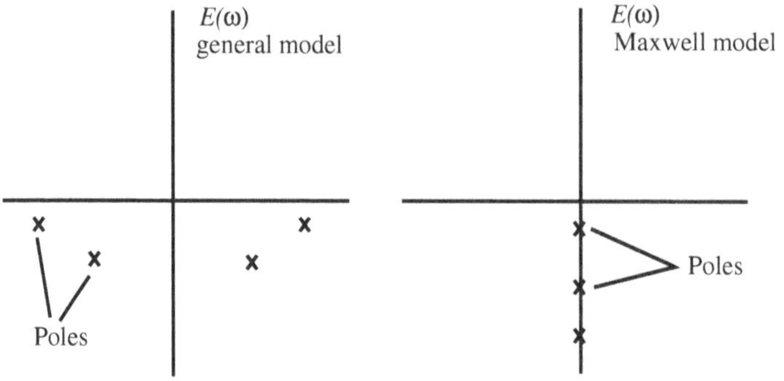

FIGURE 5.20
Poles for a general frequency-dependent material versus poles for a generalized Maxwell model.

of length l is deformed by a fixed amount. We want to see how the stress changes with time. Recall, that if the bar is extended by an amount $u(x)$ where $x = 0$ at one end of the bar, then the one-dimensional strain is defined as

$$\varepsilon = \frac{du}{dx}.$$

Therefore, the displacement in the bar can be expressed in terms of the strain as

$$u(x) = \varepsilon\, x \qquad \Longrightarrow \qquad \varepsilon = \frac{u(\ell)}{\ell} = \frac{\Delta\ell}{\ell}.$$

Also, if F is the applied force on the bar and A is its cross-sectional area, then the stress is given by

$$\sigma = \frac{F}{A}.$$

Let us now apply a strain to the bar at time $t = 0$ and hold the strain fixed. Due to the initial application of the strain, the stress reaches a value K_0 and then relaxes as time increases (due to the relaxation of polymer chains for instance). If the strain is applied by the superposition of two stepped strains as shown in Figure 5.21, we have

$$\frac{d\varepsilon}{dt} = a\,\delta(t - t_1) + b\,\delta(t - t_2)\,.$$

The stress is then given by

$$\sigma = a\,K(t - t_1) + b\,K(t - t_2)\,.$$

If the strain is applied by a series of infinitesimal steps, then we get a more general form for the stress:

$$\sigma(t) = \int_{-\infty}^{t} K(t - t')\,\frac{d\varepsilon(t')}{dt'}\,\mathrm{d}t' \qquad (5.67)$$

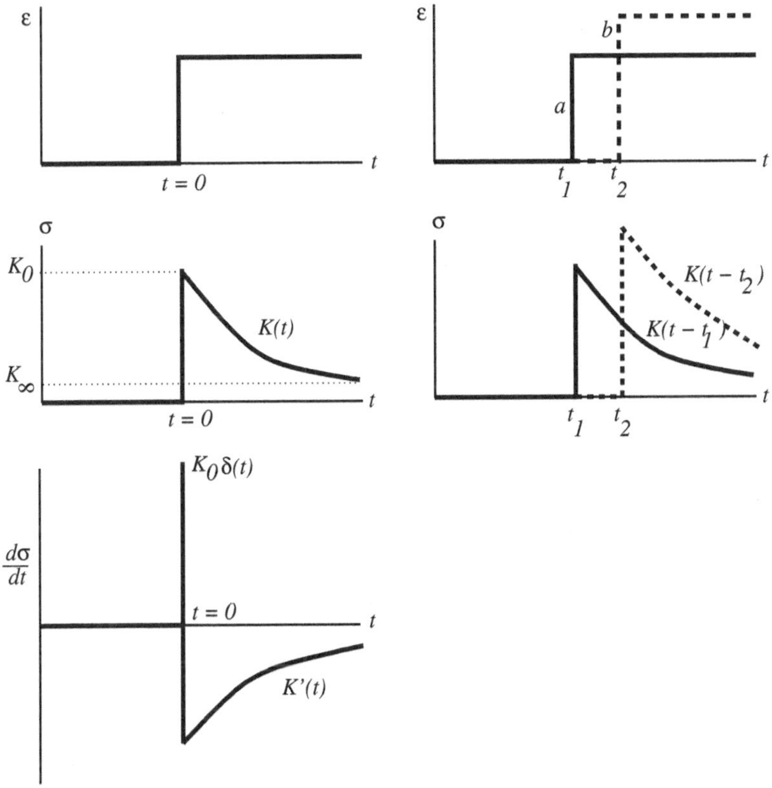

FIGURE 5.21
One-dimensional stress relaxation curves for a viscoelastic material and the linear superposition of strains.

where the integral should be interpreted in the distributional sense. Integrating by parts (and assuming that $\varepsilon = 0$ at $t = -\infty$), we get

$$\sigma(t) = K_0\ \varepsilon(t) + \int_{-\infty}^{t} K(t-t')\ \varepsilon(t')\ \mathrm{d}t'\ .$$

Now, $\sigma(t)$ clearly depends on past values of $d\varepsilon/dt$. We expect $\sigma(t)$ should have a stronger dependence on $d\varepsilon/dt$ in the recent past than in the distant past. More precisely, the dependence should decrease monotonically as $\tau = t - t'$ increases. This implies that $K(\tau)$ should decrease as τ increases, i.e.,

$$\frac{dK(\tau)}{d\tau} < 0 \quad \forall \tau > 0 \ \text{ and } \ K(\tau) > 0\,.$$

This is the assumption of *fading memory*. From equation (5.67) the rate of change of σ is given by

$$\frac{d\sigma(t)}{dt} = \int_{-\infty}^{t} \left.\frac{dK(\tau)}{d\tau}\right|_{\tau=t-t'} \frac{d\varepsilon(t')}{dt'}\,\mathrm{d}t'\,.$$

Again, we expect $d\sigma/dt$ to have a stronger dependence on $d\varepsilon(t')/dt'$ in the recent past than in the far past, i.e.,

$$\left|\frac{dK(\tau)}{d\tau}\right| \quad \text{should decrease as } \tau \text{ increases.}$$

Now, for all $\tau > 0$, the requirement of fading memory implies that

$$\frac{dK(\tau)}{d\tau} < 0 \implies \frac{d^2K(\tau)}{d\tau^2} > 0 \implies \frac{d^3K(\tau)}{d\tau^3} < 0 \implies \frac{d^4K(\tau)}{d\tau^4} > 0 \quad \cdots$$

Such functions are said to be *completely monotonic*. An example is

$$K(\tau) = e^{-\tau/\tau'} \qquad \implies \qquad \frac{dK(\tau)}{d\tau} < 0\,,\ \frac{d^2K(\tau)}{d\tau^2} > 0 \quad \cdots.$$

More generally,

$$K(\tau) = K_\infty + \int_0^\infty H(\tau')\,e^{-\tau/\tau'}\,\mathrm{d}\tau' \tag{5.68}$$

is completely monotonic if $H(\tau') \geq 0$ for all τ and $K_\infty \geq 0$. The function $H(\tau')$ is called the *relaxation spectrum*. Conversely, any completely monotonic function can be written in the form given in equation (5.68) (Bernstein, 1928). Let us consider the situation where

$$\varepsilon(t) = \mathrm{Re}(\widehat{\varepsilon}\,e^{-i\omega t}) \qquad \implies \qquad \frac{d\varepsilon(t)}{dt} = \mathrm{Re}(-i\omega\,\widehat{\varepsilon}\,e^{-i\omega t})\,.$$

Then, from (5.67), we have

$$\sigma(t) = \mathrm{Re}\left(-i\omega\,\widehat{\varepsilon}\int_{-\infty}^{t} K(t-t')\,e^{-i\omega t'}\,\mathrm{d}t'\right)\,.$$

The substitution $\tau = t - t'$ in the above equation gives

$$\sigma(t) = \mathrm{Re}\left(-i\omega\,\widehat{\varepsilon}\,e^{-i\omega t}\int_0^\infty K(\tau)\,e^{-i\omega\tau}\,\mathrm{d}\tau\right)\,.$$

If we define

$$E(\omega) := -i\omega\int_0^\infty K(\tau)\,e^{-i\omega\tau}\,\mathrm{d}\tau \tag{5.69}$$

we have

$$\sigma(t) = \mathrm{Re}\left(E(\omega)\,\widehat{\varepsilon}\,e^{-i\omega t}\right) = \mathrm{Re}\left(\widehat{\sigma}\,e^{-i\omega t}\right) \tag{5.70}$$

where $\widehat{\sigma} := E(\omega)\, \widehat{\varepsilon}$. Now, let $K(\tau)$ be a completely monotonic function of the form given in (5.68). Then from equation (5.69) we get

$$E(\omega) = -i\omega \int_0^\infty K_\infty\, e^{-i\omega\tau}\, \mathrm{d}\tau - i\omega \int_0^\infty \mathrm{d}\tau'\, H(\tau') \int_0^\infty \mathrm{d}\tau\, e^{\tau(i\omega - 1/\tau')}\,.$$

Assume that ω has a very small positive imaginary part (which implies that $\varepsilon(t)$ increases very slowly as t goes to ∞). Then

$$E(\omega) = -i\omega \left(\frac{-K_\infty}{i\omega}\right) - i\omega \int_0^\infty \mathrm{d}\tau'\, H(\tau') \left(\frac{-1}{i\omega - 1/\tau'}\right)\,.$$

Making the substitution $\tau = \tau'$ and $E_0 = K_\infty$ gives us

$$E(\omega) = E_0 + \int_0^\infty \frac{H(\tau)}{1 + i/(\omega\tau)}\, \mathrm{d}\tau\,. \tag{5.71}$$

This is the generalized Maxwell model that we had used to derive equation (5.66). Is the assumption of fading memory always correct? Recall the model of the Helmholtz resonator shown in Figure 5.17. If we apply a strain in the form of a step function to this model, the resulting stress response is not a monotonically decreasing function of time. Rather it oscillates around a certain value and may damp out over time. A similar oscillatory behavior is expected in other spring-mass systems and $K(\tau)$ will, in general, not be monotonic. Hence fading memory is only applicable to a subset of possible frequency-dependent models for the elastic modulus.

5.4 Band gaps, negative index and lenses

The similarity between the equations of TE- and TM-wave equations in electrodynamics and the SH-wave equation in elastodynamics (antiplane shear) and acoustic wave equation suggests that we should be able to observe similar phenomena in these disparate fields. Recall the split-ring resonator geometry shown in Figure 4.6 (p. 136). This structure has a frequency-dependent magnetic permeability the real part of which can be negative at certain frequencies when subjected to an out-of-plane magnetic field.

For TM-mode electromagnetic waves where H_3 is the out-of-plane magnetic induction in the $\mathbf{e}_3$-direction, the governing equation is

$$\overline{\boldsymbol{\nabla}} \cdot (\frac{1}{\varepsilon}\, \overline{\boldsymbol{\nabla}} H_3) + \omega^2\, \mu\, H_3 = 0$$

where ε is the permittivity and μ is the magnetic permeability. If the value of $1/\varepsilon$ in the region of the ring is small and hence ε is large (which implies that the conductivity σ is large), then the effective permeability μ_{33}^{eff} can be negative for this material. Now consider the equivalent SH-wave (or antiplane shear wave) propagation problem. The governing equation for this problem is

$$\overline{\nabla} \cdot (\mu \, \overline{\nabla} u_3) + \omega^2 \, \rho \, u_3 = 0$$

where μ is the shear modulus, ρ is the mass density, and u_3 is the out-of-plane displacement field. The similarity between the two equations suggests that the shear modulus (μ) and the inverse of the permittivity ($1/\varepsilon$) are analogous, and the density (ρ) is analogous to the magnetic permeability (μ). Therefore, we can reasonably expect an elastic composite structure containing split rings to exhibit a negative effective mass density at certain frequencies.

Recall that if a material has a negative density but a positive shear modulus, shear waves cannot propagate through the material. Hence, at the frequencies where the effective mass density becomes negative we expect to see a *band gap*. In fact, Movchan and Guenneau (2004) showed numerically and using asymptotic homogenization that an array of double split rings does exhibit a strong band gap.

We can compare the behavior of elastic split-ring resonators with the simple spring-mass model that we discussed earlier. Consider the periodic geometry shown in Figure 5.22. The matrix material has a high value of shear modulus (μ) while the split-ring-shaped region has a low shear modulus or is a void. The material inside the ring has the same shear modulus as the matrix material and is connected to the matrix by a thin ligament. The system is subjected to an antiplane shear displacement u_3 in the x_3-direction (parallel to the axis of each cylindrical split ring).

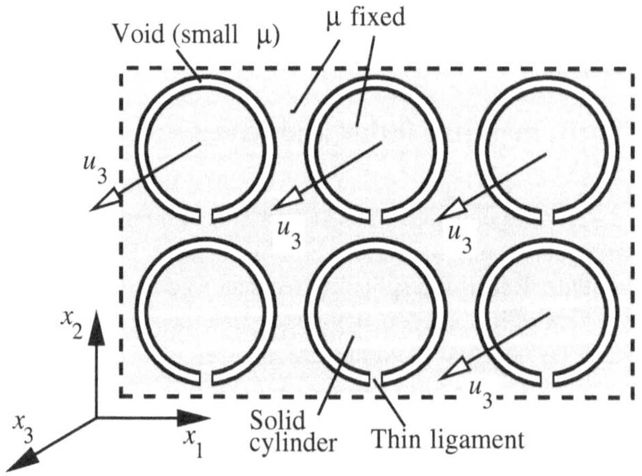

FIGURE 5.22

A periodic geometry containing split hollow cylinders of soft material in a matrix of stiff material. The x_3-direction is parallel to the axis of each cylinder.

Clearly, each periodic component of the system behaves like a mass attached to a spring. This is a resonant structure and the effective density $\rho_{33}^{\text{eff}}(\omega)$ can be negative. Therefore, we can expect to create elastic band-gap materials with such microstructures. However, for transmission through the material at the resonant frequencies we also need a negative shear modulus. But materials with negative shear modulus have been found to be unstable (see Jaglinski et al. (2007) and references cited in that paper) and therefore negative refractive index elastic materials are unlikely to be found.

On the other had, we have seen that effective negative bulk moduli can actually be achieved using Helmholtz resonator structures. This suggests that materials with simultaneously negative mass density and bulk modulus may actually be manufactured. The obvious candidate for such materials is in the context of acoustic waves and these unusual composites have been called *acoustic metamaterials*.

Recall that the acoustic wave equation at fixed frequency ω can be written as

$$\boldsymbol{\nabla} \cdot \left(\frac{1}{\rho} \boldsymbol{\nabla} p \right) + \frac{\omega^2}{\kappa} p = 0$$

where ρ is the mass density, κ is the bulk modulus, and p is the pressure. If $\rho = \rho(x_1, x_2)$ and $\kappa = \kappa(x_1, x_2)$ only depend on x_1 and x_2 and p is independent of x_3, then the three-dimensional gradient operator $\boldsymbol{\nabla}$ can be replaced with the two-dimensional gradient operator $\overline{\boldsymbol{\nabla}}$, and we get

$$\overline{\boldsymbol{\nabla}} \cdot \left(\frac{1}{\rho(\omega)} \overline{\boldsymbol{\nabla}} p \right) + \frac{\omega^2}{\kappa(\omega)} p = 0 \quad \text{and} \quad \mathbf{u}(\omega) = \frac{1}{\omega^2 \rho(\omega)} \overline{\boldsymbol{\nabla}} p$$

where we have explicitly shown the dependence of the material parameters on the frequency. Therefore, by analogy with the results from antiplane shear elasticity and TM (or TE) electromagnetism, if $\rho(\omega)$ and $\kappa(\omega)$ are both negative, we get a negative refractive index material for acoustics. The speed of sound in an acoustic medium is given by $c = \sqrt{\kappa/\rho}$ and is imaginary if either κ or ρ is negative. A material with these properties appears opaque to sound waves. However, if κ and ρ are both positive or both negative, the sound speed is real and acoustic waves can propagate through the medium. Such a material is a *negative refractive index* acoustic medium. Recall the situation shown in Figure 2.9. Since this situation is analogous to the one we observe for electromagnetism, the effect of a slab of negative κ, ρ material will be to translate a source of acoustic waves and the slab will act like a perfect lens.

In 2004, Zhang and Liu (2004) showed simulations that indicated that negative acoustic refraction was possible with an array of regularly spaced cylinders. This negative refraction effect was based on Bragg diffraction and was therefore limited to wavelengths of the order of the radii of the cylinders. Experimental evidence for focusing of ultrasound using negative refraction (at approximately 1.6 MHz frequency) was demonstrated by Yang et al. (2004) using a ~8 mm thick phononic crystal consisting of 0.8 mm diameter tungsten beads in water. The problem with such an approach is that, as the frequency decreases, the size of the crystal needed to achieve the same effect increases. For example, for kHz waves one would need a crystal of the order of 1 meter thick. This makes such structures too large to be of practical use in many applications. Metamaterials that use local resonances in the microstructure to create effective negative mass and bulk modulus are not subject to such constraints. However, the needed effective properties in metamaterials occur only over narrow frequency ranges and are accompanied by large losses.

The possibility of materials with simultaneously negative density and bulk modulus appears unlikely at first glance. However, Li and Chan (2004) showed, using a coherent potential approximation estimate of the effective properties of a rubber-water composite, that a negative refractive index acoustic metamaterial was possible. Another possible architecture, proposed by Ding et al. (2007), consists of two lattices in a zinc blende matrix; one lattice containing water-in-air spheres and the other shifted lattice containing gold-in-rubber spheres. The first demonstration of focusing of acoustic waves with negative refractive index locally resonant structure was by Zhang et al. (2009) who used an array of Helmholtz resonators with period $\sim$3.2 mm to focus a $\sim$60 kHz acoustic wave generated from a localized source. The sample periodicity was approximately 1/8th the incident wavelength. A transmission line model of the system was used to explain the frequency-dependent behavior of the mass density and the bulk modulus (see Zhang (2010) for details). However, the quality of the focus was not found to be of a high quality.

An alternative approach to achieving focusing and magnification has been experimentally demonstrated by Li et al. (2009). Instead of using resonances to achieve negative density and bulk modulus, a highly anisotropic dynamic mass density was used to achieve focus.

5.5 Anisotropic density

We have seen in Section 5.1.3 that a frequency-dependent anisotropic dynamic mass density can be observed in composites that can be approximated by spring-mass architectures. Geometries of such composites were suggested in Milton and Willis (2007) and one such model can be seen in Figure 5.11 (p. 162). In fact, the idea of tensorial anisotropic densities was discussed considerably earlier in a paper by Schoenberg and Sen (1983) in the context of layered media. In this section we will review some models of materials with anisotropic density and examine the status of experimental verification of these ideas.

The Schoenberg-Sen model is discussed in detail in Section 8.5.1. The primary observation of this model is that, for a periodic layered acoustic medium, the low-frequency effective response of the medium is indistinguishable for a homogeneous medium with a scalar bulk modulus and a transversely isotropic density. The effective bulk modulus (κ^{eff}) and the effective density $\boldsymbol{\rho}^{\text{eff}}$ for such a layered medium are given by

$$\kappa^{\text{eff}} = \frac{1}{\langle 1/\kappa \rangle} \quad \text{and} \quad \boldsymbol{\rho}^{\text{eff}} \equiv \begin{bmatrix} \langle \rho \rangle & 0 & 0 \\ 0 & 1/\langle 1/\rho \rangle & 0 \\ 0 & 0 & 1/\langle 1/\rho \rangle \end{bmatrix} \quad \text{with} \quad \langle x \rangle := \sum_{j=1}^{N} f_j x_j$$

where x_j is the value of the quantity x in layer j and the volume fraction occupied by layer j in a period containing N layers is f_j.

Structures made of layered fluids can be difficult to manufacture, even when the layers are separated by thin membranes similar to biological cell walls.** Recently, Mei et al. (2006) and Torrent and Sánchez-Dehesa (2008b) have shown that we need not limit ourselves to layered media. They show, using multiple scattering calculations, that periodic structures can also be designed to exhibit anisotropic density in the low-frequency limit.

In the approach taken by Torrent and Sánchez-Dehesa (2008b), the medium is taken to consist of a periodic array of isotropic circular cylinders embedded in an isotropic fluid. The circular cylinders can be rigid or elastic. The effective acoustic medium is assumed to be governed by the acoustic equations:

$$\boldsymbol{\rho}_{\text{eff}}^{-1} \cdot \boldsymbol{\nabla} p + \dot{\mathbf{v}} = \mathbf{0} \quad \text{and} \quad \boldsymbol{\nabla} \cdot \mathbf{v} + \kappa_{\text{eff}}^{-1} \dot{p} = 0 .$$

Taking the divergence of the first equation, and plugging in the time derivative of the second equation into the result, gives

$$\boldsymbol{\rho}_{\text{eff}}^{-1} : \boldsymbol{\nabla}(\boldsymbol{\nabla} p) - \kappa_{\text{eff}}^{-1} \ddot{p} = 0 .$$

If we look for time harmonic plane wave solutions of the form $p(\mathbf{x},t) = \widehat{p} \exp[i(\mathbf{k} \cdot \mathbf{x} - \omega t)]$ we have

$$\boldsymbol{\rho}_{\text{eff}}^{-1} : (\mathbf{k} \otimes \mathbf{k}) - \omega^2 \kappa_{\text{eff}}^{-1} = 0 .$$

If $\mathbf{k} = \|\mathbf{k}\| \, \widehat{\mathbf{k}}$ we can write

$$\kappa_{\text{eff}} \boldsymbol{\rho}_{\text{eff}}^{-1} : (\widehat{\mathbf{k}} \otimes \widehat{\mathbf{k}}) = \omega^2 / \|\mathbf{k}\|^2 \quad \Longrightarrow \quad \boxed{(\boldsymbol{C}_{\text{eff}})^2 : (\widehat{\mathbf{k}} \otimes \widehat{\mathbf{k}}) = c^2}$$

where $\boldsymbol{C}_{\text{eff}}$ is the effective sound speed tensor and we have used $\omega / \|\mathbf{k}\| = c$ with c being the phase speed. The above relation is matched with solutions derived using multiple scattering to determine the effective anisotropic density tensor. The bulk modulus can be calculated using Hashin's relation (7.24) (p. 262) discussed in Section 7.2.2.

**The effective density of such a fluid-membrane structure has not been calculated yet.

5.6 Willis materials in elastodynamics

Many of the ideas discussed so far can be interpreted as special cases of a general theory for the dynamics of elastic composites developed in the early 1980s in a series of papers by Willis (Willis, 1980, 1981a,b). A summary was provided in 1997 (Willis, 1997) and further developments and implications for metamaterials were presented in Milton and Willis (2007). These developments have led to the *Willis equations* discussed in this section. These equations hint at some unusual dynamic properties of certain classes of composite materials which we call *Willis materials*.

5.6.1 Ensemble averaging

At present, the Willis equations depend crucially on the assumption that an ensemble average is a good descriptor of behavior of individual realizations of the microstructure of a medium. Before we proceed to describe the Willis equations, let us distinguish between ensemble averaging and volume averaging.

Consider a property $\phi(\mathbf{x},t)$ that varies within a body and is a function of space ($\mathbf{x}$) and time (t). We will avoid any time averages at this point and concentrate on spatial averages. The volume average of the property ϕ is defined as

$$\langle\phi(t)\rangle_{\text{vol}} = \lim_{N\to\infty}\frac{1}{N}\sum_{i=1}^{N}\phi(\mathbf{x}_i,t) = \frac{1}{V}\int_{\Omega}\phi(\mathbf{x},t)\,\mathrm{dV}$$

where V is the volume of the body and N is the number of points in the body. In this case the sum is over all points in the body and such an average is quite convenient if the geometry of the body is known. For bodies which have a microstructure that may not be known exactly, but for which some microstructural statistics are known, an ensemble average is often more convenient. The ensemble average of ϕ is defined as

$$\langle\phi(\mathbf{x},t)\rangle = \lim_{n\to\infty}\frac{1}{n}\sum_{i=1}^{n}\phi_i(\mathbf{x},t) = \int_{\mathcal{A}}\phi(\mathbf{x},t;\alpha)\,\mathrm{d}[p(\alpha)].$$

In this case n is the number of times that the value of ϕ is sampled at the location $\mathbf{x}$. In the integral form, $\mathcal{A}$ is the sample space of possible realizations, α is a parameter such that $\alpha \in \mathcal{A}$, and $p(\alpha)$ is the probability of realizing α. For an ergodic ensemble we get the surprising result that the volume and ensemble averages are identical. See Zohdi and Wriggers (2008) for further discussion of these ideas in the context of micromechanics. Another good resource is Kleinstreuer (2003) (in the context of two-phase flows).

Some examples of ensembles are:

- Periodic media with a period δ where the fields are not necessarily periodic (see Figure 5.23). The ensembles is the material and all translations of it. Of course, a translation that is equal to the period gives back the same material.

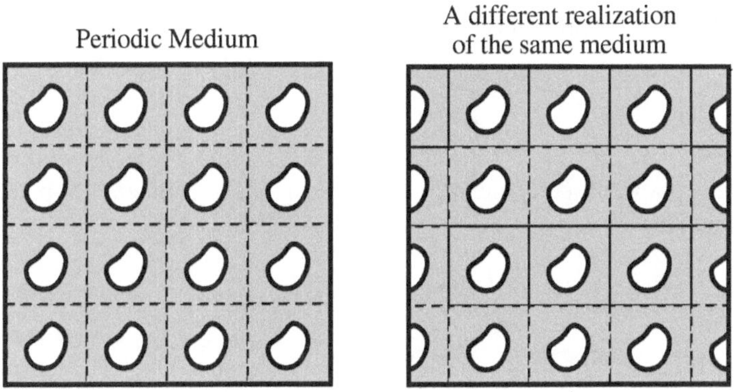

FIGURE 5.23
An example of a periodic medium and its translated version.

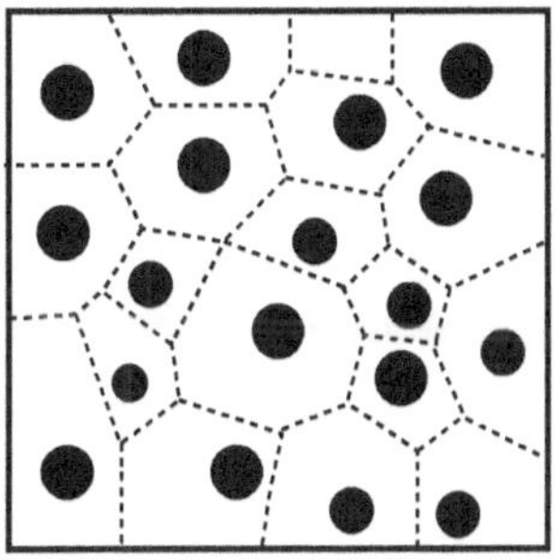

FIGURE 5.24
An example of a medium that can be described by a statistical process.

- Media generated by some translation invariant statistical process. This means that a particular realization and its translations are equally likely to occur (roughly speaking). An example is a medium generated by a Poisson process. We can represent the ensemble by constructing a Voronoi tessellation and assigning constants to each cell at random (see Figure 5.24).
- Media generated by some statistical process where the statistics vary slowly with position.

Properties of such media, if they are composed of m constituents, can be expressed in terms of characteristic functions χ_j where

$$\chi_j(\mathbf{x};\alpha) = \begin{cases} 1 & \text{if } \mathbf{x} \in \text{ constituent } j \text{ for realization } \alpha \\ 0 & \text{otherwise} \end{cases} .$$

Only one constituent is present at each point, hence

$$\sum_{j=1}^{m} \chi_j(\mathbf{x};\alpha) = 1 .$$

Because of the above property, the probability of finding constituent j at point $\mathbf{x}$ (for all realizations) is

$$p_j(\mathbf{x}) = \int_{\mathcal{A}} \chi_j(\mathbf{x};\alpha)\, \mathrm{d}[p(\alpha)] = \langle \chi_j(\mathbf{x}) \rangle$$

where, in accordance with linear elasticity, we have assumed that $\mathbf{x}$ is the position in the reference configuration of the body. For ensembles that are translation invariant (or for those in which the statistics vary sufficiently slowly), $p_j(\mathbf{x}) = p_j$, i.e., the dependence on $\mathbf{x}$ is lost and $p_j = f_j$, where f_j is the volume fraction of phase j.

With the above definitions we can proceed to find ensemble averages of quantities that are of interest in elastodynamics. Thus, the stiffness tensor of an m-component composite, with homogeneous individual phases, is given by

$$\mathsf{C}(\mathbf{x};\alpha) = \sum_{j=1}^{m} \chi_j(\mathbf{x};\alpha)\, \mathsf{C}_j$$

where C_j is the stiffness tensor of the j-th phase. The ensemble averaged stiffness tensor in this case is

$$\langle \mathsf{C}(\mathbf{x}) \rangle = \left\langle \sum_{j=1}^{m} \chi_j(\mathbf{x};\alpha) \right\rangle \mathsf{C}_j = \sum_{j=1}^{m} p_j(\mathbf{x}) \mathsf{C}_j .$$

Similarly, if the displacement field is $\mathbf{u}(\mathbf{x},t;\alpha)$, the ensemble average at a location $\mathbf{x}$ at time t is defined via

$$p_j(\mathbf{x})\, \langle \mathbf{u}(\mathbf{x},t) \rangle_j = \int_{\mathcal{A}} \chi_j(\mathbf{X};\alpha)\, \mathbf{u}(\mathbf{x},t;\alpha)\, \mathrm{d}[p(\alpha)] .$$

We will have more to say about the ensemble averaged displacement later.

5.6.2 The Willis equations

Recall that the equations governing the balance of momentum of a linear elastic body are

$$\boldsymbol{\nabla}\cdot\boldsymbol{\sigma} + \mathbf{f} = \dot{\mathbf{p}} \qquad \text{where} \quad \dot{\mathbf{p}} := \frac{\partial \mathbf{p}}{\partial t} \tag{5.72}$$

and $\boldsymbol{\sigma}$ is the Cauchy stress, $\mathbf{f}$ is the body force, and $\mathbf{p}$ is the momentum. We assume that the body force is independent of the realization. The microscopic constitutive relations are assumed to be

$$\boldsymbol{\sigma} = \int_{-\infty}^{t} \mathsf{C}(t-\tau) : \boldsymbol{\varepsilon}(\tau)\, \mathrm{d}\tau \quad \text{and} \quad \mathbf{p} = \rho\, \dot{\mathbf{u}} \qquad \text{where} \quad \dot{\mathbf{u}} := \frac{\partial \mathbf{u}}{\partial t} \tag{5.73}$$

and $\mathbf{C}$ is the stiffness tensor, $\boldsymbol{\varepsilon}$ is the strain tensor, ρ is the mass density, and $\mathbf{u}$ is the displacement vector. Note that the spatial dependence on $\mathbf{x}$ is implicit in the above equations. We will simplify the notation in the following discussion by defining

$$\mathbf{C}\star\boldsymbol{\varepsilon} \equiv \int_{-\infty}^{t} \mathbf{C}(t-\tau) : \boldsymbol{\varepsilon}(\tau)\, d\tau .$$

In particular, the convolutions of tensors of equal order are defined as

$$\mathbf{a}\star\mathbf{b} = \int_{-\infty}^{t} \mathbf{a}(t-\tau)\cdot\mathbf{b}(\tau)\, d\tau\,;\quad \boldsymbol{A}\star\boldsymbol{B} = \int_{-\infty}^{t} \boldsymbol{A}(t-\tau) : \boldsymbol{B}(\tau)\, d\tau$$

where $\mathbf{a}, \mathbf{b}$ are vectors and $\boldsymbol{A}, \boldsymbol{B}$ are second-order tensors.

By ensemble averaging (5.72) we get

$$\boldsymbol{\nabla}\cdot\langle\boldsymbol{\sigma}\rangle + \mathbf{f} = \langle\dot{\mathbf{p}}\rangle . \tag{5.74}$$

Recall that in linear elasticity the strain-displacement relation is

$$\boldsymbol{\varepsilon} = \tfrac{1}{2}\left[\boldsymbol{\nabla}\mathbf{u} + (\boldsymbol{\nabla}\mathbf{u})^T\right] \tag{5.75}$$

An ensemble average of the above equation gives

$$\langle\boldsymbol{\varepsilon}\rangle = \tfrac{1}{2}\left[\boldsymbol{\nabla}\langle\mathbf{u}\rangle + (\boldsymbol{\nabla}\langle\mathbf{u}\rangle)^T\right]. \tag{5.76}$$

However, we cannot ensemble average (5.73) since a product of averages is not in the general equation for the average of the product, i.e.,

$$\langle\boldsymbol{\sigma}\rangle \neq \langle\mathbf{C}\rangle\star\langle\boldsymbol{\varepsilon}\rangle \quad \text{and} \quad \langle\mathbf{p}\rangle \neq \langle\rho\rangle\ \langle\dot{\mathbf{u}}\rangle .$$

We therefore need some effective constitutive relation. These are the *Willis equations* and can be written as

$$\boxed{\begin{aligned} \langle\boldsymbol{\sigma}\rangle &= \mathbf{C}_{\text{eff}}\star\langle\boldsymbol{\varepsilon}\rangle + \mathcal{S}_{\text{eff}}\star\langle\dot{\mathbf{u}}\rangle \\ \langle\mathbf{p}\rangle &= \mathcal{S}^{\dagger}_{\text{eff}}\star\langle\boldsymbol{\varepsilon}\rangle + \rho_{\text{eff}}\star\langle\dot{\mathbf{u}}\rangle \end{aligned}} \tag{5.77}$$

where the operator $\star$ represents a convolution over time, i.e.,

$$\begin{aligned} \mathbf{C}\star\boldsymbol{\varepsilon} &\equiv \int_{-\infty}^{t} \mathbf{C}(t-\tau) : \boldsymbol{\varepsilon}(\tau)\, d\tau\,; \quad \mathcal{S}\star\dot{\mathbf{u}} \equiv \int_{-\infty}^{t} \mathcal{S}(t-\tau)\cdot\dot{\mathbf{u}}(\tau)\, d\tau \\ \mathcal{S}^{\dagger}\star\boldsymbol{\varepsilon} &\equiv \int_{-\infty}^{t} \mathcal{S}^{\dagger}(t-\tau)\cdot\boldsymbol{\varepsilon}(\tau)\, d\tau\,; \quad \rho\star\dot{\mathbf{u}} \equiv \int_{-\infty}^{t} \rho(t-\tau)\cdot\dot{\mathbf{u}}(\tau)\, d\tau . \end{aligned}$$

These operators are non-local in time and, in general, also non-local in space. The adjoint operator (represented by the superscript $\dagger$) is defined via

$$\int \mathbf{a}\star(\mathcal{S}^{\dagger}_{\text{eff}}\star\boldsymbol{A})\, d\mathbf{x} = \int \boldsymbol{A}\star(\mathcal{S}_{\text{eff}}\star\mathbf{a})\, d\mathbf{x}$$

for all vector fields $\mathbf{a}$ and second-order tensor fields $\boldsymbol{A}$ and at time t. The coupling terms $\mathcal{S}_{\text{eff}}$ and $\mathcal{S}^{\dagger}_{\text{eff}}$ are third-order tensors.

5.6.3 Derivation of the Willis equations

Let us introduce a homogeneous reference medium with properties $\mathbf{C}_0$ and ρ_0. Let us define *polarization fields*

$$\begin{aligned} \boldsymbol{S}(\mathbf{x},t) &:= [\mathbf{C}(\mathbf{x},t)-\mathbf{C}_0]\star\boldsymbol{\varepsilon}(\mathbf{x},t) = \boldsymbol{\sigma}(\mathbf{x},t)-\mathbf{C}_0\star\boldsymbol{\varepsilon}(\mathbf{x},t) \\ \mathbf{m}(\mathbf{x},t) &:= [\rho(\mathbf{x},t)-\rho_0]\,\dot{\mathbf{u}}(\mathbf{x},t) = \mathbf{p}(\mathbf{x},t)-\rho_0\,\dot{\mathbf{u}}(\mathbf{x},t)\,. \end{aligned} \tag{5.78}$$

Then, dropping the explicit dependence on $\mathbf{x}$ and t for convenience, we have

$$\boldsymbol{\sigma} = \boldsymbol{S}+\mathbf{C}_0\star\boldsymbol{\varepsilon} \qquad \text{and} \qquad \mathbf{p} = \mathbf{m}+\rho_0\,\dot{\mathbf{u}}\,. \tag{5.79}$$

Taking the divergence of the first of equations (5.79), we get

$$\boldsymbol{\nabla}\cdot\boldsymbol{\sigma} = \boldsymbol{\nabla}\cdot\boldsymbol{S}+\boldsymbol{\nabla}\cdot(\mathbf{C}_0\star\boldsymbol{\varepsilon})\,. \tag{5.80}$$

Also, taking the time derivative of the second of equations (5.79), we have

$$\dot{\mathbf{p}} = \dot{\mathbf{m}}+\rho_0\,\ddot{\mathbf{u}}\,. \tag{5.81}$$

Recall that the equation of motion is

$$\boldsymbol{\nabla}\cdot\boldsymbol{\sigma}+\mathbf{f} = \dot{\mathbf{p}}\,. \tag{5.82}$$

Plugging (5.80) and (5.81) into (5.82), and rearranging, gives

$$\boldsymbol{\nabla}\cdot(\mathbf{C}_0\star\boldsymbol{\varepsilon})+\mathbf{f}+\boldsymbol{\nabla}\cdot\boldsymbol{S}-\dot{\mathbf{m}} = \rho_0\,\ddot{\mathbf{u}}\,. \tag{5.83}$$

In the reference medium, $\boldsymbol{S}=\mathbf{0}$ and $\mathbf{m}=\mathbf{0}$. Let $\mathbf{u}_0$ be the solution in the reference medium in the presence of the body force $\mathbf{f}$ and with the same boundary conditions and initial conditions as in the actual body. For example, if the actual body has $\mathbf{u}\rightarrow\mathbf{0}$ as $t\rightarrow-\infty$, then $\mathbf{u}_0\rightarrow\mathbf{0}$ as $t\rightarrow-\infty$ in the reference medium. Then, in the reference medium, we have

$$\boldsymbol{\nabla}\cdot(\mathbf{C}_0\star\boldsymbol{\varepsilon}_0)+\mathbf{f} = \rho_0\,\ddot{\mathbf{u}}_0 \qquad \text{where} \qquad \boldsymbol{\varepsilon}_0 = \tfrac{1}{2}[\boldsymbol{\nabla}\mathbf{u}_0+(\boldsymbol{\nabla}\mathbf{u}_0)^T]\,. \tag{5.84}$$

If we want our effective stress-strain relations to be independent of the body force $\mathbf{f}$, all we have to do is subtract (5.84) from (5.83) to get

$$\boldsymbol{\nabla}\cdot[\mathbf{C}_0\star(\boldsymbol{\varepsilon}-\boldsymbol{\varepsilon}_0)]+\boldsymbol{\nabla}\cdot\boldsymbol{S}-\dot{\mathbf{m}} = \rho_0\,[\ddot{\mathbf{u}}-\ddot{\mathbf{u}}_0]\,.$$

or,

$$\boxed{-\boldsymbol{\nabla}\cdot(\mathbf{C}_0\star\boldsymbol{\varepsilon}')+\rho_0\,\ddot{\mathbf{u}}' = \mathbf{h}} \tag{5.85}$$

where we have defined

$$\mathbf{u}' := \mathbf{u}-\mathbf{u}_0\,; \quad \boldsymbol{\varepsilon}' := \tfrac{1}{2}[\boldsymbol{\nabla}\mathbf{u}'+(\boldsymbol{\nabla}\mathbf{u}')^T] = \boldsymbol{\varepsilon}-\boldsymbol{\varepsilon}_0\,; \quad \mathbf{h} := \boldsymbol{\nabla}\cdot\boldsymbol{S}-\dot{\mathbf{m}}\,.$$

If we assume that $\mathbf{h}$ is independent of $\mathbf{u}'$, then (5.85) can be written as

$$\mathcal{L}\,\mathbf{u}' = \mathbf{h} \tag{5.86}$$

where $\mathcal{L}$ is a linear operator. The solution of this equation is

$$\mathbf{u}' = \boldsymbol{G}\star\mathbf{h} \tag{5.87}$$

where $\boldsymbol{G}$ is the Green's function associated with the operator $\mathcal{L}$. Plugging back our definitions of $\mathbf{u}'$ and $\mathbf{h}$, we get the solution

$$\boxed{\mathbf{u} = \mathbf{u}_0 + \boldsymbol{G}\star(\boldsymbol{\nabla}\cdot\boldsymbol{S} - \dot{\mathbf{m}}) = \mathbf{u}_0 + \boldsymbol{G}\star(\boldsymbol{\nabla}\cdot\boldsymbol{S}) - \boldsymbol{G}\star\dot{\mathbf{m}}\,.} \tag{5.88}$$

Plugging the solution (5.88) into the strain-displacement relation (5.75) gives

$$\begin{aligned}\boldsymbol{\varepsilon} = \boldsymbol{\varepsilon}_0 &+ \tfrac{1}{2}\boldsymbol{\nabla}[\boldsymbol{G}\star(\boldsymbol{\nabla}\cdot\boldsymbol{S})] + \tfrac{1}{2}\left[\boldsymbol{\nabla}\{\boldsymbol{G}\star(\boldsymbol{\nabla}\cdot\boldsymbol{S})\}\right]^T \\ &- \tfrac{1}{2}\boldsymbol{\nabla}(\boldsymbol{G}\star\dot{\mathbf{m}}) - \tfrac{1}{2}\left[\boldsymbol{\nabla}(\boldsymbol{G}\star\dot{\mathbf{m}})\right]^T\,.\end{aligned} \tag{5.89}$$

Define the fourth-order tensor S_x and the third-order tensor $\mathcal{M}_x$ via

$$\begin{aligned}\mathsf{S}_x\star\boldsymbol{S} &= -\tfrac{1}{2}\left\{\boldsymbol{\nabla}[\boldsymbol{G}\star(\boldsymbol{\nabla}\cdot\boldsymbol{S})] + [\boldsymbol{\nabla}[\boldsymbol{G}\star(\boldsymbol{\nabla}\cdot\boldsymbol{S})]]^T\right\} \\ \mathcal{M}_x\star\mathbf{m} &= \tfrac{1}{2}\left\{\boldsymbol{\nabla}(\boldsymbol{G}\star\dot{\mathbf{m}}) + \tfrac{1}{2}\left[\boldsymbol{\nabla}(\boldsymbol{G}\star\dot{\mathbf{m}})\right]^T\right\}\,.\end{aligned}$$

Then we can write (5.89) as

$$\boxed{\boldsymbol{\varepsilon} = \boldsymbol{\varepsilon}_0 - \mathsf{S}_x\star\boldsymbol{S} - \mathcal{M}_x\star\mathbf{m}\,.} \tag{5.90}$$

Also, taking the time derivative of (5.88), we get

$$\dot{\mathbf{u}} = \dot{\mathbf{u}}_0 + \frac{d}{dt}[\boldsymbol{G}\star(\boldsymbol{\nabla}\cdot\boldsymbol{S})] - \frac{d}{dt}[\boldsymbol{G}\star\dot{\mathbf{m}}]\,. \tag{5.91}$$

Define the third-order tensor $\mathcal{S}_t$ and the second-order tensor $\boldsymbol{M}_t$ via

$$\begin{aligned}\mathcal{S}_t\star\boldsymbol{S} &= -\frac{d}{dt}[\boldsymbol{G}\star(\boldsymbol{\nabla}\cdot\boldsymbol{S})] \\ \boldsymbol{M}_t\star\mathbf{m} &= \frac{d}{dt}[\boldsymbol{G}\star\dot{\mathbf{m}}]\,.\end{aligned}$$

Then we can write (5.91) as

$$\boxed{\dot{\mathbf{u}} = \dot{\mathbf{u}}_0 - \mathcal{S}_t\star\boldsymbol{S} - \boldsymbol{M}_t\star\mathbf{m}\,.} \tag{5.92}$$

Willis (1981b) has shown that $\mathcal{S}_t$ and $\mathcal{M}_x$ are formal adjoints, i.e., $\mathcal{S}_t = \mathcal{M}_x^\dagger$, in the sense that

$$\int \mathbf{m}\star(\mathcal{S}_t\star\boldsymbol{S})\,d\mathbf{x} = \int \boldsymbol{S}\star(\mathcal{M}_x\star\mathbf{m})\,d\mathbf{x} \qquad \forall\,\mathbf{m},\boldsymbol{S},t\,.$$

If we eliminate $\boldsymbol{\varepsilon}$ and $\dot{\mathbf{u}}$ from equations (5.78) using (5.90) and (5.92), we get

$$\begin{aligned}(\mathsf{C}-\mathsf{C}_0)^{-1}\star\boldsymbol{S} + \mathsf{S}_x\star\boldsymbol{S} + \mathcal{M}_x\star\mathbf{m} &= \boldsymbol{\varepsilon}_0 \\ (\rho-\rho_0)^{-1}\,\mathbf{m} + \mathcal{S}_t\star\boldsymbol{S} + \boldsymbol{M}_t\star\mathbf{m} &= \dot{\mathbf{u}}_0\,.\end{aligned} \tag{5.93}$$

Also, ensemble averaging equations (5.90) and (5.92), we have

$$\begin{aligned}\langle\boldsymbol{\varepsilon}\rangle &= \boldsymbol{\varepsilon}_0 - \mathbf{S}_x \star \langle \boldsymbol{S}\rangle - \mathcal{M}_x \star \langle \mathbf{m}\rangle \\ \langle\dot{\mathbf{u}}\rangle &= \dot{\mathbf{u}}_0 - \mathcal{S}_t \star \langle \boldsymbol{S}\rangle - \boldsymbol{M}_t \star \langle \mathbf{m}\rangle \ .\end{aligned} \tag{5.94}$$

From (5.93) and (5.94), eliminating $\boldsymbol{\varepsilon}_0$ and $\dot{\mathbf{u}}_0$, we get

$$\begin{aligned}(\mathbf{C}-\mathbf{C}_0)^{-1} \star \boldsymbol{S} + \mathbf{S}_x \star (\boldsymbol{S} - \langle \boldsymbol{S}\rangle) + \mathcal{M}_x \star (\mathbf{m} - \langle \mathbf{m}\rangle) &= \langle\boldsymbol{\varepsilon}\rangle \\ (\rho-\rho_0)^{-1}\, \mathbf{m} + \mathcal{S}_t \star (\boldsymbol{S} - \langle \boldsymbol{S}\rangle) + \boldsymbol{M}_t \star (\mathbf{m} - \langle \mathbf{m}\rangle) &= \langle\dot{\mathbf{u}}\rangle \ .\end{aligned} \tag{5.95}$$

Equations (5.95) are linear in $\boldsymbol{S}$ and $\mathbf{m}$. Therefore, formally these equations have the form[††]

$$\begin{bmatrix}\boldsymbol{S}\\ \mathbf{m}\end{bmatrix} = \mathcal{T} \star \begin{bmatrix}\langle\boldsymbol{\varepsilon}\rangle \\ \langle\dot{\mathbf{u}}\rangle\end{bmatrix} . \tag{5.96}$$

From the definition of $\boldsymbol{S}$ and $\mathbf{m}$, taking the ensemble average gives us

$$\langle \boldsymbol{S}\rangle = \langle\boldsymbol{\sigma}\rangle - \mathbf{C}_0 \star \langle\boldsymbol{\varepsilon}\rangle \ ; \ \ \langle\mathbf{m}\rangle = \langle\mathbf{p}\rangle - \rho_0\ \langle\dot{\mathbf{u}}\rangle \ . \tag{5.97}$$

Also, from (5.96), taking the ensemble average leads to

$$\begin{bmatrix}\langle\boldsymbol{S}\rangle\\ \langle\mathbf{m}\rangle\end{bmatrix} = \langle\mathcal{T}\rangle \star \begin{bmatrix}\langle\boldsymbol{\varepsilon}\rangle \\ \langle\dot{\mathbf{u}}\rangle\end{bmatrix} .$$

Plugging in the relations (5.97) in these equations gives us

$$\begin{bmatrix}\langle\boldsymbol{\sigma}\rangle - \mathbf{C}_0 \star \langle\boldsymbol{\varepsilon}\rangle\\ \langle\mathbf{p}\rangle - \rho_0\ \langle\dot{\mathbf{u}}\rangle\end{bmatrix} = \langle\mathcal{T}\rangle \star \begin{bmatrix}\langle\boldsymbol{\varepsilon}\rangle \\ \langle\dot{\mathbf{u}}\rangle\end{bmatrix}$$

or

$$\begin{bmatrix}\langle\boldsymbol{\sigma}\rangle\\ \langle\mathbf{p}\rangle\end{bmatrix} = \langle\mathcal{T}\rangle \star \begin{bmatrix}\langle\boldsymbol{\varepsilon}\rangle \\ \langle\dot{\mathbf{u}}\rangle\end{bmatrix} + \begin{bmatrix}\mathbf{C}_0 & 0\\ 0 & \rho_0\,\boldsymbol{1}\end{bmatrix} \star \begin{bmatrix}\langle\boldsymbol{\varepsilon}\rangle \\ \langle\dot{\mathbf{u}}\rangle\end{bmatrix}$$

or

$$\boxed{\begin{bmatrix}\langle\boldsymbol{\sigma}\rangle\\ \langle\mathbf{p}\rangle\end{bmatrix} = \begin{bmatrix}\mathbf{C}_{\text{eff}} & \mathcal{S}_{\text{eff}}\\ \mathcal{S}^{\dagger}_{\text{eff}} & \rho_{\text{eff}}\end{bmatrix} \star \begin{bmatrix}\langle\boldsymbol{\varepsilon}\rangle \\ \langle\dot{\mathbf{u}}\rangle\end{bmatrix}\ .} \tag{5.98}$$

These are the Willis equations.

5.6.4 Extension to composites with voids

Sometimes the quantity $\langle\mathbf{u}\rangle$ is not an appropriate macroscopic variable. For example, in materials with voids the displacement (and therefore the velocity) is undefined inside the voids. Even if the voids are filled with an elastic material with modulus tending to zero, the value of $\langle\mathbf{u}\rangle$ will depend on the way this limit is taken. Also, for materials such as the rigid matrix filled with rubber and lead (see Figure 5.25), it makes senses to average $\mathbf{u}$ only over the deformable material phase.

[††]That such an argument can be made has been rigorously shown for low-contrast media but not for high-contrast media. Hence, these ideas work for composites that are close to homogeneous.

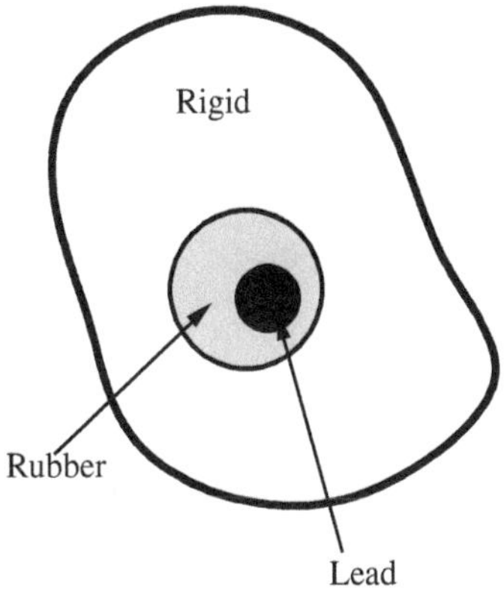

FIGURE 5.25
A composite consisting of a rigid matrix and deformable phases.

Therefore it makes sense to look for equations for $\langle \mathbf{u}_w \rangle$ where

$$\mathbf{u}_w(\mathbf{x},t) = w(\mathbf{x})\ \mathbf{u}(\mathbf{x},t) \tag{5.99}$$

where $w(\mathbf{x})$ is a weight which could be zero in the region where there are voids. Also, the weights could vary from realization to realization. Also, if we have $\dot{\mathbf{u}}$ we can recover $\mathbf{u}$ by integrating over time, i.e.,

$$\mathbf{u}(t) = \int_{-\infty}^{t} \dot{\mathbf{u}}(\tau)\ \mathrm{d}\tau = \int_{-\infty}^{\infty} W(t-\tau)\ \dot{\mathbf{u}}(\tau)\ \mathrm{d}\tau$$

where

$$W(v) = \begin{cases} 1 & \text{for } v > 0 \\ 0 & \text{for } v < 0 \end{cases}.$$

Hence we can write

$$\mathbf{u} = W \star \dot{\mathbf{u}}\ . \tag{5.100}$$

So, from the definitions of $\boldsymbol{S}$ and $\mathbf{m}$ and using the relation (5.100), we have

$$\begin{bmatrix} \boldsymbol{\varepsilon} \\ \mathbf{u} \end{bmatrix} = \begin{bmatrix} (\mathbf{C} - \mathbf{C}_0)^{-1} & 0 \\ 0 & W \star (\rho - \rho_0)^{-1} \end{bmatrix} \star \begin{bmatrix} \boldsymbol{S} \\ \mathbf{m} \end{bmatrix}.$$

From the Willis equations (5.96) we have

$$\begin{bmatrix} \boldsymbol{S} \\ \mathbf{m} \end{bmatrix} = \mathcal{T} \star \begin{bmatrix} \langle \boldsymbol{\varepsilon} \rangle \\ \langle \dot{\mathbf{u}} \rangle \end{bmatrix}.$$

Therefore,

$$\begin{bmatrix} \boldsymbol{\varepsilon} \\ \mathbf{u} \end{bmatrix} = \begin{bmatrix} (\mathbf{C} - \mathbf{C}_0)^{-1} & 0 \\ 0 & W \star (\rho - \rho_0)^{-1} \end{bmatrix} \star \mathcal{T} \star \begin{bmatrix} \langle \boldsymbol{\varepsilon} \rangle \\ \langle \dot{\mathbf{u}} \rangle \end{bmatrix}. \tag{5.101}$$

Now, if the weighted strain is defined as

$$\boldsymbol{\varepsilon}_w = \tfrac{1}{2}\left[\boldsymbol{\nabla}\mathbf{u}_w + (\boldsymbol{\nabla}\mathbf{u}_w)^T\right]$$

then, taking the ensemble average, we have

$$\langle\boldsymbol{\varepsilon}_w\rangle = \tfrac{1}{2}\left\langle[\boldsymbol{\nabla}\mathbf{u}_w + (\boldsymbol{\nabla}\mathbf{u}_w)^T]\right\rangle .$$

Using equation (5.99) we can show that

$$\langle\boldsymbol{\varepsilon}_w\rangle = \langle w\,\boldsymbol{\varepsilon}\rangle + \tfrac{1}{2}\,\langle\boldsymbol{\nabla}w\otimes\mathbf{u} + \mathbf{u}\otimes\boldsymbol{\nabla}w\rangle . \tag{5.102}$$

Using (5.101) we can express (5.102) in terms of $\langle\boldsymbol{\varepsilon}\rangle$ and $\dot{\mathbf{u}}$, and hence also in terms of $\dot{\mathbf{u}}_w$. After some algebra (see Milton and Willis (2007) for details), we can show that

$$\begin{bmatrix}\langle\boldsymbol{\varepsilon}_w\rangle\\ \langle\dot{\mathbf{u}}_w\rangle\end{bmatrix} = \boldsymbol{R}_w\star\begin{bmatrix}\langle\boldsymbol{\varepsilon}\rangle\\ \langle\dot{\mathbf{u}}\rangle\end{bmatrix}$$

where $\boldsymbol{R}_w = \boldsymbol{1}$ when $w(\mathbf{x}) = 1$. Taking the inverse, we can express the Willis equations (5.98) in terms of $\langle\boldsymbol{\varepsilon}_w\rangle$ and $\langle\dot{\mathbf{u}}_w\rangle$ as

$$\begin{bmatrix}\langle\boldsymbol{\sigma}\rangle\\ \langle\mathbf{p}\rangle\end{bmatrix} = \begin{bmatrix}\mathbf{C}_{\text{eff}} & \mathcal{S}_{\text{eff}}\\ \mathcal{S}^{\dagger}_{\text{eff}} & \boldsymbol{\rho}_{\text{eff}}\end{bmatrix}\star\boldsymbol{R}_w^{-1}\star\begin{bmatrix}\langle\boldsymbol{\varepsilon}_w\rangle\\ \langle\dot{\mathbf{u}}_w\rangle\end{bmatrix}$$

or

$$\boxed{\begin{bmatrix}\langle\boldsymbol{\sigma}\rangle\\ \langle\mathbf{p}\rangle\end{bmatrix} = \begin{bmatrix}\mathbf{C}^w_{\text{eff}} & \mathcal{S}^w_{\text{eff}}\\ \mathcal{D}^w_{\text{eff}} & \boldsymbol{\rho}^w_{\text{eff}}\end{bmatrix}\star\begin{bmatrix}\langle\boldsymbol{\varepsilon}_w\rangle\\ \langle\dot{\mathbf{u}}_w\rangle\end{bmatrix} .}$$

These equations have the same form as the Willis equations. However, $\mathcal{D}^w_{\text{eff}} \neq (\mathcal{S}^w_{\text{eff}})^{\dagger}$. We now have a means of using the Willis equations even in the case where there are voids.

5.6.5 Willis equations for electromagnetism

For electromagnetism, we can use similar arguments to obtain

$$\begin{aligned}\langle\mathbf{D}\rangle &= \boldsymbol{\varepsilon}_{\text{eff}}\star\langle\mathbf{E}\rangle + \boldsymbol{\alpha}_{\text{eff}}\star\langle\mathbf{B}\rangle\\ \langle\mathbf{H}\rangle &= \boldsymbol{\alpha}_{\text{eff}}\star\langle\mathbf{E}\rangle + (\boldsymbol{\mu}_{\text{eff}})^{-1}\star\langle\mathbf{B}\rangle\end{aligned}$$

where $\boldsymbol{\alpha}_{\text{eff}}$ is a coupling term. In particular, if the fields are time harmonic with non-local operators being approximated by local ones, then

$$\begin{aligned}\left\langle\widehat{\mathbf{D}}\right\rangle &= \boldsymbol{\varepsilon}_{\text{eff}}\cdot\left\langle\widehat{\mathbf{E}}\right\rangle + \boldsymbol{\lambda}_{\text{eff}}\cdot\left\langle\widehat{\mathbf{H}}\right\rangle\\ \left\langle\widehat{\mathbf{B}}\right\rangle &= \overline{\boldsymbol{\lambda}_{\text{eff}}}\cdot\left\langle\widehat{\mathbf{E}}\right\rangle + \boldsymbol{\mu}_{\text{eff}}\cdot\left\langle\widehat{\mathbf{H}}\right\rangle .\end{aligned}$$

If the operators are local, then $\boldsymbol{\varepsilon}_{\text{eff}}, \boldsymbol{\lambda}_{\text{eff}}, \boldsymbol{\mu}_{\text{eff}}$ will just be matrices that depend on the frequency ω. A medium that has these properties is called *bi-anisotropic*. If the composite material is isotropic, then

$$\boldsymbol{\varepsilon}_{\text{eff}} = \varepsilon_{\text{eff}}\,\boldsymbol{1} ;\quad \boldsymbol{\lambda}_{\text{eff}} = \lambda_{\text{eff}}\,\boldsymbol{1} ;\quad \boldsymbol{\mu}_{\text{eff}} = \mu_{\text{eff}}\,\boldsymbol{1} .$$

Under reflection, $\left\langle \widehat{\mathbf{E}} \right\rangle$ reflects like a normal vector. However, $\left\langle \widehat{\mathbf{H}} \right\rangle$ reflects like an axial vector (i.e., it changes direction). Hence λ_{eff} would have to change sign under a reflection. Therefore, with λ_{eff} fixed, the constitutive relations are not invariant with respect to reflections! This means that if $\lambda_{\text{eff}} \neq 0$ the medium has a certain handedness and is called a *chiral medium.*

5.7 A Milton-Willis model material

Milton has devised a number of microstructures that exhibited coupling between the stress and the momentum and, in some cases, an unsymmetric stress (Milton, 2007). These microstructures do not require the assumption of ensemble averaging that the original Willis equations require. We discuss one of these microstructures in this section.

Consider the two-dimensional spring-mass model shown in Figure 5.26. This is a unit cell in an infinite network of springs and masses. The springs have stiffnesses hk where $2h$ is the length of a diagonal of the unit cell. The spring stiffnesses are scaled by h so that they have the correct behavior as $h \to 0$. The masses m_5 and m_6 are connected to nodes 2 and 4 by rigid bars.

Let us first find the equation of motion of the model. To do that it is convenient to focus on a particular node and take advantage of the periodicity of the structure. Let us choose node 4 as the point of interest. Figure 5.27 shows the nodes required

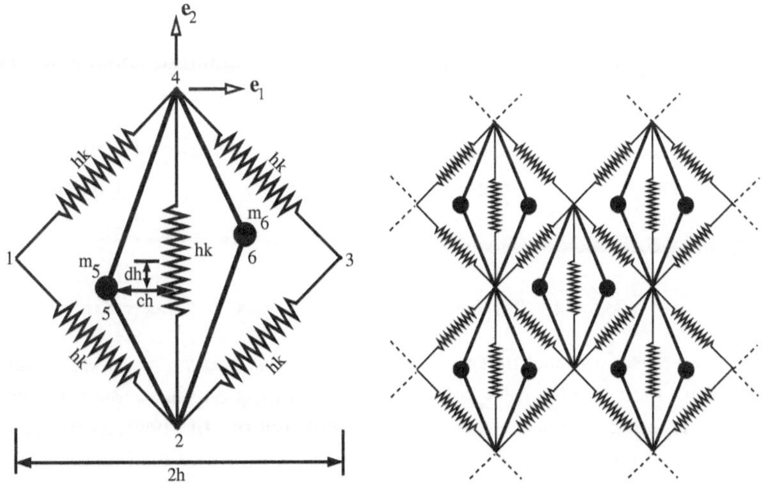

FIGURE 5.26

A unit cell (left) of a two-dimensional Milton-Willis spring-mass network (right).

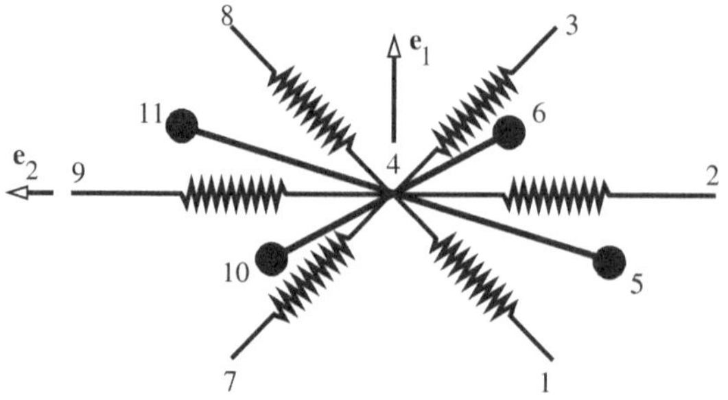

FIGURE 5.27
Elements needed to find equation of motion of spring model with rigid bars.

to find the equation of motion. The origin is chosen to be at node 4.

Equation of motion

The Lagrangian for the spring-mass system around node 4 can be written as

$$\begin{aligned} \mathcal{L} = \frac{hk}{2}\Big[&\|\mathbf{u}_1-\mathbf{u}_4\|^2+\|\mathbf{u}_2-\mathbf{u}_4\|^2+\|\mathbf{u}_3-\mathbf{u}_4\|^2 \\ &+\|\mathbf{u}_7-\mathbf{u}_4\|^2+\|\mathbf{u}_8-\mathbf{u}_4\|^2+\|\mathbf{u}_9-\mathbf{u}_4\|^2\Big] \\ -\frac{m_5}{2}&\left[\|\dot{\mathbf{u}}_5-\dot{\mathbf{u}}_4\|^2+\|\dot{\mathbf{u}}_{10}-\dot{\mathbf{u}}_4\|^2\right]-\frac{m_6}{2}\left[\|\dot{\mathbf{u}}_6-\dot{\mathbf{u}}_4\|^2+\|\dot{\mathbf{u}}_{11}-\dot{\mathbf{u}}_4\|^2\right] \end{aligned} \tag{5.103}$$

where $\mathbf{u}_i$ is the displacement of node i. We will express the Lagrangian in terms of the displacement at node 4 by taking advantage of the rigid rods and the periodicity of the structure. The coordinates of the points 2, 5, 6, 9, 10, and 11 are

$$\begin{aligned} &\mathbf{x}_2=-2h\mathbf{e}_2\ ;\ \mathbf{x}_5=-ch\mathbf{e}_1-(1+d)h\mathbf{e}_2\ ;\ \mathbf{x}_6=ch\mathbf{e}_1-(1-d)h\mathbf{e}_2 \\ &\mathbf{x}_9=2h\mathbf{e}_2\ ;\ \mathbf{x}_{10}=-ch\mathbf{e}_1+(1-d)h\mathbf{e}_2\ ;\ \mathbf{x}_{11}=ch\mathbf{e}_1+(1+d)h\mathbf{e}_2\,. \end{aligned}$$

The lengths of the rigid bars are

$$l_1=\|\mathbf{x}_5-\mathbf{x}_4\|=\|\mathbf{x}_6-\mathbf{x}_2\|\ ;\quad l_2=\|\mathbf{x}_5-\mathbf{x}_2\|=\|\mathbf{x}_6-\mathbf{x}_4\|$$

where we have used the fact that $\mathbf{x}_2=-\mathbf{x}_9$, $\mathbf{x}_{10}=-\mathbf{x}_6$, and $\mathbf{x}_{11}=-\mathbf{x}_5$. Because the bars are rigid, their lengths must remain constant during any deformation. Consider the bars connecting nodes $4-5-2$. After deformation, the new positions of the nodes are $\mathbf{x}'_4=\mathbf{x}_4+\mathbf{u}_4$, $\mathbf{x}'_5=\mathbf{x}_5+\mathbf{u}_5$ and $\mathbf{x}'_2=\mathbf{x}_2+\mathbf{u}_2$. Since the lengths of the bar remain constant, we must have

$$\begin{aligned} l_1&=\|\mathbf{x}'_5-\mathbf{x}'_4\|=\|(\mathbf{x}_5-\mathbf{x}_4)+(\mathbf{u}_5-\mathbf{u}_4)\| \\ l_2&=\|\mathbf{x}'_5-\mathbf{x}'_2\|=\|(\mathbf{x}_5-\mathbf{x}_2)+(\mathbf{u}_5-\mathbf{u}_2)\|\,. \end{aligned}$$

For small displacements we can ignore quadratic displacement terms, and we are left with

$$l_1 = \sqrt{(\mathbf{x}_5 - \mathbf{x}_4)\cdot(\mathbf{x}_5 - \mathbf{x}_4) + (\mathbf{u}_5 - \mathbf{u}_4)\cdot(\mathbf{x}_5 - \mathbf{x}_4)}$$
$$l_2 = \sqrt{(\mathbf{x}_5 - \mathbf{x}_2)\cdot(\mathbf{x}_5 - \mathbf{x}_2) + (\mathbf{u}_5 - \mathbf{u}_2)\cdot(\mathbf{x}_5 - \mathbf{x}_2)}\,.$$

This implies that $(\mathbf{u}_5 - \mathbf{u}_4)\cdot\mathbf{x}_5 = 0$ and $(\mathbf{u}_5 - \mathbf{u}_2)\cdot(\mathbf{x}_5 - \mathbf{x}_2) = 0$. Similarly, we can show that $(\mathbf{u}_6 - \mathbf{u}_4)\cdot\mathbf{x}_6 = 0$, $(\mathbf{u}_6 - \mathbf{u}_2)\cdot(\mathbf{x}_6 - \mathbf{x}_2) = 0$, $(\mathbf{u}_{10} - \mathbf{u}_4)\cdot\mathbf{x}_6 = 0$, $(\mathbf{u}_{10} - \mathbf{u}_9)\cdot(\mathbf{x}_6 - \mathbf{x}_2) = 0$, $(\mathbf{u}_{11} - \mathbf{u}_4)\cdot\mathbf{x}_5 = 0$, and $(\mathbf{u}_{11} - \mathbf{u}_9)\cdot(\mathbf{x}_5 - \mathbf{x}_2) = 0$. Solving for $\mathbf{u}_5$, $\mathbf{u}_6$, $\mathbf{u}_{10}$, and $\mathbf{u}_{11}$ in terms of $\mathbf{u}_2$, $\mathbf{u}_4$, and $\mathbf{u}_9$, we get relations of the form (in matrix notation)

$$\underline{\underline{u}}_5 = \frac{1}{2c}\begin{bmatrix} c(1+d) & -(1-d^2) & c(1-d) & (1-d^2) \\ -c^2 & c(1-d) & c^2 & c(1+d) \end{bmatrix} \begin{bmatrix} \underline{\underline{u}}_2 \\ \underline{\underline{u}}_4 \end{bmatrix} =: \begin{bmatrix} \underline{\underline{B}}_1 & \underline{\underline{B}}_2 \end{bmatrix} \begin{bmatrix} \underline{\underline{u}}_2 \\ \underline{\underline{u}}_4 \end{bmatrix}.$$

We can express these relations in direct notation as

$$\begin{aligned} \mathbf{u}_5 &= \boldsymbol{B}_1\cdot\mathbf{u}_2 + \boldsymbol{B}_2\cdot\mathbf{u}_4\ ;\quad \mathbf{u}_6 = \boldsymbol{B}_2\cdot\mathbf{u}_2 + \boldsymbol{B}_1\cdot\mathbf{u}_4 \\ \mathbf{u}_{10} &= \boldsymbol{B}_1\cdot\mathbf{u}_4 + \boldsymbol{B}_2\cdot\mathbf{u}_9\ ;\quad \mathbf{u}_{11} = \boldsymbol{B}_2\cdot\mathbf{u}_4 + \boldsymbol{B}_1\cdot\mathbf{u}_9\,. \end{aligned} \tag{5.104}$$

The periodicity of the unit cell indicates that we can use Bloch-Floquet theory to relate the displacements at certain nodes for plane-wave propagation through the structure. Without going into specifics at this stage, let us assume that

$$\begin{aligned} \mathbf{u}_1 &= A_1\mathbf{u}_4\ ;\quad \mathbf{u}_2 = A_2\mathbf{u}_4\ ;\quad \mathbf{u}_3 = A_3\mathbf{u}_4 \\ \mathbf{u}_7 &= A_3^{-1}\mathbf{u}_4\ ;\quad \mathbf{u}_8 = A_1^{-1}\mathbf{u}_4\ ;\quad \mathbf{u}_9 = A_2^{-1}\mathbf{u}_4 \end{aligned} \tag{5.105}$$

where A_j is a constant that depends on the location of node j. Plugging equations (5.104) and (5.105) into (5.103) gives

$$\begin{aligned} \mathcal{L} = &\frac{hk}{2}(A_1 + A_1^{-1} + A_2 + A_2^{-1} + A_3 + A_3^{-1} - 6)\,\|\mathbf{u}_4\|^2 + \\ &-\frac{m_5}{2}\left[\|(A_2\boldsymbol{B}_1 + \boldsymbol{B}_2 - \boldsymbol{1})\cdot\dot{\mathbf{u}}_4\|^2 + \|(A_2^{-1}\boldsymbol{B}_2 + \boldsymbol{B}_1 - \boldsymbol{1})\cdot\dot{\mathbf{u}}_4\|^2\right] \\ &-\frac{m_6}{2}\left[\|(A_2\boldsymbol{B}_2 + \boldsymbol{B}_1 - \boldsymbol{1})\cdot\dot{\mathbf{u}}_4\|^2 + \|(A_2^{-1}\boldsymbol{B}_1 + \boldsymbol{B}_2 - \boldsymbol{1})\cdot\dot{\mathbf{u}}_4\|^2\right]. \end{aligned}$$

In abbreviated form,

$$\mathcal{L} = \frac{k\alpha}{2}\|\mathbf{u}_4\|^2 - \frac{m_5}{2}\left[\|\boldsymbol{\beta}_1\cdot\dot{\mathbf{u}}_4\|^2 + \|\boldsymbol{\beta}_2\cdot\dot{\mathbf{u}}_4\|^2\right] - \frac{m_6}{2}\left[\|\boldsymbol{\beta}_3\cdot\dot{\mathbf{u}}_4\|^2 + \|\boldsymbol{\beta}_4\cdot\dot{\mathbf{u}}_4\|^2\right]. \tag{5.106}$$

The Euler-Lagrange equations for the system are given by

$$\frac{\partial\mathcal{L}}{\partial\mathbf{u}_4} - \frac{d}{dt}\left(\frac{\partial\mathcal{L}}{\partial\dot{\mathbf{u}}_4}\right) = 0\,.$$

Plugging (5.106) into the Euler-Lagrange equations leads to[‡‡]

$$\boxed{k\alpha\mathbf{u}_4 + \left[m_5\left(\boldsymbol{\beta}_1^T\cdot\boldsymbol{\beta}_1 + \boldsymbol{\beta}_2^T\cdot\boldsymbol{\beta}_2\right) + m_6\left(\boldsymbol{\beta}_3^T\cdot\boldsymbol{\beta}_3 + \boldsymbol{\beta}_4^T\cdot\boldsymbol{\beta}_4\right)\right]\cdot\ddot{\mathbf{u}}_4 = 0.} \tag{5.107}$$

This is the equation of motion of the system. For time harmonic $\mathbf{u}_4$, using the standard approach, we get

$$\left[k\alpha\boldsymbol{1} - \omega^2\left\{m_5\left(\boldsymbol{\beta}_1^T\cdot\boldsymbol{\beta}_1 + \boldsymbol{\beta}_2^T\cdot\boldsymbol{\beta}_2\right) + m_6\left(\boldsymbol{\beta}_3^T\cdot\boldsymbol{\beta}_3 + \boldsymbol{\beta}_4^T\cdot\boldsymbol{\beta}_4\right)\right\}\right]\cdot\mathbf{u}_4 = 0.$$

This is an eigenvalue problem and a solution exists if

$$\det\left[k\alpha\boldsymbol{1} - \omega^2\left\{m_5\left(\boldsymbol{\beta}_1^T\cdot\boldsymbol{\beta}_1 + \boldsymbol{\beta}_2^T\cdot\boldsymbol{\beta}_2\right) + m_6\left(\boldsymbol{\beta}_3^T\cdot\boldsymbol{\beta}_3 + \boldsymbol{\beta}_4^T\cdot\boldsymbol{\beta}_4\right)\right\}\right] = 0\,. \tag{5.108}$$

For plane waves, the above is a dispersion relation.

Forces

Let us now look at the forces in the system. For time harmonic motions, using (5.104), the inertial forces at nodes 5, 6, 10, and 11 are given by

$$\begin{aligned}\mathbf{f}_5 &= -m_5\omega^2(\boldsymbol{B}_1\cdot\mathbf{u}_2 + \boldsymbol{B}_2\cdot\mathbf{u}_4)\;;\;\; \mathbf{f}_6 = -m_6\omega^2(\boldsymbol{B}_2\cdot\mathbf{u}_2 + \boldsymbol{B}_1\cdot\mathbf{u}_4)\\ \mathbf{f}_{10} &= -m_5\omega^2(\boldsymbol{B}_1\cdot\mathbf{u}_4 + \boldsymbol{B}_2\cdot\mathbf{u}_9)\;;\;\; \mathbf{f}_{11} = -m_6\omega^2(\boldsymbol{B}_2\cdot\mathbf{u}_4 + \boldsymbol{B}_1\cdot\mathbf{u}_9)\,.\end{aligned} \tag{5.109}$$

These inertial forces will generate loads at node 4 which will need to be balanced. The components of the forces at nodes 5, 6, 10, and 11 along the bars are

$$\mathbf{f}_5 = \mathbf{f}_{5;2} + \mathbf{f}_{5;4}\;;\;\; \mathbf{f}_6 = \mathbf{f}_{6;2} + \mathbf{f}_{6;4}\;;\;\; \mathbf{f}_{10} = \mathbf{f}_{10;4} + \mathbf{f}_{10;9}\;;\;\; \mathbf{f}_{11} = \mathbf{f}_{11;4} + \mathbf{f}_{11;9}\,.$$

Since the resolved components of force are aligned along the bars we have $\mathbf{f}_{i;j} = \alpha_{i;j}\,(\mathbf{x}_j - \mathbf{x}_i)$ where $\alpha_{i;j}$ are scaling factors. Using (5.109) and the above relations we can solve for $\alpha_{i;j}$ in terms of the displacements. Ignoring the springs for now, the forces $\mathbf{f}_5$ and $\mathbf{f}_6$ will be balanced by $\mathbf{f}_{10}$ and $\mathbf{f}_{11}$. Hence we only need to know the factors $\alpha_{5;4}$ and $\alpha_{6;4}$,

$$\begin{aligned}\alpha_{5;4} &= \frac{m_5\omega^2}{4c^2h}\left[-(c^2+d^2-1)\left[-c \quad (1-d)\right]\underline{\underline{u}}_2 - \{c^2+(1-d)^2\}\left[c \quad (1+d)\right]\underline{\underline{u}}_4\right]\\ \alpha_{6;4} &= \frac{m_6\omega^2}{4c^2h}\left[-(c^2+d^2-1)\left[c \quad (1+d)\right]\underline{\underline{u}}_2 - \{c^2+(1+d)^2\}\left[-c \quad (1-d)\right]\underline{\underline{u}}_4\right].\end{aligned}$$

In direct notation,

$$\alpha_{5;4} = -\frac{m_5\omega^2}{4c^2h}(C_1\mathbf{c}_1\cdot\mathbf{u}_2 + C_2\mathbf{c}_2\cdot\mathbf{u}_4)\;;\;\; \alpha_{6;4} = -\frac{m_6\omega^2}{4c^2h}(C_1\mathbf{c}_2\cdot\mathbf{u}_2 + C_3\mathbf{c}_1\cdot\mathbf{u}_4)\,.$$

[‡‡]Note that

$$\frac{\partial}{\partial u_i}(A_{k\ell}u_\ell A_{km}u_m) = A_{k\ell}A_{km}(u_m\delta_{\ell i} + u_\ell\delta_{mi}) = 2A_{ki}A_{kj}u_j\,.$$

The inertial force at node 4 for a unit cell is

$$\mathbf{f}_4^{\text{inertial}} = \mathbf{f}_{5;4} + \mathbf{f}_{6;4} = -\alpha_{5;4}\mathbf{x}_5 - \alpha_{6;4}\mathbf{x}_6 \,.$$

We can use the periodicity of the structure to express the forces in term of $\mathbf{u}_4$. Using the relation $\mathbf{u}_2 = A_2\mathbf{u}_4$ we can write

$$\begin{aligned}
\alpha_{5;4} &= -\frac{m_5\omega^2}{4c^2h}(A_2C_1\mathbf{c}_1 + C_2\mathbf{c}_2)\cdot\mathbf{u}_4 \\
\alpha_{6;4} &= -\frac{m_6\omega^2}{4c^2h}(C_3\mathbf{c}_1 + A_2C_1\mathbf{c}_2)\cdot\mathbf{u}_4 \,.
\end{aligned}$$

Therefore the inertial force can be expressed as

$$\boxed{\begin{aligned}
\mathbf{f}_4^{\text{inertial}} &= \frac{m_5\omega^2}{4c^2h}(A_2C_1\,\mathbf{x}_5\otimes\mathbf{c}_1 + C_2\,\mathbf{x}_5\otimes\mathbf{c}_2)\cdot\mathbf{u}_4 \\
&+ \frac{m_6\omega^2}{4c^2h}(C_3\,\mathbf{x}_6\otimes\mathbf{c}_1 + A_2C_1\,\mathbf{x}_6\otimes\mathbf{c}_2)\cdot\mathbf{u}_4 \,.
\end{aligned}} \tag{5.110}$$

Let us now look at the elastic forces at node 4 due to the springs. Once again, we concentrate on the contribution from the unit cell. For equilibrium at node 4, the inertial forces must be balanced by the forces in the springs. For the model in Figure 5.27 we have

$$\mathbf{f}_4^{\text{elastic}} = hk\,\boldsymbol{D}_{41}\cdot(\mathbf{u}_4 - \mathbf{u}_1) + hk\,\boldsymbol{D}_{42}\cdot(\mathbf{u}_4 - \mathbf{u}_2) + hk\,\boldsymbol{D}_{43}\cdot(\mathbf{u}_4 - \mathbf{u}_3)$$

where the transformation $\boldsymbol{D}_{ij}$ is needed to orient the forces and the displacements along the springs and has the form (in matrix notation)

$$\underline{\underline{\mathsf{D}}}_{ij} = \begin{bmatrix} \cos^2\theta_{ij} & \cos\theta_{ij}\sin\theta_{ij} \\ \cos\theta_{ij}\sin\theta_{ij} & \sin^2\theta_{ij} \end{bmatrix}$$

where θ_{ij} is the angle made by the line from point i to point j with the $\mathbf{e}_1$-axis. We also have $\boldsymbol{D}_{41} + \boldsymbol{D}_{43} = \boldsymbol{1}$. Using these relations and the periodicity conditions that we had introduced earlier, we get

$$\boxed{\mathbf{f}_4^{\text{elastic}} = -hk\left[(1-A_3)\boldsymbol{1} + (A_3 - A_1)\boldsymbol{D}_{41} + (1-A_2)\boldsymbol{D}_{42}\right]\cdot\mathbf{u}_4 \,.} \tag{5.111}$$

Bloch wave solutions

Earlier we had proposed that we could take advantage of the periodicity of the structure to express the displacements in the form

$$\begin{aligned}
&\mathbf{u}_1 = A_1\mathbf{u}_4 \,;\;\; \mathbf{u}_2 = A_2\mathbf{u}_4 \,;\;\; \mathbf{u}_3 = A_3\mathbf{u}_4 \\
&\mathbf{u}_7 = A_3^{-1}\mathbf{u}_4 \,;\;\; \mathbf{u}_8 = A_1^{-1}\mathbf{u}_4 \,;\;\; \mathbf{u}_9 = A_2^{-1}\mathbf{u}_4 \,.
\end{aligned}$$

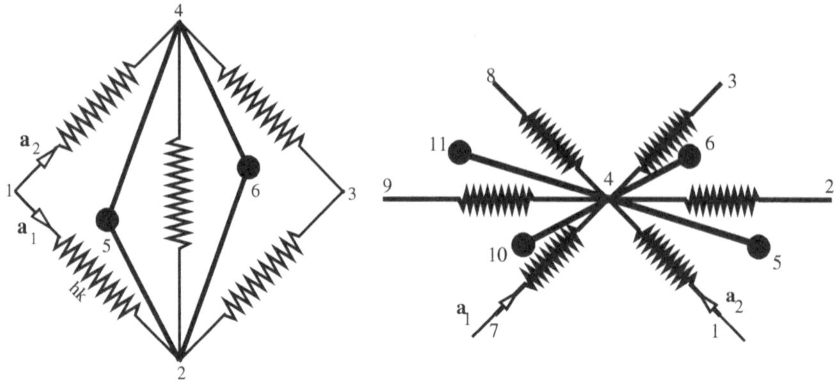

FIGURE 5.28
Lattice vectors for the unit cell of the Milton-Willis model.

Bloch-Floquet theory indicates that, for an infinite periodic structure, we can look for quasi-periodic Bloch wave solutions of the form

$$\mathbf{u}_i(\mathbf{x}) = \mathbf{u}_4 \, e^{i\mathbf{k}\cdot\mathbf{r}}$$

where $\mathbf{k}$ is a plane-wave wave vector, and $\mathbf{r}$ is a lattice parameter. We will say more about Bloch waves in Chapter 7.

Let the lattice vectors of the model be $\mathbf{a}_1$ (oriented from node 1 to node 2) and $\mathbf{a}_2$ (oriented from node 1 to node 4) as shown in Figure 5.28. Then we have

$$\mathbf{u}_1 = \mathbf{u}_4 \, e^{-i\mathbf{k}\cdot\mathbf{a}_2} = A_1\mathbf{u}_4 \; ; \; \mathbf{u}_2 = \mathbf{u}_4 \, e^{i\mathbf{k}\cdot(\mathbf{a}_1-\mathbf{a}_2)} = A_2\mathbf{u}_4 \; ; \; \mathbf{u}_3 = \mathbf{u}_4 \, e^{i\mathbf{k}\cdot\mathbf{a}_1} = A_3\mathbf{u}_4 \, . \tag{5.112}$$

We can see from the geometry of the structure that the expressions for $\mathbf{u}_7$, $\mathbf{u}_8$, and $\mathbf{u}_9$ in terms of $\mathbf{u}_4$ must also hold and hence the relations in equations (5.110) and (5.111) must hold under Bloch wave conditions. For our choice of basis $(\mathbf{e}_1, \mathbf{e}_2)$ and origin in this problem we have $\mathbf{x}_1 = -h\mathbf{e}_1 - h\mathbf{e}_2$. Recalling that $\mathbf{x}_2 = -2h\mathbf{e}_2$, we can express the lattice vectors of the unit cell as

$$\mathbf{a}_1 = \mathbf{x}_2 - \mathbf{x}_1 = h\mathbf{e}_1 - h\mathbf{e}_2 \quad \text{and} \quad \mathbf{a}_2 = \mathbf{x}_4 - \mathbf{x}_1 = h\mathbf{e}_1 + h\mathbf{e}_2 \, .$$

If the wave vector $\mathbf{k} = k_1\mathbf{e}_1 + k_2\mathbf{e}_2$, we have

$$A_1 = e^{-i\mathbf{k}\cdot\mathbf{a}_2} = e^{-ih(k_1+k_2)} \; ; \; A_2 = e^{-i\mathbf{k}\cdot(\mathbf{a}_1-\mathbf{a}_2)} = e^{ihk_2} \; ; \; A_3 = e^{-i\mathbf{k}\cdot\mathbf{a}_1} = e^{-ih(k_1-k_2)} \, .$$

We can use these factors to solve the equation of motion and find the inertial and elastic forces for particular choices of the wave vector $\mathbf{k}$.

A simpler model

Let us now consider the situation (Milton, 2007) where $m_5 = hm$, $m_6 = -hm + \delta h^2$, and $d = 0$. The quantity δ is possibly complex with a non-negative imaginary part.

In that case

$$\underline{\underline{\mathsf{B}}}_1 = \frac{1}{2c}\begin{bmatrix} c & -1 \\ -c^2 & c \end{bmatrix} \; ; \; \underline{\underline{\mathsf{B}}}_2 = \frac{1}{2c}\begin{bmatrix} c & 1 \\ c^2 & c \end{bmatrix} .$$

Using the above and the expressions for A_i from the previous section in the dispersion relations (5.108) gives

$$\begin{aligned} &-2k[-3+\{1+2\cos(hk_1)\}\cos(hk_2)]\left[6k+\delta h^2\omega^2-2k\cos\{h(k_1-k_2)\}\right. \\ &\left.-2(k+\delta h^2\omega^2)\cos(hk_2)+\delta h^2\omega^2\cos(2hk_2)-2k\cos\{h(k_1+k_2)\}\right]=0. \end{aligned}$$

Notice that if $\delta = 0$ the two masses cancel each other and the dispersion relation is degenerate. The inertial force for this situation has the form

$$\begin{aligned} \mathbf{f}_4^{\text{inertial}} = & \frac{\omega^2 h}{4c}\left[-c\delta h\left\{(1+c^2)+(1-c^2)e^{ihk_2}\right\}u_{4x}\right. \\ & \left.+\left\{(1+c^2)(\delta h-2m)-(1-c^2)(\delta h-2m)e^{ihk_2}\right\}u_{4y}\right]\mathbf{e}_1 \\ & +\frac{\omega^2 h}{4c^2}\left[\left\{c(1+c^2)(\delta h-2m)+c(1-c^2)(\delta h-2m)e^{ihk_2}\right\}u_{4x}\right. \\ & \left.+\delta h\left\{-(1+c^2)+(1-c^2)e^{ihk_2}\right\}u_{4y}\right]\mathbf{e}_2 \end{aligned}$$

where we have used $\mathbf{u}_4 = u_{4x}\mathbf{e}_1 + u_{4y}\mathbf{e}_2$. A force $\mathbf{t}_4^{\text{inertial}} = \mathbf{f}_4^{\text{inertial}}/(2h)$ needs to act on node 4 to balance the inertial forces. Dividing the above expression by $2h$ and taking the limit as $h \to 0$ (the continuum limit) gives us the traction,

$$\boxed{\mathbf{t}_4^{\text{inertial}} = -\frac{m\omega^2}{2}\left(cu_{4y}\,\mathbf{e}_1 + \frac{u_{4x}}{c}\mathbf{e}_2\right).} \qquad (5.113)$$

This expression differs by a factor of $1/2$ from that derived using a simpler approach by Milton (2007). Similarly, using

$$\boldsymbol{D}_{41} \equiv \underline{\underline{\mathsf{D}}}_{41} = \begin{bmatrix} 1/2 & 1/2 \\ 1/2 & 1/2 \end{bmatrix} \quad \text{and} \quad \boldsymbol{D}_{42} \equiv \underline{\underline{\mathsf{D}}}_{42} = \begin{bmatrix} 0 & 0 \\ 0 & 1 \end{bmatrix}$$

we can express the elastic force due to the springs at node 4 as

$$\begin{aligned} \mathbf{f}_4^{\text{elastic}} = & hk\left[\{1-e^{-ihk_1}\cos(hk_2)\}u_{4x}+\{1-e^{-ihk_1}\cos(hk_2)\}u_{4y}\right]\mathbf{e}_1 \\ & +hk\left[(1-e^{-ihk_1}\cos(hk_2)\}u_{4x}+\{2-e^{ihk_2}-e^{-ihk_1}\cos(hk_2)\}u_{4y}\right]\mathbf{e}_2. \end{aligned}$$

The traction at node 4 due to these elastic forces, $\mathbf{f}_4^{\text{elastic}}/(2h)$, goes to *zero* as $h \to 0$.

Momentum and stress

To find the momentum density and inertial and elastic stress-strain relations it is convenient to follow Milton (2007) and express the displacements at the nodes to first order in h in terms of the displacement, $\mathbf{u}(\mathbf{x}_0)$, at the center of the unit cell ($\mathbf{x}_0$) as

$$\mathbf{u}_1 = \mathbf{u}(\mathbf{x}_0) - h\,\mathbf{q}\,; \quad \mathbf{u}_2 = \mathbf{u}(\mathbf{x}_0) - h\,\mathbf{w}\,; \quad \mathbf{u}_3 = \mathbf{u}(\mathbf{x}_0) + h\,\mathbf{q}\,; \quad \mathbf{u}_4 = \mathbf{u}(\mathbf{x}_0) + h\,\mathbf{w} \tag{5.114}$$

where

$$\mathbf{q} = \lim_{h\to 0} \frac{\mathbf{u}_3 - \mathbf{u}_1}{2h} =: \left.\frac{\partial \mathbf{u}}{\partial x_1}\right|_{\mathbf{x}=\mathbf{x}_0} \quad \text{and} \quad \mathbf{w} = \lim_{h\to 0} \frac{\mathbf{u}_4 - \mathbf{u}_2}{2h} =: \left.\frac{\partial \mathbf{u}}{\partial x_2}\right|_{\mathbf{x}=\mathbf{x}_0}.$$

In the above $\mathbf{u}(\mathbf{x})$ is a sufficiently smooth displacement field that can represent the displacements in the unit cell.

The momentum *density* in the unit cell for time harmonic motions is

$$\mathbf{p} = \frac{1}{2h^2}(-i\omega m_5 \mathbf{u}_5 - i\omega m_6 \mathbf{u}_6) = -\frac{i\omega}{2h^2}\left[m_5(A_2\boldsymbol{B}_1 + \boldsymbol{B}_2)\cdot\mathbf{u}_4 + m_6(A_2\boldsymbol{B}_2 + \boldsymbol{B}_1)\cdot\mathbf{u}_4\right].$$

For the simple model from the previous section,

$$\begin{aligned}\mathbf{p} = {} & \frac{i\omega}{4ch}\left[c(e^{ihk_2}+1)\delta h\, u_{4x} + (e^{ihk_2}-1)(\delta h - 2m)u_{4y}\right]\mathbf{e}_1 \\ & + \frac{i\omega}{4h}\left[c(e^{ihk_2}-1)(\delta h - 2m)u_{4x} + (e^{ihk_2}+1)\delta h\, u_{4y}\right]\mathbf{e}_2\,.\end{aligned}$$

In the limit $h \to 0$ we have $\mathbf{u}_4 \to \mathbf{u}(\mathbf{x}_0)$ and we can write the momentum density as

$$\mathbf{p} = \frac{i\omega}{2c}\left[(c\delta\, u_1 - ik_2 m\, u_2)\,\mathbf{e}_1 + c^2\left(-ik_2 m\, u_1 + \frac{\delta}{c}u_2\right)\right]\mathbf{e}_2 \tag{5.115}$$

where we have used $\mathbf{u}_4 \to \mathbf{u}(\mathbf{x}_0) = u_1\mathbf{e}_1 + u_2\mathbf{e}_2$. At this stage it is worth recalling that the displacement field has been assumed to be plane harmonic in the preceding discussion, i.e., $\mathbf{u}(\mathbf{x},t) = \widehat{\mathbf{u}}\exp[i(\mathbf{k}\cdot\mathbf{x} - \omega t)]$. Therefore,

$$u_{1,2} = ik_2\widehat{u}_1\, e^{i(\mathbf{k}\cdot\mathbf{x}-\omega t)} \quad \text{and} \quad u_{2,2} = ik_2\widehat{u}_2\, e^{i(\mathbf{k}\cdot\mathbf{x}-\omega t)}$$

and we can write equation (5.115) as

$$\mathbf{p} = \frac{i\omega}{2c}\left[(c\delta\, u_1 - m\, u_{2,2})\,\mathbf{e}_1 + c^2\left(-m\, u_{1,2} + \frac{\delta}{c}u_2\right)\right]\mathbf{e}_2.$$

Using the relation $\mathbf{v} = -i\omega\mathbf{u}$ between velocity and displacement, we can write the above equation in the form

$$\boxed{\mathbf{p} = \frac{1}{2}\left(-\frac{im\omega}{c}u_{2,2} - \delta\, v_1\right)\mathbf{e}_1 + \frac{1}{2}(-im\omega c\, u_{1,2} - \delta\, v_2)\,\mathbf{e}_2\,.} \tag{5.116}$$

In matrix form,

$$\underline{\underline{p}} = \begin{bmatrix} p_1 \\ p_2 \end{bmatrix} = \frac{1}{2}\begin{bmatrix} 0 & 0 & 0 & -im\omega c^{-1} & -\delta & 0 \\ 0 & -im\omega c & 0 & 0 & 0 & -\delta \end{bmatrix} \begin{bmatrix} u_{1,1} \\ u_{1,2} \\ u_{2,1} \\ u_{2,2} \\ v_1 \\ v_2 \end{bmatrix}. \tag{5.117}$$

Let us now examine the Cauchy stress in the system in the continuum limit, $h \to 0$. The total homogeneous stress in a unit cell is the superposition of the elastic component ($\boldsymbol{\sigma}^E$, due to the springs) and the inertial component ($\boldsymbol{\sigma}^I$, due to the masses) and is given by

$$\boldsymbol{\sigma} = \boldsymbol{\sigma}^E + \boldsymbol{\sigma}^I.$$

Recall that the inertial component of the traction at node 4 is given by equation (5.113). There is an equal and opposite traction at node 2, and nodes 1 and 3 do not have any inertial traction components. Let us define the *effective stress* in terms of the boundary tractions using the relation $\mathbf{t} = \mathbf{n} \cdot \boldsymbol{\sigma}$, where $\mathbf{n}$ is the normal to the boundary. Then,

$$\mathbf{t} = (n_1\sigma_{11} + n_2\sigma_{21})\mathbf{e}_1 + (n_1\sigma_{12} + n_2\sigma_{22})\mathbf{e}_2.$$

At node 4, $n_1 = 0$ and $n_2 = 1$. Hence $\mathbf{t}_4 = \sigma_{21}\mathbf{e}_1 + \sigma_{22}\mathbf{e}_2 = -\mathbf{t}_2$. Similarly, at node 3, $n_1 = 1$ and $n_2 = 0$ which implies that $\mathbf{t}_3 = \sigma_{11}\mathbf{e}_1 + \sigma_{12}\mathbf{e}_2 = -\mathbf{t}_2$. From the expression in equation (5.113) and noting that $\mathbf{t}_3^{\text{inertial}} = \mathbf{t}_2^{\text{inertial}} = 0$, we have $\sigma^I_{11} = 0$, $\sigma^I_{12} = 0$, $\sigma^I_{21} = -m\omega^2 c u_{4y}/2$, and $\sigma^I_{22} = -m\omega^2 u_{4x}/(2c)$. In the limit of small h, $\mathbf{u}_4 \to \mathbf{u}_0 = \mathbf{u}(\mathbf{x}_0)$ and the inertial component of stress is given by

$$\boxed{\boldsymbol{\sigma}^I \equiv \underline{\underline{\boldsymbol{\sigma}}}^I = \frac{1}{2}\begin{bmatrix} 0 & 0 \\ -m\omega^2 c\, u_2 & -m\omega^2\, u_1/c \end{bmatrix}} \tag{5.118}$$

where $\mathbf{u}(\mathbf{x}) = u_1(\mathbf{x})\mathbf{e}_1 + u_2(\mathbf{x})\mathbf{e}_2$. We can express the above relation for the inertial stress in the alternative form

$$\underline{\underline{\boldsymbol{\sigma}}}^I = \begin{bmatrix} \sigma^I_{11} \\ \sigma^I_{12} \\ \sigma^I_{21} \\ \sigma^E_{22} \end{bmatrix} = \frac{1}{2}\begin{bmatrix} 0 \\ 0 \\ -m\omega^2 c\, u_2 \\ -m\omega^2 c^{-1}\, u_1 \end{bmatrix} = \frac{1}{2}\begin{bmatrix} 0 & 0 \\ 0 & 0 \\ 0 & -i\omega m\, c \\ -i\omega m\, c^{-1} & 0 \end{bmatrix}\begin{bmatrix} v_1 \\ v_2 \end{bmatrix}$$

where $v_1 = -i\omega u_1$ and $v_2 = -i\omega u_2$.

Next, we need a relationship for the elastic part of the stress of the form

$$\boldsymbol{\sigma}^E = \mathbf{C} : [\nabla\mathbf{u} + (\nabla\mathbf{u})^T].$$

To get this relation we can ignore the masses and the rigid bars and look only at the spring network in our model. The relations between forces and displacements in the individual springs are

$$\begin{aligned}
\mathbf{f}_2^{21} &= hk(\boldsymbol{D}_{21}\cdot\mathbf{u}_2 - \boldsymbol{D}_{21}\cdot\mathbf{u}_1)\ ; \quad \mathbf{f}_1^{21} = hk(-\boldsymbol{D}_{21}\cdot\mathbf{u}_2 + \boldsymbol{D}_{21}\cdot\mathbf{u}_1) \\
\mathbf{f}_2^{23} &= hk(\boldsymbol{D}_{23}\cdot\mathbf{u}_2 - \boldsymbol{D}_{23}\cdot\mathbf{u}_3)\ ; \quad \mathbf{f}_3^{23} = hk(-\boldsymbol{D}_{23}\cdot\mathbf{u}_2 + \boldsymbol{D}_{23}\cdot\mathbf{u}_3) \\
\mathbf{f}_4^{41} &= hk(\boldsymbol{D}_{41}\cdot\mathbf{u}_4 - \boldsymbol{D}_{41}\cdot\mathbf{u}_1)\ ; \quad \mathbf{f}_1^{41} = hk(-\boldsymbol{D}_{41}\cdot\mathbf{u}_4 + \boldsymbol{D}_{41}\cdot\mathbf{u}_1) \\
\mathbf{f}_4^{42} &= hk(\boldsymbol{D}_{42}\cdot\mathbf{u}_4 - \boldsymbol{D}_{42}\cdot\mathbf{u}_2)\ ; \quad \mathbf{f}_2^{42} = hk(-\boldsymbol{D}_{42}\cdot\mathbf{u}_4 + \boldsymbol{D}_{42}\cdot\mathbf{u}_2) \\
\mathbf{f}_4^{43} &= hk(\boldsymbol{D}_{43}\cdot\mathbf{u}_4 - \boldsymbol{D}_{43}\cdot\mathbf{u}_3)\ ; \quad \mathbf{f}_3^{43} = hk(-\boldsymbol{D}_{43}\cdot\mathbf{u}_4 + \boldsymbol{D}_{43}\cdot\mathbf{u}_3).
\end{aligned}$$

Adding force contributions from each spring at each node gives

$$\begin{aligned}
\mathbf{f}_1 &= hk[(\boldsymbol{D}_{21}+\boldsymbol{D}_{41})\cdot\mathbf{u}_1 - \boldsymbol{D}_{21}\cdot\mathbf{u}_2 - \boldsymbol{D}_{41}\cdot\mathbf{u}_4] \\
\mathbf{f}_2 &= hk[-\boldsymbol{D}_{21}\cdot\mathbf{u}_1 + (\boldsymbol{D}_{21}+\boldsymbol{D}_{23}+\boldsymbol{D}_{42})\cdot\mathbf{u}_2 - \boldsymbol{D}_{23}\cdot\mathbf{u}_3 - \boldsymbol{D}_{42}\cdot\mathbf{u}_4] \\
\mathbf{f}_3 &= hk[-\boldsymbol{D}_{23}\cdot\mathbf{u}_2 + (\boldsymbol{D}_{23}+\boldsymbol{D}_{43})\cdot\mathbf{u}_3 - \boldsymbol{D}_{43}\cdot\mathbf{u}_4] \\
\mathbf{f}_4 &= hk[-\boldsymbol{D}_{41}\cdot\mathbf{u}_1 - \boldsymbol{D}_{42}\cdot\mathbf{u}_2 - \boldsymbol{D}_{43}\cdot\mathbf{u}_3 + (\boldsymbol{D}_{41}+\boldsymbol{D}_{42}+\boldsymbol{D}_{43})\cdot\mathbf{u}_4].
\end{aligned}$$

In terms of the expressions for the nodal displacements from equation (5.114) we have

$$\begin{aligned}
\mathbf{f}_1 &= -h^2k\,[\boldsymbol{D}_{21}\cdot(\mathbf{q}-\mathbf{w}) + \boldsymbol{D}_{41}\cdot(\mathbf{q}+\mathbf{w})] \\
\mathbf{f}_2 &= h^2k\,[(\boldsymbol{D}_{21}-\boldsymbol{D}_{23})\cdot\mathbf{q} - (\boldsymbol{D}_{21}+\boldsymbol{D}_{23}+2\boldsymbol{D}_{42})\cdot\mathbf{w}] \\
\mathbf{f}_3 &= h^2k\,[\boldsymbol{D}_{43}\cdot(\mathbf{q}-\mathbf{w}) + \boldsymbol{D}_{23}\cdot(\mathbf{q}+\mathbf{w})] \\
\mathbf{f}_4 &= h^2k\,[(\boldsymbol{D}_{41}-\boldsymbol{D}_{43})\cdot\mathbf{q} + (\boldsymbol{D}_{41}+2\boldsymbol{D}_{42}+\boldsymbol{D}_{43})\cdot\mathbf{w}]\ .
\end{aligned}$$

From the geometry of the unit cell,

$$\boldsymbol{D}_{21} = \frac{1}{2}\begin{bmatrix} 1 & -1 \\ -1 & 1 \end{bmatrix} \quad \text{and} \quad \boldsymbol{D}_{23} = \frac{1}{2}\begin{bmatrix} 1 & 1 \\ 1 & 1 \end{bmatrix}$$

and we have already seen the matrix form of $\boldsymbol{D}_{41}$, $\boldsymbol{D}_{42}$, and $\boldsymbol{D}_{43}$. Using these we find that

$$\begin{aligned}
\mathbf{f}_1 &= -h^2k\,[(q_1+w_2)\mathbf{e}_1 + (q_2+w_1)\mathbf{e}_2] = -\mathbf{f}_3 \\
\mathbf{f}_2 &= -h^2k\,[(q_2+w_1)\mathbf{e}_1 + (q_1+3w_2)\mathbf{e}_2] = -\mathbf{f}_4.
\end{aligned}$$

If we express the components of $\mathbf{q}$ and $\mathbf{w}$ in terms of the smooth displacement field $\mathbf{u}(\mathbf{x})$, we have

$$\begin{aligned}
\mathbf{f}_1 &= -h^2k\,[(u_{1,1}+u_{2,2})\mathbf{e}_1 + (u_{2,1}+u_{1,2})\mathbf{e}_2] = -\mathbf{f}_3 \\
\mathbf{f}_2 &= -h^2k\,[(u_{2,1}+u_{1,2})\mathbf{e}_1 + (u_{1,1}+3u_{2,2})\mathbf{e}_2] = -\mathbf{f}_4\,.
\end{aligned}$$

If we tile the plane with these unit cells, the only non-zero forces will be on the edges where nodes are shared by two springs. Let us consider the edges of such a tile. If there are n nodes along and edge, the line containing these nodes has length $2nh$ and the total force is $n\mathbf{f}_3$ (right side) and $n\mathbf{f}_4$ (top side). The two-dimensional tractions are given by

$$\mathbf{t}_3^{\text{elastic}} = \mathbf{f}_3/(2h) \quad \text{and} \quad \mathbf{t}_4^{\text{elastic}} = \mathbf{f}_4/(2h)\,.$$

If we assume that the stress in the unit cell is homogeneous in the limit $h \to 0$, the relation $\mathbf{t} = \mathbf{n} \cdot \boldsymbol{\sigma}$ gives us $\sigma_{11}^E = (hk/2)(u_{1,1} + u_{2,2})$, $\sigma_{12}^E = (hk/2)(u_{2,1} + u_{1,2}) = \sigma_{21}^E$, and $\sigma_{22}^E = (hk/2)(u_{1,1} + 3u_{2,2})$. In matrix notation,

$$\boxed{\boldsymbol{\sigma}^E \equiv \underline{\underline{\boldsymbol{\sigma}}}^E = \frac{hk}{2}\begin{bmatrix} u_{1,1}+u_{2,2} & u_{2,1}+u_{1,2} \\ u_{2,1}+u_{1,2} & u_{1,1}+3u_{2,2} \end{bmatrix}.} \tag{5.119}$$

Alternatively, we can write the above equation as

$$\underline{\underline{\boldsymbol{\sigma}}}^E = \begin{bmatrix} \sigma_{11}^E \\ \sigma_{12}^E \\ \sigma_{21}^E \\ \sigma_{22}^E \end{bmatrix} = \frac{hk}{2}\begin{bmatrix} 1 & 0 & 0 & 1 \\ 0 & 1 & 1 & 0 \\ 0 & 1 & 1 & 0 \\ 1 & 0 & 0 & 3 \end{bmatrix}\begin{bmatrix} u_{1,1} \\ u_{1,2} \\ u_{2,1} \\ u_{2,2} \end{bmatrix}.$$

Therefore, the total stress ($\boldsymbol{\sigma} = \boldsymbol{\sigma}^I + \boldsymbol{\sigma}^E$) is

$$\underline{\underline{\boldsymbol{\sigma}}} = \begin{bmatrix} \sigma_{11} \\ \sigma_{12} \\ \sigma_{21} \\ \sigma_{22} \end{bmatrix} = \frac{1}{2}\begin{bmatrix} hk & 0 & 0 & hk & 0 & 0 \\ 0 & hk & hk & 0 & 0 & 0 \\ 0 & hk & hk & 0 & 0 & -i\omega mc \\ hk & 0 & 0 & 3hk & -i\omega mc^{-1} & 0 \end{bmatrix}\begin{bmatrix} u_{1,1} \\ u_{1,2} \\ u_{2,1} \\ u_{2,2} \\ v_1 \\ v_2 \end{bmatrix}.$$

Combining the above relation with the momentum relation (5.117) gives us

$$\boxed{\begin{bmatrix} \sigma_{11} \\ \sigma_{21} \\ \sigma_{12} \\ \sigma_{22} \\ p_1 \\ p_2 \end{bmatrix} = \frac{1}{2}\begin{bmatrix} hk & 0 & 0 & hk & 0 & 0 \\ 0 & hk & hk & 0 & 0 & -i\omega mc \\ 0 & hk & hk & 0 & 0 & 0 \\ hk & 0 & 0 & 3hk & -i\omega mc^{-1} & 0 \\ 0 & 0 & 0 & -im\omega c^{-1} & -\delta & 0 \\ 0 & -im\omega c & 0 & 0 & 0 & -\delta \end{bmatrix}\begin{bmatrix} u_{1,1} \\ u_{1,2} \\ u_{2,1} \\ u_{2,2} \\ v_1 \\ v_2 \end{bmatrix}.} \tag{5.120}$$

Note that we have switched the rows corresponding to σ_{12} and σ_{21} in the above equation to bring out the symmetries that we seek. Also note that we have retained the spring stiffnesses in the above equation because though $h \to 0$, hk is never equal to zero for composites of this nature. The above equation can be written in compact form as

$$\begin{bmatrix} \underline{\underline{\boldsymbol{\sigma}}} \\ \underline{\underline{\mathbf{p}}} \end{bmatrix} = \begin{bmatrix} \underline{\underline{\mathbf{C}}} & \underline{\underline{S}} \\ \underline{\underline{S}}^T & \underline{\underline{\boldsymbol{\rho}}} \end{bmatrix}\begin{bmatrix} \underline{\underline{\boldsymbol{\varepsilon}}} \\ \underline{\underline{\mathbf{v}}} \end{bmatrix}.$$

These are similar to the Willis equations. However, the additional momentum terms that contribute to the stress make the stress unsymmetric. Milton has proposed a microstructure that has the characteristics of the Milton model. The physical realization of such a composite and a more rigorous justification of the approach are open questions.

5.8 Extremal materials

In this section we will discuss the class of extremal materials introduced by Milton and Cherkaev (1995) in the context of linear elasticity. Though these extremal materials are not strictly metamaterials, in the sense that resonances are not exploited, they have potentially important applications in acoustic cloaking and in linear elastic composite design in general. In fact, hierarchical laminates made of these materials can be used to realize almost any positive definite fourth-order elastic stiffness tensor $\mathbf{C}$.* Alternative extremal material designs are discussed in Sigmund (2000).

The first extremal materials to be designed were materials with negative Poisson's ratio (also called auxetic materials) such as foams (Lakes, 1987) ($\nu \approx -0.7$), designed structures that unfold when compressed (Larsen et al., 2002), or structures that exhibit a negative Poisson's ratio due to internal buckling (Bertoldi et al., 2010). Milton (2002) found, surprisingly, that negative Poisson's ratio materials could be built using hierarchical lamination. Here, we focus on *pentamode materials*, composites which can behave like fluids (in the sense that they cannot support shear) even though they have solid components. These materials can form the building blocks of hierarchical laminates which can be used to realize a large range of composite behaviors. Unfortunately, we are not aware of any experimental confirmation of the properties of these proposed structures.

The starting point in the design of pentamode and other extremal materials of this class is the assumption that we can find two isotropic materials; one perfectly rigid and the other perfectly compliant. Since an isotropic elastic modulus can be characterized by its bulk and shear moduli, we need one material with bulk modulus $\kappa \equiv \psi\widehat{\kappa}$ and shear modulus $\mu \equiv \psi\widehat{\mu}$ where $\psi \to \infty$. This is the stiff material. The compliant material has a bulk modulus $\kappa \equiv \delta\widehat{\kappa}$ and a shear modulus $\mu \equiv \delta\widehat{\mu}$ where $\delta \to 0$. In practice, one can only hope to find materials that approximate these requirements. But the richness of the possible range of composite behaviors, even though imperfect from the standpoint of theory, is worth exploring.

5.8.1 Pentamode materials

A rigorous definition of pentamode and other extremal materials of that class can be found in Milton and Cherkaev (1995). We will instead work with the definition that these are composites for which the eigenvalues of the effective stiffness tensor ($\mathbf{C}$) are either very large or very small. Of course, the same composite can have some eigenvalues that are very large and others that are very small. Detailed discussions of the eigenvalues and the associated spectral decomposition of fourth-order tensors can be found in Itskov (2000) and Moakher (2008) and the references cited there.

---

*A brief description of hierarchical laminates is given in Section 8.6.5 (p. 331).

The eigenvalues of a fourth-order tensor $\mathbf{C}$ can be found by solving the eigenvalue problem

$$\mathbf{C} : \boldsymbol{N} = \lambda \boldsymbol{N}$$

where $\boldsymbol{N}$ is a second-order tensor. For a general fourth-order tensor we can find nine eigenvalues and nine principal invariants. For a symmetric fourth-order tensor such as the stiffness tensor of elasticity, the number of eigenvalues reduces to six, which means that at least three eigenvalues and three invariants are zero. Important invariants of a fourth-order tensor are its trace and its determinant which are defined as

$$\mathrm{tr}(\mathbf{C}) = \sum_{j=1}^{6} \lambda_j \quad \text{and} \quad \det(\mathbf{C}) = \prod_{j=1}^{6} \lambda_j \,.$$

The spectral decomposition of $\mathbf{C}$ is[†]

$$\mathbf{C} = \sum_{j=1}^{6} \lambda_j \boldsymbol{N}_j \otimes \boldsymbol{N}_j \quad \text{with} \quad \|\boldsymbol{N}_j\| = 1\,.$$

The eigenvalues in the above decomposition are also called the *Kelvin moduli*. The positive definiteness of $\mathbf{C}$ implies that $\lambda_j > 0$. Special treatment is needed when the eigenvalues are not distinct and the spectral decomposition may not be convenient in those situations.

Any n-dimensional symmetric fourth-order tensor can also be viewed as a $\frac{1}{2}n(n+1)$-dimensional second-order tensor. This leads to the Voigt notation representation of the stress-strain relations for a linear elastic material:

$$\begin{bmatrix} \sigma_{11} \\ \sigma_{22} \\ \sigma_{33} \\ \sqrt{2}\sigma_{23} \\ \sqrt{2}\sigma_{13} \\ \sqrt{2}\sigma_{12} \end{bmatrix} = \begin{bmatrix} C_{1111} & C_{1122} & C_{1133} & \sqrt{2}C_{1123} & \sqrt{2}C_{1113} & \sqrt{2}C_{1112} \\ & C_{2222} & C_{2233} & \sqrt{2}C_{2223} & \sqrt{2}C_{2213} & \sqrt{2}C_{2212} \\ & & C_{3333} & \sqrt{2}C_{3323} & \sqrt{2}C_{3313} & \sqrt{2}C_{3312} \\ & & & 2C_{2323} & 2C_{2313} & 2C_{2312} \\ & \text{symm.} & & & 2C_{1313} & 2C_{1312} \\ & & & & & 2C_{1212} \end{bmatrix} \begin{bmatrix} \varepsilon_{11} \\ \varepsilon_{22} \\ \varepsilon_{33} \\ \sqrt{2}\varepsilon_{23} \\ \sqrt{2}\varepsilon_{13} \\ \sqrt{2}\varepsilon_{12} \end{bmatrix}.$$

In compact form

$$\underline{\underline{\sigma}} = \underline{\underline{\mathsf{C}}}\,\underline{\underline{\varepsilon}}.$$

We can use the eigen decomposition theorem to write

$$\underline{\underline{\mathsf{C}}} = \underline{\underline{\mathsf{V}}}\,\underline{\underline{\Lambda}}\,\underline{\underline{\mathsf{V}}}^{-1}$$

where $\underline{\underline{\Lambda}}$ is a diagonal matrix whose diagonal elements are the eigenvalues λ_i of $\underline{\underline{\mathsf{C}}}$, and $\underline{\underline{\mathsf{V}}}$ is the matrix whose ℓ-th column is the eigenvector $\underline{\mathsf{n}}^{\ell}$ that corresponds to the

[†]In a Cartesian basis the tensor product $\otimes$ of two second-order tensors has the form $\boldsymbol{A} \otimes \boldsymbol{B} = A_{ij}B_{k\ell}\mathbf{e}_i \otimes \mathbf{e}_j \otimes \mathbf{e}_k \otimes \mathbf{e}_\ell$. The inner product of two second-order tensors is given by $\boldsymbol{A} : \boldsymbol{B} = A_{ij}B_{ij}$. The norm of the tensor $\boldsymbol{N}$ is $\|\boldsymbol{N}\| = \sqrt{\boldsymbol{N} : \boldsymbol{N}}$.

eigenvalue λ_ℓ. Since $\underline{\underline{\mathbf{C}}}$ is symmetric and positive definite, we have $\underline{\underline{\mathsf{V}}}^{-1} = \underline{\underline{\mathsf{V}}}^T$ and $\lambda_\ell > 0$. Let $\underline{\underline{\mathsf{Q}}}$ be the orthogonalized form of $\underline{\underline{\mathsf{V}}}$. Then,

$$\underline{\underline{\mathbf{C}}} = \underline{\underline{\mathsf{Q}}}\,\underline{\underline{\boldsymbol{\Lambda}}}\,\underline{\underline{\mathsf{Q}}}^T .$$

Notice that, in general, the stress-strain relation can be expressed as

$$\underline{\boldsymbol{\sigma}} = (\underline{\underline{\mathsf{Q}}}\,\underline{\underline{\boldsymbol{\Lambda}}}\,\underline{\underline{\mathsf{Q}}}^T)\underline{\boldsymbol{\varepsilon}} \quad \Longrightarrow \quad \underline{\underline{\mathsf{Q}}}^T \underline{\boldsymbol{\sigma}} = \underline{\underline{\boldsymbol{\Lambda}}}(\underline{\underline{\mathsf{Q}}}^T \underline{\boldsymbol{\varepsilon}}) \quad \Longrightarrow \quad \widehat{\underline{\boldsymbol{\sigma}}} = \underline{\underline{\boldsymbol{\Lambda}}}\widehat{\underline{\boldsymbol{\varepsilon}}}.$$

For an isotropic material, the matrix $\underline{\underline{\mathbf{C}}}$ has the form

$$\underline{\underline{\mathbf{C}}} = \begin{bmatrix} \kappa + \frac{4}{3}\mu & \kappa - \frac{2}{3}\mu & \kappa - \frac{2}{3}\mu & 0 & 0 & 0 \\ & \kappa + \frac{4}{3}\mu & \kappa - \frac{2}{3}\mu & 0 & 0 & 0 \\ & & \kappa + \frac{4}{3}\mu & 0 & 0 & 0 \\ & & & 2\mu & 0 & 0 \\ & \text{symm.} & & & 2\mu & 0 \\ & & & & & 2\mu \end{bmatrix}.$$

The eigen decomposition of $\underline{\underline{\mathbf{C}}}$ then has the form

$$\underline{\underline{\mathbf{C}}} = \begin{bmatrix} \frac{1}{\sqrt{3}} & \sqrt{\frac{2}{3}} & 0 & 0\,0\,0 \\ \frac{1}{\sqrt{3}} & -\frac{1}{\sqrt{6}} & -\frac{1}{\sqrt{2}} & 0\,0\,0 \\ \frac{1}{\sqrt{3}} & -\frac{1}{\sqrt{6}} & \frac{1}{\sqrt{2}} & 0\,0\,0 \\ 0 & 0 & 0 & 1\,0\,0 \\ 0 & 0 & 0 & 0\,1\,0 \\ 0 & 0 & 0 & 0\,0\,1 \end{bmatrix} \begin{bmatrix} 3\kappa & 0 & 0 & 0 & 0 & 0 \\ 0 & 2\mu & 0 & 0 & 0 & 0 \\ 0 & 0 & 2\mu & 0 & 0 & 0 \\ 0 & 0 & 0 & 2\mu & 0 & 0 \\ 0 & 0 & 0 & 0 & 2\mu & 0 \\ 0 & 0 & 0 & 0 & 0 & 2\mu \end{bmatrix} \begin{bmatrix} \frac{1}{\sqrt{3}} & \frac{1}{\sqrt{3}} & \frac{1}{\sqrt{3}} & 0\,0\,0 \\ \sqrt{\frac{2}{3}} & -\frac{1}{\sqrt{6}} & -\frac{1}{\sqrt{6}} & 0\,0\,0 \\ 0 & -\frac{1}{\sqrt{2}} & \frac{1}{\sqrt{2}} & 0\,0\,0 \\ 0 & 0 & 0 & 1\,0\,0 \\ 0 & 0 & 0 & 0\,1\,0 \\ 0 & 0 & 0 & 0\,0\,1 \end{bmatrix}.$$

The $\underline{\underline{\boldsymbol{\Lambda}}}$ matrices for stiff and soft isotropic materials can then be expressed in the form

$$\underline{\underline{\boldsymbol{\Lambda}}}^{\text{stiff}} = \begin{bmatrix} \frac{3}{2}r\psi & 0 & 0 & 0 & 0 & 0 \\ 0 & \psi & 0 & 0 & 0 & 0 \\ 0 & 0 & \psi & 0 & 0 & 0 \\ 0 & 0 & 0 & \psi & 0 & 0 \\ 0 & 0 & 0 & 0 & \psi & 0 \\ 0 & 0 & 0 & 0 & 0 & \psi \end{bmatrix} \quad \text{and} \quad \underline{\underline{\boldsymbol{\Lambda}}}^{\text{soft}} = \begin{bmatrix} \frac{3}{2}r\delta & 0 & 0 & 0 & 0 & 0 \\ 0 & \delta & 0 & 0 & 0 & 0 \\ 0 & 0 & \delta & 0 & 0 & 0 \\ 0 & 0 & 0 & \delta & 0 & 0 \\ 0 & 0 & 0 & 0 & \delta & 0 \\ 0 & 0 & 0 & 0 & 0 & \delta \end{bmatrix}$$

where $r := \kappa/\mu > 0$. Extremally stiff isotropic materials are those where $\psi \to \infty$ while r is fixed. On the other hand, extremally soft isotropic materials are those for which $\delta \to 0$ while r is fixed. A large class of composite materials can be obtained by mixing only these two soft and stiff isotropic phases.

For anisotropic materials, we get more complicated expressions for the eigenvalues of $\underline{\underline{\mathbf{C}}}$. However, the general form is

$$\underline{\underline{\Lambda}} = \begin{bmatrix} \lambda_1 & 0 & 0 & 0 & 0 & 0 \\ 0 & \lambda_2 & 0 & 0 & 0 & 0 \\ 0 & 0 & \lambda_3 & 0 & 0 & 0 \\ 0 & 0 & 0 & \lambda_4 & 0 & 0 \\ 0 & 0 & 0 & 0 & \lambda_5 & 0 \\ 0 & 0 & 0 & 0 & 0 & \lambda_6 \end{bmatrix}.$$

Extremal anisotropic composite materials are those where the λs are either extremally stiff or extremally soft. A pentamode material is one for which λ_1 is large while the other values of $\lambda_j, j = 2\ldots 6$ are small. In that case the stress-strain relation can be written in matrix form as

$$\begin{bmatrix} \widehat{\sigma}_1 \\ \widehat{\sigma}_2 \\ \widehat{\sigma}_3 \\ \widehat{\sigma}_4 \\ \widehat{\sigma}_5 \\ \widehat{\sigma}_6 \end{bmatrix} \approx \begin{bmatrix} \psi & 0 & 0 & 0 & 0 & 0 \\ 0 & \delta & 0 & 0 & 0 & 0 \\ 0 & 0 & \delta & 0 & 0 & 0 \\ 0 & 0 & 0 & \delta & 0 & 0 \\ 0 & 0 & 0 & 0 & \delta & 0 \\ 0 & 0 & 0 & 0 & 0 & \delta \end{bmatrix} \begin{bmatrix} \widehat{\varepsilon}_1 \\ \widehat{\varepsilon}_2 \\ \widehat{\varepsilon}_3 \\ \widehat{\varepsilon}_4 \\ \widehat{\varepsilon}_5 \\ \widehat{\varepsilon}_6 \end{bmatrix}.$$

Therefore, there are five strains for which the material is compliant and one for which the material is stiff. These five strains are called the "easy strains" and since there are five easy strains the material is called pentamode. A unimodal material, in contrast, will have only one easy strain. Clearly, the stress field in a pentamode material is dominated by the single mode in which the material remains stiff. This stress is called the "supporting stress."

For the special case where the five easy modes correspond to the shear moduli and the stiff mode corresponds to the bulk modulus, we get a pentamode material that behaves like a fluid. Then the stiffness matrix can be expressed as

$$\underline{\underline{\mathbf{C}}} = \underline{\underline{\mathbf{Q}}}\,\underline{\underline{\Lambda}}\,\underline{\underline{\mathbf{Q}}}^T = \lambda_1 \begin{bmatrix} Q_{11}^2 & Q_{11}Q_{21} & Q_{11}Q_{31} & Q_{11}Q_{41} & Q_{11}Q_{51} & Q_{11}Q_{61} \\ Q_{11}Q_{21} & Q_{21}^2 & Q_{21}Q_{31} & Q_{21}Q_{41} & Q_{21}Q_{51} & Q_{21}Q_{61} \\ Q_{11}Q_{31} & Q_{21}Q_{31} & Q_{31}^2 & Q_{31}Q_{41} & Q_{31}Q_{51} & Q_{31}Q_{61} \\ Q_{11}Q_{41} & Q_{21}Q_{41} & Q_{31}Q_{41} & Q_{41}^2 & Q_{41}Q_{51} & Q_{41}Q_{61} \\ Q_{11}Q_{51} & Q_{21}Q_{51} & Q_{31}Q_{51} & Q_{41}Q_{51} & Q_{51}^2 & Q_{51}Q_{61} \\ Q_{11}Q_{61} & Q_{21}Q_{61} & Q_{31}Q_{61} & Q_{41}Q_{61} & Q_{51}Q_{61} & Q_{61}^2 \end{bmatrix}. \tag{5.121}$$

In compact form,

$$\underline{\underline{\mathbf{C}}} = \lambda_1\, \underline{\mathbf{n}}\,\underline{\mathbf{n}}^T =: 3\widehat{\kappa}\,\underline{\mathbf{n}}\,\underline{\mathbf{n}}^T$$

where $\underline{\mathbf{n}}^T = [Q_{11}\ Q_{21}\ Q_{31}\ Q_{41}\ Q_{51}\ Q_{61}]$ and we have identified the non-zero eigenvalue with a "bulk modulus," $\widehat{\kappa}$, in analogy with an isotropic material. Notice that the vector $\underline{\mathbf{n}}$ is an eigenvector of $\underline{\underline{\mathbf{C}}}$. Also, since the trace of the matrix $\mathbf{C}$ is invariant, we have

$$\lambda_1 = 3\widehat{\kappa} = \text{tr}(\mathbf{C}) = C_{1111} + C_{2222} + C_{3333} + 2C_{2323} + 2C_{1313} + 2C_{1212}.$$

Following Norris (2008), we can define a suitably normalized vector $\widehat{\underline{\mathsf{n}}}$ to express $\underline{\underline{\mathbf{C}}}$ as

$$\underline{\underline{\mathbf{C}}} = \widehat{\kappa}\widehat{\underline{\mathsf{n}}}\widehat{\underline{\mathsf{n}}}^T \quad \text{where} \qquad \widehat{\underline{\mathsf{n}}}^T\widehat{\underline{\mathsf{n}}} = 3. \tag{5.122}$$

Therefore,

$$\underline{\boldsymbol{\sigma}} = \underline{\underline{\mathbf{C}}}\underline{\boldsymbol{\varepsilon}} = \widehat{\kappa}\widehat{\underline{\mathsf{n}}}\widehat{\underline{\mathsf{n}}}^T\underline{\boldsymbol{\varepsilon}} = \widehat{\kappa}\varepsilon\widehat{\underline{\mathsf{n}}} = \sigma\widehat{\underline{\mathsf{n}}} \tag{5.123}$$

where $\varepsilon := \widehat{\underline{\mathsf{n}}}^T\underline{\boldsymbol{\varepsilon}}$ and $\sigma := \widehat{\kappa}\varepsilon$. In direct tensor notation,

$$\boldsymbol{\sigma} = \mathbf{C} : \boldsymbol{\varepsilon} = \widehat{\kappa}(\widehat{\boldsymbol{N}}\otimes\widehat{\boldsymbol{N}}) : \boldsymbol{\varepsilon} = \widehat{\kappa}\varepsilon\widehat{\boldsymbol{N}} = \sigma\widehat{\boldsymbol{N}} \quad \text{where} \quad \widehat{\boldsymbol{N}}^T : \widehat{\boldsymbol{N}} = \text{tr}(\widehat{\boldsymbol{N}}^2) = 3 \tag{5.124}$$

where $\varepsilon = \widehat{\boldsymbol{N}} : \boldsymbol{\varepsilon}$. Note that there is no unique inverse relation between $\boldsymbol{\varepsilon}$ and $\boldsymbol{\sigma}$ because the tensor $\mathbf{C}$ is rank deficient. For the situation where $Q_{41} = Q_{51} = Q_{61} = 0$, equation (5.121) gives

$$\underline{\underline{\mathbf{C}}} = \lambda_1 \begin{bmatrix} Q_{11}^2 & Q_{11}Q_{21} & Q_{11}Q_{31} & 0\,0\,0 \\ Q_{11}Q_{21} & Q_{21}^2 & Q_{21}Q_{31} & 0\,0\,0 \\ Q_{11}Q_{31} & Q_{21}Q_{31} & Q_{31}^2 & 0\,0\,0 \\ 0 & 0 & 0 & 0\,0\,0 \\ 0 & 0 & 0 & 0\,0\,0 \\ 0 & 0 & 0 & 0\,0\,0 \end{bmatrix}.$$

This is an example on an orthotropic pentamode material with $C_{2323} = C_{1313} = C_{1212} = 0$. For a given strain $\underline{\boldsymbol{\varepsilon}}$, the shear stress components in this case are zero. Hence the 3×3 stress matrix must be diagonal in the chosen Cartesian coordinate system, i.e.,

$$\boldsymbol{\sigma} \equiv \begin{bmatrix} \sigma_1 & 0 & 0 \\ 0 & \sigma_2 & 0 \\ 0 & 0 & \sigma_3 \end{bmatrix}, \quad \widehat{\kappa} = \tfrac{1}{3}(C_{1111} + C_{2222} + C_{3333}) \quad \text{and} \quad \widehat{\boldsymbol{N}} = \widehat{\kappa}^{-\frac{1}{2}} \sum_{j=1}^{3} C_{jjjj}^{\frac{1}{2}} \mathbf{e}_j \otimes \mathbf{e}_j .$$

A pentamode structure that meets these requirements in shown in Figure 5.29. However, the material may not be very stable unless there are additional stiffeners. A more stable structure will have imperfect pentamode behavior. Details of the design of similar structures can be found in Milton and Cherkaev (1995) and Milton (2002).

Using equation (5.124) as the basis, for an inhomogeneous pentamode material, we have

$$\mathbf{C}(\mathbf{x}) = \kappa(\mathbf{x})\boldsymbol{N}(\mathbf{x})\otimes\boldsymbol{N}(\mathbf{x}) \quad \Longrightarrow \quad \boldsymbol{\sigma}(\mathbf{x}) = \kappa(\mathbf{x})\left[\boldsymbol{N}(\mathbf{x})\otimes\boldsymbol{N}(\mathbf{x})\right] : \boldsymbol{\varepsilon}(\mathbf{x})$$

where we have chosen not to normalize the tensor $\boldsymbol{N}$. Since the components of the strain tensor that are not oriented along the eigenvector $\boldsymbol{Q}$ do not have any effect on the stress, we can write $\boldsymbol{\varepsilon}(\mathbf{x}) = \varepsilon(\mathbf{x})\boldsymbol{N}(\mathbf{x})$ where $\varepsilon(\mathbf{x})$ is a scalar function. Then we have

$$\boldsymbol{\sigma}(\mathbf{x}) = \kappa(\mathbf{x})\varepsilon(\mathbf{x})\left[\boldsymbol{N}(\mathbf{x})\otimes\boldsymbol{N}(\mathbf{x})\right] : \boldsymbol{N}(\mathbf{x}) = \kappa(\mathbf{x})\varepsilon(\mathbf{x})[\boldsymbol{N}(\mathbf{x}) : \boldsymbol{N}(\mathbf{x})]\boldsymbol{N}(\mathbf{x}) =: \sigma(\mathbf{x})\boldsymbol{N}(\mathbf{x}).$$

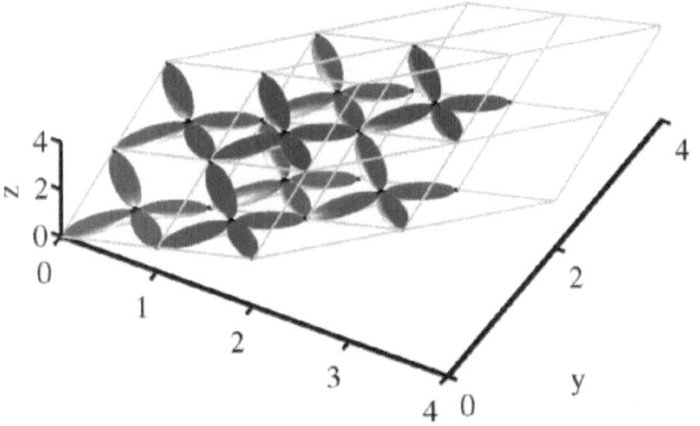

FIGURE 5.29
A pentamode structure containing stiff and compliant regions. The shaded regions are stiff while the blank regions are compliant. The light gray lines show the boundaries of the unit cell. This structure is based on an example in Milton and Cherkaev (1995).

If we define $\boldsymbol{S} := \sigma \boldsymbol{N}$, we observe that

$$\mathbf{C} = \kappa \boldsymbol{N} \otimes \boldsymbol{N} = \frac{\kappa}{\sigma^2} \boldsymbol{S} \otimes \boldsymbol{S} =: K \boldsymbol{S} \otimes \boldsymbol{S} \quad \Longrightarrow \quad \boxed{\boldsymbol{\sigma} = K(\boldsymbol{S} : \boldsymbol{\varepsilon})\boldsymbol{S}} \tag{5.125}$$

and

$$\boldsymbol{\nabla} \cdot \boldsymbol{\sigma} = 0 \quad \Longrightarrow \quad \boxed{\boldsymbol{\nabla} \cdot \boldsymbol{S} = 0}\,. \tag{5.126}$$

Equations (5.125) are a convenient form of the stiffness tensor of a pentamode material and equation (5.126) provides a constraint on the values that the tensor can take. Note that $\boldsymbol{S}$ is a symmetric second-order tensor. The "easy" strains in a pentamode material can be identified using a spectral decomposition of $\boldsymbol{S}$. Let

$$\boldsymbol{S} = \sum_{j=1}^{3} \lambda_j \mathbf{s}_j \otimes \mathbf{s}_j \tag{5.127}$$

where λ_j are the eigenvalues and $\mathbf{s}_j$ are the corresponding orthonormal eigenvectors. We assume that there are no repeated eigenvalues. The strain energy in a pentamode material is given by

$$W = \boldsymbol{\sigma} : \boldsymbol{\varepsilon} = K(\boldsymbol{S} : \boldsymbol{\varepsilon})^2\,. \tag{5.128}$$

The strain energy is zero for the five easy strains. Plugging in the spectral decomposition for $\boldsymbol{S}$ into the expression for W and equating the result to zero gives us an equation that can be used to determine the strains that are "easy." Let us examine the case of pure shear to see whether such strains are "easy." In the basis $(\mathbf{s}_1, \mathbf{s}_2, \mathbf{s}_3)$, pure shear strains can be expressed in the form (see, for example, Ogden (1997))

$$\boldsymbol{\varepsilon}^{(12)} = \gamma(\mathbf{s}_1 \otimes \mathbf{s}_2 + \mathbf{s}_2 \otimes \mathbf{s}_1)\,, \; \boldsymbol{\varepsilon}^{(23)} = \gamma(\mathbf{s}_2 \otimes \mathbf{s}_3 + \mathbf{s}_3 \otimes \mathbf{s}_2)\,, \; \boldsymbol{\varepsilon}^{(31)} = \gamma(\mathbf{s}_3 \otimes \mathbf{s}_1 + \mathbf{s}_1 \otimes \mathbf{s}_2)$$

where γ is the magnitude of the strain. If we plug these strains into (5.128) we see that $W = 0$ because $\mathbf{s}_i \cdot \mathbf{s}_j = 0, i \neq j$. Therefore, these shear strains are easy strains. If we use the same approach for normal strains of the form

$$\boldsymbol{\varepsilon} = \varepsilon_1 \mathbf{s}_1 \otimes \mathbf{s}_1 + \varepsilon_2 \mathbf{s}_2 \otimes \mathbf{s}_2 + \varepsilon_3 \mathbf{s}_3 \otimes \mathbf{s}_3$$

we find two other independent easy strains (Norris, 2009) of the form

$$\boldsymbol{\varepsilon} = \lambda_3 \mathbf{s}_2 \otimes \mathbf{s}_2 - \lambda_2 \mathbf{s}_3 \otimes \mathbf{s}_3 \quad \text{and} \quad \boldsymbol{\varepsilon} = \lambda_1 \mathbf{s}_2 \otimes \mathbf{s}_2 - \lambda_2 \mathbf{s}_1 \otimes \mathbf{s}_1 \,.$$

All other easy strains are linear combinations of the above five easy strains.

Norris (2008, 2009) has discovered several relations pertaining to the static and dynamic behavior of pentamode materials and the related acoustic metafluids. An interesting feature of pentamode materials with $\boldsymbol{\nabla} \cdot \boldsymbol{S} = 0$ is that the traction-free surface of such a material is not horizontal when it is at rest under gravity. To examine the dynamic dynamic behavior of such materials let us define a pseudo-pressure, p, such that

$$p := -K(\boldsymbol{S} : \boldsymbol{\varepsilon}) = -K(\boldsymbol{S} : \boldsymbol{\nabla}\mathbf{u}) \quad \Longrightarrow \quad \boldsymbol{\sigma} = -p\boldsymbol{S} \tag{5.129}$$

where $\mathbf{u}$ is the displacement. Then, if $\mathbf{v}$ is the velocity, we have

$$\dot{p} = -K(\boldsymbol{S} : \boldsymbol{\nabla}\mathbf{v}) \,. \tag{5.130}$$

This equation is the pentamode equivalent of the pressure constitutive relation in linear acoustics. The balance of momentum in the absence of body forces is given by

$$\boldsymbol{\nabla} \cdot \boldsymbol{\sigma} = \boldsymbol{\rho} \cdot \dot{\mathbf{v}} \tag{5.131}$$

where $\boldsymbol{\rho}$ is an anisotropic density. Taking the time derivative of (5.130), plugging the result into (5.131), assuming that condition (5.126) holds, and noting that $\boldsymbol{S}$ is symmetric gives us the wave equation for the pseudo-pressure in pentamode materials:

$$\boxed{K\boldsymbol{S} : \boldsymbol{\nabla}[\boldsymbol{\rho}^{-1} \cdot (\boldsymbol{S} \cdot \boldsymbol{\nabla} p)] - \ddot{p} = 0.} \tag{5.132}$$

For a plane harmonic wave displacement with amplitude $\widehat{\mathbf{u}}$ and wave vector $\mathbf{k}$ incident upon a pentamode material, we can show that (5.131) implies that

$$\left[K\boldsymbol{A} - \omega^2 \boldsymbol{\rho}\right] \cdot \widehat{\mathbf{u}} = \mathbf{0} \quad \text{where} \quad \boldsymbol{A} := (\boldsymbol{S} \cdot \mathbf{k}) \otimes (\boldsymbol{S} \cdot \mathbf{k}) \,. \tag{5.133}$$

Comparison with equation (2.2) shows that the quantity $\boldsymbol{A}$ is the acoustic tensor $\mathbf{k} \cdot \mathsf{C} \cdot \mathbf{k}$. Since C is rank 1, the tensor $\boldsymbol{A}$ must also be rank 1, i.e., only one of the three solutions of the generalized eigenvalue problem in (5.133) is non-zero. Therefore,

$$\omega^2 = K\boldsymbol{\rho}^{-1} : \boldsymbol{A} = K(\mathbf{k} \cdot \boldsymbol{S}) \cdot (\boldsymbol{\rho}^{-1} \cdot \boldsymbol{S} \cdot \mathbf{k}) = K(\mathbf{k} \cdot \boldsymbol{S}) \cdot \widehat{\mathbf{u}} \,.$$

Notice that the above is also an expression for the phase velocity (c) in the pentamode material because $c = \omega / \|\mathbf{k}\|$. This implies that the slowness surface is an ellipsoid. An excellent exploration of elastic waves in anisotropic solids can be found in Musgrave (1970) and we direct the interested reader to that book for further information on waves in various types of anisotropic media.

Exercises

Problem 5.1 Plot the effective mass and transmission loss for a bar of mass $M_0 = 2$ kg with a cavity containing a mass $m_1 = 2$ kg and two springs with $k_1 = k_2 = 10^5(1+0.5i)$ N/m. Compare the transmission loss with the mass law for the system. The impedance used to calculate the mass law is $Z = i\,\omega\,(M_0 + m_1)$.

What happens when the mass m_1 in the system is also frequency dependent and has the form

$$m_1 = m_0 + \frac{\omega_0^2\, m_2}{\omega_0^2 - \omega^2}\,; \quad \omega_0 := \sqrt{\frac{2\,G}{m_2}}$$

with $m_1 = m_2 = 1$ kg and $G = 10^4$ N/m?

Problem 5.2 Consider the modified rigid bar with voids shown in the figure below.

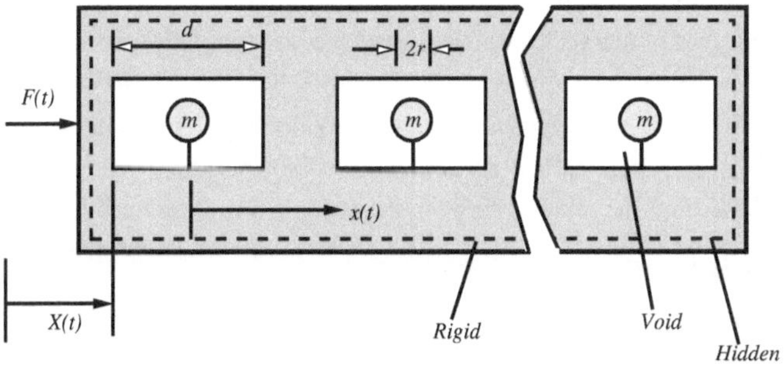

Each ball is attached to the bar by a massless beam with a circular cross-section and radius h. Calculate the effective dynamic mass of the system. Plot the effective mass as a function of frequency. Are there regions where the effective mass is negative?

Problem 5.3 Consider the model of an elastic bar containing hidden springs and masses. Plot the effective mass of the bar as a function of frequency and compare the result to the rigid case. Can you think of other representations of an elastic bar with hidden springs and masses?

Extend the the model of the elastic bar containing springs and masses to two dimensions and plot the effective mass of the system in polar coordinates as a function of angle.

Problem 5.4 Show that the effective Young's modulus function in equation (5.65) is analytic in the entire complex plane except for isolated singularities in the negative $\text{Im}(\omega)$ part of the plane. Draw a schematic of the locations of the singularities. Also show that $\overline{E}(\omega) = E(-\overline{\omega})$ for this model, where $\overline{(\bullet)}$ indicates the complex conjugate.

Problem 5.5 Examine the effective mass as a function of frequency when there is dissipation in the system. What is the predominant effect that you observe as the amount of dissipation increases?

Problem 5.6 Starting from equation (5.22) show that

$$\widehat{\mathbf{P}}(\omega) - \boldsymbol{M}(\omega) \cdot \widehat{\mathbf{V}}(\omega)$$

where

$$\boldsymbol{M}(\omega) := \widehat{\boldsymbol{H}}(\omega) = \int_{-\infty}^{\infty} \boldsymbol{H}(s)\, e^{i\omega s}\, \mathrm{d}s\,.$$

Problem 5.7 Starting from the relation

$$\mathbf{q}_\mathrm{r}(t) = \int_0^{2\pi} \overline{\mathbf{x}}_\mathrm{r}(\varphi,t) \times d\mathbf{p}_\mathrm{r}(\varphi,t)$$

show that the angular momentum of the ring in Figure 5.13 can be expressed in the form

$$\mathbf{q}_\mathrm{r}(t) = \mathbf{q}_0 + \mathrm{Re}(\widehat{\mathbf{q}}_\mathrm{r} e^{-i\omega t})$$

where

$$\mathbf{q}_0 = -m r_0^2 \omega_r \mathbf{e}_3$$

and

$$\widehat{\mathbf{q}}_\mathrm{r} = \frac{\varepsilon m r_0^2}{2}\left[-\left(\frac{2i\omega_r\widehat{\Omega}_2}{\omega} + 3\widehat{\Omega}_1\right)\mathbf{e}_1 + \left(\frac{2i\omega_r\widehat{\Omega}_1}{\omega} - 3\widehat{\Omega}_2\right)\mathbf{e}_2\right].$$

Problem 5.8 Using the Helmholtz model as an analogy, come up with another model system that can have a negative effective elastic modulus.

Problem 5.9 Verify that the ensemble averaged equations (5.94) are correct, i.e., the quantities $\mathbf{C}_x$, $\mathcal{M}_x$, $\mathcal{S}_t$, and $\boldsymbol{M}_t$ can be taken outside the integrals during averaging.

Problem 5.10 Given what you know about the Willis equations, develop a simple model that exhibits coupling between the elastic response and density.

Problem 5.11 Find the relation between the stress-momentum vector and the displacement gradient-velocity vector for the Milton-Willis model for the case where $c = 1/2$ and $d = 1/4$. Comment on your findings.

Problem 5.12 Show that the wave equation for pentamode materials can be expressed as

$$K\boldsymbol{S} : \boldsymbol{\nabla}[\boldsymbol{\rho}^{-1} \cdot (\boldsymbol{S} \cdot \boldsymbol{\nabla} p)] - \ddot{p} = 0.$$

For a plane harmonic wave with wave vector $\mathbf{k}$ incident upon a pentamode material, show that

$$[(\boldsymbol{S}\cdot\mathbf{k}) \otimes (\boldsymbol{S}\cdot\mathbf{k})]\cdot\mathbf{u} - \frac{\omega^2}{K}\boldsymbol{\rho}\cdot\mathbf{u} = \mathbf{0}.$$

6

Transformation-based Methods and Cloaking

> Facts which at first seem improbable will, even on scant explanation, drop the cloak which has hidden them and stand forth in naked and simple beauty.
>
> GALILEO GALILEI, *Dialogues concerning two new sciences*, 1638.

We are able to hear and see objects in our environment when these objects interact with elastic, sound, and light waves. Even if our senses are not quite up to the task, we have ultrasonic sensors or infrared sensors that can be used to identify objects. Elastic waves are used routinely in seismic exploration and to determine whether complicated machine parts are in working order. But our senses, natural or artificial, can be tricked into making the wrong identification of an object because different material properties in an object can lead to the same behavior at a detection point.

Cloaking is the latest of many attempts to hide an object from prying waves. Several approaches can be used to cloak objects and coatings that reduce the radar cross-section of fighter jets have been in use for many years. However, the choice of cloaking material has always been difficult. A new approach for designing materials that can cloak an object arose following a discovery by Greenleaf et al. in 2003 (Greenleaf et al., 2003b).

The new idea was that certain singular spatial transformations, if chosen appropriately, could lead to cloaking for electrical conductivity. Leonhardt (Leonhardt, 2006b) and Pendry et al. (Pendry et al., 2006) soon showed that Greenleaf's idea could be applied to general electromagnetic fields by taking advantage of the invariance of Maxwell's electromagnetic wave equations under spatial transformations. Acoustic cloaks were described in 2007 by Chen and Chan (Chen and Chan, 2007a) and Cummer and Schurig (Cummer and Schurig, 2007).

The transformation-based cloaking idea provides a straightforward formula for designing cloak materials. There is an extensive literature on the application of the idea to electromagnetic problems and a growing literature on acoustic cloaking.

In this chapter we will explore the transformation-based approach to the design of materials that can be used to guide waves around an object. The invariance of Maxwell's equations under coordinate transformations means that cloaks for electromagnetic waves can be designed quite readily. Experimental observation of electromagnetic cloaking effects have been observed both in two dimensions (see, for example, Schurig et al. (2006a), Liu et al. (2009)) and in three dimensions (Ergin et al., 2010).

Acoustic cloaks can be designed either by analogy with TE-waves, the electrical conductivity equations, or by direct manipulation of the acoustic wave equation. Inertial cloaks have anisotropic density and isotropic bulk modulus. Such cloaks are limited by the requirement of infinite mass if the transformation is singular. Also, acoustic cloaking experiments are more difficult to design than electromagnetic cloaks because of the requirement of anisotropic density. Acoustic cloaking has been demonstrated experimentally by Zhang et al. (2010) who showed that a layered structure could cloak ultrasonic waves in water. Other experiments using pentamode materials can be envisaged.

Elastodynamic cloaking is more difficult than electromagnetic or acoustic cloaking because the governing equations change into the Willis equations under coordinate transformations. However, the elastodynamic equations retain their form when the reference material is a pentamode material. Transformation elastodynamics is an active area of research and has the potential for a broad range of applications, not limited to cloaking.

Instead of the passive cloaking methods discussed in this chapter we may also seek to create active cloaking devices. The interested reader can find examples of active cloaking in Miller (2006) for cloaking of the interior of a region and Vasquez et al. (2009) for cloaking of part of the exterior of a region.

6.1 Transformations

Let us first examine the transformation of a vector $\mathbf{u}$ from an orthonormal basis $(\mathbf{e}_i)$ to another $(\mathbf{e}'_i)$ as shown in Figure 6.1. We want to find the relation between the components of the vector in the two bases. The vector can be expressed as

$$\mathbf{u} = u_j \mathbf{e}_j = u'_i \mathbf{e}'_i \qquad \Longrightarrow \qquad u_j (\mathbf{e}'_i \cdot \mathbf{e}_j) = u'_i \,.$$

If we define $A_{ij} := \mathbf{e}'_i \cdot \mathbf{e}_j$, we can write

$$u'_i = A_{ij} u_j \qquad \Longrightarrow \qquad u_j = A^{-1}_{ji} u'_i = A^T_{ji} u'_i = A_{ij} u'_i \,. \tag{6.1}$$

This is the coordinate transformation rule for vectors.

A second-order tensor maps a vector $\mathbf{u}$ to another vector $\mathbf{v}$. Let us examine how the components of such a tensor transform under change of basis. Let $\mathbf{v} = \boldsymbol{S} \cdot \mathbf{u}$. We can write this relation in component form as

$$v_i \mathbf{e}_i = (S_{ij} \mathbf{e}_i \otimes \mathbf{e}_j) \cdot (u_k \mathbf{e}_k) = S_{ij} u_j \mathbf{e}_i \quad \text{and} \quad v'_i \mathbf{e}'_i = (S'_{ij} \mathbf{e}'_i \otimes \mathbf{e}'_j) \cdot (u'_k \mathbf{e}'_k) = S'_{ij} u'_j \mathbf{e}'_i \,.$$

Using the vector transformation rule (6.1) in the second of the above relations gives us

$$A_{ik} v_k = S'_{ij} u'_j \qquad \Longrightarrow \qquad A_{ik} S_{kl} u_l = S'_{ij} u'_j \qquad \Longrightarrow \qquad A_{ik} S_{kl} A^T_{lj} u'_j = S'_{ij} u'_j \,.$$

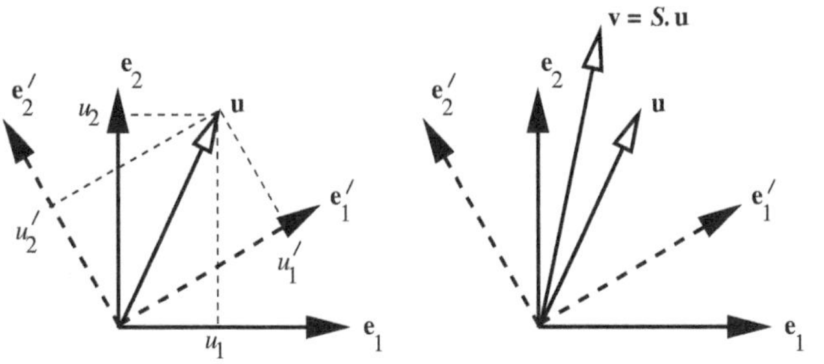

FIGURE 6.1
Components of a vector in two different Cartesian coordinate systems. The transformation rule for vectors gives us a relation between these components.

Therefore the second-order tensor transformation rule is

$$S'_{ij} = A_{ik} S_{kl} A^T_{lj} . \tag{6.2}$$

Similar relations can be derived for higher-order tensors. What about curvilinear coordinate systems?

Let us consider two curvilinear coordinate systems θ^i and $\theta^{i'}$ which are related to each other by

$$\theta'^i \equiv \theta'^i(\theta^1, \theta^2, \theta^3) \qquad \text{and} \qquad \theta^i \equiv \theta^i(\theta'^1, \theta'^2, \theta'^3) .$$

The Jacobian matrix of the transformation and its inverse are

$$F^i_{.j} = \frac{\partial \theta'^i}{\partial \theta^j} \qquad \text{and} \qquad (F^{-1})^i_{.j} = \frac{\partial \theta^i}{\partial \theta'^j} .$$

The natural basis vectors at the point $\mathbf{x}$ are (see Figure 6.2)

$$\mathbf{g}_i = \frac{\partial \mathbf{x}}{\partial \theta^i} \quad \text{and} \quad \mathbf{g}'_i = \frac{\partial \mathbf{x}}{\partial \theta'^i} .$$

Therefore, the transformation rules for the basis vectors are

$$\mathbf{g}_i = \frac{\partial \theta'^j}{\partial \theta^i} \frac{\partial \mathbf{x}}{\partial \theta'^j} = \frac{\partial \theta'^j}{\partial \theta^i} \mathbf{g}'_j \qquad \text{and} \qquad \mathbf{g}'_i = \frac{\partial \theta^j}{\partial \theta'^i} \mathbf{g}_j .$$

A vector $\mathbf{u}$ can be expressed in these bases as

$$\mathbf{u} = u^i \mathbf{g}_i = u'^i \mathbf{g}'_i = u'^i \frac{\partial \theta^j}{\partial \theta'^i} \mathbf{g}_j .$$

Therefore,

$$u^i \mathbf{g}_i \cdot \mathbf{g}^k = u'^i \frac{\partial \theta^j}{\partial \theta'^i} \mathbf{g}_j \cdot \mathbf{g}^k \qquad \Longrightarrow \qquad u^k = u'^i \frac{\partial \theta^k}{\partial \theta'^i} .$$

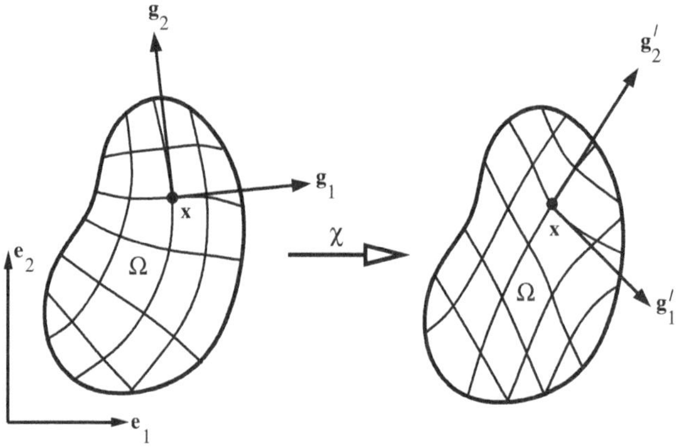

FIGURE 6.2
Transformation between two sets of curvilinear coordinates for the same region.

Similarly, we can show that,

$$u'^i = u^j \frac{\partial \theta'^i}{\partial \theta^j} = F^i_{\cdot j}\, u^j . \tag{6.3}$$

This is the transformation rule for vectors in curvilinear coordinates. For second-order tensor fields, a similar process leads to the relation

$$S'^{ij} = \frac{\partial \theta'^i}{\partial \theta^k} \frac{\partial \theta'^j}{\partial \theta^p} S^{kp} = F^i_{\cdot k} F^j_{\cdot p} S^{kp} . \tag{6.4}$$

Now consider the mapping shown in Figure 6.3. This mapping takes points $\mathbf{X}$ in the region Ω to points $\mathbf{x}$ in Ω_x.* In elastodynamics, such a mapping is expressed in the form $\mathbf{x} = \boldsymbol{\chi}(\mathbf{X})$. The mapping is one to one and invertible except, possibly, for a small number of points. If the points $\mathbf{X}$ are mapped to $\mathbf{x}$ by a linear transformation $\boldsymbol{F}$, we can write

$$\mathbf{x} = \boldsymbol{F} \cdot \mathbf{X} .$$

The quantity $\boldsymbol{F}$ is called the deformation gradient and, for sufficiently smooth transformations, is given by

$$\boldsymbol{F} = \frac{\partial \mathbf{x}}{\partial \mathbf{X}} = \boldsymbol{\nabla}_X \mathbf{x} .$$

The Jacobian of the transformation $\mathbf{X} \to \mathbf{x}$ relates infinitesimal volume elements in Ω to volume elements in Ω_x and is written as

$$J := \det \boldsymbol{F} \qquad \Longrightarrow \qquad d\Omega_x = J\, d\Omega .$$

*See Ogden (1997) for more details on such transformations in the context of nonlinear elasticity. Keep in mind that in this section Ω_x represents a configuration of the body and not the frequency.

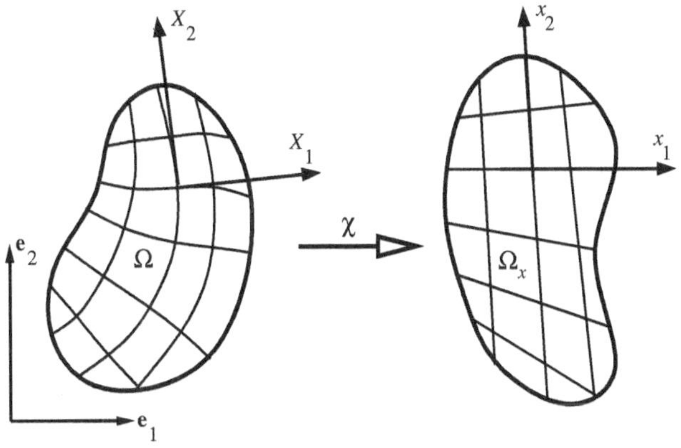

FIGURE 6.3
Transformation from a reference to a deformed configuration.

In terms of components in rectangular Cartesian coordinates, we can write

$$x_i = F_{iJ} X_J \; ; \;\; F_{iJ} = \frac{\partial x_i}{\partial X_J} = x_{i,J} \; ; \;\; J = e_{ijk} F_{i1} F_{j2} F_{k3} \, .$$

The special situation $\det \boldsymbol{F} = 1$ corresponds to volume preserving motions which include shear and rigid body rotations and translations. If we also have $\boldsymbol{F}^T = \boldsymbol{F}^{-1}$ then motion is an orthogonal transformation and is usually written as $\boldsymbol{Q}$. The form of the transformation also suggests that such deformations may be treated in a manner analogous to curvilinear deformations, i.e., equations (6.3) and (6.4) hold.

6.2 Cloaking of electrical conductivity

The problem of cloaking in electrical conductivity derives from the inverse problem of trying to find the internal structure of a body using electrical impedance tomography (Greenleaf et al., 2003a). One finds that the anisotropic conductivity of a medium cannot be found uniquely from measurements of voltages and currents at the boundary. This is because any smooth spatial transformation that changes the geometry of the interior without changing the boundary cannot be distinguished from another that similarly preserves the boundary. In this section we describe how such a transformation works.

6.2.1 Coordinate transformation for electrical conductivity

Let us consider a mapping that takes a reference configuration of a body ($\mathbf{X}$) to a deformed configuration ($\mathbf{x}$). The Jacobian of the transformation $\mathbf{X} \to \mathbf{x}$ is given by

$$J = \det(\boldsymbol{F}) \; ; \; F_{ij} := \frac{\partial x_i}{\partial X_j} \, .$$

Recall from equation (1.108) (p. 42) that the power dissipated by an electrically conducting body Ω in the reference configuration can be expressed as

$$W(\Phi) = \int_{\Omega} \boldsymbol{\nabla}_{\mathbf{X}} \Phi \cdot \boldsymbol{\Sigma} \cdot \boldsymbol{\nabla}_{\mathbf{X}} \Phi \mathrm{d}\Omega$$

where the electrical field is derived from the potential Φ via $\mathbf{E} = -\boldsymbol{\nabla}_{\mathbf{X}} \Phi$, $\boldsymbol{\Sigma}$ is the second-order electrical conductivity tensor, and $\boldsymbol{\nabla}_{\mathbf{X}}(\bullet)$ denotes a gradient with respect to the $\mathbf{X}$-coordinates. In terms of components with respect to a rectangular Cartesian basis,

$$W(\Phi) = \int_{\Omega} \frac{\partial \Phi}{\partial X_i} \, \Sigma_{ij} \, \frac{\partial \Phi}{\partial X_j} \, \mathrm{d}\Omega \, .$$

To get the expression for W in the deformation configuration we use the chain rule and the relation $\mathrm{d}\Omega_x = J\mathrm{d}\Omega$ to get

$$W(\phi) = \int_{\Omega_x} \left(\frac{\partial \phi}{\partial x_m} \frac{\partial x_m}{\partial X_i} \right) \Sigma_{ij} \left(\frac{\partial x_l}{\partial X_j} \frac{\partial \phi}{\partial x_l} \right) \frac{1}{J} \mathrm{d}\Omega_x$$

or

$$W(\phi) = \int_{\Omega_x} \frac{\partial \phi}{\partial x_m} \, \sigma_{ml} \, \frac{\partial \phi}{\partial x_l} \, \mathrm{d}\Omega_x$$

where

$$\sigma_{ml} = \frac{1}{J} \frac{\partial x_m}{\partial X_i} \, \Sigma_{ij} \, \frac{\partial x_l}{\partial X_j} = \frac{1}{J} F_{mi} \, \Sigma_{ij} \, F_{lj} \, .$$

Hence, in the transformed coordinates, the functional $W(\phi)$ takes the form

$$W(\phi) = \int_{\Omega_x} \boldsymbol{\nabla}_{\mathbf{x}} \phi \cdot \boldsymbol{\sigma} \cdot \boldsymbol{\nabla}_{\mathbf{x}} \phi \, \mathrm{d}\Omega_x \tag{6.5}$$

where $\boldsymbol{\nabla}_{\mathbf{x}}(\bullet)$ denotes a gradient with respect to the $\mathbf{x}$-coordinates. In direct notation we can write the transformation rule for the conductivity tensor as

$$\boxed{\boldsymbol{\sigma}(\mathbf{x}) = \frac{1}{J} \boldsymbol{F}(\mathbf{X}) \cdot \boldsymbol{\Sigma}(\mathbf{X}) \cdot \boldsymbol{F}^T(\mathbf{X}) \, .} \tag{6.6}$$

The form of equation (6.5) suggests that the function $\phi(\mathbf{x})$ minimizes W in a body Ω_x containing material with conductivity $\boldsymbol{\sigma}(\mathbf{x})$.[†] Therefore, for W to remain positive

[†]Note that an isotropic material transforms to an anisotropic material via the transformation equation (6.6) for conductivity.

(see Section 1.4.9), we must have

$$\mathbf{j}(\mathbf{x}) = \boldsymbol{\sigma}(\mathbf{x}) \cdot \boldsymbol{\nabla}_{\mathbf{x}} \phi(\mathbf{x}) = \frac{1}{J} [\boldsymbol{F}(\mathbf{X}) \cdot \boldsymbol{\Sigma}(\mathbf{X}) \cdot \boldsymbol{F}^T(\mathbf{X})] \cdot \boldsymbol{\nabla}_{\mathbf{x}} \phi(\mathbf{x}) \,.$$

Now, in Cartesian coordinates,

$$\boldsymbol{\nabla}_{\mathbf{X}} \Phi(\mathbf{X}) = \frac{\partial \Phi}{\partial X_i} \mathbf{e}_i = \frac{\partial x_m}{\partial X_i} \frac{\partial \phi}{\partial x_m} \mathbf{e}_i = F_{mi} \frac{\partial \phi}{\partial x_m} \mathbf{e}_i = \boldsymbol{F}^T(\mathbf{X}) \cdot \boldsymbol{\nabla}_{\mathbf{x}} \phi(\mathbf{x}) \,.$$

Hence,

$$\mathbf{j}(\mathbf{x}) = \frac{1}{J} \boldsymbol{F}(\mathbf{X}) \cdot \boldsymbol{\Sigma}(\mathbf{X}) \cdot \boldsymbol{\nabla}_{\mathbf{X}} \Phi(\mathbf{X}) = \frac{1}{J} \boldsymbol{F}(\mathbf{X}) \cdot \mathbf{J}(\mathbf{X}) \,.$$

or,

$$\boxed{\mathbf{j}(\mathbf{x}) = \frac{\boldsymbol{F} \cdot \mathbf{J}(\mathbf{X})}{\det(\boldsymbol{F})} \,.} \tag{6.7}$$

This is the transformation rule for currents. Using the same arguments as e used in Section 1.4.9, we can show that

$$\boxed{\boldsymbol{\nabla}_{\mathbf{x}} \cdot \mathbf{j}(\mathbf{x}) = 0 \,.} \tag{6.8}$$

Also, since the electric field $\mathbf{E}$ is derived from the potential Φ, the fields $\mathbf{E}(\mathbf{X}) = \boldsymbol{\nabla}_{\mathbf{X}} \Phi(\mathbf{X})$ and $\mathbf{e}(\mathbf{x}) = \boldsymbol{\nabla}_{\mathbf{x}} \phi(\mathbf{x})$ are related via

$$\boxed{\mathbf{e}(\mathbf{x}) = (\boldsymbol{F}^T)^{-1} \cdot \mathbf{E}(\mathbf{X}) \,.} \tag{6.9}$$

The two transformation rules (for the current and for the electric field) are equivalent. We can show that identical relations are obtained if we work in curvilinear coordinates rather than Cartesian coordinates.

6.2.2 Electrical tomography

Consider the body Ω with boundary Γ shown in Figure 6.4. Let the conductivity of the body be $\boldsymbol{\sigma}(\mathbf{x})$ and let us require that $\boldsymbol{\nabla} \cdot (\boldsymbol{\sigma} \cdot \boldsymbol{\nabla} \phi) = 0$ inside the body. In electrical tomography one measures the current flux $\mathbf{n} \cdot \mathbf{J}(\mathbf{x})$ at the surface for all choices of the potential ϕ_0.

Suppose one knows the Dirchlet to Neumann map (φ)

$$\varphi : \phi_0 \rightarrow \mathbf{n} \cdot \mathbf{J}(\mathbf{x}) = g(\mathbf{x}) \,.$$

Can one find $\boldsymbol{\sigma}(\mathbf{x})$ uniquely if one knows the boundary conditions at all points on the boundary? Greenleaf et al. (2003a) showed, using a counterexample, that even though there were cases in which it was possible to find the conductivity of the interior, in general $\boldsymbol{\sigma}(\mathbf{x})$ could not be determined uniquely. Figure 6.5 illustrates why that is the case. For the body in the figure, the transformation is $\mathbf{x} = \mathbf{X}$ outside the gray region while inside the gray region $\mathbf{x} \neq \mathbf{X}$. Also, outside the gray region, $\Omega_x = \Omega$, $\mathbf{j} =$

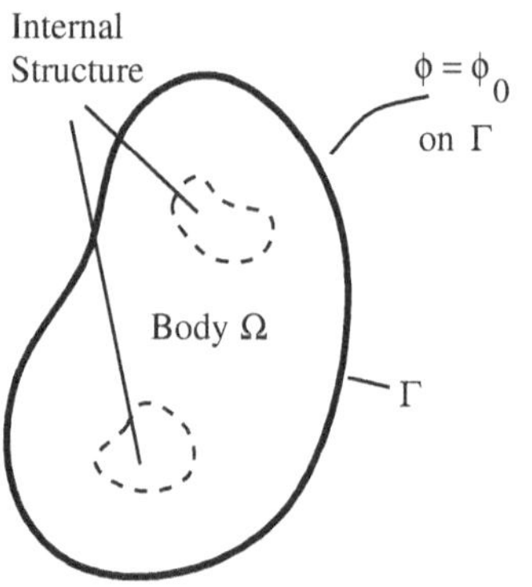

FIGURE 6.4
A region Ω that has a boundary Γ and a specified potential $\phi = \phi_0$ on the boundary.

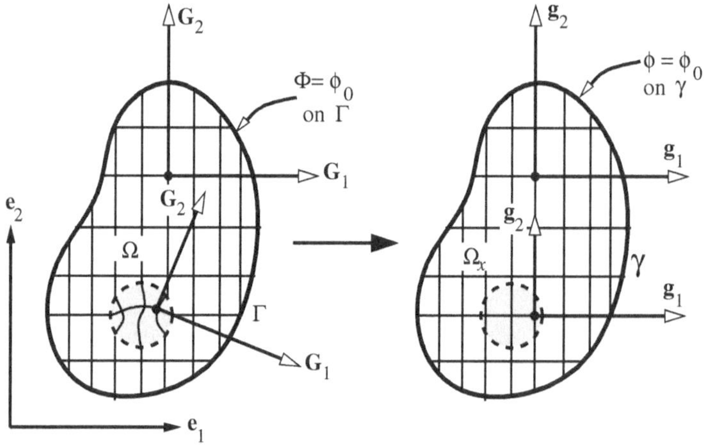

FIGURE 6.5
Illustration of why the Dirchlet to Neumann map on the surface may not, in general, be used to determine the conductivity inside a body.

$\mathbf{J}$, and $\phi = \Phi$. Inside the gray region $\Omega_x \neq \Omega$ and $\mathbf{j}$ is obtained via the transformation rule.

From the figure we can see that the Dirichlet-Neumman map will remain unchanged on the boundary Γ. Hence, the body appears to be exactly the same in $\mathbf{x}$-space as in $\mathbf{X}$-space, but has a different conductivity. Even though this fact has been known for a while, there was still hope that you could determine $\boldsymbol{\sigma}(\mathbf{x})$ uniquely, modulo a coordinate transformation. However, such hopes were dashed when Greenleaf, Lassas, and Uhlmann provided their counterexample in 2003.

6.2.3 Examples of transformation-based cloaking

Greenleaf et al. (Greenleaf et al., 2003b) provided the first example of transformation-based cloaking. They considered a singular transformation

$$\mathbf{x}(\mathbf{X}) = \begin{cases} \left(\dfrac{|\mathbf{X}|}{2}+1\right)\dfrac{\mathbf{X}}{|\mathbf{X}|} & \text{if } |\mathbf{X}| < 2 \\ \mathbf{X} & \text{if } |\mathbf{X}| > 2\,. \end{cases} \tag{6.10}$$

The effect of this mapping is shown in the schematic in Figure 6.6(a). An epsilon ball at the center of Ω is mapped into a sphere of radius 1 in Ω_x. The value of $\boldsymbol{\sigma}(\mathbf{x})$ is singular at the boundary of this sphere. Inside the sphere of radius 1, the transformed conductivity has the form $\boldsymbol{\sigma}(\mathbf{x}) = h(\mathbf{x})$. Therefore we can put a small body inside and the potential outside will be undisturbed by the presence of the body in the cloaking region.

Another unusual mapping that can be used to achieve cloaking is to fold back space upon itself (Leonhardt and Philbin, 2006, Milton et al., 2008). An example of such a mapping is

$$\mathbf{x} = \begin{cases} \mathbf{X} & \text{if } X_1 < 0 \\ (-X_1, X_2, X_3) & \text{if } d > X_1 > 0 \\ \mathbf{X} - (2d, 0, 0) & \text{if } X_1 > d\,. \end{cases} \tag{6.11}$$

The effect of this folding transformation is shown in Figure 6.6(b). Note that there is a sharp (discontinuous) fold and the separation shown in the thickness direction is simply for the purpose of illustration. In reality, space is folded upon itself and the determination of the Jacobian inside the fold is -1.

6.3 Cloaking for electromagnetism

Pendry, Schurig, and Smith (Pendry et al., 2006) were the first to show that cloaking could be achieved for electromagnetic waves. Their concept of cloaking follows from the observation that Maxwell's equations keep their form under coordinate transformations. This observation, and the extended range of permittivity and permeability made possible by the development of metamaterials, has led to the creation of a new field of study called *transformation optics*. A review of recent developments in the field can be found in Chen et al. (2010).

The Maxwell's equations at fixed frequency (ω) are

$$\boldsymbol{\nabla}_X \times \mathbf{E} + i\omega\boldsymbol{\mu}\cdot\mathbf{H} = \mathbf{0} \quad \text{and} \quad \boldsymbol{\nabla}_X \times \mathbf{H} - i\omega\boldsymbol{\varepsilon}\cdot\mathbf{E} = \mathbf{0}$$

where the spatial variable has been assumed to be $\mathbf{X}$ and the gradient operator is to be interpreted accordingly. A coordinate transformation ($\mathbf{X} \rightarrow \mathbf{x}$) gives us the equivalent

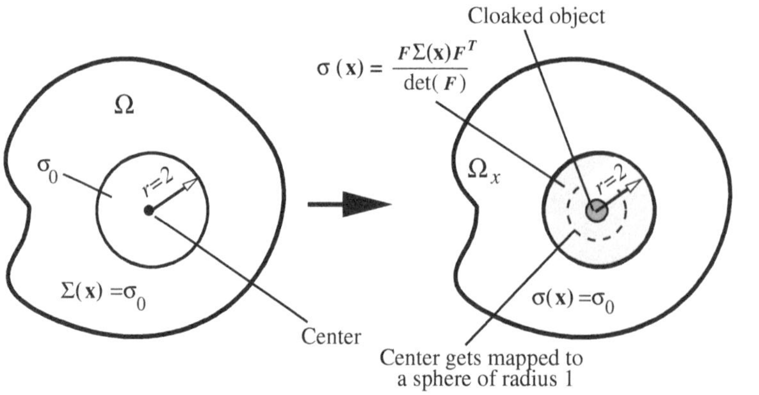

(a) Greenleaf-Lassas-Uhlman map.

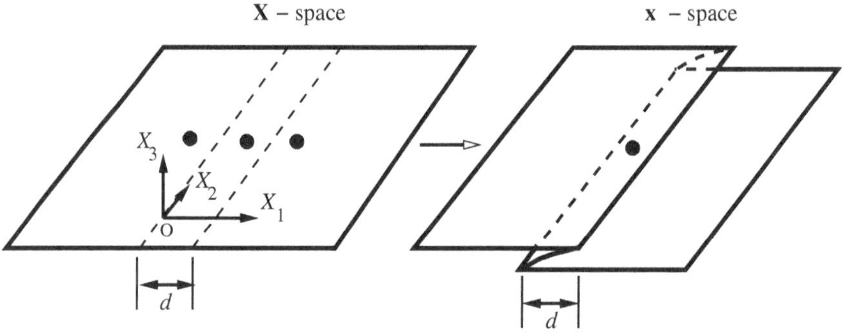

(b) A space folding map.

FIGURE 6.6
Examples of transformations that can be used for cloaking and lensing.

relations

$$\nabla_{\mathbf{x}} \times \mathbf{e} + i\omega\boldsymbol{\mu}' \cdot \mathbf{h} = \mathbf{0} \quad \text{and} \quad \nabla_{\mathbf{x}} \times \mathbf{h} - i\omega\boldsymbol{\varepsilon}' \cdot \mathbf{e} = \mathbf{0}$$

with $\mathbf{x} = \mathbf{x}(\mathbf{X})$. The vector fields $\mathbf{E}$ and $\mathbf{H}$ transform as[‡]

$$\mathbf{e}(\mathbf{x}) = (\boldsymbol{F}^T)^{-1} \cdot \mathbf{E}(\mathbf{X}) \,, \quad \mathbf{h}(\mathbf{x}) = (\boldsymbol{F}^T)^{-1} \cdot \mathbf{H}(\mathbf{X})$$

[‡]Notice that these are pull-back transformations. See Marsden and Ratiu (1999) (p. 71) for a concise definition of pull-back and push-forward operations and the conditions under which such transformations are valid operations.

where $\boldsymbol{F} = \partial\mathbf{x}/\partial\mathbf{X}$ and $\boldsymbol{F}^{-1} = \partial\mathbf{X}/\partial\mathbf{x}$. The material properties transform as

$$\boxed{\boldsymbol{\mu}'(\mathbf{x}) = \frac{\boldsymbol{F}\cdot\boldsymbol{\mu}(\mathbf{X})\cdot\boldsymbol{F}^T}{\det(\boldsymbol{F})}\,;\;\; \boldsymbol{\varepsilon}'(\mathbf{x}) = \frac{\boldsymbol{F}\cdot\boldsymbol{\varepsilon}(\mathbf{X})\cdot\boldsymbol{F}^T}{\det(\boldsymbol{F})}.} \tag{6.12}$$

When the reference permittivity and magnetic permeability are isotropic, i.e., $\boldsymbol{\mu}(\mathbf{X}) = \mu(\mathbf{X})\boldsymbol{1}$ and $\boldsymbol{\varepsilon}(\mathbf{X}) = \varepsilon(\mathbf{X})\boldsymbol{1}$, the transformation relations are[§]

$$\boxed{\boldsymbol{\mu}'(\mathbf{x}) = \mu(\mathbf{X})\frac{\boldsymbol{F}\cdot\boldsymbol{F}^T}{\det(\boldsymbol{F})}\,;\;\; \boldsymbol{\varepsilon}'(\mathbf{x}) = \varepsilon(\mathbf{X})\frac{\boldsymbol{F}\cdot\boldsymbol{F}^T}{\det(\boldsymbol{F})}.}$$

To see that the invariance of form of Maxwell's equations under coordinate transformations does indeed hold, observe that

$$-i\omega\boldsymbol{\mu}'\cdot\mathbf{h} = \frac{-i\omega\boldsymbol{F}\cdot\boldsymbol{\mu}\cdot\mathbf{H}}{\det(\boldsymbol{F})} = \frac{\boldsymbol{F}\cdot(\boldsymbol{\nabla}_{\mathbf{X}}\times\mathbf{E})}{\det(\boldsymbol{F})}. \tag{6.13}$$

Let us work in rectangular Cartesian coordinates to show that equation (6.13) equals $\boldsymbol{\nabla}_{\mathbf{x}}\times\mathbf{e}$. In index notation with respect to a Cartesian basis $(\mathbf{e}_1,\mathbf{e}_2,\mathbf{e}_3)$ (not to be confused with the transformed electric field, $\mathbf{e}$), equation (6.13) can be written as

$$-i\omega\boldsymbol{\mu}'\cdot\mathbf{h} = \frac{1}{\det(\boldsymbol{F})}\frac{\partial x_p}{\partial X_j}e_{jmk}\frac{\partial E_k}{\partial X_m}\mathbf{e}_p = \frac{1}{\det(\boldsymbol{F})}e_{jmk}\frac{\partial x_p}{\partial X_j}\frac{\partial x_\ell}{\partial X_m}\frac{\partial E_k}{\partial x_\ell}\mathbf{e}_p$$

where e_{ijk} is the permutation symbol. On the other hand,

$$\begin{aligned}\boldsymbol{\nabla}_{\mathbf{x}}\times\mathbf{e} = \boldsymbol{\nabla}_{\mathbf{x}}\times[(\boldsymbol{F}^T)^{-1}\cdot\mathbf{E}] &= e_{p\ell m}\frac{\partial}{\partial x_\ell}\left(\frac{\partial X_k}{\partial x_m}E_k\right)\mathbf{e}_p\\ &= \underbrace{e_{p\ell m}\frac{\partial^2 X_k}{\partial x_\ell\partial x_m}}_{0}E_k\,\mathbf{e}_p + e_{p\ell m}\frac{\partial X_k}{\partial x_m}\frac{\partial E_k}{\partial x_\ell}\mathbf{e}_p.\end{aligned}$$

For the first term above, $[\boldsymbol{\nabla}_{\mathbf{x}}\times(\boldsymbol{F}^T)^{-1}]\cdot\mathbf{E}$ is zero because of an important identity which states that if a second-order tensor is derived from the gradient of a vector, $\mathbf{x}$, then $\boldsymbol{\nabla}\times(\boldsymbol{\nabla}\mathbf{x}) = \mathbf{0}$. Note that the curl of a second-order tensor is also a second-order tensor.

Now we have to show that

$$e_{p\ell m}\frac{\partial X_k}{\partial x_m}\frac{\partial E_k}{\partial x_\ell} = \frac{1}{\det(\boldsymbol{F})}e_{jmk}\frac{\partial x_p}{\partial X_j}\frac{\partial x_\ell}{\partial X_m}\frac{\partial E_k}{\partial x_\ell}$$

or that

$$\det(\boldsymbol{F})\,e_{p\ell m}\frac{\partial X_k}{\partial x_m} = e_{jmk}\frac{\partial x_p}{\partial X_j}\frac{\partial x_\ell}{\partial X_m}. \tag{6.14}$$

[§]The simpler form of the transformation relations has been used widely because it is easier to begin a design with an isotropic material.

Multiplying both sides of (6.14) by $\boldsymbol{F}$ and summing over k gives

$$\det(\boldsymbol{F})\, e_{p\ell m}\, \frac{\partial X_k}{\partial x_m}\, \frac{\partial x_q}{\partial X_k} = e_{jmk}\, \frac{\partial x_p}{\partial X_j}\, \frac{\partial x_\ell}{\partial X_m}\, \frac{\partial x_q}{\partial X_k}$$

or

$$\det(\boldsymbol{F})\, e_{p\ell m}\, \delta_{qm} = \det(\boldsymbol{F})\, e_{p\ell q} = e_{jmk}\, \frac{\partial x_p}{\partial X_j}\, \frac{\partial x_\ell}{\partial X_m}\, \frac{\partial x_q}{\partial X_k}\,.$$

Therefore,

$$\det(\boldsymbol{F})\, e_{p\ell q} = e_{jmk} F_{pj} F_{\ell m} F_{qk}\,. \tag{6.15}$$

But, from the expression for the determinant of a second-order tensor (see Ogden (1997)),

$$\det \boldsymbol{F} = e_{ijk} F_{1i} F_{2j} F_{3k} = e_{ijk} F_{i1} F_{j2} F_{k3}$$

we can deduce directly that

$$e_{ijk} F_{pi} F_{qj} F_{rk} = \det(\boldsymbol{F})\, e_{pqr}\,.$$

Therefore, equation (6.15) is an identity, i.e.,

$$\boldsymbol{\nabla}_{\mathbf{x}} \times \mathbf{e} = -i\omega \boldsymbol{\mu}' \cdot \mathbf{h}\,.$$

This means that the first of the transformed Maxwell equations holds. We can follow the same procedure to show that the second Maxwell's equation also maintains its form under coordinate transformations. Therefore, Maxwell's equations are invariant with respect to coordinate transformations. It can be shown that the same invariance holds in spatial curvilinear coordinates or space-time coordinates (see Leonhardt and Philbin (2006)).

Pendry et al. (2006) observed that a material with spatially varying permittivity and magnetic permeability could be designed using equations (6.12) as the basis. A judicious choice of the transformation $\boldsymbol{F}$ could then be chosen such that the resulting material would guide incident electromagnetic waves around a region in space. An example of such a cloak is shown in Figure 6.7. The original Pendry et al. cloaking transformation, where a concentric spherical (or cylindrical) cloaking region with radially varying properties is mapped to the surface of a sphere, has been quite popular because of its simplicity. The situation is similar to that shown in Figure 6.7. The transformation is, in spherical coordinates ($R \to r, \Theta \to \theta, \Phi \to \phi$),

$$r = R_1 + t R\,, \quad \theta = \Theta\,, \quad \phi = \Phi \tag{6.16}$$

where $t := (R_2 - R_1)/R_2$. R_1 is the internal radius of the cloak and R_2 is the external radius of the cloak.

Let us use curvilinear coordinates to find out the expression for $\boldsymbol{F}$ for this situation. A similar approach can be used for more complicated maps. The relation between Cartesian and spherical coordinates is assumed to be

$$\mathbf{x} \equiv \mathbf{x}(\theta^1, \theta^2, \theta^3) \quad \text{where} \quad (\theta^1, \theta^2, \theta^3) \equiv (r, \theta, \phi)\,.$$

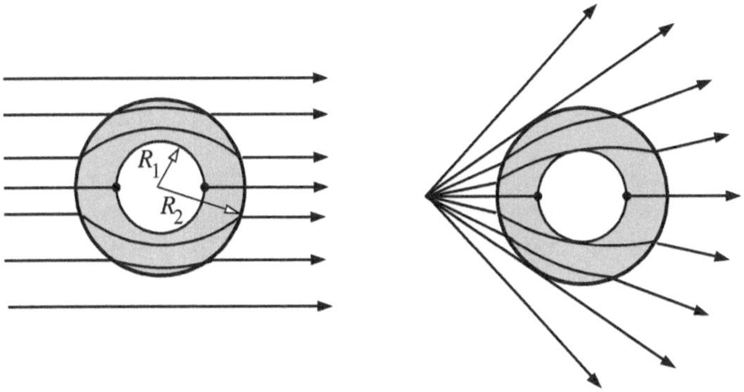

FIGURE 6.7
Schematic of the path of rays in a cloak around an object. Notice that to achieve a perfect cloak we will need to bend the rays significantly around the object. This implies that we will need a composite with a refractive index that varies strongly with spatial position. Notice also that the difference in the lengths of ray paths around the object implies that there will be phase differences in the observed field. For certain frequencies, a conformal map reduces those differences (Leonhardt, 2006a). Schematic adapted from Pendry et al. (2006).

In explicit form,

$$x_1 = r\cos\theta\sin\phi\,,\quad x_2 = r\sin\theta\sin\phi\,,\quad x_3 = r\cos\phi\,.$$

If $(\mathbf{e}_1, \mathbf{e}_2, \mathbf{e}_3)$ is a background Cartesian frame, the orthogonal basis vectors for the spherical coordinate system are determined from

$$\begin{aligned}
\mathbf{g}_1 &= \frac{\partial\mathbf{x}}{\partial\theta^1} = \cos\theta\sin\phi\,\mathbf{e}_1 + \sin\theta\sin\phi\,\mathbf{e}_2 + \cos\phi\,\mathbf{e}_3 = \mathbf{e}_r \\
\mathbf{g}_2 &= \frac{\partial\mathbf{x}}{\partial\theta} = -r\sin\theta\sin\phi\,\mathbf{e}_1 + r\cos\theta\sin\phi\,\mathbf{e}_2 = r\sin\phi\,\mathbf{e}_\theta \\
\mathbf{g}_3 &= \frac{\partial\mathbf{x}}{\partial\phi} = r\cos\theta\cos\phi\,\mathbf{e}_1 + r\sin\theta\cos\phi\,\mathbf{e}_2 - r\sin\phi\,\mathbf{e}_3 = r\mathbf{e}_\phi\,.
\end{aligned}$$

We have normalized the covariant basis vectors to get the spherical basis vectors $(\mathbf{e}_r, \mathbf{e}_\theta, \mathbf{e}_\phi)$. Using $\mathbf{g}_i \cdot \mathbf{g}^j = \delta_i^j$ and solving for the components of $\mathbf{g}^j$, we have

$$\begin{aligned}
\mathbf{g}^1 &= \cos\theta\sin\phi\,\mathbf{e}_1 + \sin\theta\sin\phi\,\mathbf{e}_2 + \cos\phi\,\mathbf{e}_3 = \mathbf{e}_r \\
\mathbf{g}^2 &= -\frac{1}{r}\frac{\sin\theta}{\sin\phi}\mathbf{e}_1 + \frac{1}{r}\frac{\cos\theta}{\sin\phi}\mathbf{e}_2 = \frac{1}{r\sin\phi}\mathbf{e}_\theta \\
\mathbf{g}^3 &= \frac{1}{r}\cos\theta\cos\phi\,\mathbf{e}_1 + \frac{1}{r}\sin\theta\cos\phi\,\mathbf{e}_2 - \frac{1}{r}\sin\phi\,\mathbf{e}_3 = \frac{1}{r}\mathbf{e}_\phi\,.
\end{aligned} \tag{6.17}$$

From the definition of the gradient of a scalar function and using the above relations between $\mathbf{g}^i$ and $(\mathbf{e}_r, \mathbf{e}_\theta, \mathbf{e}_\phi)$ we get

$$\nabla f = \frac{\partial f}{\partial \theta^i}\mathbf{g}^i = \frac{\partial f}{\partial r}\mathbf{e}_r + \frac{1}{r\sin\phi}\frac{\partial f}{\partial \theta}\mathbf{e}_\theta + \frac{1}{r}\frac{\partial f}{\partial \phi}\mathbf{e}_\phi .$$

The above expression gives us a definition of the gradient operator in spherical coordinates which we can now use to calculate the gradient of $\mathbf{x}$ with respect to $\mathbf{X}$. Therefore, we have (see Bower (2010) for further details)

$$\boldsymbol{F} = \nabla\mathbf{x} = (x_r\mathbf{e}_r + x_\theta\mathbf{e}_\theta + x_\phi\mathbf{e}_\phi)\otimes\left(\mathbf{e}_R\frac{\partial}{\partial R} + \mathbf{e}_\Theta\frac{1}{R\sin\Phi}\frac{\partial}{\partial\Theta} + \mathbf{e}_\Phi\frac{1}{R}\frac{\partial}{\partial\Phi}\right) \tag{6.18}$$

where the uppercase letters represent quantities in the "reference" configuration while the lowercase letters represent the "deformed" configuration. Notice from equations (6.17) that

$$\begin{aligned}
&\frac{\partial\mathbf{e}_r}{\partial r} = \frac{\partial\mathbf{e}_\theta}{\partial r} = \frac{\partial\mathbf{e}_\phi}{\partial r} = \frac{\partial\mathbf{e}_\theta}{\partial\phi} = \mathbf{0}\,, \quad \frac{\partial\mathbf{e}_r}{\partial\phi} = \mathbf{e}_\phi\,, \quad \frac{\partial\mathbf{e}_\phi}{\partial\phi} = -\mathbf{e}_r \\
&\frac{\partial\mathbf{e}_r}{\partial\theta} = \sin\phi\,\mathbf{e}_\theta\,, \quad \frac{\partial\mathbf{e}_\phi}{\partial\theta} = \cos\phi\,\mathbf{e}_\theta\,, \quad \frac{\partial\mathbf{e}_\theta}{\partial\theta} = -\sin\phi\,\mathbf{e}_r - \cos\phi\,\mathbf{e}_\phi\,.
\end{aligned} \tag{6.19}$$

Let us now look at the case where $\mathbf{e}_r = \mathbf{e}_R$, $\mathbf{e}_\theta = \mathbf{e}_\Theta$, $\mathbf{e}_\phi = \mathbf{e}_\Phi$, and

$$\mathbf{x} = f(R)\mathbf{e}_R \quad \text{and} \quad \mathbf{X} = R\mathbf{e}_R\,.$$

Then, using (6.18) and (6.19), we have

$$\boldsymbol{F}_{RR} = \frac{\partial}{\partial R}(f\mathbf{e}_R)\otimes\mathbf{e}_R = \frac{df}{dR}\mathbf{e}_R\otimes\mathbf{e}_R\,;\ \boldsymbol{F}_{R\Theta} = \frac{1}{R\sin\phi}\frac{\partial}{\partial\Theta}(f\mathbf{e}_R)\otimes\mathbf{e}_\Theta = \frac{f}{R}\mathbf{e}_\Theta\otimes\mathbf{e}_\Theta$$

$$\boldsymbol{F}_{R\Phi} = \frac{1}{R}\frac{\partial}{\partial\Phi}(f\mathbf{e}_R)\otimes\mathbf{e}_\Phi = \frac{f}{R}\mathbf{e}_\Phi\otimes\mathbf{e}_\Phi\,.$$

In matrix form[¶]

$$\boldsymbol{F} \equiv \begin{bmatrix} \frac{df}{dR} & 0 & 0 \\ 0 & \frac{f}{R} & 0 \\ 0 & 0 & \frac{f}{R} \end{bmatrix}.$$

For the radial map in equation (6.16), we have

$$f(R) = r = R_1 + tR \quad \Longrightarrow \quad \frac{df}{dR} = t\,.$$

Therefore,

$$\boldsymbol{F}\cdot\boldsymbol{F}^T \equiv \begin{bmatrix} 0 & \frac{(R_1+tR)^2}{R^2} & 0 \\ 0 & 0 & \frac{(R_1+tR)^2}{R^2} \end{bmatrix} = \begin{bmatrix} t^2 & 0 & 0 \\ 0 & \frac{t^2r^2}{(r-R_1)^2} & 0 \\ 0 & 0 & \frac{t^2r^2}{(r-R_1)^2} \end{bmatrix} \tag{6.20}$$

[¶]Several other maps are discussed in Ogden (1997), pp. 98–118.

and

$$J = \det(\boldsymbol{F}) = \frac{t(R_1 + tR)^2}{R^2} = \frac{t^3 r^2}{(r - R_1)^2}. \tag{6.21}$$

Using the transformation rules in equation (6.12) gives us the material properties of the cloak as a function of the radius, r. Thus we have, for a homogeneous and isotropic reference medium with permittivity ε and permittivity μ,

$$\boldsymbol{\mu}'(r) \equiv \mu \begin{bmatrix} \frac{1}{t}\frac{(r-R_1)^2}{r^2} & 0 & 0 \\ 0 & \frac{1}{t} & 0 \\ 0 & 0 & \frac{1}{t} \end{bmatrix} \quad \text{and} \quad \boldsymbol{\varepsilon}'(r) \equiv \varepsilon \begin{bmatrix} \frac{1}{t}\frac{(r-R_1)^2}{r^2} & 0 & 0 \\ 0 & \frac{1}{t} & 0 \\ 0 & 0 & \frac{1}{t} \end{bmatrix}. \tag{6.22}$$

Anisotropy of the cloak region appears to be a necessity for perfect cloaking. For $r < R_1$, the medium can have any value of permittivity and permeability because the fields are guided around the region. At $r = R_1$ we need a material with $\mu'_{rr} = \varepsilon'_{rr} \approx 0$. At $r = R_2$ we have $\mu'_{rr} = t\mu$ and $\varepsilon'_{rr} = t\varepsilon$. Recall that the impedance is given by the ratio of the tangential components of the electrical and magnetic fields, i.e., $Z = \sqrt{\mu_{\theta\theta}/\varepsilon_{\theta\theta}} = \sqrt{\mu_{\phi\phi}/\varepsilon_{\phi\phi}}$. If $\mu = \mu_0$ and $\varepsilon = \varepsilon_0$ we have $Z = \sqrt{\mu_0/\varepsilon_0}$, i.e., the cloak is perfectly impedance matched with vacuum and there are no reflections from the cloak. The effectiveness of an electromagnetic cloak is limited by considerations of causality. However, such concerns are of less importance in the acoustic cloaks which we will discuss later. Numerical simulations of cylindrical cloaks that use the same type of transformation can be found in Cummer et al. (2006). Detailed single scattering calculations for a coated sphere (see, for example, Meng et al. (2009)) show that the scattered field intensity increases as the amount of loss in the material increases but can be reduced by making the coating thin.

Several other types of cloak based on the transformation of coordinates have been suggested since the initial proposal by Pendry et al. (2006). Examples include a "carpet cloak" (Li and Pendry, 2008), concentrators (Rahm et al., 2008, Luo et al., 2008), rotators (Chen and Chan, 2007b), and absorbers (Ng et al. (2009)). Many other applications of the transformation method can be envisaged.

6.3.1 The perfect lens

Let us now consider the effect on Maxwell's equations of the fold-back transformation shown in Figure 6.6(b). Recall that this transformation has the form

$$\mathbf{x} = \begin{cases} \mathbf{X} & \text{if } X_1 < 0 \\ (-X_1, X_2, X_3) & \text{if } d > X_1 > 0 \\ \mathbf{X} - (2d, 0, 0) & \text{if } X_1 > d\,. \end{cases}$$

Now, let us suppose that $\boldsymbol{\varepsilon}' = \boldsymbol{\mu}' = \boldsymbol{1}$ everywhere in the region. Let us continue to work with components in a Cartesian basis. Since the Jacobian of the transformation is $F_{ij} = \partial x_i/\partial X_j$, we have

$$\boldsymbol{F} = \boldsymbol{1} \quad \text{if } X_1 < 0 \text{ or } X_1 > d \qquad \text{and} \qquad \boldsymbol{F} \equiv \begin{bmatrix} -1 & 0 & 0 \\ 0 & 1 & 0 \\ 0 & 0 & 1 \end{bmatrix} \text{ if } d > X_1 > 0 \,.$$

This implies that in the region $d > X_1 > 0$ we have $\boldsymbol{F} \cdot \boldsymbol{F}^T = \boldsymbol{1}$ and $\det(\boldsymbol{F}) = -1$. Since the materials in the region are isotropic, i.e., $\boldsymbol{\mu} = \mu\boldsymbol{1}$ and $\boldsymbol{\varepsilon} = \varepsilon\boldsymbol{1}$, then from equation (6.12) we see that in the region $d > X_1 > 0$,

$$\mu'(\mathbf{x}) = -\mu \,; \;\; \varepsilon'(\mathbf{x}) = -\varepsilon \,.$$

If $\mu = \varepsilon = 1$, the fold-back transformation is realized in a geometry that is equivalent to the perfect lens that we discussed in Section 4.4 (p. 139). Figure 6.8 shows the geometry involved.

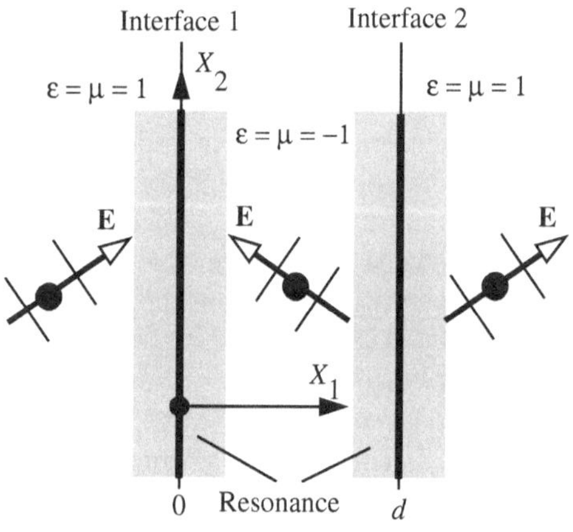

FIGURE 6.8

The "perfect" lens corresponds to a transformation that causes folding over of space. Compare this figure with Figure 6.6(b) to see the similarity between the two.

For a source that is less than a distance d from the first interface, the fields blow up to infinity and there is no solution. Recall that for a solution to exist, we need to regularize the problem and add a small loss δ, i.e., $\varepsilon = \mu = -1 + i\delta$ in the lens. In that case, the fields go to infinity in two strips of length $d - d_0$, where d_0 is the distance of the source from the first interface. The is the region of localized anomalous resonance. Outside this region, the fields converge to that expected by the Pendry solution (see Figure 4.10, p. 142, for a schematic.)

If $d - d_0 > d_0$, i.e., $d_0 < d/2$, then the sources will be in a region of enormous fields. In fact, the source produces infinite energy per unit time in such regions as the loss $\delta \rightarrow 0$. This is clearly unphysical. So any realistic point or line source with finite energy such as a polarizable particle must have an amplitude which goes to zero as $\delta \rightarrow 0$. This means that the particle will have become cloaked!

Several examples of this anomalous cloaking behavior can be found in Milton and Nicorovici (2006) and Milton et al. (2008) for polarizable dipoles close to a cylindrical negative index lens with a small loss. When the polarizable dipole is located close to the lens, the field is barely perturbed. However, when the dipole is at a distance from the lens, the field shows significant perturbations.

6.3.2 Magnification

So far we have not dealt with the issue of magnification. Is there a coordinate transformation that leads to magnification? One such possible transformation is illustrated in Figure 6.9.

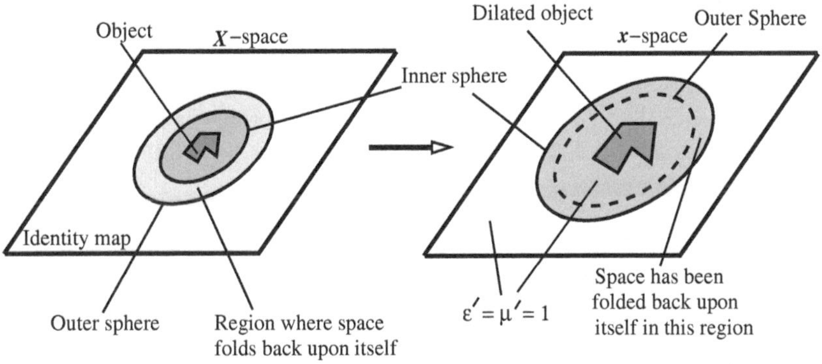

FIGURE 6.9
Magnification due to a folding-type coordinate transformation. This particular transformation was introduced by G. W. Milton.

In this case, in the region of dilation, the transformation is

$$\mathbf{x} = \beta\,\mathbf{X}\,;\ \ \beta > 1\,.$$

Therefore, the Jacobian of the transformation is

$$\boldsymbol{F} = \beta\,\boldsymbol{1}\,;\ \ \det(\boldsymbol{F}) = \beta^3\,.$$

Hence the material tensors in the region of dilation transform as

$$\boldsymbol{\varepsilon}' = \frac{\boldsymbol{\varepsilon}}{\beta}\,;\ \ \boldsymbol{\mu}' = \frac{\boldsymbol{\mu}}{\beta}\,. \tag{6.23}$$

However, in the folded region, the $\boldsymbol{\varepsilon}$ and $\boldsymbol{\mu}$ tensors are anisotropic and negative. Such a transformation acts like a magnifying lens. Similar transformations are discussed in Leonhardt and Philbin (2006), Schurig et al. (2007), Tsang and Psaltis (2008), and Milton et al. (2008).

6.4 Inertial acoustic cloaking

The acoustic cloaks discussed in this section involve anisotropic densities but isotropic bulk moduli. Norris (Norris, 2008) calls such cloaks inertial cloaks. A review of the developments in the field can be found in Chen and Chan (2010). We will now examine the basic theory behind transformation-based acoustics.

Recall from (1.50) (p. 21) that the acoustic wave equation can be expressed as

$$\boldsymbol{\nabla}\cdot(\boldsymbol{\rho}^{-1}\cdot\boldsymbol{\nabla}p)+\frac{\omega^2}{\kappa}p=0 \tag{6.24}$$

where $\boldsymbol{\rho}$ is a tensorial mass density, κ is a scalar bulk modulus, p is the acoustic pressure, and ω is the frequency. For two-dimensional problems in the x_2-x_3 plane, i.e., with x_1 invariance, we can express the above relation in the form

$$\overline{\boldsymbol{\nabla}}\cdot\left(\boldsymbol{\rho}^{-1}\cdot\overline{\boldsymbol{\nabla}}p\right)+\frac{\omega^2}{\kappa}p=0 \tag{6.25}$$

where $\overline{\boldsymbol{\nabla}}$ is the two-dimensional gradient operator. The quantity $\boldsymbol{\rho}$ in the above equation is two dimensional. Also recall from (1.95) (p. 38) that the TE-wave equation can be expressed as

$$\overline{\boldsymbol{\nabla}}\cdot\left[(\boldsymbol{R}_\perp\cdot\boldsymbol{M}^{-1}\cdot\boldsymbol{R}_\perp^T)\cdot\overline{\boldsymbol{\nabla}}E_1\right]+\omega^2\varepsilon_{11}E_1=0 \tag{6.26}$$

where $\boldsymbol{R}_\perp$ is a reflection operator and $\boldsymbol{M}$ is the two-dimensional tensor of magnetic permeability in the x_2-x_3 plane. We can see that (6.25) and (6.26) have identical forms with

$$p\leftrightarrow E_1\,,\quad \boldsymbol{\rho}^{-1}\leftrightarrow\boldsymbol{R}_\perp\cdot\boldsymbol{M}^{-1}\cdot\boldsymbol{R}_\perp\,,\quad \kappa\leftrightarrow 1/\varepsilon_{11}\,.$$

Since equation (6.26) is derived from Maxwell's equations, it must be invariant under coordinate transformations. The equivalence of (6.25) and (6.26) imply that the two-dimensional acoustic wave equation must also be invariant with respect to coordinate transformations and the material properties in the acoustic cloak also transform as

$$\boldsymbol{\rho}_x(\mathbf{x})=\frac{\boldsymbol{F}\cdot\boldsymbol{\rho}(\mathbf{X})\cdot\boldsymbol{F}^T}{\det(\boldsymbol{F})}\quad\text{and}\quad\frac{1}{\kappa_x(\mathbf{x})}=\frac{1}{\kappa(\mathbf{X})}\frac{\boldsymbol{F}\cdot\boldsymbol{F}^T}{\det(\boldsymbol{F})}. \tag{6.27}$$

This observation was used by Cummer and Schurig (2007) to explore a cylindrical acoustic cloak using the properties

$$\boldsymbol{\rho}\equiv\begin{bmatrix}\frac{r}{r-R_1} & 0\\ 0 & \frac{r-R_1}{r}\end{bmatrix}\quad\text{and}\quad\frac{1}{\kappa}=\frac{r-R_1}{t^2 r} \tag{6.28}$$

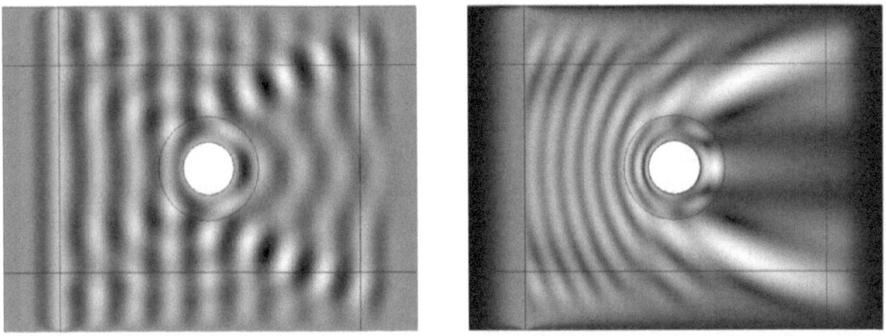

(a) Constant bulk-modulus coating (Left: $p(\mathbf{x})$ Right: $|p(\mathbf{x})|$).

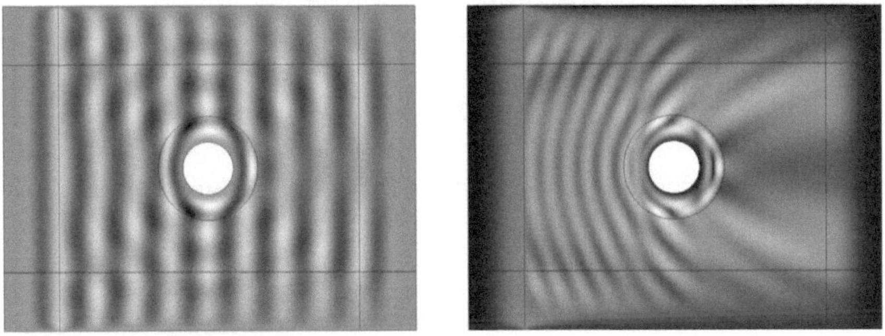

(b) Cummer-Schurig coating (Left: $p(\mathbf{x})$ Right: $|p(\mathbf{x})|$).

FIGURE 6.10

Acoustic pressure fields for two different coatings around a cylinder. In (a) the density varies according to (6.28) while the bulk modulus is constant. In (b) the full cloaking transformation given in (6.28) has been used. Simulations were performed by Bryan Smith using Comsol^{TM}.

where the quantities R_1 and t are equivalent to those discussed earlier in equation (6.16) for spherical electromagnetic cloaks.

Plots of the pressure field from finite element simulations of the above cloak is shown in Figure 6.10. If we look at the pressure fields on the images on the left, we see that the Cummer-Schurig cloak appears to distort the field far less than the constant bulk modulus coat. However, such plots can be deceptive and we also need to check the pressure amplitudes. For a perfect cloak we should see a constant value of the pressure amplitude in the region outside the cloak. Instead we observe a significant amount of back scatter from both coatings. The Cummer-Schurig cloak has a smaller shadow zone but the simulation does not indicate perfect cloaking. One of the reasons is that the cloaking transformation requires point-to-point variation for perfect cloaking. Therefore we need a large number of radial layers in the finite element model of the cloak. Numerical errors are also introduced by the boundary conditions. For instance, the incident plane wave is uniform and special considerations are needed in the absorbing layers around the boundary of the domain. The infinite mass that is needed as $r \to R_1$ (see Section 6.4.1) is also an issue. A discussion of some of these issues can be found in Cai and Sánchez-Dehesa (2007). From a practical point of view, Figure 6.10 indicates that the effectiveness of the cloak is quite sensitive to the material parameters.

Soon after Cummer and Schurig, Chen and Chan (2007a) showed that acoustic cloaking was not necessarily limited to two dimensions by appealing to the invariance of the three-dimensional electrical conductivity equations which have the form (see equation (1.106), p. 42)

$$\boldsymbol{\nabla}\cdot(\boldsymbol{\sigma}\cdot\boldsymbol{\nabla}\phi) - s = 0.$$

where $\boldsymbol{\sigma}$ is the electrical conductivity tensor, ϕ is the potential, and s is a current source. Comparing the above to the acoustic wave equation,

$$\boldsymbol{\nabla}\cdot(\boldsymbol{\rho}^{-1}\cdot\boldsymbol{\nabla}p) + \frac{\omega^2}{\kappa}p = 0$$

gives us the correspondence $\boldsymbol{\rho}^{-1} \leftrightarrow \boldsymbol{\sigma}$, $p \leftrightarrow \phi$, $\omega^2 p/\kappa \leftrightarrow -s$. Since the conductivity equations are invariant under the transformations

$$\boldsymbol{\sigma}(\mathbf{x}) = \frac{1}{J}\boldsymbol{F}\cdot\boldsymbol{\Sigma}(\mathbf{X})\cdot\boldsymbol{F}^T \quad \text{and} \quad s(\mathbf{x}) = \frac{1}{J}S(\mathbf{X})$$

we can explore an equivalent acoustic material with

$$\boldsymbol{\rho}^{-1}(\mathbf{x}) = \boldsymbol{\rho}_X^{-1}(\mathbf{X})\frac{\boldsymbol{F}\cdot\boldsymbol{F}^T}{J} \quad \text{and} \quad \kappa(\mathbf{x}) = J\kappa_X(\mathbf{X}). \tag{6.29}$$

Compare (6.27) and (6.29) to see the difference between the two transformations. For the Pendry et al. linear radial map, we have the relations (6.20) and (6.21) which can be used directly for the acoustic case. Chen and Chan (2007a) and Cummer et al. (2008) have shown analytically that such a cloak does not scatter incident plane waves.

An explanation of the cloaking effect in terms of the relation between material properties and the metric tensor can be found in Greenleaf et al. (2007). Recall that the equation for acoustic pressure in curvilinear coordinates ((1.45), p. 20), is given by

$$\frac{\partial^2 p}{\partial t^2} - \frac{c_0^2}{\sqrt{g}}\frac{\partial}{\partial\theta^i}\left[\sqrt{g}g^{ji}\frac{\partial p}{\partial\theta^j}\right] = 0$$

where $g = \det(g_{ij})$. For plane harmonic waves, and in the presence of a source (s), the above equation takes the form of a Helmholtz equation with source,

$$\frac{1}{\sqrt{g}}\frac{\partial}{\partial\theta^i}\left[\sqrt{g}g^{ji}\frac{\partial p}{\partial\theta^j}\right] + k^2 p = s \qquad \text{with} \qquad k^2 = \frac{\omega^2}{c_0^2}.$$

If we compare this relation with the acoustic wave equation for spatially varying κ and $\boldsymbol{\rho}$,

$$\kappa\boldsymbol{\nabla}\cdot\left(\boldsymbol{\rho}^{-1}\cdot\boldsymbol{\nabla}p\right)+\omega^2 p=0$$

we get the connection

$$\kappa\leftrightarrow\frac{1}{\sqrt{g}},\quad (\rho^{-1})^{ij}\leftrightarrow\sqrt{g}g^{ji},\quad k\leftrightarrow\omega .$$

Greenleaf et al. (2007) showed that cloaking could be achieved for electrostatics problems that satisfy the Helmholtz equation in curvilinear coordinates. Therefore we can design acoustic cloaks using the above connection if we know the metric tensor (g_{ij}) of the transformation (see Greenleaf et al. (2008) for an example). The connection between the SH-wave equation and the acoustic wave equation suggests that similar approaches can be used for cloaking elastodynamic SH-waves.

The transformations discussed so far are difficult to realize in practice and simpler designs have been proposed, for example, a multilayered cloak consisting of alternating layers of fluids composed of two primary materials (Torrent and Sánchez-Dehesa, 2008a). This approach requires the design of a large number (100–400) of separate structures to achieve the distribution of density and bulk modulus needed to approximate the Cummer-Schurig cloak. Norris and Nagy (2009) have devised an approach that reduces the number of required microstructures to three. An alternative approach based on a periodic arrangement of plates with infinite mass has been proposed by Pendry and Li (2008). Experimental validation of these ideas is yet to be performed.

The ideas behind the cloaking transforms discussed so far have been generalized by Norris (2008) who devised a transformation starting with the uniform wave equation (1.43) and comparing the result to the general acoustic wave equation (6.24). The uniform wave equation can be expressed as

$$\kappa_X\boldsymbol{\nabla}_{\mathbf{X}}\cdot\left[\rho_X\boldsymbol{\nabla}_{\mathbf{X}}p(\mathbf{X},t)\right]-\ddot{p}(\mathbf{X},t)=0\qquad\text{where}\quad\mathbf{X}\in\Omega \tag{6.30}$$

where κ_X,ρ_X are the spatially uniform bulk modulus and mass density, respectively. In the following discussion the units have been chosen so $\kappa_X=\rho_X=1$.

The goal is to devise a coating whose properties mimic that of the homogeneous medium that is described by (6.30). A schematic of the transformation is shown in Figure 6.11. We start with the Laplacian term in equation (6.30). The gradient term can be expressed in **x**-space by observing that

$$\boldsymbol{\nabla}_{\mathbf{X}}p\equiv\frac{\partial p}{\partial X_i}=\frac{\partial p}{\partial x_m}\frac{\partial x_m}{\partial X_i}\equiv\boldsymbol{F}^T\cdot\boldsymbol{\nabla}_{\mathbf{x}}p$$

and the divergence of the gradient term can be expressed in **x**-space using the relation

$$\boldsymbol{\nabla}_{\mathbf{X}}\cdot\mathbf{a}\equiv\frac{\partial a_i}{\partial X_i}=\frac{\partial a_i}{\partial x_m}\frac{\partial x_m}{\partial X_i}\equiv\boldsymbol{\nabla}_{\mathbf{x}}\mathbf{a}:\boldsymbol{F}^T$$

where **a** is a vector field. Then we can write

$$\boldsymbol{\nabla}_{\mathbf{X}}\cdot\left[\boldsymbol{\nabla}_{\mathbf{X}}p\right]=\boldsymbol{\nabla}_{\mathbf{x}}\left(\boldsymbol{F}^T\cdot\boldsymbol{\nabla}_{\mathbf{x}}p\right):\boldsymbol{F}^T . \tag{6.31}$$

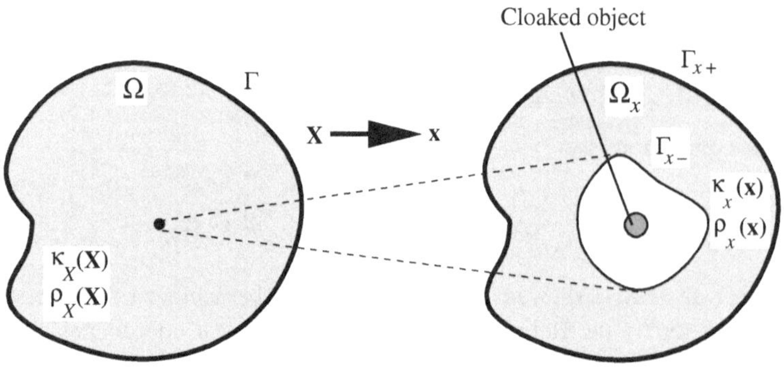

FIGURE 6.11
A singular cloaking transformation in acoustics. The simply connected region Ω is transformed into a multiply connected region Ω_x with inner boundary Γ_{x-} and outer boundary Γ_{x+}. The mapping is a diffeomorphism except for one point.

Using the identity

$$\nabla \cdot [(\alpha \mathbf{a}) \cdot (\beta \boldsymbol{B})] = \nabla(\alpha \mathbf{a}) : (\beta \boldsymbol{B}) + (\alpha \mathbf{a}) \cdot \nabla \cdot (\beta \boldsymbol{B}^T)$$

where α, β are scalars, $\mathbf{a}$ is a vector field, and $\boldsymbol{B}$ is a second-order tensor field, we have

$$\begin{aligned}\nabla_{\mathbf{x}} \left(\boldsymbol{F}^T \cdot \nabla_{\mathbf{x}} p\right) : \boldsymbol{F}^T &= J \left[\nabla_{\mathbf{x}} \left(\boldsymbol{F}^T \cdot \nabla_{\mathbf{x}} p\right)\right] : \left[J^{-1} \boldsymbol{F}^T\right] = \\ & J \nabla_{\mathbf{x}} \cdot \left[\left(\boldsymbol{F}^T \cdot \nabla_{\mathbf{x}} p\right) \cdot (J^{-1} \boldsymbol{F}^T)\right] - J \left(\boldsymbol{F}^T \cdot \nabla_{\mathbf{x}} p\right) \cdot \left[\nabla_{\mathbf{x}} \cdot (J^{-1} \boldsymbol{F})\right]\end{aligned} \tag{6.32}$$

where $J = \det \boldsymbol{F}$, From Nanson's formula (see Ogden (1997), p. 88), we have

$$\mathbf{n}\,\mathrm{d}a = J \boldsymbol{F}^{-T} \cdot \mathbf{N}\,\mathrm{d}A \quad \Longrightarrow \quad J^{-1} \boldsymbol{F}^T \cdot \mathbf{n}\,\mathrm{d}a = \mathbf{N}\,\mathrm{d}A \tag{6.33}$$

where $\mathrm{d}a, \mathrm{d}A$ are surface areas in $\mathbf{x}$- and $\mathbf{X}$-space, respectively, and $\mathbf{n}, \mathbf{N}$ are the corresponding outward normals. For a domain $\mathrm{d}\Omega$ with boundary $d\Gamma$ that maps to a region $\mathrm{d}\Omega_x$ with boundary $\mathrm{d}\Gamma_x$, we can write

$$\int_{\Gamma_x} J^{-1} \boldsymbol{F}^T \cdot \mathbf{n}\,\mathrm{d}a = \int_{\Gamma} \mathbf{N}\,\mathrm{d}A = 0.$$

From the divergence theorem and the arbitrariness of Ω_x, we then have

$$\nabla_{\mathbf{x}} \cdot (J^{-1} \boldsymbol{F}) = 0. \tag{6.34}$$

Comparing (6.31) and (6.32), and using (6.34) gives

$$\nabla_{\mathbf{X}} \cdot [\nabla_{\mathbf{X}} p] = J \nabla_{\mathbf{x}} \cdot \left[\left(\boldsymbol{F}^T \cdot \nabla_{\mathbf{x}} p\right) \cdot (J^{-1} \boldsymbol{F}^T)\right] = J \nabla_{\mathbf{x}} \cdot \left[J^{-1} \boldsymbol{F} \cdot \boldsymbol{F}^T \cdot \nabla_{\mathbf{x}} p\right]. \tag{6.35}$$

Plugging (6.35) into (6.30) gives us the transformed wave equation

$$\boxed{J \nabla_{\mathbf{x}} \cdot \left[J^{-1} \boldsymbol{F} \cdot \boldsymbol{F}^T \cdot \nabla_{\mathbf{x}} p\right] - \ddot{p} = 0.} \tag{6.36}$$

If we compare the above with the acoustic wave equation

$$\kappa(\mathbf{x})\boldsymbol{\nabla}_{\mathbf{x}} \cdot \left[\boldsymbol{\rho}^{-1}(\mathbf{x}) \cdot \boldsymbol{\nabla}_{\mathbf{x}} p(\mathbf{x},t)\right] - \ddot{p}(\mathbf{x},t) = 0 \qquad \text{where} \quad \mathbf{x} \in \Omega_x$$

we see the connections

$$\kappa(\mathbf{x}) \leftrightarrow J \quad \text{and} \quad \boldsymbol{\rho}^{-1}(\mathbf{x}) \leftrightarrow J^{-1}\boldsymbol{F} \cdot \boldsymbol{F}^T .$$

This is similar to the result obtained by Greenleaf et al. (2008) and can be used to determine the parameters of an inertial cloak. Note that from the polar decomposition theorem we have $\boldsymbol{F} = \boldsymbol{V} \cdot \boldsymbol{R}$ where $\boldsymbol{V}$ is a symmetric tensor and $\boldsymbol{R}$ is a proper orthogonal tensor with $\det \boldsymbol{R} = +1$. If the properties of the homogeneous medium are isotropic, i.e., ρ_X and κ_X, we have the relations

$$\boxed{\kappa(\mathbf{x}) = J\kappa_X \quad \text{and} \quad \boldsymbol{\rho}(\mathbf{x}) = J\rho_X \boldsymbol{V}^{-1} \cdot \boldsymbol{V}^{-1} = J\rho_X \boldsymbol{V}^{-2} .} \tag{6.37}$$

Since $J = \det \boldsymbol{F} = \det \boldsymbol{V} \det \boldsymbol{R} = \det \boldsymbol{V}$ we have, in three dimensions, $\det(\boldsymbol{\rho}/\rho_X) = J^3 J^{-2} = J = \kappa/\kappa_X$ and, in two dimensions, $\det(\boldsymbol{\rho}/\rho_X) = J^2 J^{-2} = 1$.

For perfect cloaking, the cloak and the surrounding acoustic fluid must behave as a continuous medium. The requirement that the boundaries of the uniform region and the cloak must match implies that $d\Gamma = d\Gamma_{x+}$. From Nanson's relation (6.33) we have $J^{-1}\boldsymbol{F}^T \cdot \mathbf{n}\, d\Gamma_{x+} = \mathbf{N}\, d\Gamma$. Therefore,

$$\mathbf{N} = J^{-1}\boldsymbol{F}^T \cdot \mathbf{n} = (\boldsymbol{F}^{-1} \cdot \boldsymbol{\rho}^{-1}) \cdot \mathbf{n} \quad \Longrightarrow \quad \boldsymbol{F} \cdot \mathbf{N} = \boldsymbol{\rho}^{-1} \cdot \mathbf{n} .$$

This relation and the conditions that the pressure and normal component of the velocity (or acceleration) must be continuous across the boundary lead to the physical conditions

$$[p] = 0 \quad \text{and} \quad [(\boldsymbol{\rho}^{-1} \cdot \boldsymbol{\nabla} p) \cdot \mathbf{n}] = 0$$

where the square brackets indicate jumps of the quantities across the interface.

6.4.1 The infinite mass requirement

It turns out that an infinite mass is needed for perfect cloaking with cloaks made of materials with anisotropic density and isotropic bulk modulus. Norris (2008) has shown that this is true for arbitrary geometries. Let us first examine the Cummer-Schurig cloaking parameters from (6.28),

$$\rho_r = \frac{r}{r - R_1} , \quad \rho_\perp = \frac{r - R_1}{r} , \quad \kappa = \frac{t^2 r}{r - R_1} .$$

The radial and circumferential phase speeds are

$$c_r^2 = \frac{\kappa}{\rho_r} = t^2 \quad \text{and} \quad c_\perp^2 = \frac{\kappa}{\rho_\perp} = \frac{t^2 r^2}{(r - R_1)^2} = c_r^2 \rho_r^2 .$$

Note that perfect cloaking requires that the phase speed $c_\perp$ be infinite at $r = R_1$. Since c_r^2 is always finite, the only way $c_\perp$ can be infinite as $r \to R_1$ is if $\rho_r \to \infty$. This is the infinite mass argument. Norris has pointed out that cloaking is more difficult to achieve in two dimensions than in three.

Let us now look at the three-dimensional situation. Consider a radially symmetric cloak with anisotropic density and isotropic bulk modulus given by

$$\boldsymbol{\rho}(\mathbf{x}) = \rho_r(r)\boldsymbol{P}_{\parallel} + \rho_{\perp}(r)\boldsymbol{P}_{\perp} \; ; \quad \kappa(\mathbf{x}) = \kappa(r)\boldsymbol{1}$$

where $\mathbf{x} = r\mathbf{e}_r$ and the projection operators $\boldsymbol{P}_{\parallel}, \boldsymbol{P}_{\perp}$ are given by

$$\boldsymbol{P}_{\parallel} = \mathbf{e}_r \otimes \mathbf{e}_r \quad \text{and} \quad \boldsymbol{P}_{\perp} = \boldsymbol{1} - \mathbf{e}_r \otimes \mathbf{e}_r$$

where $\boldsymbol{1} = \mathbf{e}_r \otimes \mathbf{e}_r + \mathbf{e}_\theta \otimes \mathbf{e}_\theta + \mathbf{e}_\phi \otimes \mathbf{e}_\phi$. Consider the inverse map

$$\mathbf{X} = f(r)\mathbf{e}_r = \frac{f(r)}{r}\mathbf{x} \quad \Longrightarrow \quad \mathbf{x} = \frac{r}{f(r)}\mathbf{X}.$$

Then we can show, using arguments similar to those on p. 230, that

$$\boldsymbol{F} = \boldsymbol{\nabla}_X \mathbf{x} = \left(\frac{df}{dr}\right)^{-1} \boldsymbol{P}_{\parallel} + \frac{r}{f)} \boldsymbol{P}_{\perp}$$

Therefore,

$$J = \det \boldsymbol{F} = \left(\frac{df}{dr}\right)^{-1} \left(\frac{r}{f}\right)^2 \quad \text{and} \quad \boldsymbol{V}^2 = \boldsymbol{F} \cdot \boldsymbol{F}^T = \left(\frac{df}{dr}\right)^{-2} \boldsymbol{P}_{\parallel} + \left(\frac{r}{f}\right)^2 \boldsymbol{P}_{\perp}.$$

From the relations in (6.37), we then have

$$\frac{\kappa}{\kappa_X} = \left(\frac{df}{dr}\right)^{-1} \left(\frac{r}{f}\right)^2 \quad \text{and} \quad \frac{\boldsymbol{\rho}}{\rho_X} = \left(\frac{r}{f}\right)^2 \left[\frac{df}{dr}\boldsymbol{P}_{\parallel} + \left(\frac{f}{r}\right)^2 \left(\frac{df}{dr}\right)^{-1} \boldsymbol{P}_{\perp}\right].$$

Hence,

$$\frac{\rho_r}{\rho_X} = \left(\frac{r}{f}\right)^2 \frac{df}{dr} \quad \text{and} \quad \frac{\rho_{\perp}}{\rho_X} = \left(\frac{df}{dr}\right)^{-1}. \tag{6.38}$$

From the relation $\det(\boldsymbol{\rho}/\rho_X) = \kappa/\kappa_X$, we have $\rho_r\rho_{\perp}^2/\rho_X^3 = \kappa/\kappa_X$. Then the phase speeds c_r and $c_{\perp}$ are given by

$$\frac{c_r^2}{c_X^2} = \frac{\kappa}{\rho_r} = \left(\frac{df}{dr}\right)^{-2} \quad \text{and} \quad \frac{c_{\perp}^2}{c_X^2} = \frac{\kappa}{\rho_{\perp}} = \left(\frac{r}{f}\right)^2 \tag{6.39}$$

where $c_X^2 = \kappa_X/\rho_X$. Then the material parameters in terms of the phase speeds are

$$\rho_r/\rho_X = c_r^{-1}c_{\perp}^2/c_X \; , \quad \rho_{\perp}/\rho_X = c_r/c_X \; , \quad \kappa/\kappa_X = c_r c_{\perp}^2/c_X^3 \, .$$

Consider a medium with $\kappa_X = 1, \rho_X = 1$ containing an embedded cloak that occupies the region $R_1 \leq r \leq R_2$. Then the requirement

$$\boldsymbol{F} \cdot \boldsymbol{N} = \boldsymbol{\rho}^{-1} \cdot \mathbf{n} \quad \text{with} \quad \boldsymbol{N} = \mathbf{n} = \mathbf{e}_r \quad \text{on} \quad d\Gamma = d\Gamma_{x+} \quad \Longrightarrow \quad f(R_2) = R_2\,.$$

Similarly, the continuity conditions at $r = R_2$ are

$$[p] = 0 \quad \text{and} \quad [(\boldsymbol{\rho}^{-1} \cdot \boldsymbol{\nabla} p) \cdot \mathbf{n}] = \left[\left(\frac{df}{dr} \right)^{-1} \frac{\partial p}{\partial r} \right] = 0\,.$$

For perfect cloaking we need $f(R_1) = 0$. From (6.39) we see that for the limit $r \to R_1$ we have $c_\perp \to \infty$ and $\kappa\rho_r \to \infty$. We could try to choose the function f so that both κ and ρ_r remain bounded as $r \to R_1$. However, Norris (2008) has found that no analytic function f leads to both quantities being bounded.

In fact, even for an imperfect cloak with $f(R_1) \neq 0$ we get a blow-up in the total mass as $f(R_1) \to 0$. This can be seen by calculating the nondimensional radial mass

$$m_r = \frac{1}{V} \int_{\Omega_x} \rho_r d\Omega_x \quad \text{where} \quad V := \int_{\Omega_x} d\Omega_x\,.$$

Then, using (6.38), we have

$$m_r = \frac{3}{R_2^3 - R_1^3} \left[\frac{R_1^4}{f(R_1)} - \frac{R_2^4}{f(R_2)} + 4 \int_{R_1}^{R_2} \frac{r^3}{f(r)} dr \right]\,.$$

Clearly the mass becomes singular as $f(R_1) \to 0$. A similar analysis shows that the same behavior can be expected in two dimensions and for more general geometries. This is a severe problem with inertial cloaks except for special maps where we can choose $f(R_1) < \lambda$ where λ is the wavelength of the incident wave.

6.5 Transformation-based cloaking in elastodynamics

Let us now examine whether the equations of elastodynamics also transform in an invariant manner under spatial transformations. As before, we will look at transformations that take a body from a reference configuration Ω to a transformed configuration Ω_x. Let the Jacobian ($\boldsymbol{F}$) and the Jacobian determinant (J) of the transformation be

$$\boldsymbol{F} = \frac{\partial \mathbf{x}}{\partial \mathbf{X}} \quad \text{and} \quad J = \det(\boldsymbol{F})\,.$$

The governing equations of elastodynamics in the reference configuration (at a fixed frequency) can be expressed as

$$\boldsymbol{\nabla}_X \cdot \boldsymbol{\Sigma} = -\omega^2 \rho_X \mathbf{U} \quad \text{where} \quad \boldsymbol{\Sigma} = \mathbf{C}_X : \boldsymbol{\nabla}_X \mathbf{U} \tag{6.40}$$

where $\mathbf{\Sigma}$ is the Cauchy stress, ρ_X is the mass density, $\mathbf{U}$ is the displacement vector, and $\mathbf{C}_X$ is the stiffness tensor. Instead, using the usual transformation rule for vectors (6.3) for the displacement, following Milton et al. (2006), we will use a pull-back transformation similar to that used for electromagnetism, i.e.,

$$\mathbf{u}(\mathbf{x}) = \boldsymbol{F}^{-T} \cdot \mathbf{U}(\mathbf{X}) \quad \Longrightarrow \quad \mathbf{u}(\mathbf{x}) \cdot \boldsymbol{F} = \mathbf{U}(\mathbf{X})$$

where $\mathbf{u}$ is the displacement in the transformed coordinates. This form of the pull-back operation is needed to make sure that the transformed stress tensor is symmetric and that the total power in the two configurations is the same.

Let us now find the transformed form of $\boldsymbol{\nabla}_{\mathbf{X}} \mathbf{U}$. Using Cartesian components for simplicity, we have

$$\boldsymbol{\nabla}_{\mathbf{X}} \mathbf{U} \equiv \frac{\partial U_i}{\partial X_j} = \frac{\partial}{\partial X_j}(u_k F_{ki}) = \frac{\partial u_k}{\partial x_m}\frac{\partial x_m}{\partial X_j} F_{ki} + u_k \frac{\partial F_{ki}}{\partial X_j}$$

or

$$\boldsymbol{\nabla}_{\mathbf{X}} \mathbf{U} = \boldsymbol{F}^T \cdot \boldsymbol{\nabla}_{\mathbf{x}} \mathbf{u} \cdot \boldsymbol{F} + \mathbf{u} \cdot \boldsymbol{\nabla}_{\mathbf{X}} \boldsymbol{F}. \tag{6.41}$$

The work done by the internal elastic forces and the inertial forces can be obtained by integrating the first equation in (6.40) over the volume Ω after multiplying the momentum equation with a vector-valued, displacement-like, test function $\mathbf{V}(\mathbf{X})$ which has the value $\mathbf{V} = \mathbf{0}$ on the boundary Γ. Then we have

$$W = \int_\Omega (\boldsymbol{\nabla}_{\mathbf{X}} \cdot \mathbf{\Sigma} + \omega^2 \rho_X \mathbf{U}) \cdot \mathbf{V} \, \mathrm{d}\Omega .$$

Using the identity

$$\mathbf{v} \cdot (\boldsymbol{\nabla} \cdot \boldsymbol{S}) = \boldsymbol{\nabla} \cdot (\boldsymbol{S}^T \cdot \mathbf{v}) - \boldsymbol{S} : \boldsymbol{\nabla} \mathbf{v}$$

where $\mathbf{v}$ is a vector field and $\boldsymbol{S}$ is a second-order tensor field, we have

$$W = \int_\Omega [\boldsymbol{\nabla}_{\mathbf{X}} \cdot (\mathbf{\Sigma}^T \cdot \mathbf{V}) - \mathbf{\Sigma} : \boldsymbol{\nabla}_{\mathbf{X}} \mathbf{V} + \omega^2 \rho_X \mathbf{U} \cdot \mathbf{V}] \, \mathrm{d}\Omega .$$

From the divergence theorem and using the fact that $\mathbf{V} = \mathbf{0}$ on the boundary, we have

$$W = \int_\Omega (-\mathbf{\Sigma} : \boldsymbol{\nabla}_{\mathbf{X}} \mathbf{V} + \omega^2 \rho_X \mathbf{U} \cdot \mathbf{V}) \, \mathrm{d}\Omega .$$

Substituting the second equation in (6.40) into the above expression gives us

$$W = \int_\Omega [-(\mathbf{C}_X : \boldsymbol{\nabla}_{\mathbf{X}} \mathbf{U}) : \boldsymbol{\nabla}_{\mathbf{X}} \mathbf{V} + \omega^2 \rho_X \mathbf{U} \cdot \mathbf{V}) \, \mathrm{d}\Omega . \tag{6.42}$$

We can now transform equation (6.42) into $\mathbf{x}$-space by using (6.41) and noting that $J \mathrm{d}\Omega = \mathrm{d}\Omega_x$ where $J = \det \boldsymbol{F}$,

$$W = \int_{\Omega_x} \frac{1}{J}\Big[-\{\mathbf{C}_X : (\boldsymbol{F}^T \cdot \boldsymbol{\nabla}_{\mathbf{x}} \mathbf{u} \cdot \boldsymbol{F} + \mathbf{u} \cdot \boldsymbol{\nabla}_{\mathbf{X}} \boldsymbol{F})\} : (\boldsymbol{F}^T \cdot \boldsymbol{\nabla}_{\mathbf{x}} \mathbf{v} \cdot \boldsymbol{F} + \mathbf{v} \cdot \boldsymbol{\nabla}_{\mathbf{X}} \boldsymbol{F}) \\ + \omega^2 \rho_X (\mathbf{u} \cdot \boldsymbol{F}) \cdot (\mathbf{v} \cdot \boldsymbol{F}) \Big] \mathrm{d}\Omega_x . \tag{6.43}$$

Let us expand this expression term by term. With the substitutions $\boldsymbol{A} := \boldsymbol{\nabla}_{\mathbf{x}}\mathbf{u}$ and $\boldsymbol{B} := \boldsymbol{\nabla}_{\mathbf{x}}\mathbf{v}$, the first term is

$$\left[\mathbf{C}_X : \left(\boldsymbol{F}^T \cdot \boldsymbol{A} \cdot \boldsymbol{F}\right)\right] : \left(\boldsymbol{F}^T \cdot \boldsymbol{B} \cdot \boldsymbol{F}\right) \equiv C_{ijk\ell} F^T_{kp} A_{pq} F_{q\ell} F^T_{ir} B_{rs} F_{sj}$$
$$= (F_{ri} F_{sj} C_{ijk\ell} F^T_{kp} F^T_{\ell q}) A_{pq} B_{rs} .$$

If we define the product $\boldsymbol{F} \boxtimes \boldsymbol{F}$ as $(\boldsymbol{F} \boxtimes \boldsymbol{F})_{ijk\ell} = F_{ik} F_{j\ell}$, we can write

$$\left[\mathbf{C}_X : \left(\boldsymbol{F}^T \cdot \boldsymbol{A} \cdot \boldsymbol{F}\right)\right] : \left(\boldsymbol{F}^T \cdot \boldsymbol{B} \cdot \boldsymbol{F}\right) = \left[\left\{(\boldsymbol{F} \boxtimes \boldsymbol{F}) : \mathbf{C}_X : (\boldsymbol{F}^T \boxtimes \boldsymbol{F}^T)\right\} : \boldsymbol{A}\right] : \boldsymbol{B} .$$

For the second term, using the substitution $\mathcal{G} := \boldsymbol{\nabla}_{\mathbf{X}} \boldsymbol{F}$, we have

$$[\mathbf{C}_X : (\mathbf{u} \cdot \mathcal{G})] : \left(\boldsymbol{F}^T \cdot \boldsymbol{B} \cdot \boldsymbol{F}\right) \equiv (F_{ri} F_{sj} C_{ijk\ell} G_{pk\ell} u_p) B_{rs} .$$

Noting that

$$\mathcal{G} \equiv G_{pk\ell} = \frac{\partial F_{pk}}{\partial X_\ell} = \frac{\partial F^T_{kp}}{\partial X_\ell} = \frac{\partial}{\partial X_\ell}\left(\frac{\partial x_k}{\partial X_p}\right)^T = \frac{\partial}{\partial X_p}\left(\frac{\partial x_k}{\partial X_\ell}\right)^T = \frac{\partial F^T_{k\ell}}{\partial X_p} =: G^T_{k\ell p} \equiv \mathcal{G}^T$$

we have

$$[\mathbf{C}_X : (\mathbf{u} \cdot \mathcal{G})] : \left(\boldsymbol{F}^T \cdot \boldsymbol{B} \cdot \boldsymbol{F}\right) = [(\boldsymbol{F} \boxtimes \boldsymbol{F}) : \mathbf{C}_X : (\mathcal{G}^T \cdot \mathbf{u})] : \boldsymbol{B} .$$

Similarly, the third term can be written as

$$\left[\mathbf{C}_X : \left(\boldsymbol{F}^T \cdot \boldsymbol{A} \cdot \boldsymbol{F}\right)\right] : (\mathbf{v} \cdot \mathcal{G}) = [\mathcal{G} : \mathbf{C}_X : (\boldsymbol{F}^T \boxtimes \boldsymbol{F}^T) : \boldsymbol{A}] \cdot \mathbf{v}$$

and the fourth term is

$$[\mathbf{C}_X : (\mathbf{u} \cdot \mathcal{G})] : (\mathbf{v} \cdot \mathcal{G}) = [\mathcal{G} : \mathbf{C}_X : \mathcal{G}^T \cdot \mathbf{u}) \cdot \mathbf{v} .$$

The inertial term can also be expressed as

$$\omega^2 \rho_X (\mathbf{u} \cdot \boldsymbol{F}) \cdot (\mathbf{v} \cdot \boldsymbol{F}) = \omega^2 \rho_X [(\boldsymbol{F} \cdot \boldsymbol{F}^T) \cdot \mathbf{u}] \cdot \mathbf{v} .$$

Plugging these expressions back into equation (6.43) gives

$$W = \int_{\Omega_x} \left[-\frac{1}{J}[\{(\boldsymbol{F} \boxtimes \boldsymbol{F}) : \mathbf{C}_X : (\boldsymbol{F}^T \boxtimes \boldsymbol{F}^T)\} : \boldsymbol{\nabla}_{\mathbf{x}}\mathbf{u}] : \boldsymbol{\nabla}_{\mathbf{x}}\mathbf{v} \right.$$
$$- \frac{1}{J}[(\boldsymbol{F} \boxtimes \boldsymbol{F}) : \mathbf{C}_X : (\mathcal{G}^T \cdot \mathbf{u})] : \boldsymbol{\nabla}_{\mathbf{x}}\mathbf{v} - \frac{1}{J}[\mathcal{G} : \mathbf{C}_X : (\boldsymbol{F}^T \boxtimes \boldsymbol{F}^T) : \boldsymbol{\nabla}_{\mathbf{x}}\mathbf{u}] \cdot \mathbf{v}$$
$$\left. - \frac{1}{J}[\mathcal{G} : \mathbf{C}_X : \mathcal{G}^T \cdot \mathbf{u}) \cdot \mathbf{v} + \frac{\omega^2 \rho_X}{J}[(\boldsymbol{F} \cdot \boldsymbol{F}^T) \cdot \mathbf{u}] \cdot \mathbf{v} \right] \mathrm{d}\Omega_x .$$

Let us define

$$\mathbf{C}_x := \frac{1}{J}(\boldsymbol{F} \boxtimes \boldsymbol{F}) : \mathbf{C}_X : (\boldsymbol{F}^T \boxtimes \boldsymbol{F}^T) , \ \boldsymbol{\rho}_x := \frac{\rho_X}{J}(\boldsymbol{F} \cdot \boldsymbol{F}^T) - \frac{1}{J\omega^2}(\mathcal{G} : \mathbf{C}_X : \mathcal{G}^T)$$
$$\mathcal{S}_x := \frac{1}{J}(\boldsymbol{F} \boxtimes \boldsymbol{F}) : \mathbf{C}_X : \mathcal{G}^T , \ \mathcal{D}_x := \frac{1}{J}\mathcal{G} : \mathbf{C}_X : (\boldsymbol{F}^T \boxtimes \boldsymbol{F}^T) = \mathcal{S}_x^T . \tag{6.44}$$

Then we can write the expression for W as

$$W = \int_{\Omega_x} \left[-[\mathbf{C}_x : \nabla_x \mathbf{u}] : \nabla_x \mathbf{v} - [\mathcal{S}_x \cdot \mathbf{u}] : \nabla_x \mathbf{v} - [\mathcal{D}_x : \nabla_x \mathbf{u}] \cdot \mathbf{v} + \omega^2 [\boldsymbol{\rho}_x \cdot \mathbf{u}] \cdot \mathbf{v} \right] d\Omega_x$$

or,

$$\boxed{W = \int_{\Omega_x} \left[-[\mathbf{C}_x : \nabla_x \mathbf{u} + \mathcal{S}_x \cdot \mathbf{u}] : \nabla_x \mathbf{v} - [\mathcal{D}_x : \nabla_x \mathbf{u} - \omega^2 \boldsymbol{\rho}_x \cdot \mathbf{u}] \cdot \mathbf{v} \right] d\Omega_x .} \tag{6.45}$$

If we compare (6.45) with equation (6.42), we see that the change of coordinates has transformed the equations of elastodynamics (6.40) to

$$\nabla_x \cdot \boldsymbol{\sigma} = \mathcal{D}_x : \nabla_x \mathbf{u} - \omega^2 \boldsymbol{\rho}_x \cdot \mathbf{u} \quad \text{where} \quad \boldsymbol{\sigma} = \mathbf{C}_x : \nabla_x \mathbf{u} + \mathcal{S}_x \cdot \mathbf{u} . \tag{6.46}$$

Therefore, unlike Maxwell's equations, the *equations of elastodynamics change form under a coordinate transformation*. However, that does not negate the utility of transformation methods for elasticity in special situations and for problems other than cloaking, such as diversion of elastic waves.

There is one obvious situation where the equations (6.46) and (6.40) have the same form. This is the situation where $\mathcal{G} = 0$. Milton et al. (2006) show that pentamode materials satisfy the condition $G_{pjj} = 0$. However, even in this situation the mass density transforms into a tensorial quantity. We shall explore this issue a bit further in the next section.

Notice that equations (6.46) have a form that is similar to the Willis equations![‖] This suggests that the Willis equations may be invariant with respect to coordinate transformations. In fact, as shown in Milton et al. (2006), it turns out that the Willis equations in elastodynamics also transform in a manner that is very similar to the Maxwell equations in electromagnetism. If we relax the requirement that the transformed stress be symmetric, e.g., the coating is a Cosserat type continuum, then cloaking can be achieved (see Brun et al. (2009) for an example).

6.5.1 Energy considerations

Norris (2009) has pointed out that energy considerations may be used to find materials that can lead to the cloaking of elastic waves. The strain energy density (W) and the kinetic energy density (T) in a linear elastic material with anisotropic stiffness and density can be expressed as

$$W = (\mathbf{C} : \nabla \mathbf{u}) : \nabla \mathbf{u} \quad \text{and} \quad T = (\boldsymbol{\rho} \cdot \dot{\mathbf{u}}) \cdot \dot{\mathbf{u}} .$$

If we look for transformations of the form $\mathbf{X} \rightarrow \mathbf{x}$ and $\mathbf{U} \rightarrow \boldsymbol{G}^T \cdot \mathbf{u}$ where $\boldsymbol{G} \equiv \boldsymbol{G}(\mathbf{x})$ is a gauge transformation that is independent of time, the total energy in the reference

[‖]The Willis equations are discussed in Section 5.6 and the equations have the form shown in equation (5.77) (p. 191).

configuration is

$$\int_\Omega E_X \,\mathrm{d}\Omega = \int_\Omega (W_X + T_X)\,\mathrm{d}\Omega = \int_\Omega \left[(\mathbf{C}_X : \nabla_X \mathbf{U}) : \nabla_X \mathbf{U} + (\boldsymbol{\rho}_X \cdot \dot{\mathbf{U}}) \cdot \dot{\mathbf{U}}\right] \mathrm{d}\Omega\,.$$

Transforming the above to the deformed configuration gives

$$\begin{aligned}\int_{\Omega_x} E_x \,\mathrm{d}\Omega_x &= \int_{\Omega_x} \frac{1}{J}(W_x + T_x)\,\mathrm{d}\Omega_x \\ &= \int_{\Omega_x} \frac{1}{J}\Big[\{\mathbf{C}_X : [\nabla_\mathbf{x}(\boldsymbol{G}^T \cdot \mathbf{u}) \cdot \boldsymbol{F}]\} : [\nabla_\mathbf{x}(\boldsymbol{G}^T \cdot \mathbf{u}) \cdot \boldsymbol{F}] \\ &\qquad + [\boldsymbol{\rho}_X \cdot (\boldsymbol{G}^T \cdot \dot{\mathbf{u}})] \cdot (\boldsymbol{G}^T \cdot \dot{\mathbf{u}})\Big]\,\mathrm{d}\Omega_x\,.\end{aligned}$$

Therefore the kinetic energy density in the deformed coordinates is

$$T_x = (\boldsymbol{\rho}_x \cdot \dot{\mathbf{u}}) \cdot \dot{\mathbf{u}} \quad \text{where} \quad \boldsymbol{\rho}_x := \frac{1}{J}\boldsymbol{G} \cdot \boldsymbol{\rho}_X \cdot \boldsymbol{G}^T \tag{6.47}$$

which has the same structure as the undeformed kinetic energy density. However, we see that the strain energy density in the deformed configuration has the form

$$W_x = \frac{1}{J}\Big[\{\mathbf{C}_X : [\nabla_\mathbf{x}(\boldsymbol{G}^T \cdot \mathbf{u}) \cdot \boldsymbol{F}]\} : [\nabla_\mathbf{x}(\boldsymbol{G}^T \cdot \mathbf{u}) \cdot \boldsymbol{F}]\Big] \tag{6.48}$$

which can be expressed in the form given in equation (6.45). Since the expression for W_x contains terms involving $\mathbf{u}$ in addition to terms involving $\nabla_\mathbf{x} \mathbf{u}$, it reduces only under special circumstances to the form

$$W_x = (\mathbf{C}_x : \nabla_\mathbf{x} \mathbf{u}) : \nabla_\mathbf{x} \mathbf{u}$$

with $\mathbf{C}_x$ maintaining the symmetries of $\mathbf{C}_X$. If we limit ourselves to standard linear elasticity and do not consider the possibility of Willis elasticity, the objective of elastodynamic cloak design is to find these special transformations under which the contributions of the $\mathbf{u}$ terms are zero. The design of such elastodynamic cloaks is still at an early stage of development (see, for example, Urzhumov et al. (2010) for the effect of shear stiffness on the effectiveness of cloaking).

6.6 Acoustic metafluids and pentamode materials

Milton et al. (2006) had indicated that a transformation with $G_{pjj} = 0$ would cause the reduced elastodynamic wave equations (in acoustic wave equation form) to transform to the governing equations for a pentamode material. But the chosen transformation is only achievable for identity maps and hence not useful for cloaking. Norris

(2008, 2009) found that such a restrictive transformation was not necessary if one were to start with the scalar wave equation instead of the full equations of elastodynamics.

While exploring maps from the scalar acoustic wave equation, Norris also discovered that cloaks with isotropic bulk modulus would lead to singularities in the total cloak mass if perfect cloaking was desired. This prompted his discovery of a broader class of cloaking transformations that involve pentamode materials with anisotropic moduli. The theory behind these ideas is detailed in Norris (2008). We shall discuss only the main results in this section.

Recall from equations (6.31) and (6.32) that

$$\nabla_{\mathbf{X}} \cdot (\nabla_{\mathbf{X}} p) = J \nabla_{\mathbf{x}} \left(\boldsymbol{F}^T \cdot \nabla_{\mathbf{x}} p \right) : \left(J^{-1} \boldsymbol{F}^T \right).$$

In the above equation we have chosen to multiply the right-hand term by JJ^{-1} to get the uniform wave equation into a form analogous to the acoustic wave equation with spatially inhomogeneous density and bulk modulus. In fact, the above equation can be generalized to

$$\boxed{\nabla_{\mathbf{X}} \cdot (\nabla_{\mathbf{X}} p) = J \nabla_{\mathbf{x}} \left(J^{-1} \boldsymbol{S}^{-1} \cdot \boldsymbol{V}^2 \cdot \nabla_{\mathbf{x}} p \right) : \boldsymbol{S} \quad \text{with} \quad \nabla_{\mathbf{x}} \cdot \boldsymbol{S} = 0} \tag{6.49}$$

where $\boldsymbol{S}$ is a symmetric and nonsingular second-order tensor and $\boldsymbol{V}^2 = \boldsymbol{F} \cdot \boldsymbol{F}^T$. Then the uniform wave equation can be expressed as

$$J\boldsymbol{S} : \nabla_{\mathbf{x}} \left(J^{-1} \boldsymbol{S}^{-1} \cdot \boldsymbol{V}^2 \cdot \nabla_{\mathbf{x}} p \right) - \ddot{p} = 0 \quad \text{for} \quad \mathbf{x} \in \Omega_x .$$

Recall also from equation (5.132) (p. 214) that the wave equation for the pseudo-pressure in pentamode materials has the form

$$K\boldsymbol{S} : \nabla_{\mathbf{x}} \left(\rho^{-1} \cdot \boldsymbol{S} \cdot \nabla_{\mathbf{x}} p \right) - \ddot{p} = 0 \quad \text{with} \quad \nabla_{\mathbf{x}} \cdot \boldsymbol{S} = 0 \text{ and } \mathbf{C} = K\boldsymbol{S} \otimes \boldsymbol{S}$$

where $\mathbf{C}$ is the stiffness tensor. A comparison of these two equations gives us the connections between an acoustic fluid and a pentamode material under a transformation of coordinates

$$K \leftrightarrow J \quad \text{and} \quad \rho^{-1} \leftrightarrow J^{-1} \boldsymbol{S}^{-1} \cdot \boldsymbol{V}^2 \cdot \boldsymbol{S}^{-1} .$$

Therefore,

$$\boxed{K_x = JK_X \,, \quad \mathbf{C}_x = JK_X \boldsymbol{S} \otimes \boldsymbol{S} \quad \text{and} \quad \rho_x = (J\boldsymbol{S} \cdot \boldsymbol{V}^{-2} \cdot \boldsymbol{S}) \rho_X} \tag{6.50}$$

where K_X and ρ_X are the pseudo bulk modulus and density in the uniform medium. For the situation where $\boldsymbol{S} = \boldsymbol{1}$ we recover the inertial acoustic cloaks discussed in Section 6.4. This discovery indicates that acoustic cloaks may be constructed from pentamode materials, thus extending the range of potential cloaking materials. Norris calls these unusual materials *acoustic metafluids*.

Milton, Briane, and Willis (Milton et al., 2006) show that the "harmonic" map, $G_{pjj} = 0$ for elastodynamics leads to transformed quantities

$$\rho_x = \frac{\rho_X}{J} \boldsymbol{F} \cdot \boldsymbol{F}^T = \frac{\rho_X}{J} \boldsymbol{V}^2 \quad \text{and} \quad \mathbf{C}_x = \frac{\kappa_X}{J} (\boldsymbol{F} \cdot \boldsymbol{F}^T) \otimes (\boldsymbol{F} \cdot \boldsymbol{F}^T) = \frac{\kappa_X}{J} \boldsymbol{V}^2 \otimes \boldsymbol{V}^2 .$$

The above relations are identical to (6.50) for $\boldsymbol{S} = J^{-1}\boldsymbol{V}^2$. We can also show that for harmonic mappings $\boldsymbol{\nabla} \cdot (J^{-1} \cdot \boldsymbol{V}^2) = 0$. Therefore, the harmonic map is a special case of (6.50). Note that, unlike in the general theory for elastodynamics, there is no requirement that $\mathbf{u}(\mathbf{x}) \to \boldsymbol{F}^{-T} \cdot \mathbf{U}(\mathbf{X})$ in the Norris theory. Also note that the freedom to choose the quantity $\boldsymbol{S}$ implies that numerous transformations may be used to achieve cloaking. In fact, any $\boldsymbol{S} \to \boldsymbol{S} + \boldsymbol{\nabla}\mathbf{v}$ can be chosen, where $\mathbf{v}$ is a vector field.

For a pure stretch, $\boldsymbol{F} = \boldsymbol{V} \cdot \boldsymbol{R} = \boldsymbol{V}$. In that situation, with $\boldsymbol{S} = J^{-1}\boldsymbol{V}$, we have

$$\boldsymbol{\nabla}_{\mathbf{X}} \cdot (\boldsymbol{\nabla}_{\mathbf{X}}\, p) = \boldsymbol{V} : \boldsymbol{\nabla}_{\mathbf{x}}(\boldsymbol{V} \cdot \boldsymbol{\nabla}_{\mathbf{x}}\, p) \quad \Longrightarrow \quad \rho = \frac{1}{J}\rho_X \quad \text{and} \quad K = JK_X\,.$$

Norris has shown that adding a layer of pentamode material with isotropic density to the interior of the cloak (in a region $R_2 \geq R_\beta > r \geq R_1$) can be used to remove the infinite mass problem in inertial acoustic cloaks. However, the unstable nature of pentamode materials has made it difficult to realize such cloaks in experiment.

6.6.1 A pentamode to pentamode transformation

Consider a reference pentamode material with

$$\mathbf{C}_X = K_X \boldsymbol{S}_X \otimes \boldsymbol{S}_x \quad \text{with} \quad \boldsymbol{\nabla}_{\mathbf{X}} \cdot \boldsymbol{S}_X = 0\,.$$

Note that $\boldsymbol{S}_X$ is symmetric and positive definite. From the expression from the strain energy density we have

$$W_X = (\mathbf{C}_X : \boldsymbol{\nabla}_X \mathbf{U}) : \boldsymbol{\nabla}_X \mathbf{U} = K_X(\boldsymbol{S}_X : \boldsymbol{\nabla}_{\mathbf{X}} \mathbf{U})^2 \tag{6.51}$$

where we have use the notation $\boldsymbol{A}^2 = \boldsymbol{A} : \boldsymbol{A}$. Also, from equation (6.48), we have

$$W_x = \frac{K_X}{J}\left[\{(\boldsymbol{S}_X \otimes \boldsymbol{S}_X) : [\boldsymbol{\nabla}_{\mathbf{x}}(\boldsymbol{G}^T \cdot \mathbf{u}) \cdot \boldsymbol{F}]\} : [\boldsymbol{\nabla}_{\mathbf{x}}(\boldsymbol{G}^T \cdot \mathbf{u}) \cdot \boldsymbol{F}]\right]. \tag{6.52}$$

Note that for symmetric $\boldsymbol{S}$

$$[(\boldsymbol{S} \otimes \boldsymbol{S}) : \boldsymbol{\nabla}\mathbf{u}] : \boldsymbol{\nabla}\mathbf{u} \equiv S_{ij} S_{kl} u_{k,l} u_{i,j} \equiv (\boldsymbol{S} : \boldsymbol{\nabla}\mathbf{u})^2 = [\boldsymbol{S} : (\boldsymbol{\nabla}\mathbf{u})^T]^2\,.$$

Then equation (6.52) can be written as

$$W_x = \frac{K_X}{J}\left[\boldsymbol{S}_X : \left[\boldsymbol{\nabla}_{\mathbf{x}}(\boldsymbol{G}^T \cdot \mathbf{u}) \cdot \boldsymbol{F}\right]\right]^2 = \frac{K_X}{J}\left[\boldsymbol{S}_X : \left[(\mathbf{u} \cdot \boldsymbol{\nabla}_{\mathbf{x}} \boldsymbol{G} + \boldsymbol{G}^T \cdot \boldsymbol{\nabla}_{\mathbf{x}} \mathbf{u}) \cdot \boldsymbol{F}\right]\right]^2.$$

To achieve the form $W = (\mathbf{C} : \boldsymbol{\nabla}\mathbf{u}) : \boldsymbol{\nabla}\mathbf{u}$ we can retain only terms containing $\boldsymbol{\nabla}_{\mathbf{x}}\mathbf{u}$ in the above expression, i.e.,

$$W_x = \frac{K_X}{J}\left[\boldsymbol{S}_X : \left[\boldsymbol{G}^T \cdot \boldsymbol{\nabla}_{\mathbf{x}} \mathbf{u}) \cdot \boldsymbol{F}\right]\right]^2. \tag{6.53}$$

Therefore, we need the constraint

$$\boldsymbol{S}_X : (\mathbf{u}\cdot\boldsymbol{\nabla}_{\mathbf{X}}\boldsymbol{G}\cdot\boldsymbol{F}) = \boldsymbol{S}_X : (\mathbf{u}\cdot\boldsymbol{\nabla}_{\mathbf{X}}\boldsymbol{G}) = 0 \quad \Longrightarrow \quad \boldsymbol{S}_X : (\boldsymbol{\nabla}_{\mathbf{X}}\boldsymbol{G}^T) = \mathbf{0} \quad \forall \quad \mathbf{u}.$$

Let us choose $\boldsymbol{G}^T = \boldsymbol{S}_X^{-1}\cdot\boldsymbol{T}$, where $\boldsymbol{T}$ is a second-order tensor. Then,

$$\boldsymbol{S}_X : (\boldsymbol{\nabla}_{\mathbf{X}}\boldsymbol{G}^T) = \boldsymbol{S}_X : \left[\boldsymbol{S}_X^{-1} : \boldsymbol{\nabla}_{\mathbf{X}}\boldsymbol{T} + \boldsymbol{\nabla}_{\mathbf{X}}(\boldsymbol{S}_X^{-1})\cdot\boldsymbol{T}\right] = \boldsymbol{\nabla}_{\mathbf{X}}\cdot\boldsymbol{T} + \boldsymbol{S}_X : \left[\boldsymbol{\nabla}_{\mathbf{X}}(\boldsymbol{S}_X^{-1})\cdot\boldsymbol{T}\right].$$

Using the identity (see Norris (2009))

$$\frac{\partial}{\partial\alpha}(\boldsymbol{A}^{-1}) = -\boldsymbol{A}^{-1}\cdot\frac{\partial\boldsymbol{A}}{\partial\alpha}\cdot\boldsymbol{A}^{-1}$$

where α is any parameter, we have

$$(\boldsymbol{\nabla}_{\mathbf{X}}\boldsymbol{G}^T) : \boldsymbol{S}_X = \boldsymbol{\nabla}_{\mathbf{X}}\cdot\boldsymbol{T} - \boldsymbol{S}_X : \left[\left(\boldsymbol{S}_X^{-1}\cdot\frac{\partial\boldsymbol{S}_X}{\partial\mathbf{X}}\cdot\boldsymbol{S}_X^{-1}\right)\cdot\boldsymbol{T}\right].$$

We have used the partial derivative notation to indicate that the last term is equivalent to $T_{ij}(\partial S_{jp}^{-1}/\partial X_q)S_{pq}$. Simplifying further, we get

$$(\boldsymbol{\nabla}_{\mathbf{X}}\boldsymbol{G}^T) : \boldsymbol{S}_X = \boldsymbol{\nabla}_{\mathbf{X}}\cdot\boldsymbol{T} - (\boldsymbol{\nabla}_{\mathbf{X}}\cdot\boldsymbol{S}_X)\cdot\boldsymbol{S}_X^{-1}\cdot\boldsymbol{T} = \boldsymbol{\nabla}_{\mathbf{X}}\cdot\boldsymbol{T}$$

where we have used $\boldsymbol{\nabla}_{\mathbf{X}}\cdot\boldsymbol{S}_X = 0$. Therefore, $\boldsymbol{\nabla}_{\mathbf{X}}\cdot\boldsymbol{T} = 0$. We can now write the strain energy density (6.53) as

$$W_x = \frac{K_X}{J}\left[\boldsymbol{S}_X : \left[\boldsymbol{S}_X^{-1}\cdot\boldsymbol{T}\cdot\boldsymbol{\nabla}_{\mathbf{x}}\mathbf{u}\cdot\boldsymbol{F}\right]\right]^2 = \frac{K_X}{J}\left[(\boldsymbol{F}\cdot\boldsymbol{T}) : (\boldsymbol{\nabla}_{\mathbf{x}}\mathbf{u})^T\right]^2.$$

Clearly $\boldsymbol{F}\cdot\boldsymbol{T}$ has to be symmetric if we wish to get a strain energy density of the required form. If we define

$$\boldsymbol{S}_x := \frac{1}{J}\boldsymbol{F}\cdot\boldsymbol{T} \quad \Longrightarrow \boldsymbol{T} = J\boldsymbol{F}^{-1}\cdot\boldsymbol{S}_x$$

we have

$$W_x = JK_X(\boldsymbol{S}_x\cdot\boldsymbol{\nabla}_{\mathbf{x}}\mathbf{u})^2 \quad \text{and} \quad \boldsymbol{\nabla}\cdot\boldsymbol{S}_x = 0 \tag{6.54}$$

The divergence-free nature of $\boldsymbol{S}_x$ can be shown with a straightforward calculation. Equation (6.54) is of a form similar to the strain energy density in the reference configuration (6.51). Hence, the transformation $\boldsymbol{G}^T = J\boldsymbol{S}_X^{-1}\cdot\boldsymbol{F}^{-1}\cdot\boldsymbol{S}_x$ gives us a material that satisfies the requirement that the strain energy density retain its form. From equation (6.54) we see that $K_x = JK_X$ and from equation $(6.47)_2$ we have

$$\boxed{\boldsymbol{\rho}_x = \frac{1}{J}\boldsymbol{G}\cdot\boldsymbol{\rho}_X\cdot\boldsymbol{G}^T = J(\boldsymbol{S}_x\cdot\boldsymbol{F}^{-T}\cdot\boldsymbol{S}_X^{-1})\cdot\boldsymbol{\rho}_X\cdot(\boldsymbol{S}_X^{-1}\cdot\boldsymbol{F}^{-1}\cdot\boldsymbol{S}_x).} \tag{6.55}$$

Using the above relation, Norris (2009) has shown that for transformations of the form $\boldsymbol{F} = \alpha^{-1}\boldsymbol{S}_x\cdot\boldsymbol{R}$ (where $\boldsymbol{R}$ is the rotation in the polar decomposition $\boldsymbol{F} = \boldsymbol{V}\cdot\boldsymbol{R}$) with constant $\boldsymbol{R}$, a normal acoustic fluid can be mapped to a unique pentamode metafluid with isotropic density. For more general forms of $\boldsymbol{R}$ satisfying the condition $\boldsymbol{\nabla}_{\mathbf{X}}\times\boldsymbol{R}^T\cdot\boldsymbol{\nabla}_{\mathbf{X}}\cdot\boldsymbol{R}^T = \mathbf{0}$ we can also find unique maps from an acoustic fluid to an pentamode acoustical metafluid (see Norris (2009) for details).

Exercises

Problem 6.1 Use curvilinear coordinates to derive the transformation rules for electrical conductivity and the current:

$$\boldsymbol{\sigma}(\mathbf{x}) = \frac{1}{J}\boldsymbol{F}(\mathbf{X})\cdot\boldsymbol{\Sigma}(\mathbf{X})\cdot\boldsymbol{F}^T(\mathbf{X})$$

$$\mathbf{j}(\mathbf{x}) = \frac{\boldsymbol{F}\cdot\mathbf{J}(\mathbf{X})}{\det(\boldsymbol{F})}\,.$$

Problem 6.2 The Greenleaf-Lassas-Uhlman map has the form

$$\mathbf{x}(\mathbf{X}) = \begin{cases} \left(\dfrac{|\mathbf{X}|}{2}+1\right)\dfrac{\mathbf{X}}{|\mathbf{X}|} & \text{if } |\mathbf{X}| < 2 \\ \mathbf{X} & \text{if } |\mathbf{X}| > 2\,. \end{cases}$$

Find the expression for the electrical conductivity as a function of location for this map.

Problem 6.3 The space folding map of Leonhardt and Pendry can be expressed as

$$\mathbf{x} = \begin{cases} \mathbf{X} & \text{if } X_1 < 0 \\ (-X_1, X_2, X_3) & \text{if } d > X_1 > 0 \\ \mathbf{X} - (2d, 0, 0) & \text{if } X_1 > d\,. \end{cases}$$

Find the expression for the electrical conductivity as a function of location for this map.

Problem 6.4 Use curvilinear coordinates to verify the transformation rules for the magnetic permeability and the permittivity:

$$\boldsymbol{\mu}'(\mathbf{x}) = \frac{\boldsymbol{F}\cdot\boldsymbol{\mu}(\mathbf{X})\cdot\boldsymbol{F}^T}{\det(\boldsymbol{F})}\,; \quad \boldsymbol{\varepsilon}'(\mathbf{x}) = \frac{\boldsymbol{F}\cdot\boldsymbol{\varepsilon}(\mathbf{X})\cdot\boldsymbol{F}^T}{\det(\boldsymbol{F})}\,.$$

Then show that Maxwell's equations are invariant under these transformations.

Problem 6.5 Consider a cloak in the shape of an annulus of a circular cylinder with inner radius R_1 and outer radius R_2. Assume that the material properties of the cloak are derived from the transformation

$$r = R_1 + tR\,, \quad \theta = \Theta\,, \quad z = Z \quad \text{where} \quad t := \frac{R_2 - R_1}{R_2}\,.$$

Find the permittivity and magnetic permeability of the cloak as a function of r.

Problem 6.6 The electrical conductivity equation in the presence of a current source can be expressed as

$$\boldsymbol{\nabla}\cdot\mathbf{J}(\mathbf{X}) = S(\mathbf{X}) \quad \text{with} \quad \mathbf{J}(\mathbf{X}) = \boldsymbol{\Sigma}(\mathbf{X})\cdot\boldsymbol{\nabla}\phi(\mathbf{X})$$

where $\boldsymbol{\Sigma}(\mathbf{X})$ is the electrical conductivity tensor, $\phi(\mathbf{X})$ is a scalar potential, and $S(\mathbf{X})$ is the source term. Use the approach used to derive the transformed equations of elastodynamics to show that the conductivity equation is invariant under transformations $\mathbf{x} = \boldsymbol{F}\cdot\mathbf{X}, F_{ij} = \partial x_i/\partial X_j$ with

$$\boldsymbol{\sigma}(\mathbf{x}) = \frac{1}{J}\boldsymbol{F}(\mathbf{X})\cdot\boldsymbol{\Sigma}(\mathbf{X})\cdot\boldsymbol{F}^T(\mathbf{X}) \quad \text{and} \quad s(\mathbf{x}) = \frac{1}{J}S(\mathbf{X})\,.$$

Verify that $\boldsymbol{\nabla}_{\mathbf{x}}\cdot\mathbf{j}(\mathbf{x}) = s(\mathbf{x})$ where $\mathbf{j}(\mathbf{x}) = \boldsymbol{\sigma}(\mathbf{x})\cdot\boldsymbol{\nabla}_{\mathbf{x}}\phi$.

Problem 6.7 Chen and Chan (2007a) use a separation of variables approach to show that for a cloak based on a transformation $\boldsymbol{F}$ with

$$\boldsymbol{F}\cdot\boldsymbol{F}^T \equiv \begin{bmatrix} t^2 & 0 & 0 \\ 0 & \frac{t^2r^2}{(r-R_1)^2} & 0 \\ 0 & 0 & \frac{t^2r^2}{(r-R_1)^2} \end{bmatrix} \quad \text{and} \quad J = \det(\boldsymbol{F}) = \frac{t^3r^2}{(r-R_1)^2}.$$

The pressure inside the cloak is

$$p(r,\theta,\phi) = \text{Re}\left[p_0 \exp\left(\frac{i(r-R_1)\cos\theta}{t}\sqrt{\frac{\omega^2\rho}{\kappa}}\right)\right]$$

where p_0 is the amplitude of the incident plane wave. Verify this result.

Problem 6.8 Show that the acoustic wave equation in a medium with

$$\rho(\mathbf{x}) = \rho(\mathbf{r}) = \rho_r(r)\boldsymbol{P}_{\parallel} + \rho_{\perp}(r)\boldsymbol{P}_{\perp}\,; \quad \kappa(\mathbf{x}) = \kappa(\mathbf{r}) = \kappa(r)\boldsymbol{1}$$

can be expressed in spherical coordinates as

$$\frac{\kappa(r)}{r^2}\frac{\partial}{\partial r}\left[\frac{r^2}{\rho_r(r)}\frac{\partial p}{\partial r}\right] + \frac{\kappa(r)}{r^2\rho_{\perp}(r)}\nabla_{\perp}^2 p - \ddot{p}(r,t) = 0$$

where $\nabla_{\perp}^2(\bullet)$ is the second-order Laplace-Beltrami operator in spherical coordinates. Then show that the above equation is equivalent to the uniform wave equation

$$\boldsymbol{\nabla}_{\mathbf{X}}\cdot[\boldsymbol{\nabla}_{\mathbf{X}}\, p(\mathbf{X})] - \ddot{p}(\mathbf{X}) = 0$$

only if we have a mapping $\mathbf{X} = [f(r)/r]\,\mathbf{x}$ such that

$$\rho_r = \left(\frac{r}{f}\right)^2\frac{df}{dr}, \quad \rho_{\perp} = \left(\frac{df}{dr}\right)^{-1}, \quad \kappa = \left(\frac{r}{f}\right)^2\left(\frac{df}{dr}\right)^{-1}.$$

Problem 6.9 Find the transformed equations of elastodynamics for harmonic maps with the property $G_{pjj} = 0$. Show that for the special case when the equations of elastodynamics in the reference configuration have the form

$$-\boldsymbol{\nabla}_{\mathbf{X}}\cdot\boldsymbol{\Sigma} = \omega^2\rho_X\mathbf{U} \quad \text{with} \quad \boldsymbol{\Sigma} = \kappa_X\boldsymbol{\nabla}_{\mathbf{X}}\cdot\mathbf{U}.$$

The transformed stress field has the form

$$\boldsymbol{\sigma}(\mathbf{x}) = \alpha\boldsymbol{V} \quad \text{where} \quad \alpha := \frac{\kappa}{2J}\text{tr}\left[\boldsymbol{V}\cdot\left\{\boldsymbol{\nabla}_{\mathbf{X}}\mathbf{u} + (\boldsymbol{\nabla}_{\mathbf{X}}\mathbf{u})^T\right\}\right]$$

where $\boldsymbol{V} = \boldsymbol{F}\cdot\boldsymbol{F}^T$. Compare the above stress-strain relation to those for pentamode materials.

Problem 6.10 Verify the relation

$$\boldsymbol{\nabla}_{\mathbf{X}}\cdot(\boldsymbol{\nabla}_{\mathbf{X}}\, p) = J\boldsymbol{\nabla}_{\mathbf{x}}\left(J^{-1}\boldsymbol{S}^{-1}\cdot\boldsymbol{V}^2\cdot\boldsymbol{\nabla}_{\mathbf{x}}\, p\right) : \boldsymbol{S} \quad \text{with} \quad \boldsymbol{\nabla}_{\mathbf{x}}\cdot\boldsymbol{S} = 0.$$

7

Waves in Periodic Media

> It is known that old superficially decomposed glass presents reflected tints much brighter, and transmitted tints much purer, than any of which a single transparent film is capable ... and it is easy to see how the effect may be produced by the occurrence of nearly similar laminae at nearly equal intervals.
>
> LORD RAYLEIGH, On the maintenance of vibrations ..., 1887.

Like most of the other topics discussed in this book, the propagation of waves in periodic composite media also has a long history. Brillouin's book on periodic structures (Brillouin (2003)), first published in 1946, gives a concise history of early developments in the field. The phenomenon of dispersion in a medium, where different frequency components of a wave travel at different phase velocities, was explained by Lord Kelvin using a periodic spring-mass model. Lord Rayleigh's early work (Rayleigh, 1887) on the interaction of light with periodic layered media showed that such materials can exhibit band gaps in the frequency spectrum. Wave propagation is forbidden at these frequencies and the incident energy is usually reflected from the medium. The related phenomena of dispersion and band gaps in periodic media are responsible for a large range of interesting effects such as the iridescent colors in a peacock's plumage.

Much of the work on periodic structures until 1987 had involved lattice structures at the atomic scale or layered periodic media. This changed dramatically when, in 1987, Eli Yablonovitch (Yablonovitch, 1987) and Sajeev John (John, 1987) showed that two- and three-dimensional periodic structures, artificial crystals, could also be designed to disperse waves and exhibit band gaps. Since then there has been an explosion of research into the use of such structures to control waves, primarily electromagnetic waves but also acoustic and elastic waves. In the field of electromagnetism, these two- and three-dimensional periodic structures are called photonic crystals while in elastodynamics and acoustics they are called phononic crystals. Examples of periodic media are shown in Figure 7.1.

The study of infinite periodic media depends crucially on a theorem proved by G. Floquet (Floquet, 1880) for time-periodic media and extended by F. Bloch (Bloch, 1929) to three-dimensional spatially periodic structures.* The crucial discovery of Bloch was that waves could propagate through such media without scattering (unless there were defects in the medium which broke the periodicity).

*In 1917 P. P. Ewald had published a similar result for the dispersion of waves in crystals.

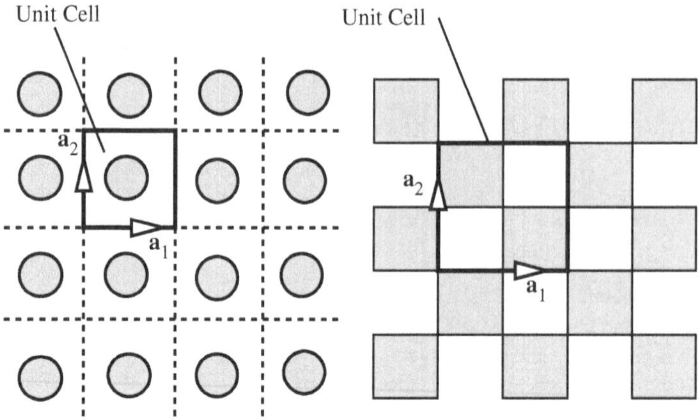

FIGURE 7.1
Examples of periodic media showing the unit cell, lattice, and primitive lattice vectors.

In this chapter we will see how the equations of acoustics, elastodynamics, and electrodynamics can be cast in the form of eigenvalue problems. We will examine the Bloch periodicity condition, a consequence of the Bloch-Floquet theorem. We will use the Bloch condition to rewrite the equations of elastodynamics and electrodynamics for a periodic unit cell. Perturbation expansions and multiple scales analysis will be used to derive the governing equations in the quasistatic limit. We will show that these equations have forms similar to the equations of elastostatics and electrostatics except that the material properties and fields are now complex. Therefore we have a straightforward means of computing the effective properties of periodic composites in the quasistatic limit; using static solutions and analytical continuation to extend those results to the complex case.

We will derive Hashin's relation for the effective elastic properties of coated sphere assemblages and show how the same results may be used to find the effective behavior of acoustic waves in a bubbly fluid. We will also show how the dispersion relation in the quasistatic limit can be used to find effective properties of electromagnetic media. Finally, will use simple lattice models to explain the concepts of band gaps and Brillouin zones.

What we will see in this chapter is a snapshot; more can be said about periodic composites. For instance, we will not discuss finite engineering structures, damping, complex material properties, complex wave vectors, and evanescent waves, all of which can be of considerable practical importance (see Hussein and Frazier (2010) and the references cited there for some examples). Another important issue is the effect of imperfections in the periodic lattice. Designed imperfections, such as the removal of parts of the lattice to create waveguides, are of technological importance. Other important issues that we will not explore are homogenization at high frequencies with the wave vector $k \to 0$ (Craster et al., 2010), and the optimization of periodic structures to maximize or minimize band gaps (Sigmund and Jensen, 2003).

Recent applications of periodic composite media range from metamaterial design for negative refraction and cloaking, guiding of electromagnetic and sonic waves, focusing and lensing applications with gradually changing periodicity, controlling vibrations of beam and plates, to blast detection. The development of nano- and micro-scale fabrication technologies has made possible a large number of other engineering uses of periodic structures. Because of the importance of the topic, the interested reader is strongly urged to explore the books by Brillouin (2003), Ashcroft and Mermin (1976), Milton (2002), Johnson and Joannopoulos (2002), and Joannopoulos and Winn (2008).

7.1 Periodic media and the Bloch condition

Let us first consider the problem of acoustic waves propagating in a periodic medium made of isotropic constituents. Suppose that the medium has a density $\rho(\mathbf{x})$ and bulk modulus $\kappa(\mathbf{x})$ which are periodic functions of $\mathbf{x}$. At fixed frequency, ω, the acoustic wave equation has the form

$$\boldsymbol{\nabla}\cdot\left(\rho^{-1}\boldsymbol{\nabla}p\right)+\frac{\omega^2}{\kappa}p=0$$

where p is the acoustic pressure. If we write the above equation in the form

$$\kappa\boldsymbol{\nabla}\cdot\left(\rho^{-1}\boldsymbol{\nabla}p\right)=-\omega^2 p$$

we can see that the left-hand side is a linear differential operator and the equation can be written in the form

$$\mathcal{L}[p(\mathbf{x})]=\lambda\, p(\mathbf{x}) \tag{7.1}$$

where

$$\mathcal{L}(\bullet):=-\kappa\boldsymbol{\nabla}\cdot\left[\rho^{-1}\boldsymbol{\nabla}(\bullet)\right] \quad \text{and} \quad \lambda:=\omega^2\,.$$

Equation (7.1) is an eigenvalue equation where λ is the eigenvalue of $\mathcal{L}$ and p is the corresponding eigenvector. Since ρ and κ are periodic, the operator $\mathcal{L}$ has the same periodicity as the medium. The periodicity of the operator becomes more obvious as we look at the lattice in Figure 7.2.

If $\mathbf{a}$ is a lattice vector, then periodicity of the medium implies that

$$\rho(\mathbf{x})=\rho(\mathbf{x}+\mathbf{a}) \quad \text{and} \quad \kappa(\mathbf{x})=\kappa(\mathbf{x}+\mathbf{a})\,.$$

Therefore,

$$\mathcal{L}[p(\mathbf{x}+\mathbf{a})]=-\kappa(\mathbf{x}+\mathbf{a})\boldsymbol{\nabla}\cdot\left[\rho^{-1}(\mathbf{x}+\mathbf{a})\boldsymbol{\nabla}p(\mathbf{x}+\mathbf{a})\right]=-\kappa(\mathbf{x})\boldsymbol{\nabla}\cdot\left[\rho^{-1}(\mathbf{x})\boldsymbol{\nabla}p(\mathbf{x}+\mathbf{a})\right].$$

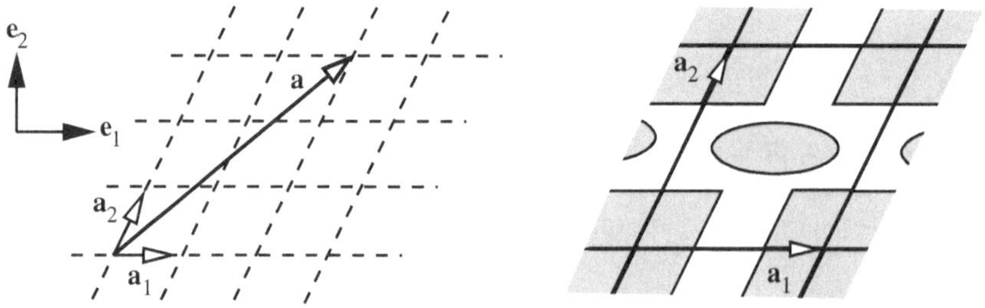

FIGURE 7.2
Lattice vector in a periodic medium.

Let $\mathcal{T}_a$ be an operator that when applied to a field $\varphi(\mathbf{x})$ shifts the argument by a vector $\mathbf{a}$. This operator is a translation operator and we may write

$$\mathcal{T}_a[\varphi(\mathbf{x})] = \varphi(\mathbf{x}+\mathbf{a})\,. \tag{7.2}$$

Notice that the operator $\mathcal{T}_a$ commutes because

$$\mathcal{T}_a\mathcal{T}_b[\varphi(\mathbf{x})] = \mathcal{T}_a[\varphi(\mathbf{x}+\mathbf{b})] = \varphi(\mathbf{x}+\mathbf{b}+\mathbf{a}) = \mathcal{T}_{a+b}[\varphi(\mathbf{x})] = \mathcal{T}_b[\varphi(\mathbf{x}+\mathbf{a})] = \mathcal{T}_b\mathcal{T}_a[\varphi(\mathbf{x})]\,.$$

We can show that the translation operator is also unitary, i.e.,

$$\mathcal{T}_a\mathcal{T}_{-a}[\varphi(\mathbf{x})] = \mathcal{T}_a[\varphi(\mathbf{x}-\mathbf{a})] = \varphi(\mathbf{x}) = I[\varphi(\mathbf{x})] = \mathcal{T}_a\mathcal{T}_a^+[\varphi(\mathbf{x})]$$

where I is the identity operator and $\mathcal{T}_a^+$ is the adjoint operator. The operators $\mathcal{L}$ and $\mathcal{T}_a$ also commute, i.e.,

$$\mathcal{T}_a\mathcal{L} = \mathcal{L}\mathcal{T}_a$$

because

$$\mathcal{T}_a[\mathcal{L}[p(\mathbf{x})]] = \mathcal{L}[p(\mathbf{x}+\mathbf{a})] = \mathcal{L}[\mathcal{T}_a[p(\mathbf{x})]] = -\kappa(\mathbf{x})\boldsymbol{\nabla}\cdot\left(\rho^{-1}(\mathbf{x})\boldsymbol{\nabla}p(\mathbf{x}+\mathbf{a})\right)\,.$$

We can write equation (7.1) in the form

$$(\mathcal{L}-\lambda I)[p(\mathbf{x})] = 0\,.$$

This equation indicates that the solutions $p(\mathbf{x})$ that we seek are in the null space of the operator $\mathcal{L}-\lambda I$.[†] Since $\mathcal{L}$ and $\mathcal{T}_a$ commute, the operators $\mathcal{L}-\lambda I$ and $\mathcal{T}_a$ must also commute. This implies that

$$\mathcal{T}_a(\mathcal{L}-\lambda I)[p(\mathbf{x})] = (\mathcal{L}-\lambda I)\mathcal{T}_a[p(\mathbf{x})] = (\mathcal{L}-\lambda I)[p(\mathbf{x}+\mathbf{a})] = 0\,.$$

[†] It is sometimes useful to think of the operator $\mathcal{L}-\lambda I$ and the solution $p(\mathbf{x})$ as vectors which are perpendicular to each other. Then the null space of $\mathcal{L}-\lambda I$ can be imagined as a plane perpendicular to the vector.

Hence the eigenstates of $\mathcal{L}-\lambda I$ and the eigenstates of $\mathcal{T}_R$ lie in the same space and these eigenstates have the property

$$\boxed{[\mathcal{T}_a - t(\mathbf{a})I][p(\mathbf{x})] = 0 \quad \Longrightarrow \quad \mathcal{T}_a[p(\mathbf{x})] = t(\mathbf{a})p(\mathbf{x})\,.} \tag{7.3}$$

Since $\mathcal{T}_a\mathcal{T}_b = \mathcal{T}_{a+b}$, we have $t(\mathbf{a})\,t(\mathbf{b}) = t(\mathbf{a}+\mathbf{b})$ which involves just a vector addition of the translation vectors. So if we know the primitive vectors of the lattice, $\mathbf{a}_1, \mathbf{a}_2, \mathbf{a}_3$, we can find the solution for an arbitrary lattice vector $\mathbf{a}$ which can be expressed as

$$\mathbf{a} = n_p\mathbf{a}_p$$

where n_p are integers.‡ Observing that for integer n, $t(n\mathbf{a}) = t^n(\mathbf{a})$, we can express $t(\mathbf{a})$ as

$$t(\mathbf{a}) = t(n_p\mathbf{a}_p) = t(n_1\mathbf{a}_1)\,t(n_2\mathbf{a}_2)\,t(n_3\mathbf{a}_3) = t^{n_1}(\mathbf{a}_1)t^{n_2}(\mathbf{a}_2)t^{n_3}(\mathbf{a}_3)\,.$$

Let us assume that, with a suitable choice of α_j, $t(\mathbf{a}_j) = \exp(2\pi i\alpha_j)$, $j = 1,2,3$. Then,

$$t(\mathbf{a}) = e^{2\pi i\alpha_p n_p} \tag{7.4}$$

Define a vector $\mathbf{k}$ in terms of the *reciprocal lattice vectors*, $\mathbf{b}_i$ such that

$$\mathbf{k} := \alpha_q\mathbf{b}_q \qquad \text{where} \qquad \mathbf{b}_q\cdot\mathbf{a}_p = 2\pi\delta_{qp}\,. \tag{7.5}$$

Then, from (7.5) and (7.4), we have

$$\mathbf{k}\cdot\mathbf{a} = 2\pi\alpha_p n_p \quad \Longrightarrow \quad \boxed{t(\mathbf{a}) = e^{i\mathbf{k}\cdot\mathbf{a}}}\,. \tag{7.6}$$

Plugging this expression into (7.3), we get

$$\mathcal{T}_a[p(\mathbf{x})] = e^{i\mathbf{k}\cdot\mathbf{a}}p(\mathbf{x})$$

or

$$\boxed{p(\mathbf{x}+\mathbf{a}) = e^{i\mathbf{k}\cdot\mathbf{a}}p(\mathbf{x})\,.} \tag{7.7}$$

Equation (7.7) is called the *Bloch condition*. The *Bloch-Floquet theorem*§ states that solutions of the form (7.7) solve the eigenvalue problem (7.3) for periodic geometries. Time harmonic solutions to the acoustic wave equations that satisfy the Bloch condition for all lattice vectors $\mathbf{a}$ are called *Bloch waves*. The vector $\mathbf{k}$ is called the *Bloch wave vector*. Note that for any vector $\mathbf{x}$, the Bloch condition (7.7) implies that

$$e^{-i\mathbf{k}\cdot(\mathbf{x}+\mathbf{a})}p(\mathbf{x}+\mathbf{a}) = e^{-i\mathbf{k}\cdot\mathbf{x}}p(\mathbf{x}).$$

Therefore the quantity $\exp(-i\mathbf{k}\cdot\mathbf{x})p(\mathbf{x})$ is periodic.

‡This is particularly useful because we usually know the periodicity of the composite in terms of the primitive lattice vectors, i.e.,$\rho(\mathbf{x}) = \rho(\mathbf{x}+\mathbf{a}_i)$ and $\kappa(\mathbf{x}) = \kappa(\mathbf{x}+\mathbf{a}_i)$.

§For a mathematical treatment of the Bloch-Floquet theorem see Kuchment (1993).

We can follow the same procedure to deduce the Bloch condition for elastodynamic waves in a periodic medium. Recall that the momentum equation at fixed frequency (ignoring body forces) can be expressed as

$$\boldsymbol{\nabla}\cdot(\boldsymbol{\mathsf{C}}:\boldsymbol{\nabla}\mathbf{u}) = -\omega^2\boldsymbol{\rho}\cdot\mathbf{u}$$

where $\mathbf{u}(\mathbf{x})$ is the displacement field, $\boldsymbol{\mathsf{C}}(\mathbf{x})$ is the stiffness tensor, and $\boldsymbol{\rho}(\mathbf{x})$ is the mass density. Expressed in the form of an eigenvalue problem, the momentum equation becomes

$$(\mathcal{L}-\lambda I)[\mathbf{u}(\mathbf{x})] = \mathbf{0}$$

where

$$\mathcal{L}[\mathbf{u}(\mathbf{x})] := -\boldsymbol{\rho}^{-1}\cdot[\boldsymbol{\nabla}\cdot(\boldsymbol{\mathsf{C}}:\boldsymbol{\nabla}\mathbf{u})] \quad \text{and} \quad \lambda := \omega^2 .$$

The Bloch condition then takes the form

$$\boxed{\mathbf{u}(\mathbf{x}+\mathbf{a}) = e^{i\mathbf{k}\cdot\mathbf{a}}\mathbf{u}(\mathbf{x}) .} \tag{7.8}$$

For a medium in which the permittivity $\varepsilon(\mathbf{x})$ and the permeability $\mu(\mathbf{x})$ are periodic, the Maxwell equations at fixed frequency are

$$\boldsymbol{\nabla}\cdot\mathbf{D} = 0\,;\;\; \boldsymbol{\nabla}\cdot\mathbf{B} = 0\,;\;\; \boldsymbol{\nabla}\times(\varepsilon^{-1}\mathbf{D}) + i\omega\mathbf{B} = 0\,;\;\; \boldsymbol{\nabla}\times(\mu^{-1}\mathbf{B}) - i\omega\mathbf{D} = 0\,.$$

These equations suggest that we should look for solutions $\mathbf{D}$ and $\mathbf{B}$ in the space of divergence-free fields such that

$$(\mathcal{L}-\omega I)\begin{bmatrix}\mathbf{D}\\ \mathbf{B}\end{bmatrix} = \mathbf{0}$$

where the operator $\mathcal{L}$ is given by

$$\mathcal{L} := \begin{bmatrix} 0 & -i\boldsymbol{\nabla}\times\mu^{-1} \\ i\boldsymbol{\nabla}\times\varepsilon^{-1} & 0 \end{bmatrix} .$$

The Bloch condition for electromagnetism then has the form

$$\boxed{\begin{bmatrix}\mathbf{D}(\mathbf{x}+\mathbf{a})\\ \mathbf{B}(\mathbf{x}+\mathbf{a})\end{bmatrix} = e^{i\mathbf{k}\cdot\mathbf{a}}\begin{bmatrix}\mathbf{D}(\mathbf{x})\\ \mathbf{B}(\mathbf{x})\end{bmatrix} .} \tag{7.9}$$

7.2 Elastodynamics in the quasistatic limit

In this section we will first determine the form of the equations of elastodynamics in the quasistatic limit and then explore the determination of the effective properties of those composites in this limit. Roughly speaking, the quasistatic limit is the situation

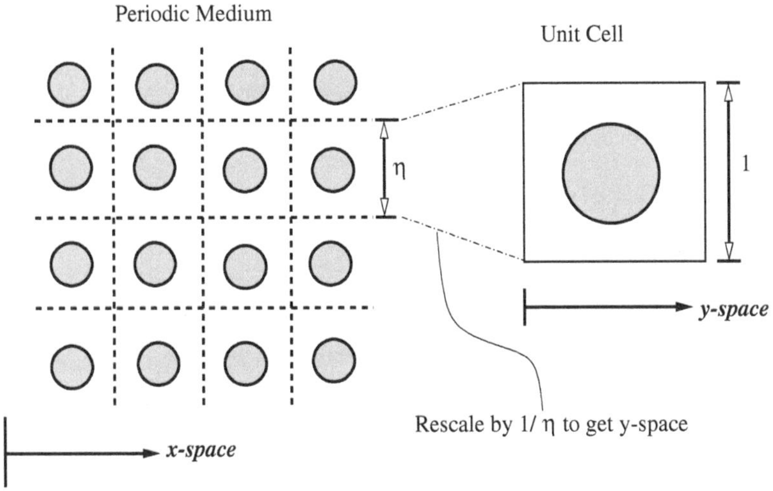

FIGURE 7.3
Periodic medium showing the unit cell and **x**- and **y**-spaces.

where the wavelength of the elastic waves is much larger than the unit cell of the periodic composite.¶

Consider the periodic medium with lattice spacing η shown in Figure 7.3. We can think of this particular composite as one of a sequence of periodic composites with successively smaller microstructure indexed by the label η. Recall that the momentum equation for linear elastodynamics at fixed frequency, in the absence of body forces, is given by

$$\boldsymbol{\nabla}\cdot\boldsymbol{\sigma}(\mathbf{x}) = -\omega^2\rho(\mathbf{x})\mathbf{u}(\mathbf{x}) \quad \text{with} \quad \boldsymbol{\sigma}(\mathbf{x}) = \mathbf{C}(\mathbf{x}) : \boldsymbol{\varepsilon}(\mathbf{x})$$

where $\boldsymbol{\sigma}$ is the stress, ρ is the mass density, $\mathbf{u}$ is the displacement, $\boldsymbol{\varepsilon}$ is the strain, and $\mathbf{C}$ is the elastic stiffness tensor. The dependence of these variables on ω has once again been kept implicit but we should keep in mind that these quantities may be complex. For infinitesimal deformations, the strain-displacement relations are

$$\boldsymbol{\varepsilon}(\mathbf{x}) = \tfrac{1}{2}\left[\boldsymbol{\nabla}\mathbf{u} + (\boldsymbol{\nabla}\mathbf{u})^T\right] .$$

Define, for a periodic medium such as the one shown in Figure 7.3, the scaled density and stiffness tensor:

$$\rho_\eta(\mathbf{x}) := \rho(\mathbf{x}/\eta) = \rho(\mathbf{y}) \quad \text{and} \quad \mathbf{C}_\eta(\mathbf{x}) := \mathbf{C}(\mathbf{x}/\eta) = \mathbf{C}(\mathbf{y})$$

where the $\mathbf{y}$-space is related to the $\mathbf{x}$-space by $\mathbf{y} = \mathbf{x}/\eta$. Because the medium is periodic, these are periodic functions, i.e.,

$$\rho(\mathbf{y} + \mathbf{a}_i) = \rho(\mathbf{y}) \quad \text{and} \quad \mathbf{C}(\mathbf{y} + \mathbf{a}_i) = \mu(\mathbf{y})$$

¶Our discussion of waves in the quasistatic limit is based on Milton (2002) (p. 230).

where $\mathbf{a}_i$ are the primitive lattice vectors. We may also write these periodicity conditions as

$$\rho_\eta(\mathbf{x}+\eta\mathbf{a}_i) = \rho_\eta(\mathbf{x}) \quad \text{and} \quad \mathbf{C}_\eta(\mathbf{x}+\eta\mathbf{a}_i) = \mathbf{C}_\eta(\mathbf{x}) \,.$$

We can similarly define the scaled stress, strain, and displacement as

$$\boldsymbol{\sigma}_\eta(\mathbf{x}) := \boldsymbol{\sigma}(\mathbf{x}/\eta) = \boldsymbol{\sigma}(\mathbf{y}) \,, \quad \boldsymbol{\varepsilon}_\eta(\mathbf{x}) := \boldsymbol{\varepsilon}(\mathbf{x}/\eta) = \boldsymbol{\varepsilon}(\mathbf{y}) \,, \quad \mathbf{u}_\eta(\mathbf{x}) := \mathbf{u}(\mathbf{x}/\eta) = \mathbf{u}(\mathbf{y}).$$

Then the governing equations can be written as

$$\boldsymbol{\nabla}\cdot\boldsymbol{\sigma}_\eta(\mathbf{x}) = -\omega^2\, \rho_\eta(\mathbf{x})\, \mathbf{u}_\eta(\mathbf{x}) \quad \text{where} \quad \boldsymbol{\sigma}_\eta(\mathbf{x}) = \mathbf{C}_\eta(\mathbf{x}) : \boldsymbol{\varepsilon}_\eta(\mathbf{x}) \tag{7.10}$$

and

$$\boldsymbol{\varepsilon}_\eta(\mathbf{x}) = \tfrac{1}{2}\left[\boldsymbol{\nabla}\mathbf{u}_\eta + (\boldsymbol{\nabla}\mathbf{u}_\eta)^T\right] . \tag{7.11}$$

Let us examine Bloch wave solutions to these equations of the form

$$\boldsymbol{\sigma}_\eta(\mathbf{x}) = e^{i\mathbf{k}\cdot\mathbf{x}}\,\widehat{\boldsymbol{\sigma}}_\eta(\mathbf{x}) \,, \quad \boldsymbol{\varepsilon}_\eta(\mathbf{x}) = e^{i\mathbf{k}\cdot\mathbf{x}}\,\widehat{\boldsymbol{\varepsilon}}_\eta(\mathbf{x}) \,, \text{ and } \mathbf{u}_\eta(\mathbf{x}) = e^{i\mathbf{k}\cdot\mathbf{x}}\,\widehat{\mathbf{u}}_\eta(\mathbf{x}) \tag{7.12}$$

where $\mathbf{k}$ is the Bloch wave vector. These variables have the same periodicity as ρ and $\mathbf{C}$, i.e.,

$$\boldsymbol{\sigma}_\eta(\mathbf{x}) = \boldsymbol{\sigma}_\eta(\mathbf{x}+\eta\mathbf{a}_i) \,, \quad \boldsymbol{\varepsilon}_\eta(\mathbf{x}) = \boldsymbol{\varepsilon}_\eta(\mathbf{x}+\eta\mathbf{a}_i) \,, \quad \text{and} \quad \mathbf{u}_\eta(\mathbf{x}) = \mathbf{u}_\eta(\mathbf{x}+\eta\mathbf{a}_i) \,.$$

Substituting (7.12) into the first equation in (7.10), taking the derivatives, assuming a symmetric stress tensor, and rearranging gives

$$\boldsymbol{\nabla}\cdot\widehat{\boldsymbol{\sigma}}_\eta + i\widehat{\boldsymbol{\sigma}}_\eta\cdot\mathbf{k} + \omega^2\rho_\eta\widehat{\mathbf{u}}_\eta = 0 \quad \text{with} \quad \widehat{\boldsymbol{\sigma}}_\eta = \mathbf{C}_\eta : \widehat{\boldsymbol{\varepsilon}}_\eta \,. \tag{7.13}$$

Similarly, the strain-displacement relation takes the form

$$\widehat{\boldsymbol{\varepsilon}}_\eta = \tfrac{i}{2}(\widehat{\mathbf{u}}_\eta\otimes\mathbf{k} + \mathbf{k}\otimes\widehat{\mathbf{u}}_\eta) + \tfrac{1}{2}\left[\boldsymbol{\nabla}\widehat{\mathbf{u}}_\eta + (\boldsymbol{\nabla}\widehat{\mathbf{u}}_\eta)^T\right] . \tag{7.14}$$

If we plug (7.14) into (7.13) and express the momentum equation in terms of $\widehat{\mathbf{u}}_\eta$ we get

$$\boldsymbol{\nabla}\cdot[\mathbf{C}_\eta : (i\mathbf{k}\otimes\widehat{\mathbf{u}}_\eta + \boldsymbol{\nabla}\widehat{\mathbf{u}}_\eta)] + i\mathbf{k}\cdot[\mathbf{C}_\eta : (i\mathbf{k}\otimes\widehat{\mathbf{u}}_\eta + \boldsymbol{\nabla}\widehat{\mathbf{u}}_\eta)] + \omega^2\rho_\eta\widehat{\mathbf{u}}_\eta = 0$$

where we have used the symmetries of $\mathbf{C}_\eta$. The above equation can be written formally in terms of a linear differential operator, $\mathcal{L}$, as

$$\mathcal{L}(\widehat{\mathbf{u}}_\eta) + \omega^2\rho_\eta\widehat{\mathbf{u}}_\eta = 0 .$$

Clearly, the above equation represents an eigenvalue problem‖ which has solutions if the frequency ω takes one of a discrete set of values, $\omega_\eta^j(\mathbf{k})$, $j = 1,2,\ldots$, which are

‖See Reddy (1998) for a discussion of the eigenvalue problem in the context of differential operators.

the dispersion relations for the periodic system. Even though the operator is continuous, the eigenvalues are discrete because the operator $\mathcal{L}$ itself depends on ω through the frequency dependence of $\mathbf{C}_\eta$.

If the wavelength $\lambda = 2\pi/\|\mathbf{k}\|$ is much larger than the size of the unit cell, the fields vary slowly in the unit cell and we can assume that the fields have perturbation expansions in powers of η. This approach is called the *multiple scales analysis* and the limit of large wavelength is called the *quasistatic limit*. The perturbation expansions of the three fields of interest are

$$\begin{aligned}\widehat{\boldsymbol{\sigma}}_\eta(\mathbf{x}) &= \boldsymbol{\sigma}_0(\mathbf{y}) + \eta\, \boldsymbol{\sigma}_1(\mathbf{y}) + \eta^2\, \boldsymbol{\sigma}_2(\mathbf{y}) + \ldots \\ \widehat{\boldsymbol{\varepsilon}}_\eta(\mathbf{x}) &= \boldsymbol{\varepsilon}_0(\mathbf{y}) + \eta\, \boldsymbol{\varepsilon}_1(\mathbf{y}) + \eta^2\, \boldsymbol{\varepsilon}_2(\mathbf{y}) + \ldots \\ \widehat{\mathbf{u}}_\eta(\mathbf{x}) &= \mathbf{u}_0(\mathbf{y}) + \eta\, \mathbf{u}_1(\mathbf{y}) + \eta^2\, \mathbf{u}_2(\mathbf{y}) + \ldots\end{aligned}$$

where the functions $\boldsymbol{\sigma}_i$, $\boldsymbol{\varepsilon}_i$, $\mathbf{u}_i$ are periodic. Let us also assume that the dispersion relations can be expressed as

$$\omega = \omega^j_\eta(\mathbf{k}) = \omega^j_0(\mathbf{y}) + \eta\, \omega^j_1(\mathbf{y}) + \eta^2\, \omega^j_2(\mathbf{y}) + \ldots .$$

We will plug these expansions into the governing equations and take the limit $\eta \to 0$. Plugging these expansions into $(7.13)_1$ gives us

$$\begin{aligned}&\tfrac{1}{\eta}\boldsymbol{\nabla}_y \cdot (\boldsymbol{\sigma}_0 + \eta\boldsymbol{\sigma}_1 + \ldots) + i(\boldsymbol{\sigma}_0 + \eta\boldsymbol{\sigma}_1 + \ldots)\cdot\mathbf{k} + \\ &\qquad \rho_\eta\left(\omega^j_0 + \eta\omega^j_1 + \ldots\right)^2 (\mathbf{u}_0 + \eta\mathbf{u}_1 + \ldots) = 0\end{aligned} \tag{7.15}$$

where we have used the subscript y to indicate derivatives with respect to $\mathbf{y}$ and

$$\boldsymbol{\nabla}\cdot\boldsymbol{\sigma}(\mathbf{x}) = \tfrac{1}{\eta}\boldsymbol{\nabla}_y \cdot \boldsymbol{\sigma}(\mathbf{y}) \quad \text{because} \quad \mathbf{y} = \frac{\mathbf{x}}{\eta}.$$

Taking the limit $\eta \to 0$ is equivalent to collecting terms of order η^{-1} and $\eta^0 = 1$ from (7.15) and equating them separately to zero. If we apply that process to (7.15) we get two equations,

$$\boxed{\boldsymbol{\nabla}_y \cdot \boldsymbol{\sigma}_0 = 0 \quad \text{and} \quad \boldsymbol{\nabla}_y \cdot \boldsymbol{\sigma}_1 + i\boldsymbol{\sigma}_0(\mathbf{y})\cdot\mathbf{k} + \rho_\eta(\omega^j_0)^2\mathbf{u}_0(\mathbf{y}) = 0.} \tag{7.16}$$

This is the momentum equation of a periodic composite in the quasistatic limit. We can follow the same procedure for the second equation in (7.13) to get

$$\boxed{\boldsymbol{\sigma}_0(\mathbf{y}) = \mathbf{C}_\eta(\mathbf{y}) : \boldsymbol{\varepsilon}_0(\mathbf{y}).} \tag{7.17}$$

Similarly, from (7.14) we have

$$\begin{aligned}&0 = \boldsymbol{\nabla}_y\mathbf{u}_0 + (\boldsymbol{\nabla}_y\mathbf{u}_0)^T \quad \text{and} \\ &\boldsymbol{\varepsilon}_0(\mathbf{y}) = \tfrac{i}{2}[\mathbf{u}_0(\mathbf{y})\otimes\mathbf{k} + \mathbf{k}\otimes\mathbf{u}_0(\mathbf{y})] + \tfrac{1}{2}[\boldsymbol{\nabla}_y\mathbf{u}_1 + (\boldsymbol{\nabla}_y\mathbf{u}_1)^T].\end{aligned} \tag{7.18}$$

From $(7.18)_1$ we see that $\mathbf{u}_0(\mathbf{y})$ must be linear in $\mathbf{y}$. We also know that from our definition that $\mathbf{u}_0(\mathbf{y})$ is periodic in $\mathbf{y}$. A function that is both linear and periodic must be constant. Therefore,

$$\mathbf{u}_0(\mathbf{y}) =: \mathbf{u}_0 = \text{ constant} .$$

Then we can write the second strain-displacement relation in (7.18) in the form

$$\boxed{\boldsymbol{\varepsilon}_0(\mathbf{y}) = \tfrac{1}{2}\left[\boldsymbol{\nabla}_y\widetilde{\mathbf{u}} + (\boldsymbol{\nabla}_y\widetilde{\mathbf{u}})^T\right] \quad \text{where} \quad \widetilde{\mathbf{u}}(\mathbf{y}) := i(\mathbf{k}\cdot\mathbf{y})\mathbf{u}_0 + \mathbf{u}_1(\mathbf{y}).} \tag{7.19}$$

This is the strain-displacement relation in the quasistatic limit.

7.2.1 Effective stiffness and dispersion relation

One of the first questions that arises in the homogenization of composites is whether determining the effective behavior of the composite by some averaging process is the right thing to do. There is a vast amount of literature on the subject and excellent sources of information are Milton (2002), Torquato (2001), and Nemat-Nasser and Hori (1993). A more mathematical treatment can be found in Zhikov et al. (1994). We will assume that there is a representative volume element (RVE) over which such an average can be obtained.

The complex effective elasticity tensor of the periodic medium in Figure 7.3 may be defined via the relation

$$\langle\boldsymbol{\sigma}_0\rangle = \mathbf{C}_{\text{eff}} : \langle\boldsymbol{\varepsilon}_0\rangle \ . \tag{7.20}$$

Since the first equation in (7.16) and equations (7.17) and (7.19) which govern these fields have a form similar to that of the quasistatic elasticity equations in the absence of inertial and body forces, we can use standard homogenization techniques to find the effective properties of the composite (see Nemat-Nasser and Hori (1993) for examples). However, the periodicity of the unit cell imposes some constraints on the allowable strain field and analytic continuation of the real-space solution is needed because the fields and material properties are now complex.

Taking the average of the second equation in (7.18) over the periodic cell and keeping in mind that $\mathbf{u}_0$ is constant, we get

$$\langle\boldsymbol{\varepsilon}_0\rangle = \tfrac{i}{2}(\mathbf{u}_0\otimes\mathbf{k}+\mathbf{k}\otimes\widehat{\mathbf{u}}_0) + \tfrac{1}{2}\left[\langle\boldsymbol{\nabla}_y\mathbf{u}_1\rangle + \langle(\boldsymbol{\nabla}_y\mathbf{u}_1)^T\rangle\right] .$$

Since the function $\mathbf{u}_1(\mathbf{y})$ is periodic, the average gradient over the unit cell is zero and we have

$$\langle\boldsymbol{\varepsilon}_0\rangle = \tfrac{i}{2}(\mathbf{u}_0\otimes\mathbf{k}+\mathbf{k}\otimes\mathbf{u}_0) . \tag{7.21}$$

Similarly, taking the average of the second equation in (7.16) we get

$$i\,\langle\boldsymbol{\sigma}_0\rangle\cdot\mathbf{k} + \langle\rho_\eta\rangle\,(\omega_0^j)^2\mathbf{u}_0 = 0 \ . \tag{7.22}$$

Plugging (7.21) into (7.20) gives

$$\langle\boldsymbol{\sigma}_0\rangle = \tfrac{i}{2}\mathbf{C}_{\text{eff}} : (\mathbf{u}_0\otimes\mathbf{k}+\mathbf{k}\otimes\mathbf{u}_0) .$$

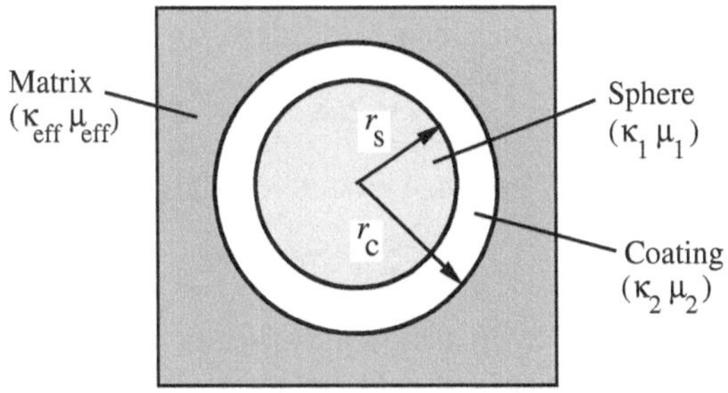

FIGURE 7.4
Elastic sphere with an elastic coating inside in an elastic matrix.

Substituting the above expression into (7.22) and rearranging gives the dispersion relation

$$\boxed{\frac{1}{\langle\rho_\eta\rangle}(\mathbf{k}\cdot\mathbf{C}_{\text{eff}}\cdot\mathbf{k})\cdot\mathbf{u}_0=(\omega_0^j)^2\mathbf{u}_0\,.} \tag{7.23}$$

The quantity

$$\boldsymbol{A}_{\text{eff}}(\mathbf{k}):=\mathbf{k}\cdot\mathbf{C}_{\text{eff}}\cdot\mathbf{k}$$

is the *effective acoustic tensor* of the periodic composite in the quasistatic limit. Equation (7.23) can be used to compute the dispersion relations for the periodic composite once the effective stiffness tensor has been calculated by quasistatic approaches. Conversely, if the dispersion relation is known then equation (7.23) can be used to compute the effective stiffness tensor.

7.2.2 Hashin's relations

Let us now derive Hashin's relations (Hashin, 1962, Hashin and Shtrikman, 1962) for the effective moduli of an assemblage of coated spheres in elastostatics and see how we can use them to find the complex dynamic moduli of a liquid containing air bubbles. We can do this because, in the quasistatic limit, the Bloch solutions satisfy equations which are directly analogous to the elasticity equations but with complex fields and complex effective tensors. This implies that if we have a formula for the effective tensor which is valid for real tensors, then, by analytic continuation, we can use the same formula when the tensors take complex values.

Consider the coated sphere shown in Figure 7.4. Let the bulk and shear moduli of the sphere be κ_1 and μ_1, let those of the coating be κ_2 and μ_2, and let the moduli of the effective medium be κ_{eff} and μ_{eff}, respectively. The radius of the sphere is r_s and the outer boundary of the coating has a radius r_c.**

**This discussion is based on Milton (2002) (p. 116) and Christensen (1984) (p. 48).

The governing equations for the linear elastostatics of isotropic media are

$$\boldsymbol{\nabla}\cdot\boldsymbol{\sigma}=0 \quad \text{where} \quad \boldsymbol{\sigma}=\lambda\,\text{tr}(\boldsymbol{\varepsilon})\boldsymbol{1}+2\mu\boldsymbol{\varepsilon} \quad \text{and} \quad \boldsymbol{\varepsilon}=\tfrac{1}{2}\left[\boldsymbol{\nabla}\mathbf{u}+(\boldsymbol{\nabla}\mathbf{u})^T\right].$$

The bulk modulus κ is related the Lamé moduli λ and μ by

$$\kappa=\lambda+\tfrac{2}{3}\mu\,.$$

Let the matrix be subject to a hydrostatic state of stress, $\boldsymbol{\sigma}=-p\boldsymbol{1}$. We will look for a solution for the radial displacement field that does not perturb this field when the coated sphere is added to the matrix. From the spherical symmetry of the problem, we know that an allowable radial displacement field, $\mathbf{u}(\mathbf{r})$, has the form

$$\begin{aligned}\mathbf{u}^{(1)}(r,\theta,\phi)&=u_r^{(1)}\mathbf{e}_r=a_1\,r\,\mathbf{e}_r && \text{in the core}\\ \mathbf{u}^{(2)}(r,\theta,\phi)&=u_r^{(2)}\mathbf{e}_r=\left(a_2r+b_2/r^2\right)\mathbf{e}_r && \text{in the coating.}\end{aligned}$$

If we replace the coated sphere with a homogeneous sphere made of the effective material, the effective displacement field is

$$\mathbf{u}^{\text{eff}}(r,\theta,\phi)=u_r^{\text{eff}}\mathbf{e}_r=a_{\text{eff}}\,r\,\mathbf{e}_r\,.$$

From the continuity of displacements at the interface $r=r_s$ and $r=r_c$, we have

$$a_1r_s=a_2r_s+b_2/r_s^2 \quad \text{and} \quad a_{\text{eff}}\,r_c=a_2r_c+b_2/r_c^2\,.$$

The continuity of radial tractions $\boldsymbol{\sigma}\cdot\mathbf{e}_r$ across the interface implies that

$$3\kappa_1a_1=3\kappa_2a_2-4b_2\mu_2/r_s^3 \quad \text{and} \quad 3\kappa_2a_2-4b_2\mu_2/r_c^3=-p=3\kappa_{\text{eff}}\,a_{\text{eff}}$$

where we have used $\text{tr}(\boldsymbol{\varepsilon})=\boldsymbol{\nabla}\cdot\mathbf{u}=du_r/dr+2u_r/r$ and $\boldsymbol{\nabla}\mathbf{u}\cdot\mathbf{e}_r=du_r/dr\,\mathbf{e}_r$. From the displacement and traction continuity relations for the coated sphere we can find expressions for a_2 and b_2 in terms of the applied pressure p. The traction continuity condition for effective medium gives us an expression for a_{eff}. Plugging these into the displacement continuity condition for the effective medium leads to the following expression for κ_{eff}:

$$\kappa_{\text{eff}}=\frac{\kappa_2(3\kappa_1+4\mu_2)r_c^3+4\mu_2(\kappa_1-\kappa_2)r_s^3}{(3\kappa_1+4\mu_2)r_c^3-3(\kappa_1-\kappa_2)r_s^3}.$$

If we express the above in terms of volume fractions using the definition

$$f_1=1-f_2=r_s^3/r_c^3$$

we can show that the κ_{eff} is given by

$$\boxed{\kappa_{\text{eff}}=\kappa_2+\cfrac{f_1}{\cfrac{1}{\kappa_1-\kappa_2}+\cfrac{f_2}{\kappa_2+\frac{4}{3}\mu_2}}}\,. \tag{7.24}$$

The effective bulk modulus in equation (7.24) is *Hashin's relation* and applies to an assemblage of coated spheres. For this choice of the bulk modulus of the matrix, a coated sphere will have no effect on the pressure field in the matrix. In such a situation, we could continue to add spheres and completely fill space (except for a set of measure zero) without affecting the field. A schematic of the situation is shown in Figure 7.5. Clearly, in that case, the effective bulk modulus of the assemblage of coated spheres is equal to that of the matrix.[††]

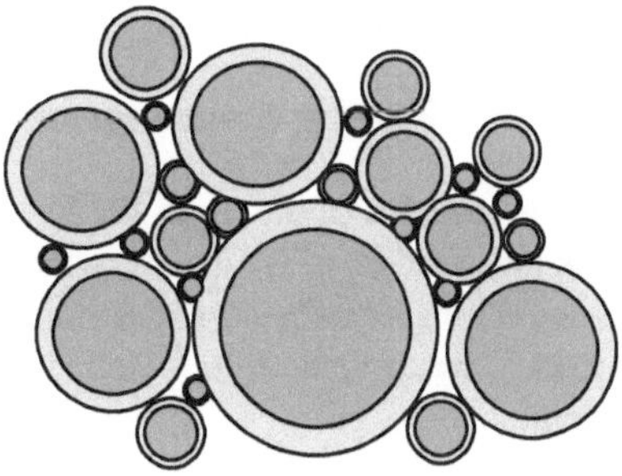

FIGURE 7.5
Assemblage of coated spheres.

For the assemblage of coated spheres, when the bulk and shear moduli of the two phases are real, the effective bulk modulus is given by equation (7.24). Let us use the Bloch wave solutions to show that the same result holds when the moduli are complex. Suppose that the spheres are made of gas and the coating is liquid so that the composite resembles bubbles in water. If we assume that the liquid is incompressible, then $\kappa_2 \to \infty$ which implies that $\boldsymbol{\nabla} \cdot \mathbf{u}^{(2)} \to 0$. If we follow the Hashin procedure again, we get an expression for the effective bulk modulus of the form

$$\kappa_{\text{eff}} = \frac{2\mu_2\left[(1+2f_1)\kappa_1 + \frac{4}{3}f_2\mu_2\right]}{3f_2\kappa_1 + 2(2+f_1)\mu_2}.$$

A simpler expression can be obtained if we invert (7.24) and take the limit $\kappa_2 \to \infty$. In that case we have

$$\kappa_{\text{eff}} \approx \frac{\kappa_1 + \frac{4}{3}f_2\mu_2}{f_1}. \tag{7.25}$$

Now consider a time harmonic plane shear wave with a real frequency, ω, and complex wave vector, $\mathbf{k}$, propagating into a fluid. Then the associated stress and strain

[††]In determining the effective properties of an assemblage of coated spheres we have assumed that the coated spheres do not overlap the boundary of the spherical unit cells and that tractions and displacements at the boundary remain unaffected by the addition of coated spheres.

fields can be expressed as

$$\boldsymbol{\sigma}(\mathbf{x},t) = \mathrm{Re}[\widehat{\boldsymbol{\sigma}}\, e^{i(\mathbf{k}\cdot\mathbf{x}-\omega t)}] \quad \text{and} \quad \boldsymbol{\varepsilon}(\mathbf{x},t) = \mathrm{Re}[\widehat{\boldsymbol{\varepsilon}}\, e^{i(\mathbf{k}\cdot\mathbf{x}-\omega t)}].$$

If the fluid is Newtonian, the stress is related to the strain by

$$\boldsymbol{\sigma} = \eta_\mu \frac{\partial \boldsymbol{\varepsilon}}{\partial t}$$

where η_μ is the shear viscosity of the fluid. Substituting the expressions for the stress and the strain into the above constitutive relation gives us

$$\boldsymbol{\sigma}(\mathbf{x},\omega) = -i\omega\eta_\mu \boldsymbol{\varepsilon}(\mathbf{x},\omega) =: \mu \boldsymbol{\varepsilon}(\mathbf{x},\omega)$$

where μ is the complex shear modulus of the fluid. We have assumed that the shear viscosity of the fluid, η_μ, is independent of frequency, which is reasonable for water. If we now consider water containing air bubbles, and assume that the bulk modulus of air is independent of frequency, we can plug $\mu_2 = -i\omega\eta_\mu$ into (7.25) (after appealing to analytic continuation arguments) to get

$$\kappa_{\text{eff}} \approx \frac{\kappa_1}{f_1} - i\omega\left(\frac{4f_2}{3f_1}\eta_\mu\right) =: \frac{\kappa_1}{f_1} - i\omega\eta_\kappa^{\text{eff}}$$

where the η_κ^{eff} is the effective bulk viscosity of the mixture. Since the imaginary part of the effective bulk modulus, and hence dissipation, is determined by the shear viscosity of water we find that sound is damped strongly in bubbly fluids. This is true even when the volume fraction of bubbles is small, in which case, $f_2 \to 1$, and we get the Taylor estimate of effective bulk viscosity,

$$\eta_\kappa^{\text{eff}} \to \frac{4}{3f_1}\eta_\mu\,.$$

7.3 Electromagnetic waves in the quasistatic limit

Let us now consider the solution of Maxwell's equation in periodic media in the quasistatic limit. We will use the notation shown in Figure 7.3 and follow the procedure used in the previous section (see Milton (2002) for further details). In this case we define the scaled permittivity and magnetic permeability as

$$\varepsilon_\eta(\mathbf{x}) := \varepsilon\left(\tfrac{\mathbf{x}}{\eta}\right) = \varepsilon(\mathbf{y}) \quad \text{and} \quad \mu_\eta(\mathbf{x}) := \mu\left(\tfrac{\mathbf{x}}{\eta}\right) = \mu(\mathbf{y})$$

where the periodicity of the composite, with lattice vector $\mathbf{a}_i$, requires that

$$\varepsilon_\eta(\mathbf{x}+\eta\mathbf{a}_i) = \varepsilon_\eta(\mathbf{x}) \quad \text{and} \quad \mu_\eta(\mathbf{x}+\eta\mathbf{a}_i) = \mu_\eta(\mathbf{x})\,.$$

Similarly, we define

$$\mathbf{D}_\eta(\mathbf{x}) := \mathbf{D}(\mathbf{y})\,;\;\; \mathbf{B}_\eta(\mathbf{x}) := \mathbf{B}(\mathbf{y})\,;\;\; \mathbf{E}_\eta(\mathbf{x}) := \mathbf{E}(\mathbf{y})\,;\;\; \mathbf{H}_\eta(\mathbf{x}) := \mathbf{H}(\mathbf{y})\,.$$

Then the Maxwell equations at fixed frequency, ω, can be written as

$$\boldsymbol{\nabla}\cdot\mathbf{D}_\eta = 0\,;\;\; \boldsymbol{\nabla}\cdot\mathbf{B}_\eta = 0\,;\;\; \boldsymbol{\nabla}\times\mathbf{E}_\eta - i\omega\,\mathbf{B}_\eta = \mathbf{0}\,;\;\; \boldsymbol{\nabla}\times\mathbf{H}_\eta + i\omega\,\mathbf{D}_\eta = \mathbf{0} \tag{7.26}$$

and from the constitutive relations we have

$$\mathbf{D}_\eta(\mathbf{x}) = \varepsilon_\eta(\mathbf{x})\mathbf{E}_\eta(\mathbf{x}) \quad \text{and} \quad \mathbf{B}_\eta(\mathbf{x}) = \mu_\eta(\mathbf{x})\mathbf{H}_\eta(\mathbf{x})\,. \tag{7.27}$$

For the **E**-field let us look for Bloch wave solutions of the form

$$\mathbf{E}_\eta(\mathbf{x}) = e^{i\mathbf{k}\cdot\mathbf{x}}\,\mathbf{e}_\eta(\mathbf{x}) \qquad \text{and} \qquad \mathbf{D}_\eta(\mathbf{x}) = e^{i\mathbf{k}\cdot\mathbf{x}}\,\mathbf{d}_\eta(\mathbf{x})$$

where $\mathbf{e}_\eta$ and $\mathbf{d}_\eta$ have the same periodicity as ε and μ, i.e.,

$$\mathbf{e}_\eta(\mathbf{x}+\eta\,\mathbf{a}_i) = \mathbf{e}(\mathbf{x}) \quad \text{and} \quad \mathbf{d}_\eta(\mathbf{x}+\eta\,\mathbf{a}_i) = \mathbf{d}(\mathbf{x})\,.$$

Similarly, for the **H**-field we look for Bloch wave solutions of the form

$$\mathbf{H}_\eta(\mathbf{x}) = e^{i\mathbf{k}\cdot\mathbf{x}}\,\mathbf{h}_\eta(\mathbf{x}) \qquad \text{and} \qquad \mathbf{B}_\eta(\mathbf{x}) = e^{i\mathbf{k}\cdot\mathbf{x}}\,\mathbf{b}_\eta(\mathbf{x})$$

where $\mathbf{h}_\eta, \mathbf{b}_\eta$ have the periodicity

$$\mathbf{h}_\eta(\mathbf{x}+\eta\,\mathbf{a}_i) = \mathbf{h}(\mathbf{x}) \qquad \text{and} \qquad \mathbf{b}_\eta(\mathbf{x}+\eta\,\mathbf{a}_i) = \mathbf{b}(\mathbf{x})\,.$$

If we substitute these solutions into equations (7.26) we get

$$\begin{aligned} i\mathbf{k}\cdot\mathbf{d}_\eta(\mathbf{x}) + \boldsymbol{\nabla}\cdot\mathbf{d}_\eta = 0, & \qquad i\mathbf{k}\cdot\mathbf{b}_\eta(\mathbf{x}) + \boldsymbol{\nabla}\cdot\mathbf{b}_\eta = 0, \\ i\mathbf{k}\times\mathbf{e}_\eta(\mathbf{x}) + \boldsymbol{\nabla}\times\mathbf{e}_\eta - i\omega\mathbf{b}_\eta(\mathbf{x}) = 0, & \quad i\mathbf{k}\times\mathbf{h}_\eta(\mathbf{x}) + \boldsymbol{\nabla}\times\mathbf{h}_\eta + i\omega\mathbf{d}_\eta(\mathbf{x}) = 0\,. \end{aligned} \tag{7.28}$$

Similarly, from (7.27) we have

$$\mathbf{d}_\eta(\mathbf{x}) = \varepsilon_\eta(\mathbf{x})\,\mathbf{e}_\eta(\mathbf{x}) \quad \text{and} \quad \mathbf{b}_\eta(\mathbf{x}) = \mu_\eta(\mathbf{x})\,\mathbf{h}_\eta(\mathbf{x})\,. \tag{7.29}$$

Using (7.29) we can organize (7.28) into an eigenvalue problem of the form

$$(\mathcal{L} - \omega\mathbf{1})\begin{bmatrix}\mathbf{d}_\eta \\ \mathbf{b}_\eta\end{bmatrix} = \mathbf{0}$$

where $\mathcal{L}$ is a linear differential operator. Then ω is an eigenvalue of $\mathcal{L}$. However, $\mathcal{L}$ depends on ω via $\varepsilon(\omega)$ and $\mu(\omega)$ and therefore Bloch wave solutions do not exist unless ω takes one of a discrete set of values. As before, let these discrete values be

$$\omega = \omega_\eta^j(\mathbf{k})$$

where the superscript j labels the solution branches. Let us see what the Bloch wave solutions reduce to as $\eta \to 0$. Following standard multiple scale analysis, let us assume that the periodic complex fields have the expansions

$$\begin{aligned}
\mathbf{e}_\eta(\mathbf{x}) &= \mathbf{e}_0(\mathbf{y}) + \eta\mathbf{e}_1(\mathbf{y}) + \eta^2\mathbf{e}_2(\mathbf{y}) + \dots \\
\mathbf{d}_\eta(\mathbf{x}) &= \mathbf{d}_0(\mathbf{y}) + \eta\mathbf{d}_1(\mathbf{y}) + \eta^2\mathbf{d}_2(\mathbf{y}) + \dots \\
\mathbf{h}_\eta(\mathbf{x}) &= \mathbf{h}_0(\mathbf{y}) + \eta\mathbf{h}_1(\mathbf{y}) + \eta^2\mathbf{h}_2(\mathbf{y}) + \dots \\
\mathbf{b}_\eta(\mathbf{x}) &= \mathbf{b}_0(\mathbf{y}) + \eta\mathbf{b}_1(\mathbf{y}) + \eta^2\mathbf{b}_2(\mathbf{y}) + \dots .
\end{aligned} \tag{7.30}$$

Let us also assume that the dependence of ω on η and $\mathbf{k}$ has an expansion of the form

$$\omega = \omega_\eta^j(\mathbf{k}) = \omega_0^j(\mathbf{y}) + \eta\, \omega_1^j(\mathbf{y}) + \eta^2\, \omega_2^j(\mathbf{y}) + \dots . \tag{7.31}$$

Plugging (7.31) and (7.30) into (7.28), and following the process that we had used to derive the quasistatic elasticity equations, we find that collecting terms of order $1/\eta$ gives

$$\boxed{\begin{aligned}
\boldsymbol{\nabla}_y \cdot \mathbf{d}_0 &= 0, & \boldsymbol{\nabla}_y \times \mathbf{e}_0 &= \mathbf{0}, \\
\boldsymbol{\nabla}_y \cdot \mathbf{b}_0 &= 0, & \boldsymbol{\nabla}_y \times \mathbf{h}_0 &= \mathbf{0}.
\end{aligned}} \tag{7.32}$$

These are the Maxwell equations in the quasistatic limit. Similarly, collecting terms of order 1 gives

$$\boxed{\begin{aligned}
i\mathbf{k} \cdot \mathbf{d}_0(\mathbf{y}) + \boldsymbol{\nabla}_y \cdot \mathbf{d}_1 &= 0, & i\mathbf{k} \times \mathbf{e}_0(\mathbf{y}) + \boldsymbol{\nabla}_y \times \mathbf{e}_1 - i\omega_0^j \mathbf{b}_0(\mathbf{y}) &= \mathbf{0} \\
i\mathbf{k} \cdot \mathbf{b}_0(\mathbf{y}) + \boldsymbol{\nabla}_y \cdot \mathbf{b}_1 &= 0, & i\mathbf{k} \times \mathbf{h}_0(\mathbf{y}) + \boldsymbol{\nabla}_y \times \mathbf{h}_1 + i\omega_0^j \mathbf{d}_0(\mathbf{y}) &= \mathbf{0}.
\end{aligned}} \tag{7.33}$$

These are additional constrains that must be satisfied by Maxwell's equations in the quasistatic limit. Also, from the constitutive equations (7.29),

$$\boxed{\mathbf{d}_0(\mathbf{y}) = \varepsilon(\mathbf{y})\, \mathbf{e}_0(\mathbf{y}) \quad \text{and} \quad \mathbf{b}_0(\mathbf{y}) = \mu(\mathbf{y})\, \mathbf{h}_0(\mathbf{y}).} \tag{7.34}$$

These are the constitutive relations in the quasistatic limit.

7.3.1 Effective properties and dispersion relations

Let us use the approach in Milton (2002) (p. 227) to find the effective permittivity and magnetic permeability of a periodic composite using Maxwell's equations in the quasistatic limit. Let $\boldsymbol{\varepsilon}_{\text{eff}}$ be an effective permittivity tensor associated with the field $\boldsymbol{\varepsilon}(\mathbf{y})$ and let $\boldsymbol{\mu}_{\text{eff}}$ be an effective permeability tensor associated with the field $\boldsymbol{\mu}(\mathbf{y})$. These effective tensors are defined through

$$\langle \mathbf{d}_0 \rangle = \boldsymbol{\varepsilon}_{\text{eff}} \cdot \langle \mathbf{e}_0 \rangle \quad \text{and} \quad \langle \mathbf{b}_0 \rangle = \boldsymbol{\mu}_{\text{eff}} \cdot \langle \mathbf{h}_0 \rangle . \tag{7.35}$$

Since $\mathbf{d}_1(\mathbf{y}), \mathbf{b}_1(\mathbf{y}), \mathbf{e}_1(\mathbf{y}), \mathbf{h}_1(\mathbf{y})$ are periodic, the volume average divergence and curl over a unit cell must be zero, i.e.,

$$\begin{aligned}
\langle \boldsymbol{\nabla}_y \cdot \mathbf{d}_1 \rangle &= 0, & \langle \boldsymbol{\nabla}_y \times \mathbf{e}_1 \rangle &= \mathbf{0} \\
\langle \boldsymbol{\nabla}_y \cdot \mathbf{b}_1 \rangle &= 0, & \langle \boldsymbol{\nabla}_y \times \mathbf{h}_1 \rangle &= \mathbf{0}
\end{aligned} \tag{7.36}$$

where $\langle \bullet \rangle$ is the volume average over the unit cell. Therefore, equations (7.33) can be written as

$$\begin{aligned} \mathbf{k} \cdot \langle \mathbf{d}_0 \rangle &= 0, \qquad \mathbf{k} \times \langle \mathbf{e}_0 \rangle - \omega_0^j \langle \mathbf{b}_0 \rangle = \mathbf{0} \\ \mathbf{k} \cdot \langle \mathbf{b}_0 \rangle &= 0, \qquad \mathbf{k} \times \langle \mathbf{h}_0 \rangle + \omega_0^j \langle \mathbf{d}_0 \rangle = \mathbf{0}\,. \end{aligned} \tag{7.37}$$

Plugging (7.35) into the two equations containing ω_0^j in (7.37) gives

$$\mathbf{k} \times \langle \mathbf{e}_0 \rangle - \omega_0^j \boldsymbol{\mu}_{\text{eff}} \cdot \langle \mathbf{h}_0 \rangle = \mathbf{0} \quad \text{and} \quad \mathbf{k} \times \langle \mathbf{h}_0 \rangle + \omega_0^j \boldsymbol{\varepsilon}_{\text{eff}} \cdot \langle \mathbf{e}_0 \rangle = \mathbf{0}\,. \tag{7.38}$$

From the first of equations (7.38) we have

$$\langle \mathbf{h}_0 \rangle = (\omega_0^j \boldsymbol{\mu}_{\text{eff}})^{-1} \cdot (\mathbf{k} \times \langle \mathbf{e}_0 \rangle)\,. \tag{7.39}$$

Plugging this expression into the second of equations (7.38) gives

$$\boldsymbol{\varepsilon}_{\text{eff}}^{-1} \cdot \left[\mathbf{k} \times \left\{\boldsymbol{\mu}_{\text{eff}}^{-1} \cdot (\mathbf{k} \times \langle \mathbf{e}_0 \rangle)\right\}\right] + (\omega_0^j)^2 \langle \mathbf{e}_0 \rangle = \mathbf{0}\,. \tag{7.40}$$

Define the matrix $\boldsymbol{A}_{\text{eff}}(\mathbf{k})$ such that for all vectors $\mathbf{v}$

$$\boldsymbol{A}_{\text{eff}}(\mathbf{k}) \cdot \mathbf{v} := \boldsymbol{\varepsilon}_{\text{eff}}^{-1} \cdot \left[\mathbf{k} \times \left\{\boldsymbol{\mu}_{\text{eff}}^{-1} \cdot (\mathbf{k} \times \mathbf{v})\right\}\right]\,.$$

Then (7.40) can be written as an eigenvalue problem

$$\boxed{\left[\boldsymbol{A}_{\text{eff}}(\mathbf{k}) + (\omega_0^j)^2 \boldsymbol{1}\right] \cdot \langle \mathbf{e}_0 \rangle = \mathbf{0}\,.} \tag{7.41}$$

The matrix $\boldsymbol{A}_{\text{eff}}$ is 3×3 for three-dimensional problems and has three eigenvalues. The eigenvalue problem (7.41) may be used to determine the dispersion relations $\omega_0^j(\mathbf{k})$ and also possible values of the eigenvectors $\langle \mathbf{e}_0 \rangle$ associated with each Bloch wave mode. Once we know $\langle \mathbf{e}_0 \rangle$ we can then find $\langle \mathbf{h}_0 \rangle$ using equation (7.39). Note that one trivially determined eigenvector/eigenvalue pair of the matrix $\boldsymbol{A}_{\text{eff}}$ is $(\langle \mathbf{e}_0 \rangle = \mathbf{k}, \omega_0^j = 0)$. So, it is necessary to examine only the other two eigenvalues of $\boldsymbol{A}_{\text{eff}}$ to find allowable average fields $\langle \mathbf{e}_0 \rangle$ and $\langle \mathbf{h}_0 \rangle$.

Isotropic constituents

When the permittivity and permeability of the medium are isotropic, equation (7.41) becomes

$$(\varepsilon_{\text{eff}} \mu_{\text{eff}})^{-1} (\mathbf{k} \times \mathbf{k} \times \langle \mathbf{e}_0 \rangle) + (\omega_0^j)^2 \langle \mathbf{e}_0 \rangle = \mathbf{0}\,.$$

Using the identity $\mathbf{a} \times \mathbf{b} \times \mathbf{c} = (\mathbf{a} \cdot \mathbf{c})\mathbf{b} - (\mathbf{a} \cdot \mathbf{b})\mathbf{c}$, we can write the above as

$$(\varepsilon_{\text{eff}} \mu_{\text{eff}})^{-1} [(\mathbf{k} \cdot \langle \mathbf{e}_0 \rangle)\mathbf{k} - (\mathbf{k} \cdot \mathbf{k}) \langle \mathbf{e}_0 \rangle] + (\omega_0^j)^2 \langle \mathbf{e}_0 \rangle = \mathbf{0}\,.$$

Further rearrangement leads to

$$(\varepsilon_{\text{eff}} \mu_{\text{eff}})^{-1} [\mathbf{k} \otimes \mathbf{k} - (\mathbf{k} \cdot \mathbf{k})\boldsymbol{1}] \cdot \langle \mathbf{e}_0 \rangle + (\omega_0^j)^2 \langle \mathbf{e}_0 \rangle = \mathbf{0}\,.$$

If we express the above relation in matrix form in a Cartesian basis, we get

$$\begin{bmatrix} k_2^2+k_3^2 & -k_1k_2 & -k_1k_3 \\ -k_1k_2 & k_1^2+k_3^2 & -k_2k_3 \\ -k_1k_3 & -k_2k_3 & k_1^2+k_2^2 \end{bmatrix} \begin{bmatrix} \langle e_0 \rangle_1 \\ \langle e_0 \rangle_2 \\ \langle e_0 \rangle_3 \end{bmatrix} = \lambda \begin{bmatrix} \langle e_0 \rangle_1 \\ \langle e_0 \rangle_2 \\ \langle e_0 \rangle_3 \end{bmatrix} \; ; \quad \lambda := (\omega_0^j)^2 \varepsilon_{\text{eff}} \mu_{\text{eff}} .$$

The eigenvalues of the matrix on the left-hand side are

$$\lambda_1 = 0 \, , \quad \lambda_2 = \lambda_3 = k_1^2 + k_2^2 + k_3^2 = \mathbf{k} \cdot \mathbf{k} .$$

Ignoring the trivial eigenvalue, we have the dispersion relation

$$\mathbf{k} \cdot \mathbf{k} = (\omega_0^j)^2 \varepsilon_{\text{eff}} \mu_{\text{eff}} . \tag{7.42}$$

We can calculate the effective phase velocity from the relation $c_{\text{eff}} = 1/\sqrt{\varepsilon_{\text{eff}}\mu_{\text{eff}}}$ and the effective group velocity by taking a derivative of the dispersion relation. If we assume that $\mu_{\text{eff}} = 1$ and

$$\mathbf{k} = \left(\tfrac{2\pi}{\lambda} + \tfrac{i}{\delta}\right) \mathbf{n}$$

where λ is the wavelength, δ is the attenuation length, and $\mathbf{n}$ is a unit vector, equation (7.42) gives

$$\varepsilon_{\text{eff}} = \frac{1}{(\omega_0^j)^2} \left(\tfrac{2\pi}{\lambda} + \tfrac{i}{\delta}\right)^2 .$$

Therefore, for a given frequency, the complex permittivity ε_{eff} is determined by the wavelength and attenuation length of the incident field.

7.4 Band gap phenomena in periodic composites

For wavelengths that are of the order of the unit cell and shorter, we can no longer use the quasistatic approximation. Band structures are useful for examining the behavior of periodic composites in those situations. Let us look at a few discrete systems to understand the basic procedure used to compute band structures. Problems involving electrodynamics are dealt with in detail by Joannopoulos and Winn (2008). We will limit ourselves to elastodynamic problems. Problems in acoustics can be solved using the same procedures.

The band structure of continuous systems can be quite difficult to calculate and several methods can be applied to achieve that goal. Popular approaches are plane wave expansions (Sigalas and Economou, 1992, Kushwaha et al., 1993, Suzuki and Yu, 1998), finite difference time domain methods (Sigalas and Garcia, 2000), finite element methods (Axmann and Kuchment, 1999, Hussein, 2009), multiple scattering theory (Liu et al., 2002), Rayleigh multipole and Green's function methods (Movchan et al., 1997, Poulton et al., 2000), and various other techniques. The aim of most of these methods is to reduce the problems into a generalized matrix eigenvalue problem of the form

$$[\underline{\underline{K}}(\mathbf{k}) - \omega^2 \underline{\underline{M}}]\underline{u} = \underline{0}$$

where $\underline{\underline{K}}$ and $\underline{\underline{M}}$ are matrices containing information about the geometry and material parameters, $\mathbf{k}$ is the wave number, ω is the frequency, and $\underline{u}$ is the vector of degrees of freedom in the system. The differences between the various methods are essentially in the choice of basis. Once a system of equations has been generated, standard numerical methods can be used to solve the system for individual values of $\mathbf{k}$ (usually chosen to be such that they lie along the borders of an irreducible Brillouin zone).

A discussion of the details of these techniques is beyond the scope of this book and excellent reviews can be found in Busch et al. (2007) and Sigalas et al. (2005). We will limit ourselves to a few simple lattice models in this section. Note that many complex structures can be adequately represented by lattice models, particularly in elastodynamics. These models have the advantage of reducing the number of degrees of freedom significantly while retaining much of the important physics. An excellent review of lattice models in micromechanics can be found in Ostoja-Starzewski (2002).

7.4.1 One-dimensional lattice models

The simplest periodic structure we can think of is a lumped mass model of a homogeneous elastic rod as shown in Figure 7.6(a). The period of the structure is a, the lumped masses are m and the spring stiffness is K. The displacements at the locations x_j are u_j and the unit cell is repeated an infinite number of times. The equation of

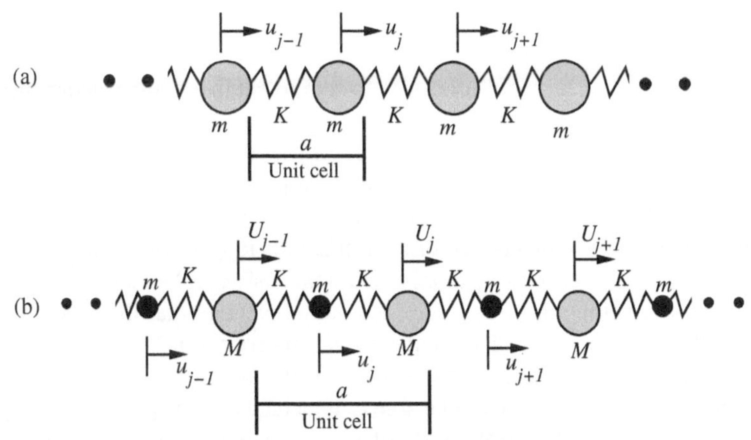

FIGURE 7.6
Lumped mass models of one-dimensional periodic media. a) A model with a single mass. b) A model with two different masses.

motion of the j-th mass is

$$K(u_{j+1} - u_j) - K(u_j - u_{j-1}) = m\ddot{u}_j\,.$$

If the motion is time harmonic with frequency ω, we have

$$K(u_{j+1} - u_j) - K(u_j - u_{j-1}) + m\omega^2 u_j = 0\,.$$

From the Bloch-Floquet theorem we have

$$u_{j-1} = e^{-ika} u_j \quad \text{and} \quad u_{j+1} = e^{ika} u_j\,.$$

Therefore we can write the equation of motion as

$$K u_j (2 - e^{ika} - e^{-ika}) - m\omega^2 u_j = 0\,.$$

Noting that $\exp(ika) + \exp(-ika) = 2\cos(ka)$ and $1 - \cos(ka) = 2\sin^2(ka/2)$, we have

$$\left[\frac{4K}{m}\sin^2(ka/2) - \omega^2\right] u_j = 0\,.$$

For real k, and since $\omega > 0$, this equation has a nontrivial solution only if

$$\omega(k) = 2\sqrt{\frac{K}{m}}\,|\sin(ka/2)| =: 2\omega_0\,|\sin(ka/2)|\,. \tag{7.43}$$

This is the dispersion relation for the structure. In the quasistatic limit, $k \to 0$, $\sin(ka/2) \to ka/2$, we have $\omega(k) = ka\sqrt{K/m}$. The group velocity is $\mathrm{d}\omega/\mathrm{d}k = a\sqrt{K/m}$. So if we know the group velocity, the lattice spacing, and the mass density, we can calculate the stiffness of the medium. But recall that the lattice model was meant to represent a homogeneous medium for which the wave number and the frequency should have a linear relationship, i.e., there should be no dispersion in such a medium. The reason for the discrepancy is that, to get to a homogeneous medium, we have to take the limit $a \to 0$. In that case we again have $\omega(k) = ka\sqrt{K/m}$, which is linear. We have to be careful when interpreting the dispersion relations for a continuous medium that has been modeled using a lattice structure.

Let us now look at a plot of the dispersion relation (7.43). The frequency is a periodic function of k with a period of $2\pi/a$ that falls to zero for certain values of k. The mode where both ω and k go to zero simultaneously is called the *acoustic mode*. Recall from the definition of the reciprocal lattice vectors that the spacing of the reciprocal lattice is $b := 1/a$. The frequency is zero every time k becomes a integer multiple of $2\pi b$. The periodicity of the plot indicates that we may concentrate on a small range of values of k and still be able to understand the behavior of the system for all k. This irreducible set of values of k is indicated by the gray region in the figure and is called the first *Brillouin zone*. We also see that at the boundaries of the Brillouin zone $\omega = 2\omega_0$, and the group velocity $\mathrm{d}\omega/\mathrm{d}k$ becomes 0. Therefore we have standing waves for values of k which are odd integer multiples of π/a. Because there is only one K and one m in the model, we see only one mode. Also, since there

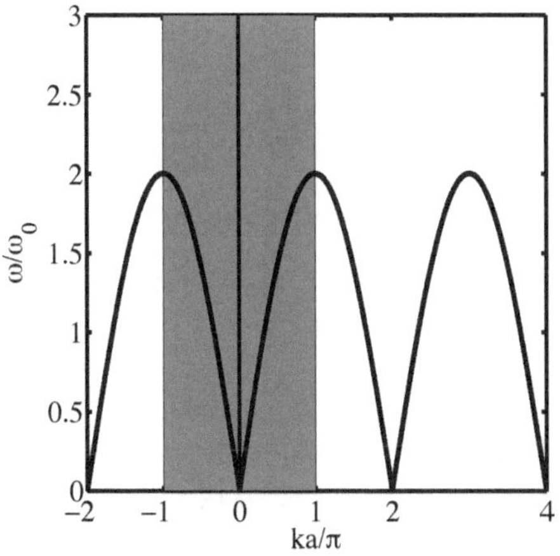

FIGURE 7.7
Dispersion plot of one-dimensional lumped-mass model of a homogeneous medium. The gray region is the first Brillouin zone.

is only one direction of wave propagation, we see only one acoustic branch. More modes and additional acoustic branches are observed as further details are included in the model.

In fact, we can use a reduced range of values of k for any periodic medium. We can see this by considering a plane wave of wavelength λ propagating in the medium with wave vector $\mathbf{k}$. Then we can superpose solutions of the form

$$u(\mathbf{x}) = \widehat{u}\,e^{i\mathbf{k}\cdot\mathbf{x}} \quad \text{where} \quad \|\mathbf{k}\| = 2\pi/\lambda\,.$$

Let $\mathbf{a}$ be a lattice vector in the medium such that $\mathbf{a} = n_j\mathbf{a}_j$ where $\mathbf{a}_j$ are the primitive basis vectors of the lattice and n_j are integers. Now consider another wave with a wave vector $\mathbf{k}' = \mathbf{k} + 2\pi\mathbf{b}$[‡‡] where $\mathbf{b} = m_j\mathbf{b}_j$ where $\mathbf{b}_j$ are the reciprocal lattice vectors and m_j are integers. Then $\mathbf{b}\cdot\mathbf{a} = m_k n_k = N$ where N is an integer. This implies that

$$e^{i\mathbf{k}'\cdot\mathbf{a}} = e^{i\mathbf{k}\cdot\mathbf{a}}\,e^{2\pi i N} = e^{i\mathbf{k}\cdot\mathbf{a}}\,.$$

Therefore, wave vectors differing by a reciprocal lattice vector have the same effect at the location $\mathbf{x} = \mathbf{a}$. Hence we only need to consider a small region of k-space.

Let us now look at the model in Figure 7.6(b). In this case we can write out two separate equations of motion for the two types of mass,

$$K(2U_j - u_j - u_{j+1}) - \omega^2 M U_j = 0 \quad \text{and} \quad K(2u_j - U_{j-1} - U_j) - \omega^2 m u_j = 0\,.$$

[‡‡]The quantity $2\pi\mathbf{b}$ is often written as $\mathbf{G}$.

From the Bloch-Floquet theorem, $u_{j+1} = u_j \exp(ika)$ and $U_{j-1} = U_j \exp(-ika)$. Therefore,

$$K[2U_j - u_j(1+e^{ika})] - \omega^2 M U_j = 0 \quad \text{and} \quad K[2u_j - U_j(1+e^{-ika})] - \omega^2 m u_j = 0\,.$$

We can write these two equation in matrix form as

$$(\underline{\underline{\mathsf{K}}} - \omega^2 \underline{\underline{\mathsf{M}}})\underline{\underline{\mathsf{u}}} = \underline{\underline{\mathsf{0}}} \quad \text{or} \quad \begin{bmatrix} 2K - M\omega^2 & -K(1+e^{ika}) \\ -K(1+e^{-ika}) & 2K - m\omega^2 \end{bmatrix} \begin{bmatrix} U_j \\ u_j \end{bmatrix} = \begin{bmatrix} 0 \\ 0 \end{bmatrix}.$$

This eigenvalue problem has a nontrivial solution if $\det(\underline{\underline{\mathsf{K}}} - \omega^2 \underline{\underline{\mathsf{M}}}) = 0$, i.e.,

$$mM\omega^4 - 2K(m+M)\omega^2 + 4K^2 \sin^2(ka/2) = 0\,.$$

The positive eigenvalues are

$$\omega(k) = \left[K\frac{m+M}{mM} \pm \frac{K}{mM}\sqrt{m^2 + M^2 + 2mM\cos(ka)}\right]^{1/2}.$$

A dispersion plot for this situation is shown in Figure 7.8. In this case we get one extra mode, called the optical mode. If we examine the value of the displacement at the Brillouin zone boundary we will see that one of the masses moves while the other stands still, i.e, if $U_j = 0$ then $u_j \neq 0$ and vice versa. Also notice that at $k = 0$ the group velocity of the optical mode is zero and standing waves appear. The appearance of a frequency band gap is also significant. This gap depends on the difference between m and M and explains why high contrast composites are more likely to exhibit band gaps.

7.4.2 Two-dimensional lattice models

The unit cell and the first Brillouin zone for one-dimensional lattices are quite straightforward to visualize and compute. But when dealing with two- and three-dimensional lattices, these structures can become quite complicated, particularly in three dimensions. Let us revisit the concept of the reciprocal lattice and examine the Brillouin zone in two dimensions.

Recall that the reciprocal lattice vectors $\mathbf{b}_j$ are related to the primitive lattice vectors $\mathbf{a}_j$ by the relation $\mathbf{a}_i \cdot \mathbf{b}_j = \delta_{ij}$.* Based on the observation of the periodicity of the dispersion relations, it is often more convenient to define the reciprocal vectors such that $\mathbf{a}_i \cdot \mathbf{b}_j = 2\pi\delta_{ij}$. Figure 7.9 shows a two-dimensional schematic of these vectors and the associated first Brillouin zone in reciprocal space. With the new definition of the reciprocal basis vectors we see that, in one dimension, the Brillouin zone will extend from $-\pi/a$ to π/a. A detailed explanation can be found in Brillouin (2003).

*Observe that the primitive lattice vectors and the reciprocal lattice vectors are similar to the covariant and contravariant basis vectors in curvilinear coordinates.

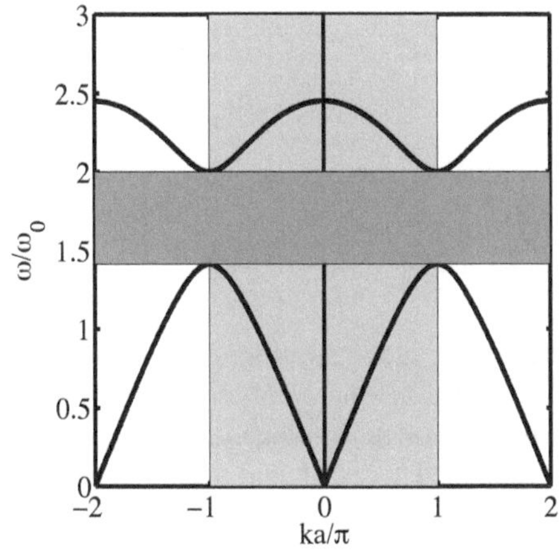

FIGURE 7.8

Dispersion plot of one-dimensional model containing two lumped masses with $M = 2m$ and $\omega_0 = \sqrt{K/mM}$. The light gray region is the first Brillouin zone and the dark gray region is the frequency band gap.

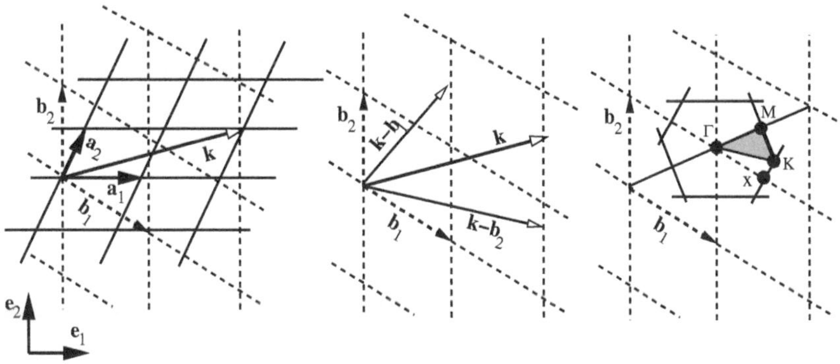

FIGURE 7.9

Lattice vectors, reciprocal vectors, and Brillouin zone for a two-dimensional lattice. The solid lines represent the physical lattice. The dashed lines show the reciprocal lattice. The Brillouin zone is the smallest closed polygon formed by the perpendicular bisectors of the unit cell in reciprocal space. The edges of the Brillouin zone are often labeled using group theoretic notation; the notation shown in the figure is for a hexagonal lattice.

The reciprocal lattice vectors can be calculated from

$$\mathbf{b}_1 = 2\pi \frac{\mathbf{a}_2 \times \mathbf{a}_3}{\mathbf{a}_1 \cdot (\mathbf{a}_2 \times \mathbf{a}_3)}, \quad \mathbf{b}_2 = 2\pi \frac{\mathbf{a}_3 \times \mathbf{a}_1}{\mathbf{a}_2 \cdot (\mathbf{a}_3 \times \mathbf{a}_1)}, \quad \mathbf{b}_3 = 2\pi \frac{\mathbf{a}_1 \times \mathbf{a}_2}{\mathbf{a}_3 \cdot (\mathbf{a}_1 \times \mathbf{a}_2)}.$$

For the lattice in Figure 5.28 (p. 202), the lattice vectors of the unit cell are

$$\mathbf{a}_1 = h\mathbf{e}_1 - h\mathbf{e}_2 \ , \ \ \mathbf{a}_2 = h\mathbf{e}_1 + h\mathbf{e}_2 \ , \ \ \mathbf{a}_3 = \mathbf{e}_3 \, .$$

The corresponding reciprocal vectors are

$$\mathbf{b}_1 = \frac{\pi}{h}\mathbf{e}_1 - \frac{\pi}{h}\mathbf{e}_1 \ , \ \ \mathbf{b}_2 = \frac{\pi}{h}\mathbf{e}_1 + \frac{\pi}{h}\mathbf{e}_2 \ , \ \ \mathbf{b}_3 = 2\pi\mathbf{e}_3 \, .$$

Consider a square lattice with lattice spacing a (Figure 7.10).* To keep things simple we will examine the situation where only an out-of-plane displacement $w(\mathbf{x})$ is allowed at the lattice nodes (but no in-plane displacements). The lattice nodes are connected with strings that have a distributed mass per unit length ρ. We may think of the lattice as a model of a membrane that is represented as a net of flexible strings.

If we look at the free-body diagram of an element of the string of length dx that is stretched under a tension T, a balance of forces in the out-of-plane direction gives us, for small deflections,

$$T\left(\frac{\partial w}{\partial x} + \frac{\partial^2 w}{\partial x^2}\mathrm{d}x\right) - T\frac{\partial w}{\partial x} = \rho \mathrm{d}x \frac{\partial^2 w}{\partial t^2}$$

or

$$\frac{\partial^2 w}{\partial x^2} = \frac{1}{c^2}\frac{\partial^2 w}{\partial t^2} \quad \text{where} \quad c^2 := \frac{T}{\rho} \, .$$

If we look for time harmonic solutions with frequency ω, we can write the above

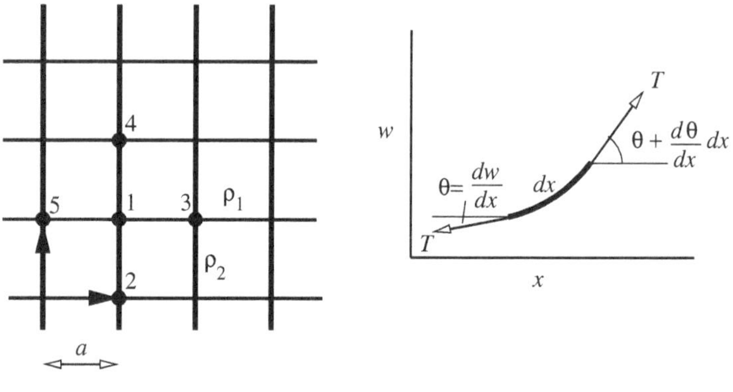

FIGURE 7.10
A square lattice made of strings with a mass densities ρ_1 and ρ_2.

*This example is based on Martinsson and Movchan (2003). Several other examples can be found in that work and in Jensen (2003).

wave equation and its solution as

$$\frac{\partial^2 w}{\partial x^2} + \frac{\omega^2}{c^2} w = 0 \quad \Longrightarrow \quad w(x) = A\,e^{i\kappa x} + B\,e^{-i\kappa x}\ , \quad \kappa := \frac{\omega}{c}\,.$$

Using the boundary conditions $w(0) = w_0$ and $w(a) = w_a$ we have the solution

$$w(x) = w_0 \cos(\kappa x) + [w_a - w_0 \cos(\kappa a)]\,\frac{\sin(\kappa x)}{\sin(\kappa a)}$$

and

$$\theta(x) = \frac{dw}{dx} = -\kappa w_0 \sin(\kappa x) + \kappa\,[w_a - w_0 \cos(\kappa a)]\,\frac{\cos(\kappa x)}{\sin(\kappa a)}\,.$$

The situation $\sin(\kappa a) = 0$ represents the normal mode of vibration of the string. We will consider the situation where $\sin(\kappa a) \neq 0$. The force that the string exerts at $x = 0$ in the out-of-plane direction is

$$F_z = -T\theta(0) = [w_0 \cos(\kappa a) - w_a]\,\frac{T\kappa}{\sin(\kappa a)}\,.$$

Therefore, if we locate $x = 0$ at node 1 in the figure and if the densities in the two directions are ρ_1 and ρ_2, a balance of forces at node 1 gives us

$$0 = [2w_1 \cos(\kappa_2 a) - w_2 - w_4]\,\frac{\kappa_2}{\sin(\kappa_2 a)} + [2w_1 \cos(\kappa_1 a) - w_3 - w_5]\,\frac{\kappa_1}{\sin(\kappa_1 a)}$$

where $\kappa_j = \omega/c_j$ and $c_j^2 = T/\rho_j$, $j = 1, 2$. Let $\mathbf{k} = k_1\mathbf{e}_1 + k_2\mathbf{e}_2$ be a Bloch wave vector. Then, from the Bloch-Floquet theorem we have

$$w_2 = w_1\,e^{-ia\mathbf{k}\cdot\mathbf{e}_2}\ , \quad w_3 = w_1\,e^{ia\mathbf{k}\cdot\mathbf{e}_1}\ , \quad w_4 = w_1\,e^{ia\mathbf{k}\cdot\mathbf{e}_2}\ , \quad w_5 = w_1\,e^{-ia\mathbf{k}\cdot\mathbf{e}_1}\,.$$

If we plug these back into the equilibrium equation, we find the resulting equation has a nontrivial solution only if we satisfy the dispersion relation

$$\kappa_1 \frac{\cos(\kappa_1 a)}{\sin(\kappa_1 a)} + \kappa_2 \frac{\cos(\kappa_2 a)}{\sin(\kappa_2 a)} - \kappa_1 \frac{\cos(k_1 a)}{\sin(\kappa_1 a)} - \kappa_2 \frac{\cos(k_2 a)}{\sin(\kappa_2 a)} = 0\,. \tag{7.44}$$

For the special situation where $\rho_1 = \rho_2$ or $\kappa_1 = \kappa_2 = \kappa = \omega/c$, the above relation simplifies to

$$2\cos(\omega a/c) - \cos(k_1 a) - \cos(k_2 a) = 0\,.$$

A plot of the reciprocal lattice vector, the first Brillouin zone, and the dispersion plot for the square lattice are shown in Figure 7.11. The calculation of the dispersion relation can be performed using bisection or with a Newton solver. However, as we can see from the figure, such methods can be inaccurate close to branch points.

Let us now consider the two-dimensional lattice model that we had discussed in Section 5.7 (p. 197). We had found that the forces at a representative node can be expressed as (see equations (5.110) and (5.111))

$$\mathbf{f}_4^{\text{inertial}} = \frac{\omega^2}{4c^2 h}\left[(m_5 A_2 C_1 \mathbf{x}_5 + m_6 C_3 \mathbf{x}_6) \otimes \mathbf{c}_1 + (m_5 C_2 \mathbf{x}_5 + m_6 A_2 C_1 \mathbf{x}_6) \otimes \mathbf{c}_2\right] \cdot \mathbf{u}_4$$

$$\mathbf{f}_4^{\text{elastic}} = -hK\,[(1 - A_3)\mathbf{1} + (A_3 - A_1)\boldsymbol{D}_{41} + (1 - A_2)\boldsymbol{D}_{42}] \cdot \mathbf{u}_4$$

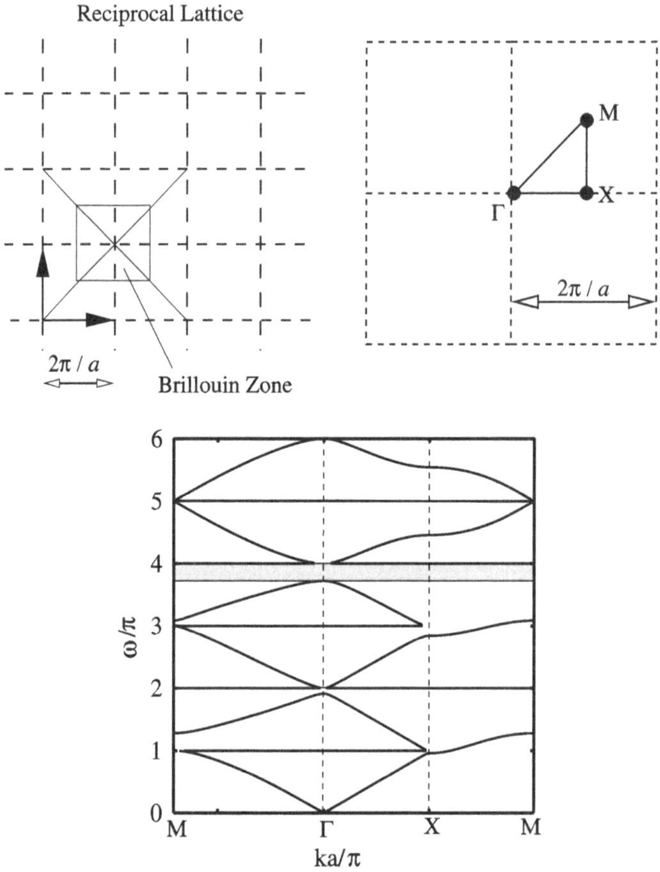

FIGURE 7.11

The reciprocal lattice and a dispersion plot for the square lattice in Figure 7.10. To generate a dispersion plot, values of **k** are chosen along the contour $\Gamma - M - X$. Along $\Gamma - X$, $k_1 \in [0, \pi/a]$ and $k_2 = 0$; along $\Gamma - M$, $k_1 = k_2 \in [0, \pi/a]$; and along $X - M$, $k_1 = 0$ and $k_2 \in [0, \pi/a]$. The dispersion plot is for the situation where the wave speeds in the two types of strings are $c_2 = 5c_1$. A bisection method was used to compute the values of ω. A band gap can be seen in the diagram.

where m_5 and m_6 are lumped masses, hK is the spring stiffness, and $\mathbf{x}_5, \mathbf{x}_6$ are the positions of the two masses. The geometric parameters C_1, C_2, C_3 are defined as

$$C_1 = c^2 + d^2 - 1 \, , \quad C_2 = c^2 + (1+d)^2 \, , \quad C_3 = c^2 + (1+d)^2$$

and the vectors $\mathbf{c}_1, \mathbf{c}_2$ are given by

$$\mathbf{c}_1 = -c\mathbf{e}_1 + (1-d)\mathbf{e}_2 \, , \quad \mathbf{c}_2 = c\mathbf{e}_1 + (1+d)\mathbf{e}_2$$

where c and d are parameters indicating the locations of the two masses. The quantities A_1, A_2, and A_3 are obtained from Bloch wave solutions (see Section 5.7):

$$A_1 = e^{-ih(k_1+k_2)} \;,\; A_2 = e^{ihk_2} \;,\; A_3 = e^{-ih(k_1-k_2)}$$

where $\mathbf{k} = k_1\mathbf{e}_1 + k_2\mathbf{e}_2$ is the Bloch wave vector. The transformations $\boldsymbol{D}_{ij}$ are

$$\boldsymbol{D}_{ij} \equiv \begin{bmatrix} \cos^2\theta_{ij} & \cos\theta_{ij}\sin\theta_{ij} \\ \cos\theta_{ij}\sin\theta_{ij} & \sin^2\theta_{ij} \end{bmatrix}$$

where i and j are the end points of the springs and θ_{ij} is the angle made by the spring with the $\mathbf{e}_1$ axis (in the counterclockwise direction).

A balance between the inertial and the elastic force at node 4 leads to a system of equations of the form

$$[\underline{\underline{\mathsf{K}}}(\mathbf{k}) - \omega^2\underline{\underline{\mathsf{M}}}(\mathbf{k})]\underline{\mathsf{u}} = \underline{\mathsf{0}} .$$

These can be solved using a numerical eigenvalue solver to find the values of ω corresponding to a particular choice of $\mathbf{k}$ within the first Brillouin zone. For this problem, in addition to a normal acoustic mode, we get a shear acoustic mode because of the extra degree of freedom in $\mathbf{u}$.

7.4.3 Dispersion in continuous elastic composites

Dispersion in periodic elastic composites has been the subject of study since at least the early 1960s (see for example Kohn et al. (1972) and the references cited there). These studies had found that band gaps typically occurred at high frequencies. The increased interest in elastic band gap materials since 2000 has been ignited by the discovery by Liu et al. (2000) that band gaps can be opened up at low frequencies by using locally resonant materials (see Chapter 5). A similar behavior is observed if the coated balls are organized in a periodic manner and a dispersion diagram is computed. It is often difficult to distinguish the regime of local resonance from the Bragg scattering regime in dispersion diagrams and additional measures, such as full wave calculations, may be needed.

Let us briefly discuss a procedure that is used to calculate dispersion curves for elastic composites. Recall that for continuous, periodic, elastic composite media, the momentum equation at fixed frequency in the absence of body forces can be expressed in the form of an eigenvalue problem:

$$(\mathcal{L} - \lambda\mathcal{I})[\mathbf{u}(\mathbf{x})] = \mathbf{0}$$

where

$$\mathcal{L}[\mathbf{u}(\mathbf{x})] := -\rho^{-1} \cdot [\boldsymbol{\nabla} \cdot (\mathbf{C} : \boldsymbol{\nabla}\mathbf{u})] \quad \text{and} \quad \lambda := \omega^2$$

and $\mathbf{u}(\mathbf{x})$ is the displacement field, $\mathbf{C}(\mathbf{x})$ is the stiffness tensor, and $\rho(\mathbf{x})$ is the mass density. We can also write the above equation as

$$(\mathcal{K} - \omega^2\mathcal{M})[\mathbf{u}(\mathbf{x})] = \mathbf{0}$$

where

$$\mathcal{K}[\mathbf{u}(\mathbf{x})] := -\boldsymbol{\nabla}\cdot(\mathbf{C}:\boldsymbol{\nabla}\mathbf{u}) \quad \text{and} \quad \mathcal{M}[\mathbf{u}(\mathbf{x})] := \boldsymbol{\rho}\cdot\mathbf{u}\,.$$

Recall from (7.8) that the Bloch condition for elastodynamics has the form

$$\mathbf{u}(\mathbf{x}+\mathbf{a}) = e^{i\mathbf{k}\cdot\mathbf{a}}\mathbf{u}(\mathbf{x})$$

where $\mathbf{a} = m_j\mathbf{a}_j$ is a lattice vector, $\mathbf{a}_j$ are the primitive lattice vectors, and m_j are integers. If we consider the unit cell in Figure 7.12, the Bloch condition can be applied directly to points on the boundary of the unit cell. For points inside the unit cell Ω it

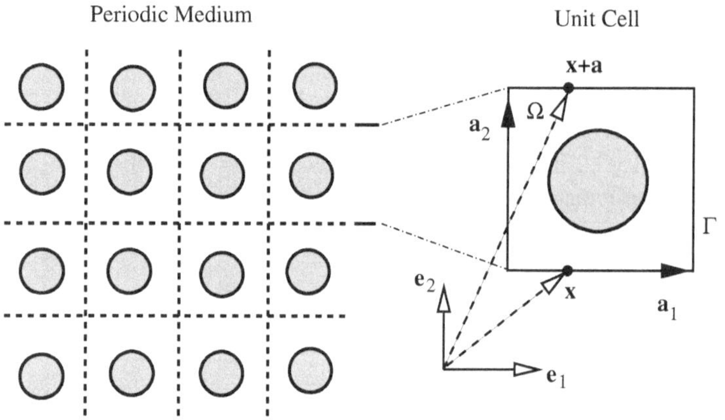

FIGURE 7.12
Periodic medium showing the unit cell Ω and periodic locations on the boundary Γ.

is more convenient to use a Bloch solution of the form

$$\mathbf{u}_\eta(\mathbf{x}) = \widehat{\mathbf{u}}_\eta(\mathbf{x})\,e^{i\mathbf{k}\cdot\mathbf{x}} \quad \Longrightarrow \quad \boldsymbol{\nabla}\mathbf{u}_\eta = [\boldsymbol{\nabla}\widehat{\mathbf{u}}_\eta + i\mathbf{k}\otimes\widehat{\mathbf{u}}_\eta]\,e^{i\mathbf{k}\cdot\mathbf{x}} \tag{7.45}$$

where the displacement $\widehat{\mathbf{u}}_\eta(\mathbf{x})$ is periodic in Ω (see Section 7.2, p. 256, for details). The momentum equation may then be expressed as

$$\boldsymbol{\nabla}\cdot[\mathbf{C}_\eta : \boldsymbol{\nabla}\mathbf{u}_\eta] + \omega^2\rho_\eta\mathbf{u}_\eta = 0\,.$$

To find the Galerkin weak form of this equation, we multiply it with a periodic test function $\mathbf{w}_\eta(\mathbf{x})$ with compact support inside the unit cell (i.e., the test function is zero on Γ). Then we get

$$\int_\Omega \left[\boldsymbol{\nabla}\cdot[\mathbf{C}_\eta : \boldsymbol{\nabla}\mathbf{u}_\eta] + \omega^2\rho_\eta\mathbf{u}_\eta\right]\cdot\mathbf{w}_\eta\,\mathrm{d}\Omega = 0\,.$$

Using the identity $\mathbf{v}\cdot(\boldsymbol{\nabla}\cdot\boldsymbol{S}) = \boldsymbol{\nabla}\cdot(\boldsymbol{S}^T\cdot\mathbf{v}) - \boldsymbol{S}:\boldsymbol{\nabla}\mathbf{v}$, where $\mathbf{v}$ is a vector field and $\boldsymbol{S}$ is a second-order tensor field, and using the compact support of the test function $\mathbf{w}$

gives us

$$\int_{\Omega} \left[\nabla \mathbf{w} : \mathbf{C}_\eta : \nabla \mathbf{u}_\eta - \omega^2 \rho_\eta \mathbf{u}_\eta \cdot \mathbf{w}_\eta \right] d\Omega = 0 \,.$$

If the periodic test function also has the form $\mathbf{w}_\eta(\mathbf{x}) = \widehat{\mathbf{w}}_\eta(\mathbf{x})\, e^{i\mathbf{k}\cdot\mathbf{x}}$, using (7.45) we can write the weak form as

$$\int_{\Omega} \left[(\nabla \widehat{\mathbf{w}}_\eta + i\mathbf{k} \otimes \widehat{\mathbf{w}}_\eta) : \mathbf{C}_\eta : (\nabla \widehat{\mathbf{u}}_\eta + i\mathbf{k} \otimes \widehat{\mathbf{u}}_\eta) - \omega^2 \rho_\eta \widehat{\mathbf{w}}_\eta \cdot \widehat{\mathbf{u}}_\eta \right] d\Omega = 0 \,. \quad (7.46)$$

Since complex quantities are involved, special care is usually needed to make the system Hermitian and keep the total energy real. For a finite element solution of the above equation,[§] we discretize the domain Ω into n_E elements and choose trial and test functions of the form

$$\mathbf{u}_\eta = \sum_j N_j \mathbf{u}_j \quad \text{and} \quad \mathbf{w}_\eta = \sum_j N_j \mathbf{w}_j$$

where N_j are finite element basis functions. Substitution of these functions into the weak form and invoking the arbitrariness of $\mathbf{w}$ gives us a system of equations

$$[\underline{\underline{\mathsf{K}}}(\mathbf{k}) - \omega^2 \underline{\underline{\mathsf{M}}}]\underline{\mathsf{u}} = \underline{\mathsf{0}} \,.$$

The eigenvalues of this system of equations give us the dispersion relations $\omega(\mathbf{k})$. As before, the wave vectors $\mathbf{k}$ are chosen from the irreducible Brillouin zone for the periodic composite. Several techniques for solving these equations can be found in Bathe (1997) and Hughes (2000). An efficient mode superposition technique for problems with real $\mathbf{C}$ and ρ has been discussed in Hussein (2009).

As we have seen before, the analogy between antiplane shear (SH-waves), acoustic waves, and transverse electromagnetic waves implies that we can use a similar approach to find the dispersion relations for a broad range of wave phenomena in composites. Efficient computational methods for the calculation of dispersion relations can be found, for example, in Figotin and Kuchment (1996), Axmann and Kuchment (1999), Dobson (1999), and in the book by Lourtioz et al. (2008).

Exercises

Problem 7.1 Show that for periodic composites in the quasistatic limit, the momentum equation, the stress-strain relations and the strain-displacements relations can be written as

$$\nabla \cdot \widehat{\boldsymbol{\sigma}}_\eta + i\widehat{\boldsymbol{\sigma}}_\eta \cdot \mathbf{k} + \omega^2 \rho_\eta \widehat{\mathbf{u}}_\eta = 0 \,; \;\; \widehat{\boldsymbol{\sigma}}_\eta = \mathbf{C}_\eta : \widehat{\boldsymbol{\varepsilon}}_\eta$$

[§] See Minagawa et al. (1984) and the references cited there for early calculations of elastic dispersion curves with finite element methods.

and

$$\widehat{\varepsilon}_\eta = \tfrac{i}{2}\left(\widehat{\mathbf{u}}_\eta \otimes \mathbf{k} + \mathbf{k} \otimes \widehat{\mathbf{u}}_\eta\right) + \tfrac{1}{2}\left[\nabla \widehat{\mathbf{u}}_\eta + (\nabla \widehat{\mathbf{u}}_\eta)^T\right].$$

Problem 7.2 Verify that the matrix $\boldsymbol{A} := \mathbf{k} \cdot \langle \mathbf{e}_0 \rangle)\mathbf{k} - (\mathbf{k} \cdot \mathbf{k})\langle \mathbf{e}_0 \rangle$ can be expressed as $\boldsymbol{A} = [\mathbf{k} \otimes \mathbf{k} - (\mathbf{k} \cdot \mathbf{k})\mathbf{1}] \cdot \langle \mathbf{e}_0 \rangle$. Show that the eigenvalues of $\boldsymbol{A}$ are

$$\lambda_1 = 0\,, \quad \lambda_2 = \lambda_3 = k_1^2 + k_2^2 + k_3^2 = \mathbf{k} \cdot \mathbf{k}.$$

What do these eigenvalues imply about the quantity $\mathbf{k} \cdot \langle \mathbf{e}_0 \rangle$?

Problem 7.3 Show that perturbation expansions of the Bloch wave equations for elastodynamics lead to the relations

$$\begin{aligned}
\boldsymbol{\sigma}_0 &= \mathbf{C}_\eta : \boldsymbol{\varepsilon}_0 \\
0 &= \nabla_y \mathbf{u}_0 + (\nabla_y \mathbf{u}_0)^T \\
\boldsymbol{\varepsilon}_0 &= \tfrac{i}{2}(\mathbf{u}_0 \otimes \mathbf{k} + \mathbf{k} \otimes \mathbf{u}_0) + \tfrac{1}{2}[\nabla_y \mathbf{u}_1 + (\nabla_y \mathbf{u}_1)^T].
\end{aligned}$$

Problem 7.4 Show that we can express the first-order term in the perturbation expansion of $\widehat{\varepsilon}_\eta$ in the form

$$\boldsymbol{\varepsilon}_0(\mathbf{y}) = \tfrac{1}{2}\left(\nabla_y \widetilde{\mathbf{u}}(\mathbf{y}) + [\nabla_y \widetilde{\mathbf{u}}(\mathbf{y})]^T\right)$$

where $\widetilde{\mathbf{u}}(\mathbf{y}) = i(\mathbf{k} \cdot \mathbf{y})\mathbf{u}_0 + \mathbf{u}_1(\mathbf{y})$.

Problem 7.5 Show that the volume averaged gradient, divergence, and curl are zero for a periodic vector-valued function $\mathbf{u}(\mathbf{x})$ which satisfies the relation $\mathbf{u}(\mathbf{y}+\mathbf{a}) = \mathbf{u}(\mathbf{y})$ within a unit cell with lattice vector $\mathbf{a}$.

Problem 7.6 For a periodic elastic composite show that

$$\frac{1}{\langle \rho_\eta \rangle}(\mathbf{k} \cdot \mathbf{C}_{\text{eff}} \cdot \mathbf{k}) \cdot \mathbf{u}_0 = (\omega_0^j)^2 \mathbf{u}_0\,.$$

Problem 7.7 Derive Hashin's estimate for the effective bulk modulus of a coated sphere:

$$\kappa_{\text{eff}} = \kappa_2 + \cfrac{f_1}{\cfrac{1}{\kappa_1 - \kappa_2} + \cfrac{f_2}{\kappa_2 + \frac{4}{3}\mu_2}}$$

after verifying the expressions for the radial displacements and the traction continuity conditions.

Problem 7.8 Show that Maxwell's equations for a Bloch periodic medium can be expressed as

$$\begin{aligned}
i\mathbf{k} \cdot \mathbf{d}_\eta(\mathbf{x}) + \nabla \cdot \mathbf{d}_\eta &= 0, & i\mathbf{k} \cdot \mathbf{b}_\eta(\mathbf{x}) + \nabla \cdot \mathbf{b}_\eta &= 0, \\
i\mathbf{k} \times \mathbf{e}_\eta(\mathbf{x}) + \nabla \times \mathbf{e}_\eta - i\omega \mathbf{b}_\eta(\mathbf{x}) &= 0, & i\mathbf{k} \times \mathbf{h}_\eta(\mathbf{x}) + \nabla \times \mathbf{h}_\eta + i\omega \mathbf{d}_\eta(\mathbf{x}) &= 0.
\end{aligned}$$

Problem 7.9 Show, using an approach similar to that used to obtain the effective bulk modulus, that Hashin's estimate for the effective electrical permittivity of a coated sphere can be expressed as

$$\varepsilon_{\text{eff}} = \varepsilon_2 + \frac{3f_1\varepsilon_2(\varepsilon_1 - \varepsilon_2)}{3\varepsilon_2 + f_2(\varepsilon_1 - \varepsilon_2)}$$

where ε_1 is the permittivity of the sphere, ε_2 is the permittivity of the coating, and $f_1 = 1 - f_2$ is the volume fraction occupied by the sphere.

Problem 7.10 Consider an infinite one-dimensional lattice of circular rings connected to each other by linear springs. The mass of each ring is M and the stiffness of each spring is K. Each ring contains an additional mass m that is connected to the ring by a spring that has a stiffness K_0. Show that the dispersion relation for the lattice is

$$mM\omega^4 - [(m+M)K_0 + 2mK(1-\cos ka)] + 2KK_0(1-\cos ka) = 0$$

where a is the lattice spacing, k is the wave number, and ω is the frequency. Plot the dispersion relation for the lattice. Hint: See Huang et al. (2009).

Problem 7.11 Plot the dispersion relation described by equation (7.44) and compare your result to the plot shown in Figure 7.11.

Problem 7.12 Find the dispersion relations for the Milton-Willis model discussed in Chapter 5, where k is the wave number. Plot the dispersion diagram for the model with $c = 1/2$.

Problem 7.13 The governing equation for antiplane shear (SH-waves) in an inhomogeneous but isotropic medium can be written as

$$\overline{\boldsymbol{\nabla}} \cdot (\mu(x_1,x_2)\overline{\boldsymbol{\nabla}} u_3(x_1,x_2)) + \omega^2\rho(x_1,x_2)u_3(x_1,x_2) = 0\,.$$

Derive the weak form of this equation as it applies to a periodic composite by taking the product with a vector test function $\mathbf{w}(x_1,x_2)$ (with compact support) and integrating over the volume of the unit cell. Discretize the resulting equation using finite element basis functions and comment on how Bloch periodic boundary conditions may be implemented in this situation. Hint: See Guenneau et al. (2007).

Problem 7.14 The weak form of the wave equation in a periodic elastic composite can be written as

$$\int_\Omega \left[\left(\boldsymbol{\nabla}\widehat{\mathbf{w}}_\eta + i\mathbf{k}\otimes\widehat{\mathbf{w}}_\eta\right) : \mathbf{C}_\eta : \left(\boldsymbol{\nabla}\widehat{\mathbf{u}}_\eta + i\mathbf{k}\otimes\widehat{\mathbf{u}}_\eta\right) - \omega^2\rho_\eta\widehat{\mathbf{w}}_\eta\cdot\widehat{\mathbf{u}}_\eta \right] d\Omega = 0\,.$$

Show that the above equation can be discretized using finite elements in system of algebraic equations of the form

$$[\underline{\underline{\mathsf{K}}}(\mathbf{k}) - \omega^2\underline{\underline{\mathsf{M}}}]\underline{\mathsf{u}} = \underline{\mathsf{0}}\,.$$

8

Waves in Layered Media

> The properties of the multilayer are not simply some combination of the properties of the bulk constituents.
>
> S. J. LLOYD AND J. M. MOLINA-ALDAREGUIA, Multilayered materials: A palette for the materials artist, 2003

Layered media form a large class of composites. Examples of layered materials include sedimentary rocks, atmospheric strata, the nacre shells of crustaceans, high-temperature copper-oxide semiconductors, martensitic twins in metals and alloys, laminated resin-fiber composites, and numerous other materials. Waves in such media have been studied extensively. Detailed expositions can be found in Brekhovskikh (1960) and Chew (1995). A good introduction to the elastodynamics of layered media can be found in Aki and Richards (1980) and a broader perspective can be found in Carcione (2007). Homogenization of layered media is covered in detail in Milton (2002) and a mathematical description with historical context can be found in Tartar (2009).

In this chapter we discuss a few topics that are useful while navigating the vast literature on layered media. We discuss the state vector approach to problems involving layered media, the approximate Wentzel-Kramers-Brillouin-Jeffreys method of solving the state equations, and the widely used propagator matrix method. We then explore the Schoenberg-Sen model that predicts an anisotropic tensorial mass density at low frequencies. We finish our discussion with the quasistatic homogenization of layered composites and hierarchical laminates. High-frequency homogenization is avoided because of the complexities involved.

8.1 Wave equations in layered media

Consider the layered medium shown in Figure 8.1. Let are assume that the material properties in each layer are scalars and locally isotropic, i.e., $\kappa \equiv \kappa(x_3)$, $\rho \equiv \rho(x_3)$, $\mu \equiv \mu(x_3)$, $\varepsilon \equiv \varepsilon(x_3)$, and so on.

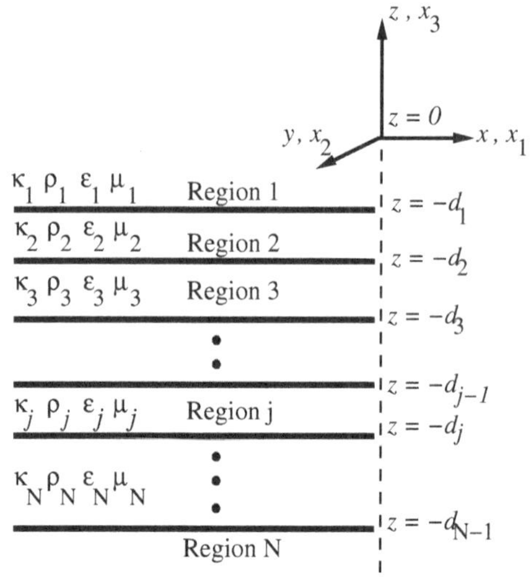

FIGURE 8.1
A medium with N parallel layers.

8.1.1 Antiplane shear elastic waves

Let us first examine the case of elastic waves in layered media. For plane harmonic waves in *elastic* layered media, a displacement field can be assumed to be of the form

$$\mathbf{u}(x_1,x_2,x_3) = \widehat{\mathbf{u}}(x_3)\, e^{i(k_1x_1-\omega t)}$$

where k_1 is the component of the wave vector in the x_1-direction, $\mathbf{x}_1 = x_1\mathbf{e}_1$ is parallel to the layers, and $\mathbf{x}_3 = x_3\mathbf{e}_3$ is perpendicular to the layers. Notice that this is a problem where the displacement is independent of x_2. Recall that the balance of momentum in the absence of body forces is given by $\sigma_{mn,m} = \rho\ddot{u}_n$ which reduces to $\sigma_{12,1}+\sigma_{23,3} = \rho\ddot{u}_2$ for antiplane shear waves. The momentum equation expressed in terms of displacements is the wave equation for antiplane shear,*

$$\rho_j\ddot{u}_2 = \mu_j(u_{2,11}+u_{2,33}) \quad \text{or} \quad \rho_j\frac{\partial^2 u_2}{\partial t^2} = \mu_j\left(\frac{\partial^2 u_2}{\partial x_1^2}+\frac{\partial^2 u_2}{\partial x_3^2}\right)$$

where μ_j is the shear modulus and ρ_j is the mass density of layer j. For the assumed displacement field we have

$$-\omega^2\rho_j\widehat{u}_2 = \mu_j(-k_1^2\widehat{u}_2+\widehat{u}_{2,33}).$$

*A commonly encountered type of antiplane shear wave is the Love wave. Love waves are antiplane shear waves in a layered medium with one free surface. For an infinite layered medium with a free surface, the boundary conditions on the displacements can be taken to be $\mathbf{n}\cdot\boldsymbol{\sigma} = \mathbf{0}$ at $x_3 = 0$ (a free surface) and $\mathbf{u} = \mathbf{0}$ at $x_3 = -\infty$.

Rearrangement of the above equation gives

$$\widehat{u}_{2,33} + k_j^2 \widehat{u}_2 = 0 \quad \text{where} \quad k_j^2 := \frac{\omega^2 \rho_j}{\mu_j} - k_1^2 .$$

The above equation has solutions of the form

$$\widehat{u}_2 = A_{1j} e^{k_j x_3} + A_{2j} e^{-k_j x_3} .$$

Hence the plane harmonic wave has the form

$$u_2(x_1, x_3, t) = (A_{1j} e^{k_j x_3} + A_{2j} e^{-k_j x_3})\, e^{i(k_1 x_1 - \omega t)} . \tag{8.1}$$

If $\text{Re}(k_j) \geq 0$ we must have $\text{Im}(k_j) \leq 0$. Therefore, for antiplane shear waves we try to find solutions of the form

$$u_1 = u_3 = 0, \quad u_2 = \widehat{u}_2(k, \omega, x_3)\, e^{i(k_1 x_1 - \omega t)} .$$

The strain components corresponding to this displacement field are

$$\varepsilon_{11} = \varepsilon_{22} = \varepsilon_{33} = \varepsilon_{13} = 0, \quad \varepsilon_{23} = \tfrac{1}{2} \widehat{u}_{2,3}\, e^{i(k_1 x_1 - \omega t)}, \quad \varepsilon_{12} = \tfrac{ik_1}{2} \widehat{u}_3\, e^{i(k_1 x_1 - \omega t)}$$

and the stress components are

$$\sigma_{11} = \sigma_{22} = \sigma_{33} = \sigma_{13} = 0, \quad \sigma_{23} = \mu \widehat{u}_{2,3}\, e^{i(k_1 x_1 - \omega t)}, \quad \sigma_{12} = ik_1 \mu \widehat{u}_2\, e^{i(k_1 x_1 - \omega t)} .$$

For a layered medium in which the shear modulus $\mu \equiv \mu(x_3)$, the momentum balance can be expressed as

$$-k_1^2 \mu \widehat{u}_2 + \frac{d}{dx_3}\left(\mu \frac{d\widehat{u}_2}{dx_3}\right) = -\rho \omega^2 \widehat{u}_2$$

or

$$\boxed{\left[\frac{1}{\mu(x_3)} \frac{d}{dx_3}\left(\mu(x_3) \frac{d\widehat{u}_2}{dx_3}\right)\right] + \left[\omega^2 \frac{\rho(x_3)}{\mu(x_3)} - k_1^2\right] \widehat{u}_2 = 0 .} \tag{8.2}$$

If $\sigma_{23} = \widehat{\sigma}_{23}\, e^{i(k_1 x_1 - \omega t)}$ we have

$$\widehat{\sigma}_{23} = \mu(x_3) \frac{d\widehat{u}_2}{dx_3}$$

and we can express equation (8.2) as

$$\frac{d\widehat{\sigma}_{23}}{dx_3} = \left[k_1^2 \mu(x_3) - \omega^2 \rho(x_3)\right] \widehat{u}_2 .$$

Thus we have reduced the second-order differential equation into two coupled first-order equations and can write the two in matrix form as

$$\boxed{\frac{d}{dx_3}\begin{bmatrix} \widehat{u}_2 \\ \widehat{\sigma}_{23} \end{bmatrix} = \begin{bmatrix} 0 & 1/\mu(x_3) \\ k_1^2 \mu(x_3) - \omega^2 \rho(x_3) & 0 \end{bmatrix} \begin{bmatrix} \widehat{u}_2 \\ \widehat{\sigma}_{23} \end{bmatrix} .} \tag{8.3}$$

The vector $[\widehat{u}_2 \ \ \widehat{\sigma}_{23}]^T$ is called the *motion-stress vector* or the *state vector* and equations (8.3) are called the *state equations*. If we define

$$\underline{\mathsf{v}} := \begin{bmatrix} \widehat{u}_2 \\ \widehat{\sigma}_{23} \end{bmatrix} \quad \text{and} \quad \underline{\underline{\mathsf{H}}} := \begin{bmatrix} 0 & 1/\mu(x_3) \\ -k_3^2\mu(x_3) & 0 \end{bmatrix}$$

where $k_3^2 := \omega^2\rho(x_3)/\mu(x_3) - k_1^2$, then equations (8.3) can be written as

$$\frac{d}{dx_3}(\underline{\mathsf{v}}) = \underline{\underline{\mathsf{H}}}\,\underline{\mathsf{v}}. \tag{8.4}$$

The above state equations form the basis of the propagator matrix methods discussed later.

8.1.2 Acoustic waves

Let us now consider the acoustic wave equation which can be expressed in the form

$$\frac{\partial^2 p}{\partial t^2} - \kappa \frac{\partial}{\partial x_m}\left(\frac{1}{\rho}\frac{\partial p}{\partial x_m}\right) = 0$$

where p is the acoustic pressure, ρ is the mass density, and κ is the bulk modulus. If we once again we assume that $p \equiv p(x_1, x_3)$ and $\rho \equiv \rho(x_3), \kappa \equiv \kappa(x_3)$ we can write

$$\frac{\partial^2 p}{\partial t^2} - \frac{\kappa(x_3)}{\rho(x_3)}\frac{\partial^2 p}{\partial x_1^2} - \kappa(x_3)\frac{\partial}{\partial x_3}\left(\frac{1}{\rho(x_3)}\frac{\partial p}{\partial x_3}\right) = 0.$$

Let us look for solutions of the form

$$p(x_1, x_3) = \widehat{p}(k, \omega, x_3)\, e^{i(k_1 x_1 - \omega t)}.$$

Then we have

$$-\omega^2\widehat{p} + k_1^2\frac{\kappa(x_3)}{\rho(x_3)}\widehat{p} - \kappa(x_3)\frac{d}{dx_3}\left(\frac{1}{\rho(x_3)}\frac{d\widehat{p}}{dx_3}\right) = 0$$

or

$$\boxed{\left[\rho(x_3)\frac{d}{dx_3}\left(\frac{1}{\rho(x_3)}\frac{d\widehat{p}}{dx_3}\right)\right] + \left[\omega^2\frac{\rho(x_3)}{\kappa(x_3)} - k_1^2\right]\widehat{p} = 0.} \tag{8.5}$$

Notice the similarity between this equation and (8.2). Now, look at the momentum balance equation in acoustics:

$$\rho\frac{\partial v_m}{\partial t} + \frac{\partial p}{\partial x_m} = 0.$$

If $\mathbf{v} = \widehat{\mathbf{v}}(x_3)\, e^{i(k_1 x_1 - \omega t)}$, we have

$$\frac{d\widehat{p}}{dx_3} = i\omega\rho(x_3)\widehat{v}_3 .$$

Therefore equation (8.5) can be written in the form

$$\frac{d\widehat{v}_3}{dx_3} = \frac{i\left[\omega^2\rho(x_3) - k_1^2\kappa(x_3)\right]}{\omega\rho(x_3)\kappa(x_3)}\widehat{p}.$$

We can write the above two equations in matrix form as

$$\boxed{\frac{d}{dx_3}\begin{bmatrix}\widehat{v}_3\\ \widehat{p}\end{bmatrix} = \begin{bmatrix} 0 & \frac{i}{\omega}\left[\omega^2/\kappa(x_3) - k_1^2/\rho(x_3)\right] \\ i\omega\rho(x_3) & 0\end{bmatrix}\begin{bmatrix}\widehat{v}_3\\ \widehat{p}\end{bmatrix}.} \tag{8.6}$$

We could also have expressed the above equation in terms of particle displacements. If we define $k_3^2 := \omega^2\rho(x_3)/\kappa(x_3) - k_1^2$, then equations (8.6) can be written as

$$\frac{d}{dx_3}(\underline{\underline{\mathsf{v}}}) = \underline{\underline{\mathsf{H}}}\,\underline{\underline{\mathsf{v}}}. \tag{8.7}$$

where

$$\underline{\underline{\mathsf{v}}} := \begin{bmatrix}\widehat{v}_3\\ \widehat{p}\end{bmatrix} \quad \text{and} \quad \underline{\underline{\mathsf{H}}} := \begin{bmatrix} 0 & \frac{ik_3^2}{\omega\rho(x_3)} \\ i\omega\rho(x_3) & 0\end{bmatrix}.$$

8.1.3 Electromagnetic waves

Assume that the permittivity and permeability are scalars and locally isotropic, i.e., $\boldsymbol{\varepsilon} = \varepsilon(x_3)\boldsymbol{1}$ and $\boldsymbol{\mu} = \mu(x_3)\boldsymbol{1}$. Let us first consider TE-waves where the electric field is $\mathbf{E} \equiv E_2(x_1,x_3)\mathbf{e}_2$. Then the TE-wave equation at a fixed frequency ω is given by

$$\frac{\partial}{\partial x_m}\left(\frac{1}{\mu(x_3)}\frac{\partial E_2}{\partial x_m}\right) + \omega^2\varepsilon(x_3)E_2(x_1,x_3) = 0, \quad m = 1,2.$$

Expanding the above equation and using our assumptions, we have

$$\frac{\partial E_2}{\partial x_1^2} + \mu(x_3)\frac{\partial}{\partial x_3}\left(\frac{1}{\mu(x_3)}\frac{\partial E_2}{\partial x_3}\right) + \omega^2\,\varepsilon(x_3)\mu(x_3)E_2 = 0.$$

As we have observed before, the above equation admits solutions of the form

$$E_2(x_1,x_3) = \widehat{E}_2(x_3)\, e^{\pm ik_1x_1}.$$

Inserting these solutions into the TE-wave equation gives us the ODE,

$$\boxed{\left[\mu(x_3)\frac{d}{dx_3}\left(\frac{1}{\mu(x_3)}\frac{d\widehat{E}_2}{dx_3}\right)\right] + \left[\omega^2\varepsilon(x_3)\mu(x_3) - k_1^2\right]\widehat{E}_2 = 0.} \tag{8.8}$$

The quantity $k_3^2 := \omega^2\varepsilon(x_3)\mu(x_3) - k_1^2$ can be less than zero, implying that k_3 may be complex. Also, at the boundary, both $\widehat{E}_2$ and $1/\mu\,\partial\widehat{E}_2/\partial x_3$ must be continuous. For a non-magnetic medium, μ is constant and we can write (8.8) as

$$\frac{d^2\widehat{E}_2}{dx^2} + \left[\omega^2\varepsilon\mu - k_1^2\right]\widehat{E}_2 = 0 \qquad \text{where} \quad x \equiv x_3\,. \tag{8.9}$$

Similarly, for TM-waves, we have

$$H_2(x_1,x_3) = \widehat{H}_2(x_3)\,e^{\pm ik_1x_1}$$

and the resulting ODE is

$$\boxed{\left[\varepsilon(x_3)\frac{d}{dx_3}\left(\frac{1}{\varepsilon(x_3)}\frac{d\widehat{H}_2}{dx_3}\right)\right] + \left[\omega^2\varepsilon(x_3)\mu(x_3) - k_1^2\right]\widehat{H}_2 = 0\,.} \tag{8.10}$$

Note that the definition of k_3 is the same for TE- and TM-waves. To reduce (8.10) to a first-order differential equation, let us introduce the quantities

$$\varphi := \widehat{H}_2 \quad \text{and} \quad \psi := \frac{1}{i\omega\varepsilon(x_3)}\frac{d\widehat{H}_2}{dx_3}\,. \tag{8.11}$$

Then

$$\frac{d\varphi}{dx_3} = i\omega\varepsilon\psi\,.$$

Clearly, ψ has to be continuous across the interface for the differential equation (8.10) to be satisfied. Plugging (8.11) into (8.10) gives

$$\frac{d\psi}{dx_3} = \frac{ik_3^2}{\omega\varepsilon}\varphi\,. \tag{8.12}$$

Therefore, the second of equations (8.11) and (8.12) form a system of differential equations which can be written as

$$\boxed{\frac{d}{dx_3}\begin{bmatrix}\varphi\\ \psi\end{bmatrix} = \begin{bmatrix}0 & i\omega\varepsilon(x_3)\\ \dfrac{ik_3^2}{\omega\varepsilon(x_3)} & 0\end{bmatrix}\begin{bmatrix}\varphi\\ \psi\end{bmatrix}\,.} \tag{8.13}$$

Similarly, for TE-waves we have

$$\boxed{\frac{d}{dx_3}\begin{bmatrix}\varphi\\ \psi\end{bmatrix} = \begin{bmatrix}0 & i\omega\mu(x_3)\\ \dfrac{ik_3^2}{\omega\mu(x_3)} & 0\end{bmatrix}\begin{bmatrix}\varphi\\ \psi\end{bmatrix}} \tag{8.14}$$

where $\varphi := \widehat{E}_2$ and $i\omega\mu\psi := d\widehat{E}/dx_3$. The above state equations can be written in the compact form

$$\frac{d}{dx_3}(\underline{\mathbf{v}}) = \underline{\underline{\mathbf{H}}}\,\underline{\mathbf{v}}\,. \tag{8.15}$$

where

$$\underline{\mathbf{v}} := \begin{bmatrix} \varphi \\ \psi \end{bmatrix} \quad \text{and} \quad \underline{\underline{\mathbf{H}}} := \begin{bmatrix} 0 & i\omega\varepsilon(x_3) \\ \frac{ik_3^2}{\omega\varepsilon(x_3)} & 0 \end{bmatrix} \text{ or } \begin{bmatrix} 0 & i\omega\mu(x_3) \\ \frac{ik_3^2}{\omega\mu(x_3)} & 0 \end{bmatrix}.$$

8.2 Piecewise-constant multilayered media

If we look at equations (8.2), (8.5), (8.8), and (8.10), we see that they all have the form

$$\left[\alpha(x_3)\frac{d}{dx_3}\left(\frac{1}{\alpha(x_3)}\frac{d\widehat{\varphi}}{dx_3}\right)\right] + \left[\omega^2\alpha(x_3)\beta(x_3) - k_1^2\right]\widehat{\varphi} = 0$$

where α, β are material properties and the variable $\widehat{\varphi}$ is related to the field φ by

$$\varphi(x_1, x_3) = \widehat{\varphi}(k, \omega, x_3)\, e^{i(k_1 x_1 - \omega t)}.$$

The subscripts in the above equation can be confusing when we are dealing with multiple layers and it is more convenient to use (x, y, z) instead of (x_1, x_2, x_3). Equation (8.16) may be written in the new notation as

$$\left[\alpha(z)\frac{d}{dz}\left(\frac{1}{\alpha(z)}\frac{d\widehat{\varphi}}{dz}\right)\right] + \left[\omega^2\alpha(z)\beta(z) - k_x^2\right]\widehat{\varphi} = 0. \tag{8.16}$$

Let us drop the hats and define $k_z^2 := \omega^2\alpha(z)\beta(z) - k_x^2$. Then the governing equation takes the form

$$\alpha(z)\frac{d}{dz}\left(\frac{1}{\alpha(z)}\frac{d\varphi}{dz}\right) + k_z^2\,\varphi = 0.$$

In Section 2.5.3 (p. 85) we took a brief look at solving the above equation for a three-layered medium where each layer is isotropic, i.e., the properties are piecewise constant. We had found the generalized reflection and transmission coefficients for a slab. Let us now try to extend those results to a medium with N layers.

We had found that the apparent reflection coefficient $\widetilde{R}_{12}$ for TE-waves incident on a slab can be expressed as†

$$\widetilde{R}_{12} = R_{12} + \frac{T_{12}T_{21}R_{23}\, e^{2ik_{z2}(d_2-d_1)}}{1 - R_{21}R_{23}\, e^{2ik_{z2}(d_2-d_1)}}. \tag{8.17}$$

If one additional layer were to be added to the slab, then we would only need to replace R_{23} in equation (8.17) with $\widetilde{R}_{23}$. This indicates that we can easily generalize the above result to a medium containing N layers (see Figure 8.2). In general, the

†This expression is given in equation (2.69). Note that we can use the same expression for acoustic waves or antiplane shear waves.

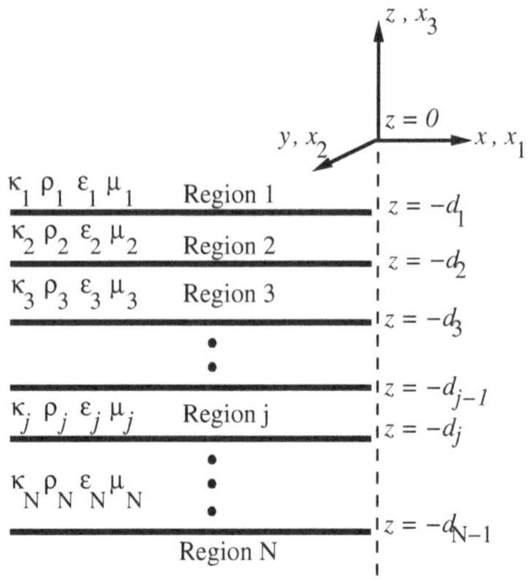

FIGURE 8.2
A medium with N parallel layers, each with a different set of material properties. The origin of the coordinate system is chosen such that the interfaces between the layers are at the locations $z = -d_j$.

wave in the j-th region takes the form

$$\varphi_j(z) = A_j \left[e^{-ik_{zj}z} + \widetilde{R}_{j,j+1}\, e^{ik_{zj}z + 2ik_{zj}d_j} \right]. \tag{8.18}$$

For the last layer,

$$\widetilde{R}_{N,N+1} = 0.$$

For all other layers we get a recursion relation

$$\widetilde{R}_{j,j+1} = R_{j,j+1} + \frac{T_{j,j+1}T_{j+1,j}\widetilde{R}_{j+1,j+2}\, e^{2ik_{z,j+1}(d_{j+1}-d_j)}}{1 - R_{j+1,j}\widetilde{R}_{j+1,j+2}\, e^{2ik_{z,j+1}(d_{j+1}-d_j)}}. \tag{8.19}$$

Generalizing equation (2.64), we have

$$T_{i,j} = 1 + R_{i,j} \qquad \text{and} \qquad R_{i,j} = -R_{j,i}. \tag{8.20}$$

We can use equation (8.20) to simplify (8.19) to

$$\boxed{\widetilde{R}_{j,j+1} = \frac{R_{j,j+1} + \widetilde{R}_{j+1,j+2}\, e^{2ik_{z,j+1}(d_{j+1}-d_j)}}{1 + R_{j,j+1}\widetilde{R}_{j+1,j+2}\, e^{2ik_{z,j+1}(d_{j+1}-d_j)}}} \tag{8.21}$$

where $R_{j,j+1}$ is the reflection coefficient for reflection at the interface between the j-th and $j+1$-th layers. For the particular case of TE-waves this is just the Fresnel reflection coefficient,

$$R_{j,j+1} = \frac{\mu_{j+1}\,k_{z,j} - \mu_j\,k_{z,j+1}}{\mu_{j+1}\,k_{z,j} + \mu_j\,k_{z,j+1}}.$$

Equation (8.18) has two unknown components, the amplitude coefficients A_j and the reflection coefficients $\widetilde{R}_{j,j+1}$. We have determined the value of $\widetilde{R}$, but how do we determine the coefficients A_j in multilayered media? Let us start with the coefficients for a slab in equation (2.68) (p. 88). We had found that

$$A_2 = \frac{T_{12}A_1\,e^{i(k_{z1}-k_{z2})d_1}}{1 - R_{21}R_{23}\,e^{2ik_{z2}(d_2-d_1)}} \tag{8.22}$$

where T_{12} is the transmission coefficient between layers 1 and 2. For TE-waves,

$$T_{12} = \frac{2\mu_2 k_{z1}}{\mu_2 k_{z1} + \mu_1 k_{z2}}.$$

We can rewrite (8.22) as

$$A_2\,e^{ik_{z2}d_1} = \frac{T_{12}A_1\,e^{ik_{z1}d_1}}{1 - R_{21}R_{23}\,e^{2ik_{z2}(d_2-d_1)}}. \tag{8.23}$$

Using the arguments that we used for the reflection coefficients, we can generalize (8.23) to a medium with N layers. Thus, for the j-th layer, we have

$$A_j\,e^{ik_{z,j}d_{j-1}} = \frac{T_{j-1,j}A_{j-1}\,e^{ik_{z,j-1}d_{j-1}}}{1 - R_{j,j-1}\widetilde{R}_{j,j+1}\,e^{2ik_{z,j}(d_j-d_{j-1})}}. \tag{8.24}$$

Let us define

$$S_{j-1,j} := \frac{T_{j-1,j}}{1 - R_{j,j-1}\,\widetilde{R}_{j,j+1}\,e^{2ik_{z,j}(d_j-d_{j-1})}}.$$

Then we can write (8.24) as

$$\begin{aligned}
A_j\,e^{ik_{z,j}d_{j-1}} &= S_{j-1,j}A_{j-1}\,e^{ik_{z,j-1}d_{j-1}} \\
&= \left(A_{j-1}\,e^{ik_{z,j-1}d_{j-2}}\right)\left(S_{j-1,j}\,e^{ik_{z,j-1}(d_{j-1}-d_{j-2})}\right).
\end{aligned}$$

The above equation gives us a recurrence relation that can be used to compute the other coefficients, i.e.,

$$\begin{aligned}
A_j\,e^{ik_{z,j}d_{j-1}} &= \left(A_{j-1}\,e^{ik_{z,j-1}d_{j-2}}\right)\left(S_{j-1,j}\,e^{ik_{z,j-1}(d_{j-1}-d_{j-2})}\right) \\
&= \left(A_{j-2}\,e^{ik_{z,j-2}d_{j-3}}S_{j-2,j-1}\,e^{ik_{z,j-2}(d_{j-2}-d_{j-3})}\right)\left(S_{j-1,j}\,e^{ik_{z,j-1}(d_{j-1}-d_{j-2})}\right) \\
&= \dots \\
&= A_1\,e^{ik_{z1}d_1}\prod_{m=1}^{j-1} S_{m,m+1}\,e^{ik_{z,m}(d_m-d_{m-1})}.
\end{aligned}$$

If we introduce a generalized transmission coefficient

$$\widetilde{T}_{1N} := \prod_{m=1}^{N-1} S_{m,m+1}\, e^{ik_{z,m}(d_m - d_{m-1})} \tag{8.25}$$

we can express the amplitudes as

$$\boxed{A_N\, e^{ik_{z,N} d_{N-1}} = \widetilde{T}_{1N} A_1\, e^{ik_{z1} d_1}\,.} \tag{8.26}$$

Therefore the downgoing wave amplitude in region N at $z = -d_{n+1}$ is $\widetilde{T}_{1N}$ times the downgoing amplitude in region 1 ($z = -d_1$). We shall use some of these results when we attempt to homogenize layered media in Section 8.6.

8.3 Smoothly varying layered media

Let us now examine a few features of waves in layered media where we can assume that the material properties are continuous functions of z. Recall that the equations governing antiplane shear waves, acoustic waves, and TE/TM electromagnetic waves in layered media with variability only in the z-direction can be expressed in the form

$$\left[\alpha(z)\frac{d}{dz}\left(\frac{1}{\alpha(z)}\frac{d\widehat{\varphi}}{dz}\right)\right] + \left[\omega^2\alpha(z)\beta(z) - k_x^2\right]\widehat{\varphi} = 0 \tag{8.27}$$

where $\widehat{\varphi}$ is the unknown field, α and β are material properties, and

$$\varphi(x,z) = \widehat{\varphi}(z)\, e^{\pm ik_x x}\,.$$

8.3.1 The thin layer limit

We can extend the approach that we have used to calculate the reflection and transmission coefficients for piecewise constant media to smoothly varying layered media by taking the limit as the layer thickness becomes small. In Section 8.2 we saw that for a piecewise constant layered medium with N layers the field in the j-th layer is given by[‡]

$$\widehat{\varphi}_j(z) = A_j\left[e^{-ik_{zj}z} + \widetilde{R}_{j,j+1}\, e^{ik_{zj}z + 2ik_{zj}d_j}\right] \tag{8.28}$$

where $\widetilde{R}_{j,j+1}$ is a generalized reflection coefficient. This coefficient can be obtained from a recursion relation of the form

$$\widetilde{R}_{j,j+1} = \frac{R_{j,j+1} + \widetilde{R}_{j+1,j+2}\, e^{2ik_{z,j+1}(d_{j+1}-d_j)}}{1 + R_{j,j+1}\widetilde{R}_{j+1,j+2}\, e^{2ik_{z,j+1}(d_{j+1}-d_j)}}. \tag{8.29}$$

[‡]For a detailed discussion see Chew (1995). Other approaches for finely layered media can be found in Brekhovskikh (1960).

The quantity $R_{j,j+1}$ is the reflection coefficient for reflection at the interface between layers j and $j+1$. In the particular case of TE-waves we have

$$R_{j,j+1} = \frac{\mu_{j+1}k_{z,j} - \mu_j k_{z,j+1}}{\mu_{j+1}k_{z,j} + \mu_j k_{z,j+1}} = \frac{k_{z,j}/\mu_j - k_{z,j+1}/\mu_{j+1}}{k_{z,j}/\mu_j + k_{z,j+1}/\mu_{j+1}} \tag{8.30}$$

where μ_j is the magnetic permeability of layer j. We can find equivalent expressions for elastic and acoustic waves. We will now proceed to determine the generalized reflection coefficient in the continuum limit.

Consider a multilayered medium where each layer has thickness Δ. For concreteness we will focus on TE-waves here. The same procedure can be used for TM-waves, acoustic waves, and antiplane SH-elastic waves. Let the generalized reflection coefficient for the medium at the interface $z = -d_j$ be $\widetilde{R}(z)$ and let the local reflection coefficient be $R(z)$. Let us write the phase velocity $k_{z,j+1}$ in the layer just below the interface as $k_z(z - \Delta/2)$. The material property of importance in TE-wave propagation is the magnetic permeability and we will assume that $\mu_{j+1} \equiv \mu(z - \Delta/2)$.[§] Then, equation (8.29) can be written as

$$\widetilde{R}(z) = \frac{R(z) + \widetilde{R}(z-\Delta)\, e^{2ik_z(z-\Delta/2)\Delta}}{1 + R(z)\widetilde{R}(z-\Delta)\, e^{2ik_z(z-\Delta/2)\Delta}} \tag{8.31}$$

where

$$R(z) = \frac{\dfrac{k_z(z+\Delta/2)}{\mu(z+\Delta/2)} - \dfrac{k_z(z-\Delta/2)}{\mu(z-\Delta/2)}}{\dfrac{k_z(z+\Delta/2)}{\mu(z+\Delta/2)} + \dfrac{k_z(z-\Delta/2)}{\mu(z-\Delta/2)}} = \frac{\widetilde{k}(z+\Delta/2) - \widetilde{k}(z-\Delta/2)}{\widetilde{k}(z+\Delta/2) + \widetilde{k}(z-\Delta/2)}$$

with $\widetilde{k} := k_z/\mu$. Expanding the $\widetilde{k}(\bullet)$ quantities in the expression for $R(z)$ in the Taylor series about z and ignoring terms containing Δ^2 and higher powers, we get

$$R(z) \approx \frac{\widetilde{k}(z) + (\Delta/2)\widetilde{k}'(z) - \widetilde{k}(z) + (\Delta/2)\widetilde{k}'(z)}{\widetilde{k}(z) + (\Delta/2)\widetilde{k}'(z) + \widetilde{k}(z) - (\Delta/2)\widetilde{k}'(z)} = \frac{\Delta\widetilde{k}'(z)}{2\widetilde{k}(z)}. \tag{8.32}$$

Similarly, ignoring powers Δ^2 and higher, we get

$$e^{2ik_z(z-\Delta/2)\Delta} \approx 1 + 2i\Delta k_z(z) \quad \text{and} \quad \widetilde{R}(z-\Delta) \approx \widetilde{R}(z) - \Delta\widetilde{R}'(z). \tag{8.33}$$

Plugging equations (8.33) into (8.31) gives

$$\widetilde{R}(z) \approx \frac{R(z) + \left[\widetilde{R}(z) - \Delta\widetilde{R}'(z)\right][1 + 2i\Delta k_z(z)]}{1 + R(z)\left[\widetilde{R}(z) - \Delta\widetilde{R}'(z)\right][1 + 2i\Delta k_z(z)]}. \tag{8.34}$$

[§]This implies that we are measuring the phase velocity and the permeability at the center of the layer. However, this is not strictly necessary and we could alternatively measure these quantities at $z - \Delta$.

Substituting (8.32) into (8.34) and dropping terms containing Δ^2 and higher leads to

$$\widetilde{R}(z) \approx \frac{\dfrac{\Delta\widetilde{k}'(z)}{2\widetilde{k}(z)} + \widetilde{R}(z) - \Delta\widetilde{R}'(z) + 2i\Delta k_z(z)\widetilde{R}(z)}{1 + \dfrac{\Delta\widetilde{k}'(z)}{2\widetilde{k}(z)}\widetilde{R}(z)}. \tag{8.35}$$

If we assume that $\Delta\widetilde{k}'$ is small such that the denominator can be expanded in a series, we get

$$\widetilde{R}(z) \approx \left[\frac{\Delta\widetilde{k}'(z)}{2\widetilde{k}(z)} + \widetilde{R}(z) - \Delta\widetilde{R}'(z) + 2i\Delta k_z(z)\widetilde{R}(z)\right]\left[1 - \frac{\Delta\widetilde{k}'(z)}{2\widetilde{k}(z)}\widetilde{R}(z) + O(\Delta^2)\right]. \tag{8.36}$$

After expanding and ignoring terms containing Δ^2, we get

$$\widetilde{R}(z) \approx \frac{\Delta\widetilde{k}'(z)}{2\widetilde{k}(z)} + \widetilde{R}(z) - \Delta\widetilde{R}'(z) + 2i\Delta k_z(z)\widetilde{R}(z) - \frac{\Delta\widetilde{k}'(z)}{2\widetilde{k}(z)}[\widetilde{R}(z)]^2. \tag{8.37}$$

Rearrangement and division by Δ gives

$$\boxed{\widetilde{R}'(z) = \frac{d\widetilde{R}}{dz} = 2ik_z(z)\widetilde{R}(z) + \frac{\widetilde{k}'(z)}{2\widetilde{k}(z)}\left\{1 - [\widetilde{R}(z)]^2\right\}.} \tag{8.38}$$

Equation (8.38) has the form of a *Riccati equation*¶ and gives a continuous representation of the generalized reflection coefficient $\widetilde{R}(z)$. The Riccati equation can be solved numerically using Runge-Kutta methods.

For example, in the situation shown in Figure 8.3(a), the generalized reflectivity coefficient at the point z_1 is $\widetilde{R}(z_1)$ while that at point z_0 is 0. If we wish to determine the value of $\widetilde{R}(z_i)$ at a point inside the smoothly varying layer, then one possibility is to assume that $\mu(z)$ and $\varepsilon(z)$ is constant for $z > z_i$ and compute the value of $\widetilde{R}$ in the usual manner. There can also be a situation where there are a few isolated strong discontinuities inside the graded layer as shown in Figure 8.3(b). If there is a discontinuity at z_c, we can use the discrete solution with layer thickness 0 at the discontinuity. Then, from (8.29), at the discontinuity

$$\widetilde{R}(z_c^+) = \frac{R(z_c) + \widetilde{R}(z_c^-)}{1 + R(z_c)\widetilde{R}(z_c^-)}. \tag{8.39}$$

¶The Riccati equation has the form

$$\frac{dy}{dx} = a_0(x) + a_1(x)y + a_2(x)y^2 \quad \text{where} \quad a_0(x) \neq 0, a_2(x) \neq 0.$$

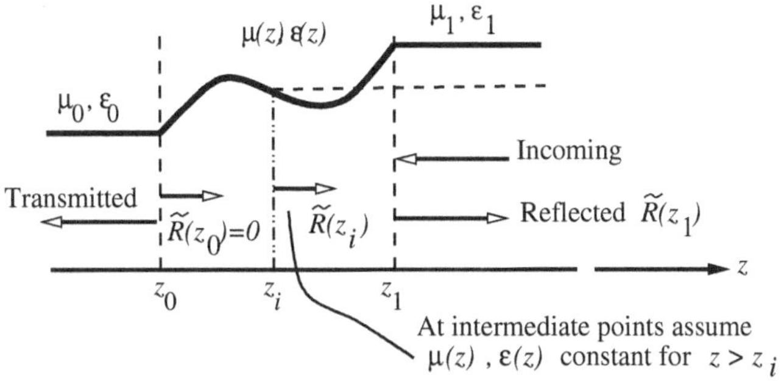

(a) Smoothly graded layered material.

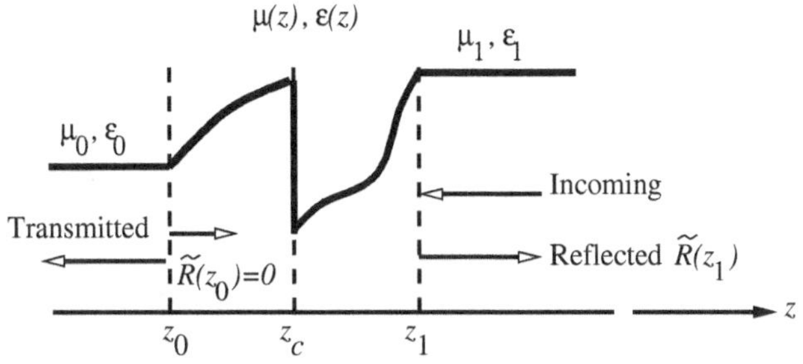

(b) Smoothly graded material with discontinuity.

FIGURE 8.3
Reflectivity in a smoothly graded layered material without and with a strong discontinuity.

Also, from (8.30)

$$R(z_c) = \frac{k_z^+/\mu^+ - k_z^-/\mu^-}{k_z^+/\mu^+ + k_z^-/\mu^-}. \tag{8.40}$$

Hence we can find the generalized reflection coefficients at isolated discontinuities within the material. Note that due to the products involved, a continuum extension of the recursion formula for the amplitudes of waves in multilayered media is not straightforward.

8.3.2 The Airy function solution

In the previous subsection we found the properties of a layered medium in the limit where the layer thickness was small. Let us now try to solve equation (8.27) directly. We will consider the simplest functional form of inhomogeneity that we can think of, a linear variation of properties with z. In particular, let us look at the situation where the material property β is a linear function of z and α is constant. The situation where α is constant can correspond to a nonmagnetic medium in electromagnetism, a constant density layered medium in acoustics, or a layered elastic medium with constant shear modulus in antiplane shear elasticity.

If β varies linearly with z, then we may write

$$\beta(z) = a + bz$$

where a and b are constants. Plugging this into (8.27) we get

$$\frac{d^2\widehat{\varphi}}{dz^2} + (A+Bz)\widehat{\varphi} = 0 \quad \text{where} \quad A := \omega^2\alpha a - k_x^2 \quad \text{and} \quad B := \omega^2\alpha b\,. \tag{8.41}$$

Let us assume that $B > 0$ (this is not strictly necessary, but simplifies things for our present analysis) and introduce a change of variables

$$\eta = B^{1/3}\left(z + \frac{A}{B}\right).$$

Then (8.41) becomes

$$\boxed{\frac{d^2\widehat{\varphi}}{d\eta^2} + \eta\widehat{\varphi} = 0.} \tag{8.42}$$

Equation (8.42) is called the *Airy equation*. The solution of this equation is

$$\widehat{\varphi}(\eta) = C_1\,\text{Ai}(-\eta) + C_2\,\text{Bi}(-\eta)$$

where Ai and Bi are *Airy functions of the first and second kind* which are related to the modified Bessel functions (see Abramowitz and Stegun (1972) for details.) A plot of the behavior of the two Airy functions as a function of real $-\eta$ is shown in Figure 8.4. As $z \to -\infty$ (i.e., as $-\eta \to \infty$), the Airy functions asymptotically approach the values

$$\begin{aligned}
\text{Ai}(-\eta) &\sim \frac{1}{2}\pi^{-1/2}\,(-\eta)^{-1/4}\,e^{-2/3\,(-\eta)^{3/2}} \\
\text{Bi}(-\eta) &\sim \pi^{-1/2}\,(-\eta)^{-1/4}\,e^{2/3\,(-\eta)^{3/2}}\,.
\end{aligned}$$

Therefore $\text{Ai}(-\eta)$ corresponds to an exponentially decaying wave as $|z| \to \infty$ and $\text{Bi}(-\eta)$ corresponds to an exponentially increasing wave at $|z| \to \infty$. A schematic of the situation is shown in Figure 8.5. If there are no sources in the region $z < 0$ then the solution $\text{Bi}(-\eta)$ is unphysical which implies that $C_2 = 0$. Therefore,

$$\widehat{\varphi}(\eta) = C_1\;\text{Ai}(-\eta)\,. \tag{8.43}$$

Now, as $z \to \infty$ (i.e., as $-\eta \to -\infty$), the Airy function $\text{Ai}(-\eta)$ takes the asymptotic form

$$\text{Ai}(-\eta) \sim \pi^{-1/2}\,\eta^{-1/4}\,\sin\left(\frac{2}{3}\eta^{3/2} + \frac{\pi}{4}\right). \tag{8.44}$$

This is a superposition of right and left traveling waves (because the sine can be decomposed into two exponentials one of which corresponds to a wave traveling in one direction and the seconds to a wave traveling in the opposite direction).

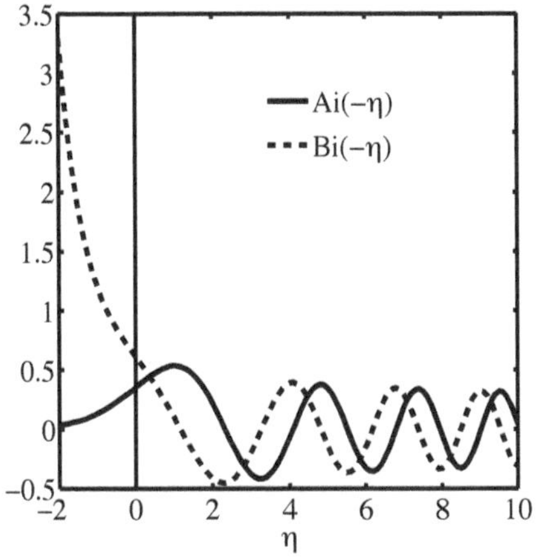

FIGURE 8.4
Plots of the Airy functions $\mathrm{Ai}(-\eta)$ and $\mathrm{Bi}(-\eta)$.

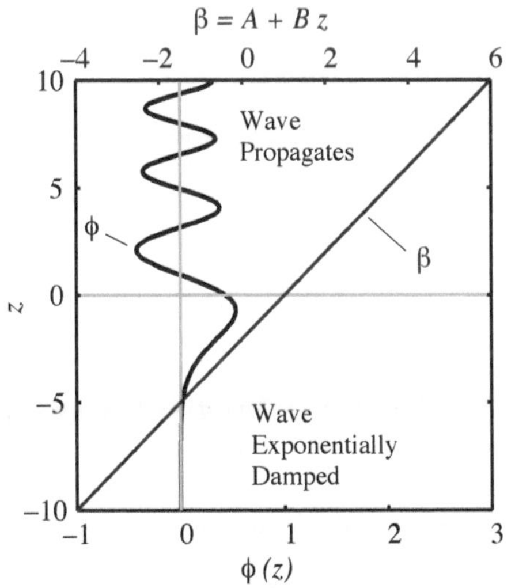

FIGURE 8.5
A schematic of $\widehat{\varphi}(z)$ and $\beta(z)$ showing the regions where the wave propagates and where it is exponentially damped.

8.3.3 The Wentzel-Kramers-Brillouin-Jeffreys method

If we don't assume any particular linear variation of the property $\beta(z)$, we can use the Wentzel-Kramers-Brillouin-Jeffreys (WKBJ) method to arrive at an approximate solution. The WKBJ method is a high-frequency method for obtaining solutions to one-dimensional (time-independent) wave equations of the form

$$\frac{d^2\varphi}{dz^2} + k_z^2(z)\,\varphi(z) = 0. \tag{8.45}$$

If α does not vary with z and if we do not assume any particular functional form for $\beta(z)$, equation (8.27) takes the form

$$\frac{d^2\widehat{\varphi}}{dz^2} + \left[\omega^2\alpha\beta(z) - k_x^2\right]\widehat{\varphi} = 0. \tag{8.46}$$

Clearly this equation can be written in the form of (8.45) by setting

$$k_z^2(z) := \omega^2\alpha\beta(z) - k_x^2.$$

In fact, we do not have to assume that α is constant to be able to express (8.27) in the form of equation (8.45). To see this let us write (8.27) in the form

$$\alpha(z)\frac{d}{dz}\left(\frac{1}{\alpha(z)}\frac{d\widehat{\varphi}}{dz}\right) + k_z^2(z)\widehat{\varphi}(z) = 0 \tag{8.47}$$

where $k_z^2(z) = \omega^2\alpha(z)\beta(z) - k_x^2$. After expanding (8.47) we get

$$\frac{d^2\widehat{\varphi}}{dz^2} = \frac{1}{\alpha(z)}\frac{d\alpha}{dz}\frac{d\widehat{\varphi}}{dz} - k_z^2(z)\,\widehat{\varphi}(z). \tag{8.48}$$

Define

$$\psi(z) := \widehat{\varphi}(z)/\sqrt{\alpha(z)}. \tag{8.49}$$

Differentiating (8.49) twice, we get

$$\frac{d^2\psi}{dz^2} = \widehat{\varphi}\frac{d^2}{dz^2}\left(\frac{1}{\sqrt{\alpha}}\right) + 2\frac{d\widehat{\varphi}}{dz}\frac{d}{dz}\left(\frac{1}{\sqrt{\alpha}}\right) + \frac{1}{\sqrt{\alpha}}\frac{d^2\widehat{\varphi}}{dz^2}. \tag{8.50}$$

Substitution of (8.48) and (8.49) into (8.50) gives

$$\frac{d^2\psi}{dz^2} = \sqrt{\alpha}\psi\frac{d^2}{dz^2}\left(\frac{1}{\sqrt{\alpha}}\right) - \frac{1}{\alpha^{3/2}}\frac{d\alpha}{dz}\frac{d\widehat{\varphi}}{dz} + \frac{1}{\alpha^{3/2}}\frac{d\alpha}{dz}\frac{d\widehat{\varphi}}{dz} - \frac{k_z^2}{\sqrt{\alpha}}\psi$$

or

$$\frac{d^2\psi}{dz^2} + \left[k_z^2 - \sqrt{\alpha}\frac{d^2}{dz^2}\left(\frac{1}{\sqrt{\alpha}}\right)\right]\psi = 0. \tag{8.51}$$

Equation (8.51) has the same form as (8.45) and therefore the WKBJ method is also applicable to the situation where α is not constant. If we assume that k_x is proportional to ω (which implies that k_z is also proportional to ω), i.e.,

$$k_z^2(z) = \omega^2\alpha(z)\beta(z) - k_x^2 =: \omega^2 s^2(z) \tag{8.52}$$

where $s(z)$ is independent of ω. In equation (8.51), if ω is large, then k_z will dominate and we will end up with exactly the same equation as (8.45), provided variations in α are smooth (and we don't get large jumps in its derivatives).

Let us now try to solve equation (8.45),

$$\varphi''(z) + k_z^2(z)\varphi(z) = 0\,. \tag{8.53}$$

When k_z is constant, the solution of the equation is a traveling wave. If we assume that k_z varies slowly with z, we can try to get solutions of the form

$$\varphi(z) = A\, e^{i\omega\tau(z)} \tag{8.54}$$

and examine the phase $\tau(z)$ rather than the solution $\varphi(z)$. Differentiating (8.54) twice, we get

$$\varphi''(z) = \left[i\omega\tau''(z) - \omega^2(\tau'(z))^2\right] A\, e^{i\omega\tau(z)}\,. \tag{8.55}$$

Plugging (8.55) into (8.53), we get

$$i\omega\tau''(z) - \omega^2[\tau'(z)]^2 + k_z^2(z) = 0\,. \tag{8.56}$$

Let us simplify the analysis slightly at this stage, even though this is not strictly necessary, by assuming that $k_z^2 > 0$ (i.e., k_z is real). For large ω, i.e., $\omega \gg 1$, we can seek a perturbation solution of equation (8.56) of the form

$$\tau(z) = \tau_0(z) + \frac{1}{\omega}\tau_1(z) + \frac{1}{\omega^2}\tau_2(z) + \ldots\,. \tag{8.57}$$

Plugging (8.57) into (8.56), using (8.52), and dividing by ω^2, we get

$$\begin{aligned} &\frac{i}{\omega}\left[\tau_0''(z) + \frac{1}{\omega}\tau_1''(z) + \frac{1}{\omega^2}\tau_2''(z) + \ldots\right] \\ &\quad - \left[\tau_0'(z) + \frac{1}{\omega}\tau_1'(z) + \frac{1}{\omega^2}\tau_2'(z) + \ldots\right]^2 + s^2(z) = 0\,. \end{aligned} \tag{8.58}$$

For large ω the above equation reduces to

$$-[\tau_0'(z)]^2 + s^2(z) = 0 \qquad \text{or} \qquad \boxed{[\tau_0'(z)]^2 = s^2(z).} \tag{8.59}$$

This is the *Eikonal equation* which is used to obtain the ray theory approximation for wave propagation at high frequencies. Therefore we have

$$\tau_0'(z) = \pm s(z)\,. \tag{8.60}$$

Integrating (8.60) from an arbitrary point z_0 to z, we get

$$\boxed{\tau_0(z) = \pm\int_{z_0}^{z} s(y)\, \mathrm{d}y + C_{0\pm}} \tag{8.61}$$

where $Co_{\pm}$ depends on the sign of the integral. Next, collecting terms of order $1/\omega$ in equation (8.58), we get

$$\frac{i}{\omega}\tau_0''(z) - \frac{2}{\omega}\tau_0'(z)\tau_1'(z) = 0.$$

We can eliminate $\tau_0'(z)$ from the above equation by substituting the result from equation (8.60). Then we have

$$\pm is'(z) \mp 2s(z)\tau_1'(z) = 0 \quad \Longrightarrow \quad \tau_1'(z) = \frac{i}{2}\frac{s'(z)}{s(z)}. \tag{8.62}$$

Integrating the second of equations (8.62), we get

$$\boxed{\tau_1(z) = \frac{i}{2}\ln[s(z)] + C_1 = i\ln[\sqrt{s(z)}] + C_1.} \tag{8.63}$$

Plugging (8.61) and (8.63) into (8.57) (and ignoring terms containing powers of ω^2 and higher) we get

$$\boxed{\tau(z) = \pm\int_{z_0}^{z} s(y)\,dy + \frac{i}{\omega}\ln[\sqrt{s(z)}] + C_{\pm}.} \tag{8.64}$$

This implies that the solution (8.54) has the form

$$\boxed{\varphi(z) = \frac{A_+}{\sqrt{s(z)}}\exp\left(i\omega\int_{z_0}^{z} s(y)\,dy\right) + \frac{A_-}{\sqrt{s(z)}}\exp\left(-i\omega\int_{z_0}^{z} s(y)\,dy\right).} \tag{8.65}$$

Equation (8.65) is the WKBJ solution assuming $k_z^2 > 0$. Note that a solution does not exist when $s(z) = k_z/\omega = (\alpha\beta - k_x^2/\omega^2)^{1/2} = 0$. Also note that since k_z^2 is proportional to ω^2,

$$|k_z^2| \gg |i\omega\tau_0''| = |2\omega\tau_0'\tau_1'| \qquad \text{for large } \omega. \tag{8.66}$$

Therefore,

$$\omega^2 s^2 \gg \omega s' \Longrightarrow \omega s \gg \frac{\omega s'}{\omega s} = \frac{d}{dz}[\ln(\omega s)] \Longrightarrow k_z \gg \frac{d}{dz}[\ln(k_z)].$$

Therefore, the restriction is that ω is large and that k_z is smooth with respect to z.

Now, consider for example the profile shown in Figure 8.6. In region I, the WKBJ solution is valid since $k_z^2 > 0$. At the point where the profile meets the z-axis, a solution does not exist since $k_z = s(z) = 0$. However, if the profile is smooth enough, we can assume that $k_z(z)$ is linear and we can use the Airy solution for the region II around this point. When the profile goes below the z-axis, $k_z^2 < 0$. However, the WKBJ solution is valid in this region (III) too as equation 8.66 can still be satisfied with $s(z) = i\,\chi(z)$.

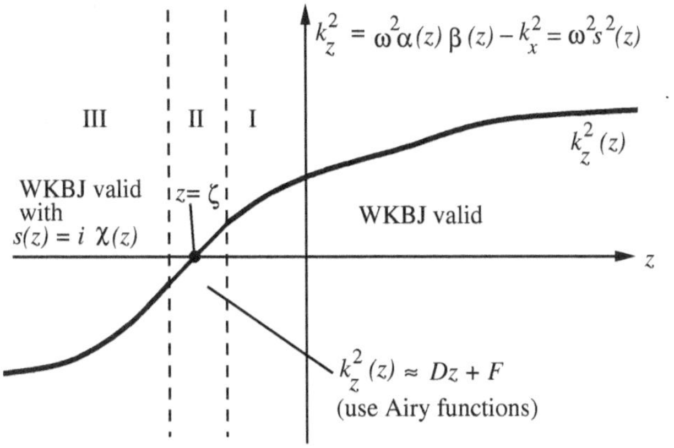

FIGURE 8.6
Regions of validity of the linear solution and the WKBJ solution for large ω.

There is an area of overlap between the regions where the WKBJ solution is valid and the region where the Airy solution is valid. In fact, the unknown parameters in the two solutions can be determined by matching the solutions at points in this region of overlap. To do this, let ζ be the point on the z-axis where $s(z) = 0$. In region I, the solution is

$$\varphi_{\mathrm{I}}(z) = \frac{A_+}{\sqrt{s(z)}} \exp\left(i\omega \int_\zeta^z s(y)\,\mathrm{d}y\right) + \frac{A_-}{\sqrt{s(z)}} \exp\left(-i\omega \int_\zeta^z s(y)\,\mathrm{d}y\right). \tag{8.67}$$

If there are no sources in region III the solution decays exponentially in the $-z$-direction. Then the WKBJ solution with $s(z) = i\chi(z)$ is

$$\varphi_{\mathrm{III}}(z) \sim \frac{B_-}{\sqrt{\chi(z)}} \exp\left(\omega \int_\zeta^z \chi(y)\,\mathrm{d}y\right) \tag{8.68}$$

where the coefficient $B_- = A_-/\sqrt{i}$. In region II, since α or β vary linearly with z, we may write

$$k_z^2 = \omega^2\alpha\beta - k_x^2 \sim D(z-\zeta). \tag{8.69}$$

Then, from (8.43),

$$\varphi_{\mathrm{II}}(z) \sim C_{\mathrm{II}}\,\mathrm{Ai}(-\eta) \qquad \text{with} \qquad \eta := D^{1/3}(z-\zeta).$$

When ω is high, regions I, II, and III overlap. Also, from (8.69) we observe that $D \propto \omega^2$. Hence, the large η expansion for the Airy function given in equation (8.44) can be used in the overlap region,

$$\varphi_{\mathrm{II}}(z) \sim C_{\mathrm{II}}\pi^{-1/2}\eta^{-1/4}\sin\left(\frac{2}{3}\eta^{3/2} + \frac{\pi}{4}\right).$$

Using the definition of η and expressing $\sin(\theta)$ in terms of exponentials, we get

$$\varphi_{\rm II}(z) \sim \frac{C_{\rm II}}{2i\pi^{1/2}D^{1/12}(z-\zeta)^{1/4}}\left\{\exp\left[\frac{2i}{3}D^{1/2}(z-\zeta)^{3/2}+\frac{\pi i}{4}\right]-\exp\left[-\frac{2i}{3}D^{1/2}(z-\zeta)^{3/2}-\frac{\pi i}{4}\right]\right\}. \tag{8.70}$$

Also, in the neighborhood of region II where $z \sim \zeta$, we have

$$\omega s(z) \sim D^{1/2}(z-\zeta)^{1/2} \quad \Longrightarrow \quad \omega\int_{\zeta}^{z} s(y)\,\mathrm{d}y \sim \frac{2}{3}D^{1/2}.$$

Therefore, $\varphi_{\rm I}$ becomes

$$\varphi_{\rm I}(z) = \frac{A_+\omega^{1/2}}{D^{1/4}(z-\zeta)^{1/4}}\exp\left[\frac{2i}{3}D^{1/2}(z-\zeta)^{3/2}\right]+\frac{A_-\omega^{1/2}}{D^{1/4}(z-\zeta)^{1/4}}\exp\left[-\frac{2i}{3}D^{1/2}(z-\zeta)^{3/2}\right]. \tag{8.71}$$

Comparing (8.71) with (8.70) we get

$$\frac{A_+}{A_-} = -i \qquad \text{and} \qquad C_{\rm II} = 2A_+\omega^{1/2}\pi^{1/2}D^{-1/6}e^{i\pi/4}. \tag{8.72}$$

Similarly, by comparing $\varphi_{\rm II}$ and $\varphi_{\rm III}$ in the region of overlap, we get

$$C_{\rm II} = 2B_-\omega^{1/2}\pi^{1/2}D^{-1/6}. \tag{8.73}$$

We can therefore express all the unknown coefficients in terms of $C_{\rm II}$ and we have an approximate solution for the entire range of values of k_z^2. These solutions are of great use in geometric optics and ray-based seismic tomography computations.

8.4 Propagator matrix approach

The *propagator matrix* relates the fields at two points in a multilayered medium.‖ This matrix is also known as the transition matrix or the transfer matrix. Since the term "matrix" is used we can presume that a matrix form of the wave equation may

‖The idea of a propagator matrix was first introduced by Volterra in 1887 in the context of systems of ordinary differential equations. In elastodynamics, the propagator matrix in its present form was introduced by Gilbert and Backus (1966).

be involved in propagator matrix methods. In fact, we have seen just such a form of the wave equations in Section 8.1. The second-order differential equation describing one-dimensional wave propagation in a layered medium was converted into two first-order differential equations, called the state equations, and expressed in matrix form as

$$\frac{d}{dz}\left(\underline{\underline{v}}\right) = \underline{\underline{H}}\,\underline{\underline{v}} \qquad \text{where} \qquad \underline{\underline{v}} := \begin{bmatrix} \varphi \\ \psi \end{bmatrix}. \tag{8.74}$$

For antiplane shear problems we had

$$\underline{\underline{H}} := \begin{bmatrix} 0 & 1/\mu(z) \\ -k_z^2\mu(z) & 0 \end{bmatrix} \quad \text{with} \quad k_z^2 := \omega^2\rho(z)/\mu(z) - k_1^2 .$$

For acoustic wave propagation we had

$$\underline{\underline{H}} := \begin{bmatrix} 0 & \frac{ik_z^2}{\omega\rho(z)} \\ i\omega\rho(z) & 0 \end{bmatrix} \quad \text{with} \quad k_z^2 := \omega^2\rho(z)/\kappa(z) - k_1^2 .$$

For TE- and TM-electromagnetic wave propagation we had

$$\underline{\underline{H}} := \begin{bmatrix} 0 & i\omega\varepsilon(z) \\ \frac{ik_z^2}{\omega\varepsilon(z)} & 0 \end{bmatrix} \text{ or } \begin{bmatrix} 0 & i\omega\mu(z) \\ \frac{ik_z^2}{\omega\mu(z)} & 0 \end{bmatrix} \quad \text{with} \quad k_z^2 := \omega^2\varepsilon(z)\mu(z) - k_1^2 .$$

Let us examine the propagation matrix for these types of waves. If $\underline{\underline{H}}$ *is constant*, which is reasonable if we are dealing with piecewise constant layer media, particular solutions to (8.74) can be sought of the form

$$\underline{\underline{v}} = \underline{\underline{v}}^0 e^{\lambda z} . \tag{8.75}$$

Plugging (8.75) into (8.74) leads to the eigenvalue problem

$$(\underline{\underline{H}} - \lambda\underline{\underline{I}})\underline{\underline{v}}^0 = \underline{\underline{0}}$$

where $\underline{\underline{I}}$ is the identity matrix. Solutions exist only if

$$\det(\underline{\underline{H}} - \lambda\underline{\underline{I}}) = 0 .$$

Notice that for all the $\underline{\underline{H}}$ matrices above, $\det(\underline{\underline{H}} - \lambda\underline{\underline{I}}) = \lambda^2 + k_z^2$. Therefore, solutions exist if

$$\lambda^2 = -k_z^2 \qquad \Longrightarrow \qquad \lambda = \pm ik_z .$$

Therefore, the general solution of (8.74) is

$$\underline{\underline{v}}(z) = A^+ e^{ik_z z}\underline{\underline{n}}^+ + A^- e^{-ik_z z}\underline{\underline{n}}^- \tag{8.76}$$

where $\underline{\underline{n}}^+$ and $\underline{\underline{n}}^-$ are the eigenvectors corresponding to the eigenvalues ik_z and $-ik_z$, respectively. Equation (8.76) can be written more compactly in the form

$$\underline{\underline{v}}(z) = \begin{bmatrix}\underline{\underline{n}}^+ & \underline{\underline{n}}^-\end{bmatrix} \begin{bmatrix} e^{ik_z z} & 0 \\ 0 & e^{-ik_z z}\end{bmatrix} \begin{bmatrix} A^+ \\ A^- \end{bmatrix}$$

or

$$\underline{\underline{v}}(z) = \underline{\underline{N}}\,\underline{\underline{K}}(z)\,\underline{\underline{a}} \tag{8.77}$$

where

$$\underline{\underline{N}} := \begin{bmatrix}\underline{\underline{n}}^+ & \underline{\underline{n}}^-\end{bmatrix} \; ; \; \underline{\underline{K}}(z) := \begin{bmatrix} e^{ik_z z} & 0 \\ 0 & e^{-ik_z z}\end{bmatrix} \; ; \; \underline{\underline{a}} := \begin{bmatrix} A^+ \\ A^- \end{bmatrix}.$$

Note that for a point z' that is different from z,

$$\underline{\underline{K}}(z-z') = \begin{bmatrix} e^{ik_z(z-z')} & 0 \\ 0 & e^{-ik_z(z-z')}\end{bmatrix} = \begin{bmatrix} e^{ik_z z} & 0 \\ 0 & e^{-ik_z z}\end{bmatrix} \begin{bmatrix} e^{-ik_z z'} & 0 \\ 0 & e^{ik_z z'}\end{bmatrix} = \underline{\underline{K}}(z)\,\underline{\underline{K}}(-z').$$

Also,

$$\underline{\underline{K}}(z-z')\,\underline{\underline{K}}(z') = \begin{bmatrix} e^{ik_z(z-z')} & 0 \\ 0 & e^{-ik_z(z-z')}\end{bmatrix} \begin{bmatrix} e^{-ik_z z'} & 0 \\ 0 & e^{ik_z z'}\end{bmatrix} = \underline{\underline{K}}(z).$$

Therefore we can write (8.77) in the form

$$\underline{\underline{v}}(z) = \underline{\underline{N}}\,\underline{\underline{K}}(z-z')\,\underline{\underline{K}}(z')\,\underline{\underline{a}} = \underline{\underline{N}}\,\underline{\underline{K}}(z-z')\,\underline{\underline{N}}^{-1}\,\underline{\underline{N}}\,\underline{\underline{K}}(z')\,\underline{\underline{a}}$$

or

$$\boxed{\underline{\underline{v}}(z) = \underline{\underline{P}}(z,z')\,\underline{\underline{v}}(z') \quad \text{where} \quad \underline{\underline{P}}(z,z') := \underline{\underline{N}}\,\underline{\underline{K}}(z-z')\,\underline{\underline{N}}^{-1}.} \tag{8.78}$$

The matrix $\underline{\underline{P}}$ is called the *propagator matrix* or the *transition matrix* that relates the fields at z and z'. Notice that we can also write the second of equations (8.78) as

$$\underline{\underline{P}}(z,z') = \underline{\underline{N}}\exp\left[i\widetilde{\underline{\underline{K}}}(z-z')\right]\underline{\underline{N}}^{-1}$$

where

$$\widetilde{\underline{\underline{K}}} := \begin{bmatrix} k_z & 0 \\ 0 & -k_z \end{bmatrix} \qquad \text{and} \qquad \exp\left[i\widetilde{\underline{\underline{K}}}z\right] = \begin{bmatrix} e^{ik_z z} & 0 \\ 0 & e^{-ik_z z}\end{bmatrix}.$$

If we substitute (8.78) into (8.74) we get

$$\frac{d}{dz}\left[\underline{\underline{P}}(z,z')\right]\underline{\underline{v}}(z') = \underline{\underline{H}}\,\underline{\underline{P}}(z,z')\,\underline{\underline{v}}(z').$$

Since z' is arbitrary, we have

$$\frac{d}{dz}\left[\underline{\underline{P}}(z,z')\right] = \underline{\underline{H}}\,\underline{\underline{P}}(z,z'). \tag{8.79}$$

Therefore $\underline{\underline{P}}(z,z')$ is an integral matrix of equation (8.74). For constant $\underline{\underline{H}}$, the solution of (8.79) is

$$\underline{\underline{P}}(z,z') = \exp\left[(z-z')\underline{\underline{H}}\right] = \underline{\underline{I}} + (z-z')\underline{\underline{H}} + \tfrac{1}{2}(z-z')^2\underline{\underline{H}}\,\underline{\underline{H}} + \ldots. \tag{8.80}$$

If the eigenvalues of $\underline{\underline{H}}$ are distinct, we can use Sylvester's formula to express the exponential function of a matrix $\underline{\underline{A}}$ in terms of the eigenvalues λ_j using

$$\exp(\underline{\underline{A}}) = \sum_{j=1}^{n} \exp(\lambda_j) \frac{\prod_{k\neq j}(\underline{\underline{A}} - \lambda_k \underline{\underline{I}})}{\prod_{k\neq j}(\lambda_j - \lambda_k)}. \tag{8.81}$$

This formula is therefore valid when both z and z' are *in the same layer* and is more convenient in situations where eigenvalues are easier to compute than eigenvectors.

Also notice from (8.78) that

$$\underline{\underline{P}}(z',z') = \underline{\underline{P}}(z,z) = \underline{\underline{I}}.$$

For any three values of z, for example z_1, z_2, z_3, we have

$$\underline{v}(z_3) = \underline{\underline{P}}(z_3,z_2)\,\underline{v}(z_2) = \underline{\underline{P}}(z_3,z_2)\,\underline{\underline{P}}(z_2,z_1)\,\underline{v}(z_1) = \underline{\underline{P}}(z_3,z_1)\,\underline{v}(z_1).$$

Therefore,

$$\underline{\underline{P}}(z_3,z_1) = \underline{\underline{P}}(z_3,z_2)\,\underline{\underline{P}}(z_2,z_1). \tag{8.82}$$

In particular, if $z_1 = z_3$, we have

$$\underline{\underline{P}}(z_1,z_1) = \underline{\underline{P}}(z_1,z_2)\,\underline{\underline{P}}(z_2,z_1) = \underline{\underline{I}}.$$

Therefore,

$$\underline{\underline{P}}(z_2,z_1) = \underline{\underline{P}}^{-1}(z_1,z_2).$$

For a multilayered medium such as the one shown in Figure 8.1 we can generalize (8.82) if we know the value of $\underline{\underline{P}}$ for each layer. If the interfaces between the layers are located at $z_1, z_2, \ldots, z_{j-1}, z_j, \ldots, z_{N-1}$, then the state vector at the location $z_{j-1} > z > z_j$ is given by

$$\boxed{\underline{v}(z) = \underline{\underline{P}}(z,z_{j-1})\,\underline{\underline{P}}(z_{j-1},z_{j-2})\ldots\underline{\underline{P}}(z_3,z_2)\,\underline{\underline{P}}(z_2,z_1)\underline{v}(z_1) = \underline{\underline{P}}(z,z_1)\underline{v}(z_1).} \tag{8.83}$$

Therefore, using $\underline{\underline{H}}^j$ to denote the $\underline{\underline{H}}$ matrix for the j-th layer and taking advantage of the solution in equation (8.80), we can write

$$\underline{\underline{P}}(z,z_1) = \exp\left[(z-z_{j-1})\underline{\underline{H}}^j\right] \prod_{k=2}^{j-1} \exp\left[(z_k - z_{k-1})\underline{\underline{H}}^k\right].$$

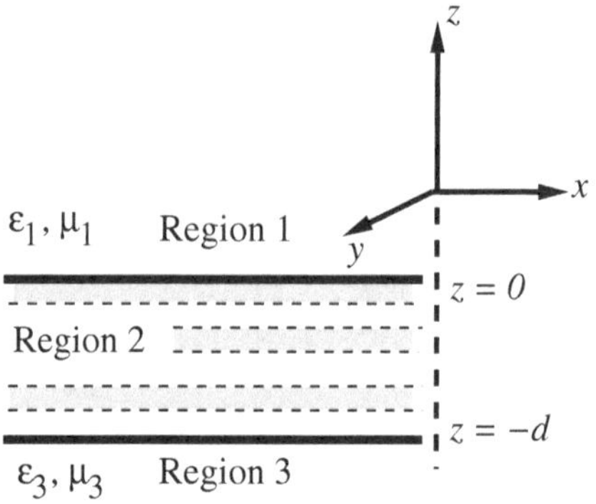

FIGURE 8.7
A multilayered medium sandwiched between two layers.

8.4.1 A three-layered sandwich medium

Let us now consider the special case of a three-layered medium in Figure 8.7. The medium consists of three layers. Regions 1 and 3 are isotropic while the sandwiched region 2 is multilayered. The interface between regions 1 and 2 is located at $z = 0$ while the interface between regions 2 and 3 is located at $z = -d$. For concreteness let us consider electromagnetic wave propagation. If region 2 is homogeneous we can use equation (8.83) directly to calculate the propagator matrix keeping in mind that the vector $\underline{\underline{v}}$ is discontinuous across the layers. Then we have

$$\underline{\underline{P}}(-d_2,0) = \underline{\underline{P}}(-d_2,-d_1)\,\underline{\underline{P}}(-d_1,0)$$

where $\underline{\underline{P}}(-d_2,-d_1)$ depends on ε_2, μ_2 and $\underline{\underline{P}}(-d_1,0)$ depends on ε_1, μ_1.

If fact we can use the same equation even if region 2 is not homogeneous because we can calculate the propagator matrix in the region using the product relation (8.83). Let the propagator matrix for region 2 be $\underline{\underline{P}}(-d,0)$. We want to find the reflection coefficient and the transmission coefficient of the system. In region 1, the state vector is given by equation (8.76),

$$\underline{\underline{v}}^1(z) = A^{1+}\, e^{ik_{z1}z}\, \underline{\underline{n}}^{1+} + A^{1-}\, e^{-ik_{z1}z}\, \underline{\underline{n}}^{1-}\,.$$

Define $A^{1+} := RA^{1-}$ where R is a scalar reflection coefficient. Then

$$\underline{\underline{v}}^1(z) = RA^{1-}\, e^{ik_{z1}z}\, \underline{\underline{n}}^{1+} + A^{1-}\, e^{-ik_{z1}z}\, \underline{\underline{n}}^{1-}\,. \tag{8.84}$$

Proceeding as before, let us define the matrices

$$\underline{\underline{N}}^1 := \begin{bmatrix} \underline{\underline{n}}^{1+} & \underline{\underline{n}}^{1-} \end{bmatrix} \qquad \text{and} \qquad \underline{\underline{K}}^1(z) = \exp\left[i\widetilde{\underline{\underline{K}}}^1 z\right] := \begin{bmatrix} e^{ik_{z1}z} & 0 \\ 0 & e^{-ik_{z1}z} \end{bmatrix}.$$

Then equation (8.84) can be written as

$$\boxed{\underline{\underline{v}}^1(z) = \underline{\underline{N}}^1 \exp\left[i\widetilde{\underline{\underline{K}}}^1 z\right] \begin{bmatrix} R \\ 1 \end{bmatrix} A^{1-} .} \tag{8.85}$$

In region 3, there is only a transmitted wave. Therefore, the state vector is given by

$$\underline{\underline{v}}^3(z) = A^{3-}\, e^{-ik_{z3}(z+d)}\, \mathbf{n}^{3-} . \tag{8.86}$$

Define $A^{3-} := TA^{1-}$ where T is a factor that behaves like a scalar transmission coefficient. Then, equation (8.86) can be written in matrix form as

$$\boxed{\underline{\underline{v}}^3(z) = \underline{\underline{N}}^3 \exp\left[i\widetilde{\underline{\underline{K}}}^3 (z+d)\right] \begin{bmatrix} 0 \\ T \end{bmatrix} A^{1-}} \tag{8.87}$$

where

$$\underline{\underline{N}}^3 := \left[\underline{\underline{n}}^{3+}\ \underline{\underline{n}}^{3-}\right] \qquad \text{and} \qquad \underline{\underline{K}}^3(z) = \exp\left[i\widetilde{\underline{\underline{K}}}^3 z\right] := \begin{bmatrix} e^{ik_{z3}z} & 0 \\ 0 & e^{-ik_{z3}z} \end{bmatrix} .$$

Since we have the propagator matrix, $\underline{\underline{P}}(-d,0)$, for region 2, we can use it to connect regions 1 and 3. The continuity of the state vector across the interfaces implies that

$$\underline{\underline{v}}^1(0) = \underline{\underline{v}}^2(0) \quad \text{at } z=0 \qquad \text{and} \qquad \underline{\underline{v}}^2(-d) = \underline{\underline{v}}^3(-d) \quad \text{at } z=-d . \tag{8.88}$$

Also, using equation (8.78) we have

$$\underline{\underline{v}}^2(-d) = \underline{\underline{P}}(-d,0)\,\underline{\underline{v}}^2(0) . \tag{8.89}$$

Therefore, using equations (8.88), we can write (8.89) as

$$\boxed{\underline{\underline{v}}^3(-d) = \underline{\underline{P}}(-d,0)\,\underline{\underline{v}}^1(0) .} \tag{8.90}$$

At $z=0$ (from (8.85)) and at $z=-d$ (from (8.87)) we have

$$\underline{\underline{v}}^1(0) = \underline{\underline{N}}^1 \begin{bmatrix} R \\ 1 \end{bmatrix} A^{1-} \qquad \text{and} \qquad \underline{\underline{v}}^3(-d) = \underline{\underline{N}}^3 \begin{bmatrix} 0 \\ T \end{bmatrix} A^{1-} .$$

Plugging these into (8.90) gives

$$\underline{\underline{N}}^3 \begin{bmatrix} 0 \\ T \end{bmatrix} A^{1-} = \underline{\underline{P}}(-d,0)\,\underline{\underline{N}}^1 \begin{bmatrix} R \\ 1 \end{bmatrix} A^{1-}$$

or

$$\boxed{\begin{bmatrix} 0 \\ T \end{bmatrix} = \left[\underline{\underline{N}}^3\right]^{-1} \underline{\underline{P}}(-d,0)\,\underline{\underline{N}}^1 \begin{bmatrix} R \\ 1 \end{bmatrix} .} \tag{8.91}$$

Equation (8.91) can then be solved to find the reflection and transmission coefficients R and T.

8.4.2 Anisotropic electromagnetic layered media

In a layered medium where each of the layers is isotropic, the TE- and TM-waves are uncoupled at the interface. However, this is not true when each of the layers is anisotropic and we have to consider the full Maxwell's equations. The state vector approach can still be used for anisotropic media by choosing the variables so that they are continuous across interfaces.

Let us start with Maxwell's equations at fixed frequency

$$\boldsymbol{\nabla} \times \mathbf{E} = i\omega\,\boldsymbol{\mu}\cdot\mathbf{H} \quad \text{and} \quad \boldsymbol{\nabla} \times \mathbf{H} = -i\omega\,\boldsymbol{\varepsilon}\cdot\mathbf{E}.$$

Recall that continuity of the fields requires that the tangential components of $\mathbf{E}$ and $\mathbf{H}$ be continuous across material interfaces. Therefore, an appropriate state vector for anisotropic media is

$$\underline{\underline{\mathsf{v}}} := \begin{bmatrix} \mathbf{E}_s \\ \mathbf{H}_s \end{bmatrix}$$

where $\mathbf{E}_s$ and $\mathbf{H}_s$ are the tangential components of $\mathbf{E}$ and $\mathbf{H}$ (i.e., the components on the surface normal to the z-direction).

Let us decompose the vector fields into a sum of the normal and tangential components:

$$\mathbf{E} = \mathbf{E}_s + \mathbf{E}_z \quad \text{and} \quad \mathbf{H} = \mathbf{H}_s + \mathbf{H}_z\,.$$

The gradient operator can also be split along the same lines, i.e.,

$$\boldsymbol{\nabla} = \boldsymbol{\nabla}_s + \frac{\partial}{\partial z}\,\mathbf{e}_z \quad \text{and} \quad \boldsymbol{\nabla}_s := \frac{\partial}{\partial x}\,\mathbf{e}_x + \frac{\partial}{\partial y}\,\mathbf{e}_y$$

where $\mathbf{e}_x, \mathbf{e}_y, \mathbf{e}_z$ are the unit vectors in the x-, y-, z-directions, respectively. Let us express the tensors $\boldsymbol{\mu}$ and $\boldsymbol{\varepsilon}$ in matrix form (in the basis $\mathbf{e}_x, \mathbf{e}_y, \mathbf{e}_z$) as

$$\boldsymbol{\mu} \equiv \begin{bmatrix} \boldsymbol{\mu}_{ss} & \boldsymbol{\mu}_{sz} \\ \boldsymbol{\mu}_{zs} & \mu_{zz} \end{bmatrix} \;;\; \boldsymbol{\varepsilon} \equiv \begin{bmatrix} \boldsymbol{\varepsilon}_{ss} & \boldsymbol{\varepsilon}_{sz} \\ \boldsymbol{\varepsilon}_{zs} & \varepsilon_{zz} \end{bmatrix}$$

where $\boldsymbol{\mu}_{ss}, \boldsymbol{\varepsilon}_{ss}$ are 2×2 matrices, $\boldsymbol{\mu}_{sz}, \boldsymbol{\varepsilon}_{sz}$ are 2×1 matrices, $\boldsymbol{\mu}_{zs}, \boldsymbol{\varepsilon}_{zs}$ are 1×2 matrices, and $\mu_{zz}, \varepsilon_{zz}$ are 1×1 matrices, i.e., scalars. Using the splits of the various quantities and the gradient operator, we can show that $\mathbf{E}_z, \mathbf{H}_z$ can be expressed in terms of $\mathbf{E}_s, \mathbf{H}_s$ as

$$\mathbf{E}_z = -\frac{1}{i\omega\varepsilon_{zz}}\boldsymbol{\nabla}_s \times \mathbf{H}_s - \frac{1}{\varepsilon_{zz}}\boldsymbol{\varepsilon}_{zs}\cdot\mathbf{E}_s \;;\; \mathbf{H}_z = \frac{1}{i\omega\mu_{zz}}\boldsymbol{\nabla}_s \times \mathbf{E}_s - \frac{1}{\mu_{zz}}\boldsymbol{\mu}_{zs}\cdot\mathbf{H}_s\,. \tag{8.92}$$

After some further manipulations, the Maxwell equations may be expressed in matrix form as (see Chew (1995) for details)

$$\boxed{\frac{\partial}{\partial z}\begin{bmatrix} \mathbf{E}_s \\ \mathbf{H}_s \end{bmatrix} = \underline{\underline{\mathsf{H}}}\begin{bmatrix} \mathbf{E}_s \\ \mathbf{H}_s \end{bmatrix}} \tag{8.93}$$

where $\underline{\underline{\mathsf{H}}}$ is a 4×4 matrix instead of the 2×2 matrix $\underline{\underline{\mathsf{H}}}$ in equation (8.13) for the isotropic case. We can now use the propagator matrix approach to calculate the generalized transmission and reflection coefficients for electromagnetic waves in a layered anisotropic medium.

8.4.3 Anisotropic elastic layered media

In anisotropic elastic layered media, a common approach is to use the Stroh formalism (see Ting (1996) for details). The direction of lamination is denoted $\mathbf{n}$ and is usually aligned with the z-direction for simplicity. Then the material properties of each layer are only functions of z, i.e., $SfC \equiv \mathbf{C}(z)$ and $\rho \equiv \rho(z)$. The governing equations of elastodynamics, in the absence of body forces, then have the form

$$\nabla \cdot \boldsymbol{\sigma} = \rho(z)\ddot{\mathbf{u}} \quad \text{and} \quad \boldsymbol{\sigma} = \mathbf{C}(z) : \nabla \mathbf{u}$$

where $\boldsymbol{\sigma}(\mathbf{r},t)$ is the stress, $\mathbf{u}(\mathbf{r},t)$ is the displacement, and $\mathbf{r}$ is the radial location of a point with respect to the origin of the coordinate system. We choose the orientation of the x-coordinate such that $\boldsymbol{\sigma}$ and $\mathbf{u}$ are independent of y, and label the unit vector along the x-direction as $\mathbf{m}$. Then $x = \mathbf{m} \cdot \mathbf{r}$ and $z = \mathbf{n} \cdot \mathbf{r}$. Note that the $\mathbf{m}$ vector is parallel to the layers.

We now consider two time harmonic plane wave vector fields (a displacement and a traction),

$$\mathbf{u}(x,z,t) = \mathbf{a}(z)\, e^{i(k_x x - \omega t)} \qquad \text{and} \qquad \mathbf{n} \cdot \boldsymbol{\sigma}(x,z,t) = ik\mathbf{f}(z)\, e^{i(k_x x - \omega t)}$$

where $\mathbf{a}(y), \mathbf{f}(y)$ are the respective amplitudes. If, following the Stroh approach, the state vector is composed of these two amplitudes:

$$\underline{\underline{\eta}}(z) = \begin{bmatrix} \mathbf{a}(z) \\ i\mathbf{f}(z) \end{bmatrix}$$

then the state equation can be expressed as

$$\boxed{\frac{d}{dz}\underline{\underline{\eta}}(z) = \underline{\underline{\mathsf{H}}}(z)\,\underline{\underline{\eta}}(z)} \tag{8.94}$$

where the matrix $\underline{\underline{\mathsf{H}}}$ has the form

$$\underline{\underline{\mathsf{H}}}(z) = i \begin{bmatrix} k_x \underline{\underline{\mathsf{N}}}^1 & \underline{\underline{\mathsf{N}}}^2 \\ k_x^2 \underline{\underline{\mathsf{N}}}^3 - \rho\omega^2 \underline{\underline{\mathsf{I}}} & k_x (\underline{\underline{\mathsf{N}}}^1)^T \end{bmatrix}.$$

The matrices $\underline{\underline{\mathsf{N}}}^j$ are 3×3 blocks of the Stroh matrix,

$$\underline{\underline{\mathsf{N}}}(z) = \begin{bmatrix} \underline{\underline{\mathsf{N}}}^1 & \underline{\underline{\mathsf{N}}}^2 \\ \underline{\underline{\mathsf{N}}}^3 & (\underline{\underline{\mathsf{N}}}^1)^T \end{bmatrix} = \begin{bmatrix} -(\underline{\underline{\mathbf{C}}}^{nn})^{-1}\underline{\underline{\mathbf{C}}}^{nm} & -(\underline{\underline{\mathbf{C}}}^{nn})^{-1} \\ \underline{\underline{\mathbf{C}}}^{mm} - \underline{\underline{\mathbf{C}}}^{mn}(\underline{\underline{\mathbf{C}}}^{nn})^{-1}\underline{\underline{\mathbf{C}}}^{nm} & -\underline{\underline{\mathbf{C}}}^{mn}(\underline{\underline{\mathbf{C}}}^{nn})^{-1} \end{bmatrix}$$

where the components of the matrices $\underline{\underline{\mathbf{C}}}$ are defined as

$$(\underline{\underline{\mathbf{C}}}^{nn})_{qr} = (\mathbf{n} \cdot \mathbf{C} \cdot \mathbf{n})_{qr}\,; \quad (\underline{\underline{\mathbf{C}}}^{mm})_{qr} = (\mathbf{m} \cdot \mathbf{C} \cdot \mathbf{m})_{qr}$$
$$(\underline{\underline{\mathbf{C}}}^{nm})_{qr} = (\mathbf{n} \cdot \mathbf{C} \cdot \mathbf{m})_{qr}\,; \quad (\underline{\underline{\mathbf{C}}}^{mn})_{qr} = (\mathbf{m} \cdot \mathbf{C} \cdot \mathbf{n})_{qr}\,.$$

The propagator matrix (also called the matricant or transfer matrix) can be found by solving (8.94). We have seen the form that the propagator matrix takes when the medium is piecewise homogeneous. In fact, that solution is a special case of a more general solution (see Aki and Richards (1980) for an example) of the form

$$\underline{\underline{P}}(z,z') = \underline{\underline{I}} + \int_{z'}^{z} \underline{\underline{H}}(\zeta_1)d\zeta_1 + \int_{z'}^{z} \underline{\underline{H}}(\zeta_1)\int_{z'}^{\zeta_1} \underline{\underline{H}}(\zeta_2)d\zeta_2 d\zeta_1 + \ldots .$$

If the matrix $\underline{\underline{H}}(z)$ is independent of z, the propagator matrix takes the simple form

$$\underline{\underline{P}}(z,z') = e^{(z-z')\underline{\underline{H}}} .$$

For periodic layered medium with a period D, the quantity $\underline{\underline{P}}(D,0)$ is called the monodromy matrix. For recent developments in the field of anisotropic layered elastic composites see Alshits and Maugin (2005) and Shuvalov et al. (2010). In fact, it can be shown that the low-frequency dispersive dynamics of such a layered medium can be described by the Willis equations.

8.5 Periodic layered media

Periodic layered media have been the object of extensive investigation since the pioneering work of Rayleigh (1887) and Maxwell (1954). In this section we will limit our study to the particular case of the Schoenberg-Sen model of a periodically layered acoustic half-space (Schoenberg and Sen, 1983). This model is of interest because it was one of the first to hint at the possibility of a tensorial anisotropic effective mass density.

8.5.1 The Schoenberg-Sen model

Consider the periodic layered medium shown in Figure 8.8. The medium occupies the region $x > 0$ and the period of the medium is D. Within each period there are n layers indexed by j with density ρ_j, bulk modulus κ_j, and thickness $h_j = f_j D$ where f_j represents the volume fraction of layer j. The region $x < 0$ is occupied by a homogeneous fluid of density ρ_0 and bulk modulus κ_0.

Recall that, for time harmonic plane acoustic wave propagation in a layered medium, the state equation (8.7) has the form

$$\frac{d}{dx}(\underline{v}) = \underline{\underline{H}}\,\underline{v}$$

where

$$\underline{v}(x) := \begin{bmatrix} v_x(x) \\ p(x) \end{bmatrix} \quad \text{and} \quad \underline{\underline{H}} := \begin{bmatrix} 0 & \frac{ik_x^2}{\omega\rho(x)} \\ i\omega\rho(x) & 0 \end{bmatrix} .$$

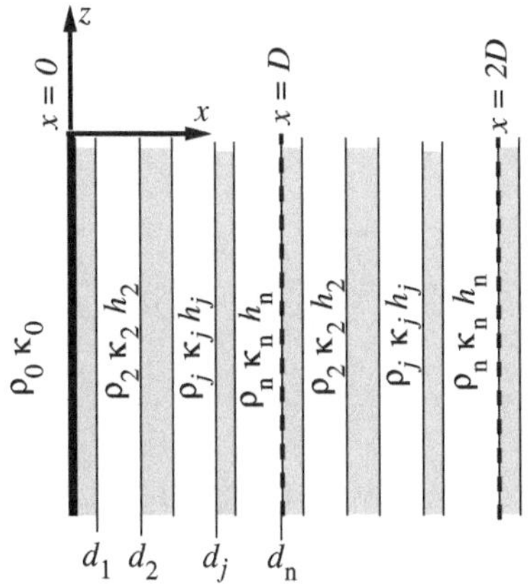

FIGURE 8.8
A periodic layered half-space of period d with lamination in the x-direction.

Here $\mathbf{k} = k_x\mathbf{e}_x + k_z\mathbf{e}_z$ is the wave vector with $k_x^2 = \omega^2\rho/\kappa - k_z^2$ and ω is the frequency. Proceeding as before, the state vector in a homogeneous region is given by

$$\underline{\underline{\mathsf{v}}}(x) = A^+ e^{ik_x x}\underline{\underline{\mathsf{n}}}^+ + A^- e^{-ik_x x}\underline{\underline{\mathsf{n}}}^-$$

where $\underline{\underline{\mathsf{n}}}^+, \underline{\underline{\mathsf{n}}}^-$ are the eigenvectors of $\underline{\underline{\mathsf{H}}}$ corresponding to the eigenvalues $\pm ik_x$. We can write the above in matrix form similar to (8.85),

$$\underline{\underline{\mathsf{v}}}(x) = \underline{\underline{\mathsf{N}}}\,\underline{\underline{\mathsf{K}}}(x)\begin{bmatrix} A^+ \\ A^- \end{bmatrix}$$

where

$$\underline{\underline{\mathsf{N}}} := \begin{bmatrix} \underline{\underline{\mathsf{n}}}^+ & \underline{\underline{\mathsf{n}}}^- \end{bmatrix} \qquad \text{and} \qquad \underline{\underline{\mathsf{K}}}(x) = \exp\left[i\underline{\underline{\widetilde{\mathsf{K}}}}x\right] := \begin{bmatrix} e^{ik_x x} & 0 \\ 0 & e^{-ik_x x} \end{bmatrix}.$$

For the acoustic problem, the matrix $\underline{\underline{\mathsf{N}}}$ is

$$\underline{\underline{\mathsf{N}}} = \begin{bmatrix} k_x/(\omega\rho) & -k_x/(\omega\rho) \\ 1 & 1 \end{bmatrix}.$$

For the incidence region we have

$$\underline{\underline{\mathsf{v}}}^0(x) = e^{ik_{x0}x}\underline{\underline{\mathsf{n}}}^{0+} + Re^{-ik_{x0}x}\underline{\underline{\mathsf{n}}}^{0-} \quad \Longrightarrow \quad \underline{\underline{\mathsf{v}}}^0(x) = \underline{\underline{\mathsf{N}}}^0\underline{\underline{\mathsf{K}}}^0(x)\begin{bmatrix} 1 \\ R \end{bmatrix} \tag{8.95}$$

where $k_{x0}^2 = \omega^2\rho_0/\kappa_0 - k_z^2$, and

$$\underline{\underline{\mathsf{N}}}^0 = \begin{bmatrix} k_{x0}/(\omega\rho_0) & -k_{x0}/(\omega\rho_0) \\ 1 & 1 \end{bmatrix} \quad \text{and} \quad \underline{\underline{\mathsf{K}}}^0(x) = \begin{bmatrix} e^{ik_{x0}x} & 0 \\ 0 & e^{-ik_{x0}x} \end{bmatrix}. \tag{8.96}$$

We have assumed that the incident wave has unit amplitude and R is the reflection coefficient. In the j-th layer of the layered region we have

$$\underline{\underline{\mathsf{v}}}^j(x) = A^{j+} e^{ik_{xj}(x-d_{j-1})}\,\underline{\underline{\mathsf{n}}}^{j+} + A^{j-}\, e^{-ik_{xj}(x-d_{j-1})}\,\underline{\underline{\mathsf{n}}}^{j-}\,.$$

In matrix form,

$$\underline{\underline{\mathsf{v}}}^j(x) = \underline{\underline{\mathsf{N}}}^j\,\underline{\underline{\mathsf{K}}}^j(x-d_{j-1})\begin{bmatrix} A^{j+} \\ A^{j-} \end{bmatrix} \tag{8.97}$$

where $d_j = \sum_{q=1}^{j} h_q = \sum_{q=1}^{j} f_q D\,,\ \ k_{xj}^2 = \omega^2\rho_j/\kappa_j - k_z^2$,

$$\underline{\underline{\mathsf{N}}}^j = \begin{bmatrix} k_{xj}/(\omega\rho_j) & -k_{xj}/(\omega\rho_j) \\ 1 & 1 \end{bmatrix} \quad \text{and} \quad \underline{\underline{\mathsf{K}}}^j(x) = \begin{bmatrix} e^{ik_{xj}(x-d_{j-1})} & 0 \\ 0 & e^{-ik_{xj}(x-d_{j-1})} \end{bmatrix}.$$

We can relate the state vector at $x = d_j$ to the state vector at $x = d_{j-1}$ with the propagator matrix (8.78), i.e.,

$$\underline{\underline{\mathsf{v}}}^j(d_j) = \underline{\underline{\mathsf{P}}}(d_j, d_{j-1})\,\underline{\underline{\mathsf{v}}}^j(d_{j-1}) \tag{8.98}$$

where

$$\underline{\underline{\mathsf{P}}}(d_j, d_{j-1}) := \underline{\underline{\mathsf{N}}}^j\,\underline{\underline{\mathsf{K}}}^j(d_j - d_{j-1})\,(\underline{\underline{\mathsf{N}}}^j)^{-1}\,.$$

Plugging in the expressions for $\underline{\underline{\mathsf{N}}}^j$ and $\underline{\underline{\mathsf{K}}}^j$, we get

$$\underline{\underline{\mathsf{P}}}^j := \underline{\underline{\mathsf{P}}}(d_j, d_{j-1}) = \begin{bmatrix} \cos(h_j k_{xj}) & -ik_{xj}\sin(h_j k_{xj})/(\rho_j\omega) \\ -i\rho_j\omega\sin(h_j k_{xj})/k_{xj} & \cos(h_j k_{xj}) \end{bmatrix}. \tag{8.99}$$

Observe that $\det(\underline{\underline{\mathsf{P}}}^j) = 1$. For one period of the layered medium, we then have

$$\underline{\underline{\mathsf{v}}}(D) = \underline{\underline{\mathsf{P}}}^n\,\underline{\underline{\mathsf{P}}}^{n-1}\ldots\underline{\underline{\mathsf{P}}}^2\,\underline{\underline{\mathsf{P}}}^1\,\underline{\underline{\mathsf{v}}}(0) =: \underline{\underline{\mathsf{P}}}(D)\,\underline{\underline{\mathsf{v}}}(0) \tag{8.100}$$

where $\underline{\underline{\mathsf{v}}}(0)$ is the state vector at the beginning of the period, $\underline{\underline{\mathsf{v}}}(D)$ is the state vector at the end of the period, and $\det[\underline{\underline{\mathsf{P}}}(D)] = 1$. The quantity $\underline{\underline{\mathsf{P}}}(D)$ is also called the monodromy matrix. Using the eigen decomposition theorem we can express $\underline{\underline{\mathsf{P}}}(D)$ in the form

$$\underline{\underline{\mathsf{P}}}(D) = \underline{\underline{\mathsf{Q}}}\,\underline{\underline{\boldsymbol{\Lambda}}}\,\underline{\underline{\mathsf{Q}}}^{-1}$$

where $\underline{\underline{\boldsymbol{\Lambda}}}$ is a diagonal matrix whose diagonal elements are the eigenvalues, λ_ℓ, of $\underline{\underline{\mathsf{P}}}(D)$ and $\underline{\underline{\mathsf{Q}}}$ is a matrix whose ℓ-th column is the eigenvector, $\underline{\mathsf{q}}^\ell$, corresponding to the eigenvalue λ_ℓ.** Plugging the above decomposition into (8.100) gives

$$\underline{\underline{\mathsf{v}}}(D) = \underline{\underline{\mathsf{Q}}}\,\underline{\underline{\boldsymbol{\Lambda}}}\,\underline{\underline{\mathsf{Q}}}^{-1}\,\underline{\underline{\mathsf{v}}}(0)\,. \tag{8.101}$$

**Observe that since $\det(\underline{\underline{\mathsf{Q}}})\det(\underline{\underline{\mathsf{Q}}}^{-1}) = 1$ and $\det[\underline{\underline{\mathsf{P}}}(D)] = 1$ we must have $\det\underline{\underline{\boldsymbol{\Lambda}}} = 1$. Therefore $\lambda_1\lambda_2 = 1$ and we can have real λ_ℓ with $|\lambda_1| < 1$ and $|\lambda_2| > 1$, or we can have complex conjugate λ_ℓ with $|\lambda_\ell| = 1$. Also note that the eigenvalues of $\underline{\underline{\mathsf{P}}}(D)$ are of the form $\exp(\pm iDk_x^{\text{eff}})$ where k_x^{eff} is the effective wave number of the layered medium. This is the expression that is expected from Bloch-Floquet theory.

Also, because the periodicity of the medium implies that each period has the same normalized eigenvectors $\underline{\underline{\mathsf{Q}}}$, the state vector at a depth of m periods is

$$\underline{\mathsf{v}}(mD) = \underline{\underline{\mathsf{Q}}}\,\underline{\underline{\boldsymbol{\Lambda}}}^m\,\underline{\underline{\mathsf{Q}}}^{-1}\,\underline{\mathsf{v}}(0)$$

where $\underline{\underline{\boldsymbol{\Lambda}}}^m = \prod_{\ell=1}^m \underline{\underline{\boldsymbol{\Lambda}}}$. Now, from equation (8.95), we have

$$\underline{\mathsf{v}}(0) = \underline{\mathsf{v}}^0(0) = \underline{\underline{\mathsf{N}}}^0\,\underline{\underline{\mathsf{K}}}^0(0)\begin{bmatrix}1\\R\end{bmatrix} = \underline{\underline{\mathsf{N}}}^0\begin{bmatrix}1\\R\end{bmatrix}.$$

Hence,

$$\underline{\mathsf{v}}(mD) = \underline{\underline{\mathsf{Q}}}\,\underline{\underline{\boldsymbol{\Lambda}}}^m\,\underline{\underline{\mathsf{Q}}}^{-1}\,\underline{\underline{\mathsf{N}}}^0\begin{bmatrix}1\\R\end{bmatrix}.$$

In expanded form,

$$\underline{\mathsf{v}}(mD) = \frac{1}{\det\underline{\underline{\mathsf{Q}}}}\begin{bmatrix}q_1^1 & q_1^2\\ q_2^1 & q_2^2\end{bmatrix}\begin{bmatrix}\lambda_1^m & 0\\ 0 & \lambda_2^m\end{bmatrix}\begin{bmatrix}q_2^2 & -q_1^2\\ -q_2^1 & q_1^1\end{bmatrix}\underline{\underline{\mathsf{N}}}^0\begin{bmatrix}1\\R\end{bmatrix} \tag{8.102}$$

where the superscripts on λ_j represent powers. The eigenvalues of λ_1, λ_2 are determined from the relation

$$\det[\underline{\underline{\mathsf{P}}}(D) - \lambda\underline{\underline{\mathsf{I}}}] = 0 \quad \text{with} \quad \det[\underline{\underline{\mathsf{P}}}(D)] = 1\,.$$

Expanding these out, we get the characteristic equation

$$\lambda^2 - \lambda(P_{11} + P_{22}) + 1 = 0\,.$$

If k_x^{eff} is the effective wave vector of the layered medium and ρ_{eff} is the effective density, then

$$\underline{\underline{\mathsf{P}}}(D) = \underline{\underline{\mathsf{P}}}^{\text{eff}} = \begin{bmatrix}\cos(Dk_x^{\text{eff}}) & -ik_x^{\text{eff}}\sin(Dk_x^{\text{eff}})/(\rho_{\text{eff}}\omega)\\ -i\rho_{\text{eff}}\omega\sin(Dk_x^{\text{eff}})/k_x^{\text{eff}} & \cos(Dk_x^{\text{eff}})\end{bmatrix}.$$

Therefore,

$$P_{11} + P_{22} = 2\cos(Dk_x^{\text{eff}}) =: 2C \tag{8.103}$$

and the solutions of the characteristic equation can be written as

$$\lambda = C \pm \sqrt{C^2 - 1}\,. \tag{8.104}$$

If the eigenvalues are real, $|C| > 1$ which means that $\cos(Dk_x^{\text{eff}}) > 1$. This implies that k_x^{eff} is imaginary which corresponds to an evanescent wave and no waves can propagate into the layered medium. On the other hand, if $|C| < 1$, the eigenvalues are complex conjugates but we have wave propagation in the medium because k_x^{eff} is real.

Interestingly, we can simplify the problem if we add a small amount of loss. Let us incorporate a small amount of loss in the system by allowing the wave number to be complex, i.e., $k_{xj} = k_{xj}(1+i\delta)$ with $0 < \delta \ll 1$. In that case, as the number of layers goes to infinity, the displacement and pressure should go to zero:

$$\lim_{m\to\infty} \underline{\underline{\mathsf{v}}}(mD) = 0\,. \tag{8.105}$$

We will now consider the situation where the eigenvalues of $\underline{\underline{\mathsf{P}}}(D)$ are complex conjugates. The eigenvalues of $\underline{\underline{\mathsf{P}}}(D)$ are of the form

$$\lambda_\ell^m = e^{\pm iDk_x m}\,, \qquad \ell = 1,2\,.$$

For $k_x = k_x(1+i\delta)$, we can express the two eigenvalues as

$$\lambda_1^m = [\cos(Dk_x m) + i\sin(Dk_x m)]e^{-Dk_x\delta m}\,,\ \lambda_2^m = [\cos(Dk_x m) - i\sin(Dk_x m)]e^{Dk_x\delta m}\,.$$

As $m \to \infty$, $\lambda_1^m \to 0$ and equation (8.105) becomes

$$\frac{1}{\det\underline{\underline{\mathsf{Q}}}}\begin{bmatrix} q_1^1 & q_1^2 \\ q_2^1 & q_2^2\end{bmatrix}\begin{bmatrix} 0 & 0 \\ 0 & \lim_{m\to\infty}\lambda_2^m\end{bmatrix}\begin{bmatrix} q_2^2 & -q_1^2 \\ -q_2^1 & q_1^1\end{bmatrix}\underline{\underline{\mathsf{N}}}^0\begin{bmatrix}1\\R\end{bmatrix} = 0$$

where we have used the expanded form from equation (8.102). Since $\exp(Dk_x\delta m) > 0 \implies \lim_{m\to\infty}\lambda_2^m \to \infty$, the above equation can be satisfied only if

$$\begin{bmatrix}-q_2^1 & q_1^1\end{bmatrix}\underline{\underline{\mathsf{N}}}^0\begin{bmatrix}1\\R\end{bmatrix} = 0. \tag{8.106}$$

Condition (8.106) implies that the expression for the state vector in equation (8.102) can be reduced to

$$\underline{\underline{\mathsf{v}}}(mD) = \frac{\lambda_1^m}{\det\underline{\underline{\mathsf{Q}}}}\begin{bmatrix}q_1^1\\q_2^1\end{bmatrix}\begin{bmatrix}q_2^2 & -q_1^2\end{bmatrix}\underline{\underline{\mathsf{N}}}^0\begin{bmatrix}1\\R\end{bmatrix}$$

After some simple manipulations we can show that the above relation can be written as

$$\underline{\underline{\mathsf{v}}}(mD) = \lambda_1^m\left\{\begin{bmatrix}1 & 0\\0 & 1\end{bmatrix} - \frac{1}{\det\underline{\underline{\mathsf{Q}}}}\begin{bmatrix}q_1^2\\q_2^2\end{bmatrix}\begin{bmatrix}-q_2^1 & q_1^1\end{bmatrix}\right\}\underline{\underline{\mathsf{N}}}^0\begin{bmatrix}1\\R\end{bmatrix}.$$

Applying (8.106), we arrive as the simple expression

$$\boxed{\underline{\underline{\mathsf{v}}}(mD) = \lambda_1^m\underline{\underline{\mathsf{N}}}^0\begin{bmatrix}1\\R\end{bmatrix}.} \tag{8.107}$$

Note that this solution is *valid for all wavelengths*. We can now calculate the reflection and transmission coefficients of the medium. Noting that $k_{x0} = \omega/c_0$ and using the definition of acoustic impedance, $Z_0 = \rho_0 c_0$, we can write $\underline{\underline{\mathsf{N}}}^0$ (8.96)$_1$ as

$$\underline{\underline{\mathsf{N}}}^0 = \begin{bmatrix}1/Z_0 & -1/Z_0\\1 & 1\end{bmatrix}$$

and the condition (8.106) can be expressed as

$$Z_0 q_1^1(1+R) = q_2^1(1-R) \quad \Longrightarrow \quad q_1^1 = q_2^1\,\frac{1-R}{Z_0(1+R)}\,. \tag{8.108}$$

Also, from the eigenvalue problem for $\underline{\underline{P}}(D)$ we have

$$\begin{bmatrix} P_{11}-\lambda_1 & P_{12} \\ P_{21} & P_{22}-\lambda_1 \end{bmatrix} \begin{bmatrix} q_1^1 \\ q_2^1 \end{bmatrix} = \begin{bmatrix} 0 \\ 0 \end{bmatrix} o \quad \Longrightarrow \quad P_{21}q_1^1 + (P_{22}-\lambda_1)q_2^1 = 0\,. \tag{8.109}$$

Eliminating q_1^1 from (8.108) and (8.109) gives us an expression for the reflection coefficient R,

$$R = \frac{P_{21}/(\lambda_1 - P_{22}) - Z_0}{P_{21}/(\lambda_1 - P_{22}) + Z_0}\,.$$

We can write the above as

$$\boxed{R = \frac{Z_{\text{eff}} - Z_0}{Z_{\text{eff}} + Z_0} \quad \text{where} \quad Z_{\text{eff}} := \frac{P_{21}}{\lambda_1 - P_{22}}\,.} \tag{8.110}$$

If we compare this expression for the reflection coefficient with the expression for R in equation (2.57), p. 82, we see that Z_{eff} can be interpreted as an effective acoustic impedance of the layered medium. To find the transmission coefficient for a layered slab with m periods, we expand equation (8.107) to get

$$\underline{\mathbf{v}}(mD) = \begin{bmatrix} v_x(mD) \\ p(mD) \end{bmatrix} = \lambda_1^m \begin{bmatrix} (1-R)/Z_0 \\ 1+R \end{bmatrix}\,.$$

Recall that the incident wave has amplitude 1 and that, for wave propagation to occur, the amplitude of λ_1 is 1. Therefore the amplitude of the transmitted wave is equal to the transmission coefficient, T, i.e.,

$$T = p(mD) = 1 + R = \frac{2Z_{\text{eff}}}{Z_{\text{eff}} + Z_0}\,.$$

8.5.2 Low-frequency effective properties

At low frequencies we can ignore the effect of dispersion[‡] and estimate the effective properties of the layered medium by expanding $\underline{\underline{P}}(D)$ in powers of ω. Recall from (8.100) and (8.99) that

$$\underline{\underline{P}}(D) = \underline{\underline{P}}^n\,\underline{\underline{P}}^{n-1}\ldots\underline{\underline{P}}^2\,\underline{\underline{P}}^1$$

where

$$\underline{\underline{P}}^j = \begin{bmatrix} \cos(h_j k_{xj}) & -i k_{xj}\sin(h_j k_{xj})/(\rho_j \omega) \\ -i\rho_j\omega\sin(h_j k_{xj})/k_{xj} & \cos(h_j k_{xj}) \end{bmatrix}\,.$$

[‡]The exact dispersion relation for P-waves in a layered elastic medium were first discovered by Rytov (1956).

For simplicity, let us look at a layered medium in which each period has two layers, i.e., n = 2. Then

$$\underline{\underline{P}}(D) = \begin{bmatrix} P_{11} & P_{12} \\ P_{21} & P_{22} \end{bmatrix}$$

where

$$\begin{aligned}
P_{11} &= \cos(h_1 k_{x1})\cos(h_2 k_{x2}) - \frac{k_{x2}\rho_1}{k_{x1}\rho_2}\sin(h_1 k_{x1})\sin(h_2 k_{x2}) \\
P_{12} &= -i\left[\frac{k_{x1}}{\rho_1\omega}\cos(h_2 k_{x2})\sin(h_1 k_{x1}) + \frac{k_{x2}}{\rho_2\omega}\cos(h_1 k_{x1})\sin(h_2 k_{x2})\right] \\
P_{21} &= -i\left[\frac{\rho_1\omega}{k_{x1}}\cos(h_2 k_{x2})\sin(h_1 k_{x1}) + \frac{\rho_2\omega}{k_{x2}}\cos(h_1 k_{x1})\sin(h_2 k_{x2})\right] \\
P_{22} &= \cos(h_1 k_{x1})\cos(h_2 k_{x2}) - \frac{k_{x1}\rho_2}{k_{x2}\rho_1}\sin(h_1 k_{x1})\sin(h_2 k_{x2}) .
\end{aligned}$$

If we use $k_{xj} = (\omega^2\rho_j/\kappa_j - k_z^2)^{1/2} =: \omega s_{xj}$, where s_{xj} is the slowness in the x-direction, expand out the cosines and sines in powers of ω, and drop terms $O(\omega^3)$ and higher, we get

$$\underline{\underline{P}}(D) \approx \begin{bmatrix} 1 - \omega^2 \sum_{j=1}^{2} \frac{h_j^2 s_{xj}^2}{2} - \omega^2 (h_1\rho_1)\frac{h_2 s_{x2}^2}{\rho_2} & -i\omega \sum_{j=1}^{2} \frac{h_j s_{xj}^2}{\rho_j} \\ -i\omega \sum_{j=1}^{2} h_j\rho_j & 1 - \omega^2 \sum_{j=1}^{2} \frac{h_j^2 s_{xj}^2}{2} - \omega^2 (h_2\rho_2)\frac{h_1 s_{x1}^2}{\rho_1} \end{bmatrix} .$$

Therefore, from the above relation and (8.103), and using $h_j = f_j D$, we have

$$\begin{aligned}
C &\approx \tfrac{1}{2}(P_{11}+P_{22}) = 1 - \omega^2 D^2 \left[\sum_{j=1}^{2} \frac{f_j s_{xj}^2}{\rho_j}\right]\left[\sum_{j=1}^{2} f_j\rho_j\right] =: 1 - \omega^2 D \left\langle \frac{s_x^2}{\rho} \right\rangle \langle\rho\rangle \\
P_{21} &\approx -i\omega D\langle\rho\rangle
\end{aligned}$$

where $\langle(\bullet)\rangle$ is the thickness weighted average. From (8.104) we know that

$$\lambda_1 = C + \sqrt{C^2 - 1} .$$

If we include the $O(\omega^4)$ terms in the series expansion and use the definition of C, we can show that

$$\lambda_1 \approx 1 + i\omega D \left\langle \frac{s_x^2}{\rho} \right\rangle^{1/2} \langle\rho\rangle^{1/2} - \frac{\omega^2 D^2}{2} \left\langle \frac{s_x^2}{\rho} \right\rangle \langle\rho\rangle + O(\omega^3) . \tag{8.111}$$

Notice that the series looks like an exponential function which is the form that we expect λ_1 to have. Recall from (8.110) that

$$Z_{\text{eff}} = \frac{P_{21}}{\lambda_1 - P_{22}} .$$

Plugging in the approximations for P_{21}, P_{22}, and λ_1, we get

$$Z_{\text{eff}} \approx \frac{\langle\rho\rangle^{1/2}}{\langle s_x^2/\rho\rangle^{1/2}} + O(\omega^2)\,.$$

From the definition of the slowness, we have

$$s_{xj}^2 = \rho_j/\kappa_j - k_z^2/\omega^2 =: 1/c_j^2 - s_z^2$$

where c_j is the phase velocity and s_z is the slowness in the z-direction. Using the above relation we can show that

$$Z_{\text{eff}} \approx \frac{[\langle\rho\rangle\,/\,\langle 1/\rho\rangle]^{1/2}}{\left[\langle 1/\kappa\rangle\,/\,\langle 1/\rho\rangle - s_z^2\right]^{1/2}} = \frac{\langle\rho\rangle\,/\,[\langle\rho\rangle\,\langle 1/\rho\rangle]^{1/2}}{\left[\langle 1/\kappa\rangle\,/\,\langle 1/\rho\rangle - s_z^2\right]^{1/2}} \tag{8.112}$$

or

$$Z_{\text{eff}} = \frac{\rho_{\text{eff}}}{(1/c_{\text{eff}}^2 - s_z^2)^{1/2}}$$

where ρ_{eff} is an effective density and c_{eff} is an effective phase velocity in the layered medium. If we define $c_{\text{eff}}^2 =: \kappa_{\text{eff}}/\rho_{\text{eff}}$ we see that

$$\boxed{\frac{1}{\kappa_{\text{eff}}} = \left\langle\frac{1}{\kappa}\right\rangle\left[\langle\rho\rangle\left\langle\frac{1}{\rho}\right\rangle\right]^{-\frac{1}{2}}\,,\ \rho_{\text{eff}} = \langle\rho\rangle\left[\langle\rho\rangle\left\langle\frac{1}{\rho}\right\rangle\right]^{-\frac{1}{2}}\,,\ c_{\text{eff}} = \left\langle\frac{1}{\rho}\right\rangle^{\frac{1}{2}}\left\langle\frac{1}{\kappa}\right\rangle^{-\frac{1}{2}}\,.}$$

Now, we can express equation (8.111) in exponential form as

$$\lambda_1 \approx \exp\left[i\omega D\left\langle\frac{s_x^2}{\rho}\right\rangle^{\frac{1}{2}}\langle\rho\rangle^{\frac{1}{2}}\right]\,.$$

We also know that the value of λ_1 for an effective homogeneous medium has the form

$$\lambda_1 = e^{iDk_x^{\text{eff}}} = e^{i\omega D s_x^{\text{eff}}}\,.$$

Comparing these two equations, we deduce that

$$s_x^{\text{eff}} = \left\langle\frac{s_x^2}{\rho}\right\rangle^{\frac{1}{2}}\langle\rho\rangle^{\frac{1}{2}} = \left[\frac{\langle 1/\kappa\rangle}{\langle 1/\rho\rangle} - s_z^2\right]^{\frac{1}{2}}\left[\langle\rho\rangle^{\frac{1}{2}}\,\langle 1/\rho\rangle\right]^{\frac{1}{2}} \tag{8.113}$$

where we have used the process that we had used earlier to arrive at equation (8.112). We can reorganize (8.113) and write it as

$$\frac{1}{\langle 1/\kappa\rangle\,\langle\rho\rangle}(s_x^{\text{eff}})^2 + \frac{\langle 1/\rho\rangle}{\langle 1/\kappa\rangle}s_z^2 = 1\,.$$

When we compare the above to the equivalent relation for a homogeneous medium, $c^2 s_x^2 + c^2 s_z^2 = 1$ (where c is the phase speed in the medium), we can make the connection

$$c_x^{\text{eff}} = \left[\frac{1}{\langle 1/\kappa\rangle \langle\rho\rangle}\right]^{\frac{1}{2}} \quad \text{and} \quad c_z^{\text{eff}} = \left[\frac{\langle 1/\rho\rangle}{\langle 1/\kappa\rangle}\right]^{\frac{1}{2}} = c_{\text{eff}} \tag{8.114}$$

where c_x^{eff} is the effective phase speed perpendicular to the layers and c_z^{eff} is the effective phase speed parallel to the layers. Equations (8.114) can be interpreted as the phase speeds in a medium with an isotropic bulk modulus and a transversely isotropic density tensor of the form

$$\boxed{\kappa^{\text{eff}} = \frac{1}{\langle 1/\kappa\rangle} \quad \text{and} \quad \boldsymbol{\rho}^{\text{eff}} \equiv \begin{bmatrix} \langle\rho\rangle & 0 & 0 \\ 0 & 1/\langle 1/\rho\rangle & 0 \\ 0 & 0 & 1/\langle 1/\rho\rangle \end{bmatrix}.} \tag{8.115}$$

We have considered a two-layer period in the above derivation. A general derivation for a period with n layers can be found in Schoenberg and Sen (1983).

8.6 Quasistatic homogenization of layered media

The effective response of a layered medium is of greater interest in many situations than the response of individual layers. We had our first encounter with homogenization of layered media in Section 8.3.1 where we found a continuum approximation of the reflection coefficient in the limit of small layer thickness. Let us now explore the effective properties of layered composite materials in the quasistatic limit.

8.6.1 Quasistatic effective properties of elastic laminates

Consider the layered medium shown in Figure 8.9. Each layer is homogeneous but has anisotropic elastic properties. Such a layered medium is called a laminate. In this case we have chosen the direction of lamination to be x_1. We will find the effective elastic properties of the laminate using an approach pioneered by Backus (1962).[††] We will assume that the layers are perfectly bonded, i.e., the displacement field is continuous across the interfaces between the layers.

Recall that the constitutive relation for an anisotropic elastic material is given by

$$\boldsymbol{\sigma} = \mathbf{C} : \boldsymbol{\varepsilon}.$$

[††] This approach is also called Backus upscaling in the geophysics literature. Further details and references can be found in Milton (2002). Similar techniques are suggested by Postma (1955) and Tartar (1976).

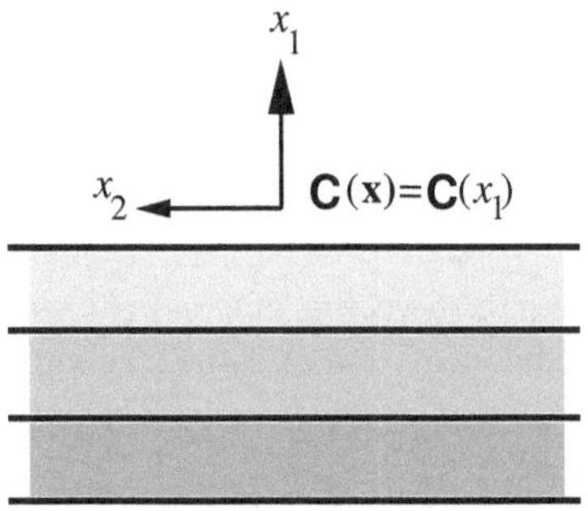

FIGURE 8.9
An elastic layered medium with direction of lamination x_1.

To find the relation between the average stress, $\langle\boldsymbol{\sigma}\rangle$, and the average strain, $\langle\boldsymbol{\varepsilon}\rangle$, we cannot just average the constitutive relation because

$$\langle\boldsymbol{\sigma}\rangle = \langle\mathbf{C} : \boldsymbol{\varepsilon}\rangle \neq \langle\mathbf{C}\rangle : \langle\boldsymbol{\varepsilon}\rangle \quad \text{where} \quad \langle f\rangle := \frac{1}{L}\int_0^L f(x_1)\,\mathrm{d}x_1$$

unless $\boldsymbol{\varepsilon}$ is constant or $\mathbf{C}$ is constant. However, there are fields which are constant in certain directions and those can be used to simplify things.

The strain field ($\boldsymbol{\varepsilon}$) in each layer is given by

$$\boldsymbol{\varepsilon}(\mathbf{x}) = \tfrac{1}{2}\left[\boldsymbol{\nabla}\mathbf{u} + (\boldsymbol{\nabla}\mathbf{u})^T\right] \quad \equiv \quad \varepsilon_{ij} = \tfrac{1}{2}\left(\frac{\partial u_i}{\partial x_j} + \frac{\partial u_j}{\partial x_i}\right)$$

where the components are with respect to a Cartesian basis $(\mathbf{e}_1, \mathbf{e}_2, \mathbf{e}_3)$ with coordinates (x_1, x_2, x_3). Because the displacements u_i are continuous across the interfaces between the layers, and each layer is homogeneous, the in-plane strain components $\varepsilon_{22}, \varepsilon_{33}, \varepsilon_{23}$ are also continuous across the interfaces. Moreover, these components of strain are also constant in each layer. Since a piecewise constant field that is also continuous must be constant, the strain components $\varepsilon_{22}, \varepsilon_{33}, \varepsilon_{23}$ must be constant throughout the laminate.

The tractions (normal components of the stress) at each interface are given by

$$\mathbf{t} = \mathbf{n}\cdot\boldsymbol{\sigma} \quad \equiv \quad t_i = n_j\sigma_{ji}$$

where $\boldsymbol{\sigma}$ is the stress in a layer and $\mathbf{n}$ is the outward unit normal at the interface (relative to the layer under consideration). Now the tractions must also be continuous at the interfaces because there are not gaps between the layers. The surface normal at an interface points in the $\pm\mathbf{e}_1$-direction and hence the components of stress that contribute to the traction are σ_{11}, σ_{21}, and σ_{31}. Since these components of the stress are piecewise constant in each layer, this implies that the stress components σ_{11}, σ_{12}, and σ_{13} must also be constant throughout the laminate.

Let us express the constitutive relation in the following matrix form:

$$\begin{bmatrix} \underline{\underline{\boldsymbol{\sigma}}}^{\rm n} \\ \underline{\underline{\boldsymbol{\sigma}}}^{\rm t} \end{bmatrix} = \begin{bmatrix} \underline{\underline{\mathbf{C}}}^{\rm nn} & \underline{\underline{\mathbf{C}}}^{\rm nt} \\ \underline{\underline{\mathbf{C}}}^{\rm tn} & \underline{\underline{\mathbf{C}}}^{\rm tt} \end{bmatrix} \begin{bmatrix} \underline{\underline{\boldsymbol{\varepsilon}}}^{\rm n} \\ \underline{\underline{\boldsymbol{\varepsilon}}}^{\rm t} \end{bmatrix} . \tag{8.116}$$

In the above we have broken the 6-vectors of stress and strain into separate 3-vectors for the out-of-plane (n) components and in-plane (t) components,

$$\underline{\underline{\boldsymbol{\sigma}}}^{\rm n} = \begin{bmatrix} \sigma_{11} \\ \sqrt{2}\,\sigma_{12} \\ \sqrt{2}\,\sigma_{13} \end{bmatrix} ; \; \underline{\underline{\boldsymbol{\sigma}}}^{\rm t} = \begin{bmatrix} \sigma_{22} \\ \sigma_{33} \\ \sqrt{2}\,\sigma_{23} \end{bmatrix} ; \; \underline{\underline{\boldsymbol{\varepsilon}}}^{\rm n} = \begin{bmatrix} \varepsilon_{11} \\ \sqrt{2}\,\varepsilon_{12} \\ \sqrt{2}\,\varepsilon_{13} \end{bmatrix} ; \; \underline{\underline{\boldsymbol{\varepsilon}}}^{\rm t} = \begin{bmatrix} \varepsilon_{22} \\ \varepsilon_{33} \\ \sqrt{2}\,\varepsilon_{23} \end{bmatrix} .$$

Accordingly, the 6×6 stiffness matrix has also been decomposed into four 3×3 matrices,

$$\underline{\underline{\mathbf{C}}}^{\rm nn} = \begin{bmatrix} C_{1111} & \sqrt{2}\,C_{1112} & \sqrt{2}\,C_{1113} \\ \sqrt{2}\,C_{1211} & 2\,C_{1212} & 2\,C_{1213} \\ \sqrt{2}\,C_{1311} & 2\,C_{1312} & 2\,C_{1313} \end{bmatrix} ; \; \underline{\underline{\mathbf{C}}}^{\rm nt} = \begin{bmatrix} C_{1122} & C_{1133} & \sqrt{2}\,C_{1123} \\ \sqrt{2}\,C_{1222} & \sqrt{2}\,C_{1233} & 2\,C_{1223} \\ \sqrt{2}\,C_{1322} & \sqrt{2}\,C_{1333} & 2\,C_{1323} \end{bmatrix}$$

$$\underline{\underline{\mathbf{C}}}^{\rm tn} = \begin{bmatrix} C_{2211} & \sqrt{2}\,C_{2212} & \sqrt{2}\,C_{2213} \\ C_{3311} & \sqrt{2}\,C_{3312} & \sqrt{2}\,C_{3313} \\ \sqrt{2}\,C_{2311} & 2\,C_{2312} & 2\,C_{2313} \end{bmatrix} ; \; \underline{\underline{\mathbf{C}}}^{\rm tt} = \begin{bmatrix} C_{2222} & C_{2233} & \sqrt{2}\,C_{2223} \\ C_{3322} & C_{3333} & \sqrt{2}\,C_{3323} \\ \sqrt{2}\,C_{2322} & \sqrt{2}\,C_{2333} & 2\,C_{2323} \end{bmatrix} .$$

From the major symmetry of $\mathbf{C}$, we see that $\underline{\underline{\mathbf{C}}}^{\rm nt} = (\underline{\underline{\mathbf{C}}}^{\rm tn})^T$. Also, $\underline{\underline{\mathbf{C}}}^{\rm nn}$ and $\underline{\underline{\mathbf{C}}}^{\rm tt}$ are symmetric. Expanding the first row of equation (8.116), we get

$$\underline{\underline{\boldsymbol{\sigma}}}^{\rm n} = \underline{\underline{\mathbf{C}}}^{\rm nn} \cdot \underline{\underline{\boldsymbol{\varepsilon}}}^{\rm n} + \underline{\underline{\mathbf{C}}}^{\rm nt} \cdot \underline{\underline{\boldsymbol{\varepsilon}}}^{\rm t}$$

or

$$\underline{\underline{\boldsymbol{\varepsilon}}}^{\rm n} = (\underline{\underline{\mathbf{C}}}^{\rm nn})^{-1} \cdot \underline{\underline{\boldsymbol{\sigma}}}^{\rm n} - \left[(\underline{\underline{\mathbf{C}}}^{\rm nn})^{-1} \cdot \underline{\underline{\mathbf{C}}}^{\rm nt} \right] \cdot \underline{\underline{\boldsymbol{\varepsilon}}}^{\rm t} . \tag{8.117}$$

From the second row of equation (8.116) we have

$$\underline{\underline{\boldsymbol{\sigma}}}^{\rm t} = (\underline{\underline{\mathbf{C}}}^{\rm nt})^T \cdot \underline{\underline{\boldsymbol{\varepsilon}}}^{\rm n} + \underline{\underline{\mathbf{C}}}^{\rm tt} \cdot \underline{\underline{\boldsymbol{\varepsilon}}}^{\rm t} .$$

Substituting the expression for $\underline{\underline{\boldsymbol{\varepsilon}}}^n$ from the first row into the above equation, we get

$$\underline{\underline{\boldsymbol{\sigma}}}^{\rm t} = (\underline{\underline{\mathbf{C}}}^{\rm nt})^T \cdot \left[(\underline{\underline{\mathbf{C}}}^{\rm nn})^{-1} \cdot \underline{\underline{\boldsymbol{\sigma}}}^{\rm n} - (\underline{\underline{\mathbf{C}}}^{\rm nn})^{-1} \cdot \underline{\underline{\mathbf{C}}}^{\rm nt} \cdot \underline{\underline{\boldsymbol{\varepsilon}}}^{\rm t} \right] + \underline{\underline{\mathbf{C}}}^{\rm tt} \cdot \underline{\underline{\boldsymbol{\varepsilon}}}^{\rm t}$$

or

$$\underline{\underline{\boldsymbol{\sigma}}}^{\rm t} = \left[(\underline{\underline{\mathbf{C}}}^{\rm nt})^T \cdot (\underline{\underline{\mathbf{C}}}^{\rm nn})^{-1} \right] \cdot \underline{\underline{\boldsymbol{\sigma}}}^{\rm n} + \left[\underline{\underline{\mathbf{C}}}^{\rm tt} - (\underline{\underline{\mathbf{C}}}^{\rm nt})^T \cdot (\underline{\underline{\mathbf{C}}}^{\rm nn})^{-1} \cdot \underline{\underline{\mathbf{C}}}^{\rm nt} \right] \cdot \underline{\underline{\boldsymbol{\varepsilon}}}^{\rm t} . \tag{8.118}$$

Collecting (8.117) and (8.118) we get

$$\begin{bmatrix} -\underline{\underline{\boldsymbol{\varepsilon}}}^{\rm n} \\ \underline{\underline{\boldsymbol{\sigma}}}^{\rm t} \end{bmatrix} = \begin{bmatrix} -(\underline{\underline{\mathbf{C}}}^{\rm nn})^{-1} & (\underline{\underline{\mathbf{C}}}^{\rm nn})^{-1} \cdot \underline{\underline{\mathbf{C}}}^{\rm nt} \\ (\underline{\underline{\mathbf{C}}}^{\rm nt})^T \cdot (\underline{\underline{\mathbf{C}}}^{\rm nn})^{-1} & \underline{\underline{\mathbf{C}}}^{\rm tt} - (\underline{\underline{\mathbf{C}}}^{\rm nt})^T \cdot (\underline{\underline{\mathbf{C}}}^{\rm nn})^{-1} \cdot \underline{\underline{\mathbf{C}}}^{\rm nt} \end{bmatrix} \begin{bmatrix} -\underline{\underline{\boldsymbol{\sigma}}}^{\rm n} \\ \underline{\underline{\boldsymbol{\varepsilon}}}^{\rm t} \end{bmatrix} . \tag{8.119}$$

Now $\underline{\underline{\boldsymbol{\sigma}}}^{\rm n}$ is constant across the layers because σ_{11}, σ_{21}, σ_{31} are constant across the layers. Also, $\underline{\underline{\boldsymbol{\varepsilon}}}^{\rm t}$ is constant across the layers because $\varepsilon_{22}, \varepsilon_{33}, \varepsilon_{23}$ are constant. Therefore, taking a volume average of equation (8.119) gives

$$\begin{bmatrix} -\left\langle \underline{\underline{\boldsymbol{\varepsilon}}}^{\rm n}\right\rangle \\ \left\langle \underline{\underline{\boldsymbol{\sigma}}}^{\rm t}\right\rangle \end{bmatrix} = \begin{bmatrix} -\left\langle (\underline{\underline{\mathbf{C}}}^{\rm nn})^{-1}\right\rangle & \left\langle (\underline{\underline{\mathbf{C}}}^{\rm nn})^{-1}\cdot \underline{\underline{\mathbf{C}}}^{\rm nt}\right\rangle \\ \left\langle (\underline{\underline{\mathbf{C}}}^{\rm nt})^T\cdot(\underline{\underline{\mathbf{C}}}^{\rm nn})^{-1}\right\rangle & \left\langle \underline{\underline{\mathbf{C}}}^{\rm tt} - (\underline{\underline{\mathbf{C}}}^{\rm nt})^T\cdot(\underline{\underline{\mathbf{C}}}^{\rm nn})^{-1}\cdot\underline{\underline{\mathbf{C}}}^{\rm nt}\right\rangle\end{bmatrix}\begin{bmatrix}-\underline{\underline{\boldsymbol{\sigma}}}^{\rm n}\\ \underline{\underline{\boldsymbol{\varepsilon}}}^{\rm t}\end{bmatrix}. \tag{8.120}$$

If we define the effective stiffness of the material by the relation[‡‡]

$$\left\langle \underline{\underline{\boldsymbol{\sigma}}}\right\rangle = \mathbf{C}_{\rm eff} : \left\langle \underline{\underline{\boldsymbol{\varepsilon}}}\right\rangle$$

we can use the same procedure to show that

$$\begin{bmatrix} -\left\langle \underline{\underline{\boldsymbol{\varepsilon}}}^{\rm n}\right\rangle \\ \left\langle \underline{\underline{\boldsymbol{\sigma}}}^{\rm t}\right\rangle \end{bmatrix} = \begin{bmatrix} -(\underline{\underline{\mathbf{C}}}^{\rm nn}_{\rm eff})^{-1} & (\underline{\underline{\mathbf{C}}}^{\rm nn}_{\rm eff})^{-1}\cdot \underline{\underline{\mathbf{C}}}^{\rm nt}_{\rm eff} \\ (\underline{\underline{\mathbf{C}}}^{\rm nt}_{\rm eff})^T\cdot(\underline{\underline{\mathbf{C}}}^{\rm nn}_{\rm eff})^{-1} & \underline{\underline{\mathbf{C}}}^{\rm tt}_{\rm eff} - (\underline{\underline{\mathbf{C}}}^{\rm nt}_{\rm eff})^T\cdot(\underline{\underline{\mathbf{C}}}^{\rm nn}_{\rm eff})^{-1}\cdot\underline{\underline{\mathbf{C}}}^{\rm nt}_{\rm eff}\end{bmatrix}\begin{bmatrix}-\underline{\underline{\boldsymbol{\sigma}}}^{\rm n}\\ \underline{\underline{\boldsymbol{\varepsilon}}}^{\rm t}\end{bmatrix}. \tag{8.121}$$

Comparing (8.120) and (8.121) we can show that

$$\boxed{\begin{aligned} \underline{\underline{\mathbf{C}}}^{\rm nn}_{\rm eff} &= \left\langle (\underline{\underline{\mathbf{C}}}^{\rm nn})^{-1}\right\rangle^{-1} \\ \underline{\underline{\mathbf{C}}}^{\rm nt}_{\rm eff} &= \left\langle (\underline{\underline{\mathbf{C}}}^{\rm nn})^{-1}\right\rangle^{-1}\cdot\left\langle (\underline{\underline{\mathbf{C}}}^{\rm nn})^{-1}\cdot\underline{\underline{\mathbf{C}}}^{\rm nt}\right\rangle \\ \underline{\underline{\mathbf{C}}}^{\rm tt}_{\rm eff} &= \left\langle \underline{\underline{\mathbf{C}}}^{\rm tt} - \underline{\underline{\mathbf{C}}}^{\rm tn}\cdot(\underline{\underline{\mathbf{C}}}^{\rm nn})^{-1}\cdot\underline{\underline{\mathbf{C}}}^{\rm nt}\right\rangle + \\ &\qquad \left\langle \underline{\underline{\mathbf{C}}}^{\rm tn}\cdot(\underline{\underline{\mathbf{C}}}^{\rm nn})^{-1}\right\rangle\cdot\left\langle (\underline{\underline{\mathbf{C}}}^{\rm nn})^{-1}\right\rangle^{-1}\cdot\left\langle (\underline{\underline{\mathbf{C}}}^{\rm nn})^{-1}\cdot\underline{\underline{\mathbf{C}}}^{\rm nt}\right\rangle . \end{aligned}} \tag{8.122}$$

If the material in each layer is isotropic, then the constitutive relation is

$$\boldsymbol{\sigma} = \mathbf{C} : \boldsymbol{\varepsilon} = \lambda(x_1)\,{\rm tr}(\boldsymbol{\varepsilon})\boldsymbol{1} + 2\mu(x_1)\boldsymbol{\varepsilon}$$

where λ is the Lamé modulus and μ is the shear modulus. In that case the effective properties of the laminate are

$$\boxed{\begin{aligned} & C^{\rm eff}_{1111} = \left\langle \tfrac{1}{\lambda+2\mu}\right\rangle^{-1} \;;\; C^{\rm eff}_{1122} = C^{\rm eff}_{1133} = \left\langle \tfrac{\lambda}{\lambda+2\mu}\right\rangle\left\langle \tfrac{1}{\lambda+2\mu}\right\rangle^{-1} \\ & C^{\rm eff}_{1212} = C^{\rm eff}_{1313} = \left\langle \tfrac{1}{\mu}\right\rangle^{-1} \;;\; C^{\rm eff}_{2323} = \langle \mu\rangle \;;\; C^{\rm eff}_{1112} = C^{\rm eff}_{1113} = C^{\rm eff}_{1123} = 0 \\ & C^{\rm eff}_{2222} = C^{\rm eff}_{3333} = \left\langle \tfrac{4\mu\lambda(\lambda+\mu)}{(\lambda+2\mu)^2}\right\rangle + \left\langle \tfrac{\lambda}{\lambda+2\mu}\right\rangle\left\langle \tfrac{1}{\lambda+2\mu}\right\rangle^{-1} \\ & C^{\rm eff}_{2233} = \left\langle \tfrac{2\mu\lambda}{\lambda+2\mu}\right\rangle + \left\langle \tfrac{\lambda}{\lambda+2\mu}\right\rangle^{2}\left\langle \tfrac{1}{\lambda+2\mu}\right\rangle^{-1} \end{aligned}} \tag{8.123}$$

[‡‡] See Hill (1963), Hill (1964), Nemat-Nasser and Hori (1993), and Milton (2002) for justifications for the definition of an effective elastic stiffness tensor in terms of the volume averaged stresses and strains.

8.6.2 Quasistatic effective properties of electromagnetic laminates

Let us now apply the Backus approach to find the effective electromagnetic properties of laminates. Similar approaches have also been used by Postma (1955) and Tartar (1976). Consider a material laminated in the x_1-direction as shown in Figure 8.10. To find the effective material properties of the laminate we take advantage

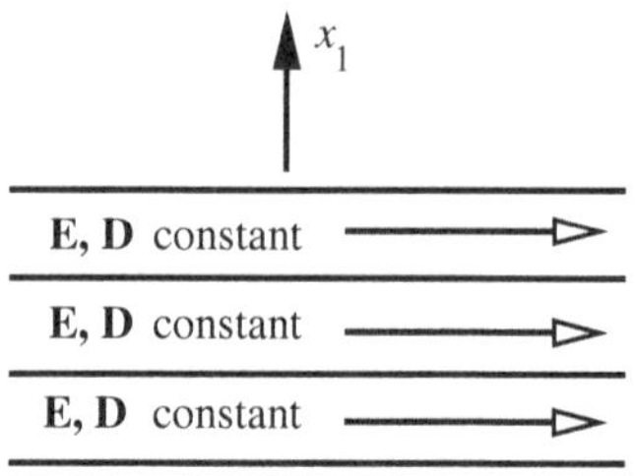

FIGURE 8.10
A laminate with direction of lamination x_1. Each layer is homogeneous.

of the fact that since the tangential components (parallel to the layers) of the electric field (**E**) are piecewise constant and continuous across the interfaces between the layers, these tangential components must be constant, i.e., E_2 and E_3 are constant in the laminate. Similarly, the continuity of the normal electric displacement field (**D**) across the interfaces and the fact that this field is constant in each layer implies that the component D_1 is constant in the laminate.

Recall that the constitutive relation between **D** and **E** is

$$\mathbf{D} = \boldsymbol{\varepsilon} \cdot \mathbf{E}\,.$$

Let us rewrite the constitutive relation in matrix form (with respect to the rectangular Cartesian basis $(\mathbf{e}_1, \mathbf{e}_2, \mathbf{e}_3)$) so that constant fields appear on the right-hand side. We start by breaking up the matrix representation of the constitutive relation into the form

$$\begin{bmatrix} D_1 \\ \underline{\mathsf{D}}^{\mathrm{t}} \end{bmatrix} = \begin{bmatrix} \varepsilon_{11} & \underline{\underline{\boldsymbol{\varepsilon}}}^{1\mathrm{t}} \\ \underline{\underline{\boldsymbol{\varepsilon}}}^{\mathrm{t}1} & \underline{\underline{\boldsymbol{\varepsilon}}}^{\mathrm{tt}} \end{bmatrix} \begin{bmatrix} E_1 \\ \underline{\mathsf{E}}^{\mathrm{t}} \end{bmatrix} \tag{8.124}$$

where

$$\underline{\mathsf{D}}^{\mathrm{t}} = \begin{bmatrix} D_2 \\ D_3 \end{bmatrix} \; ; \; \underline{\mathsf{E}}^{\mathrm{t}} = \begin{bmatrix} E_2 \\ E_3 \end{bmatrix} \tag{8.125}$$

and

$$\underline{\underline{\boldsymbol{\varepsilon}}}^{1\mathrm{t}} = \begin{bmatrix} \varepsilon_{12} & \varepsilon_{13} \end{bmatrix} \; ; \; \underline{\underline{\boldsymbol{\varepsilon}}}^{\mathrm{t}1} = \begin{bmatrix} \varepsilon_{21} \\ \varepsilon_{31} \end{bmatrix} \; ; \; \underline{\underline{\boldsymbol{\varepsilon}}}^{\mathrm{tt}} = \begin{bmatrix} \varepsilon_{22} & \varepsilon_{23} \\ \varepsilon_{32} & \varepsilon_{33} \end{bmatrix} . \tag{8.126}$$

Note that the constant fields are D_1 and $\underline{\mathsf{E}}^{\mathrm{t}}$. We want to rewrite equation (8.124) so that these constant fields appear on the right-hand side. From the first row of (8.124) we get

$$D_1 = \varepsilon_{11} E_1 + \underline{\underline{\boldsymbol{\varepsilon}}}^{1t} \cdot \underline{\underline{\mathsf{E}}}^{t}$$

or

$$E_1 = (\varepsilon_{11})^{-1} D_1 - (\varepsilon_{11})^{-1} \underline{\underline{\boldsymbol{\varepsilon}}}^{1t} \cdot \underline{\underline{\mathsf{E}}}^{t} . \tag{8.127}$$

From the second row of (8.124) we get

$$\underline{\underline{\mathsf{D}}}^{t} = \underline{\underline{\boldsymbol{\varepsilon}}}^{t1} E_1 + \underline{\underline{\boldsymbol{\varepsilon}}}^{tt} \cdot \underline{\underline{\mathsf{E}}}^{t} . \tag{8.128}$$

Substitution of (8.127) into (8.128) gives

$$\underline{\underline{\mathsf{D}}}^{t} = (\varepsilon_{11})^{-1} \underline{\underline{\boldsymbol{\varepsilon}}}^{t1} D_1 - (\varepsilon_{11})^{-1} (\underline{\underline{\boldsymbol{\varepsilon}}}^{t1} \cdot \underline{\underline{\boldsymbol{\varepsilon}}}^{1t}) \cdot \underline{\underline{\mathsf{E}}}^{t} + \underline{\underline{\boldsymbol{\varepsilon}}}^{tt} \cdot \underline{\underline{\mathsf{E}}}^{t}$$

or

$$\underline{\underline{\mathsf{D}}}^{t} = (\varepsilon_{11})^{-1} \underline{\underline{\boldsymbol{\varepsilon}}}^{t1} D_1 + \left[\underline{\underline{\boldsymbol{\varepsilon}}}^{tt} - (\varepsilon_{11})^{-1} (\underline{\underline{\boldsymbol{\varepsilon}}}^{t1} \cdot \underline{\underline{\boldsymbol{\varepsilon}}}^{1t}) \right] \cdot \underline{\underline{\mathsf{E}}}^{t} . \tag{8.129}$$

Collecting (8.127) and (8.129) gives

$$\begin{bmatrix} -E_1 \\ \underline{\underline{\mathsf{D}}}^{t} \end{bmatrix} = \frac{1}{\varepsilon_{11}} \begin{bmatrix} -1 & \underline{\underline{\boldsymbol{\varepsilon}}}^{1t} \\ \underline{\underline{\boldsymbol{\varepsilon}}}^{t1} & \varepsilon_{11} \underline{\underline{\boldsymbol{\varepsilon}}}^{tt} - \underline{\underline{\boldsymbol{\varepsilon}}}^{t1} \cdot \underline{\underline{\boldsymbol{\varepsilon}}}^{1t} \end{bmatrix} \begin{bmatrix} -D_1 \\ \underline{\underline{\mathsf{E}}}^{t} \end{bmatrix} \tag{8.130}$$

where the negative signs on E_1 and D_1 are used to make sure that the signs of the off-diagonal terms are identical. Define

$$\underline{\underline{\mathsf{L}}}(\mathbf{x}) := \frac{1}{\varepsilon_{11}} \begin{bmatrix} -1 & \underline{\underline{\boldsymbol{\varepsilon}}}^{1t} \\ \underline{\underline{\boldsymbol{\varepsilon}}}^{t1} & \varepsilon_{11} \underline{\underline{\boldsymbol{\varepsilon}}}^{tt} - \underline{\underline{\boldsymbol{\varepsilon}}}^{t1} \cdot \underline{\underline{\boldsymbol{\varepsilon}}}^{1t} \end{bmatrix} .$$

Then we have

$$\begin{bmatrix} -E_1 \\ \underline{\underline{\mathsf{D}}}^{t} \end{bmatrix} = \underline{\underline{\mathsf{L}}}(\mathbf{x}) \cdot \begin{bmatrix} -D_1 \\ \underline{\underline{\mathsf{E}}}^{t} \end{bmatrix} . \tag{8.131}$$

Since the vector on the right hand side is constant, a volume average of (8.131) gives

$$\begin{bmatrix} -\langle E_1 \rangle \\ \left\langle \underline{\underline{\mathsf{D}}}^{t} \right\rangle \end{bmatrix} = \left\langle \underline{\underline{\mathsf{L}}}(\mathbf{x}) \right\rangle \cdot \begin{bmatrix} -D_1 \\ \underline{\underline{\mathsf{E}}}^{t} \end{bmatrix} . \tag{8.132}$$

Let us define the effective permittivity of the laminate, $\underline{\underline{\boldsymbol{\varepsilon}}}^{\text{eff}}$, via the relation

$$\langle \mathbf{D} \rangle = \boldsymbol{\varepsilon}_{\text{eff}} \cdot \langle \mathbf{E} \rangle .$$

Since the tangential components of $\mathbf{E}$ are constant in the laminate, the average values $\langle E_2 \rangle$ and $\langle E_3 \rangle$ must also be constant. Similarly, the average value $\langle D_1 \rangle$ must be constant. Therefore we can use the same arguments as we used before to write the effective constitutive relation in the form

$$\begin{bmatrix} -\langle E_1 \rangle \\ \left\langle \underline{\underline{\mathsf{D}}}^{t} \right\rangle \end{bmatrix} = \underline{\underline{\mathsf{L}}}_{\text{eff}} \cdot \begin{bmatrix} -\langle D_1 \rangle \\ \left\langle \underline{\underline{\mathsf{E}}}^{t} \right\rangle \end{bmatrix} = \underline{\underline{\mathsf{L}}}_{\text{eff}} \cdot \begin{bmatrix} -D_1 \\ \underline{\underline{\mathsf{E}}}^{t} \end{bmatrix} \tag{8.133}$$

where

$$\underline{\underline{\mathsf{L}}}_{\text{eff}} := \frac{1}{\varepsilon_{11}^{\text{eff}}} \begin{bmatrix} -1 & \underline{\boldsymbol{\varepsilon}}_{\text{eff}}^{1\text{t}} \\ \underline{\boldsymbol{\varepsilon}}_{\text{eff}}^{\text{t}1} & \varepsilon_{11}^{\text{eff}} \underline{\underline{\boldsymbol{\varepsilon}}}_{\text{eff}}^{\text{tt}} - \underline{\boldsymbol{\varepsilon}}_{\text{eff}}^{\text{t}1} \cdot \underline{\boldsymbol{\varepsilon}}_{\text{eff}}^{1\text{t}} \end{bmatrix}$$

and $\boldsymbol{\varepsilon}_{\text{eff}}$ has been decomposed in exactly the same manner as $\boldsymbol{\varepsilon}$ (see equation (8.126). If we compare equations (8.132) and (8.133) we get a formula for determining the effective permittivity of the laminate.

$$\boxed{\underline{\underline{\mathsf{L}}}_{\text{eff}} = \left\langle \underline{\underline{\mathsf{L}}}(\mathbf{x}) \right\rangle .}$$

Expanding out the terms, we have

$$\boxed{\begin{aligned} \varepsilon_{11}^{\text{eff}} &= \left\langle \frac{1}{\varepsilon_{11}} \right\rangle^{-1} \\ \varepsilon_{1j}^{\text{eff}} &= \left\langle \frac{1}{\varepsilon_{11}} \right\rangle^{-1} \left\langle \frac{\varepsilon_{1j}}{\varepsilon_{11}} \right\rangle \\ \varepsilon_{ij}^{\text{eff}} &= \left\langle \varepsilon_{ij} - \frac{\varepsilon_{i1}\,\varepsilon_{1j}}{\varepsilon_{11}} \right\rangle + \left\langle \frac{1}{\varepsilon_{11}} \right\rangle^{-1} \left\langle \frac{\varepsilon_{i1}}{\varepsilon_{11}} \right\rangle \left\langle \frac{\varepsilon_{1j}}{\varepsilon_{11}} \right\rangle \quad ,i,j \neq 1. \end{aligned}} \tag{8.134}$$

When the off-diagonal elements vanish, we get

$$\begin{aligned} \varepsilon_{11}^{\text{eff}} &= \left\langle \frac{1}{\varepsilon_{11}} \right\rangle^{-1} && \text{(Harmonic average)} \\ \varepsilon_{jj}^{\text{eff}} &= \langle \varepsilon_{jj} \rangle && \text{(Arithmetic average)} \\ \varepsilon_{ab}^{\text{eff}} &= 0 && \text{if } a \neq b\,. \end{aligned}$$

The harmonic average corresponds to a situation in which each layer may be thought of as a capacitor in series while the arithmetic average corresponds to a situation where the capacitors are in parallel. We can use the same approach to determine the effective magnetic permeability of a laminate in the quasistatic limit.

8.6.3 Laminates with arbitrary lamination direction

So far we have dealt with laminates with a single direction of lamination that was oriented in the x_1-direction. Let us generalize our approach to deal with laminates with a normal, $\mathbf{n}$, which is not necessarily parallel to any coordinate axis. The manipulations necessary become a bit cumbersome for elastic laminates. Therefore we concentrate on the electric permittivity of laminates in this section. Note the same ideas can be used for elasticity or to find the effective magnetic permeability of a laminate.

Recall that, if there is only one direction of lamination, the normal component of $\mathbf{D}$, i.e., $\mathbf{D} \cdot \mathbf{n}$, is constant and the tangential components of $\mathbf{E}$ are constant throughout the entire laminate. Let us introduce the second-order tensor basis

$$\boxed{\boldsymbol{P}_{\parallel}(\mathbf{n}) = \mathbf{n} \otimes \mathbf{n}\,;\;\; \boldsymbol{P}_{\perp}(\mathbf{n}) = \boldsymbol{1} - \mathbf{n} \otimes \mathbf{n} = \boldsymbol{1} - \boldsymbol{P}_{\parallel}(\mathbf{n})\,.}$$

Notice that these also act as projection operators in the sense that $\boldsymbol{P}_{\|} \cdot \boldsymbol{P}_{\|} = \boldsymbol{P}_{\|}$ and $\boldsymbol{P}_{\perp} \cdot \boldsymbol{P}_{\perp} = \boldsymbol{P}_{\perp}$. Note that

$$\boldsymbol{P}_{\|} \cdot \mathbf{D} = (\mathbf{D} \cdot \mathbf{n})\mathbf{n} \quad \text{and} \quad \boldsymbol{P}_{\perp} \cdot \mathbf{E} = \mathbf{E} - (\mathbf{E} \cdot \mathbf{n})\mathbf{n}.$$

Therefore the operator $\boldsymbol{P}_{\|}$ can be used to extract the components of $\mathbf{D}$ in the direction of lamination while $\boldsymbol{P}_{\perp}$ can be used to extract the tangential components of the electric field along the interfaces between the layers. Because of the continuity of these components (and hence constancy across layers), we have

$$\boldsymbol{P}_{\|}(\mathbf{n}) \cdot \mathbf{D}(\mathbf{x}) = \boldsymbol{P}_{\|}(\mathbf{n}) \cdot \langle \mathbf{D} \rangle \quad \text{and} \quad \boldsymbol{P}_{\perp}(\mathbf{n}) \cdot \mathbf{E}(\mathbf{x}) = \boldsymbol{P}_{\perp}(\mathbf{n}) \cdot \langle \mathbf{E} \rangle . \tag{8.135}$$

Therefore,

$$\boldsymbol{P}_{\|}(\mathbf{n}) \cdot \mathbf{E}(\mathbf{x}) = \mathbf{E}(\mathbf{x}) - \langle \mathbf{E} \rangle + \boldsymbol{P}_{\|}(\mathbf{n}) \cdot \langle \mathbf{E} \rangle . \tag{8.136}$$

Let us now introduce a *polarization field*

$$\mathbf{p}(\mathbf{x}) = [\boldsymbol{\varepsilon}(\mathbf{x}) - \varepsilon_0 \boldsymbol{1}] \cdot \mathbf{E}(\mathbf{x}) = \mathbf{D}(\mathbf{x}) - \varepsilon_0 \mathbf{E}(\mathbf{x}) \tag{8.137}$$

where ε_0 is an arbitrary constant. The volume averaged polarization field is given by

$$\langle \mathbf{p} \rangle = (\boldsymbol{\varepsilon}_{\text{eff}} - \varepsilon_0 \boldsymbol{1}) \cdot \langle \mathbf{E} \rangle . \tag{8.138}$$

Define

$$\boldsymbol{S}(\mathbf{x}) := \varepsilon_0 [\varepsilon_0 \boldsymbol{1} - \boldsymbol{\varepsilon}(\mathbf{x})]^{-1} \quad \text{and} \quad \boldsymbol{S}_{\text{eff}} := \varepsilon_0 [\varepsilon_0 \boldsymbol{1} - \boldsymbol{\varepsilon}_{\text{eff}}]^{-1} .$$

Then,

$$\begin{aligned} \boldsymbol{S}(\mathbf{x}) \cdot \mathbf{p}(\mathbf{x}) &= -\left[\varepsilon_0 \{\varepsilon_0 \boldsymbol{1} - \boldsymbol{\varepsilon}(\mathbf{x})\}^{-1}\right] \cdot [\{\varepsilon_0 \boldsymbol{1} - \boldsymbol{\varepsilon}(\mathbf{x})\} \cdot \mathbf{E}(\mathbf{x})] = -\varepsilon_0 \mathbf{E}(\mathbf{x}) \\ \boldsymbol{S}_{\text{eff}} \cdot \langle \mathbf{p} \rangle &= -\left[\varepsilon_0 (\varepsilon_0 \boldsymbol{1} - \boldsymbol{\varepsilon}_{\text{eff}})^{-1}\right] \cdot [(\varepsilon_0 \boldsymbol{1} - \boldsymbol{\varepsilon}_{\text{eff}}) \cdot \langle \mathbf{E} \rangle] = -\varepsilon_0 \langle \mathbf{E} \rangle . \end{aligned} \tag{8.139}$$

Applying the projection $\boldsymbol{P}_{\|}(\mathbf{n})$ to the second of equations (8.137), we get

$$\boldsymbol{P}_{\|}(\mathbf{n}) \cdot \mathbf{p}(\mathbf{x}) = \boldsymbol{P}_{\|}(\mathbf{n}) \cdot \mathbf{D}(\mathbf{x}) - \varepsilon_0 \boldsymbol{P}_{\|}(\mathbf{n}) \cdot \mathbf{E}(\mathbf{x}) .$$

Substituting the first of equations (8.135) and (8.136) into the above equation, we have

$$\boldsymbol{P}_{\|}(\mathbf{n}) \cdot \mathbf{p}(\mathbf{x}) = \boldsymbol{P}_{\|}(\mathbf{n}) \cdot \langle \mathbf{D} \rangle - \varepsilon_0 \mathbf{E}(\mathbf{x}) + \varepsilon_0 \langle \mathbf{E} \rangle - \varepsilon_0 \boldsymbol{P}_{\|}(\mathbf{n}) \cdot \langle \mathbf{E} \rangle .$$

From the definitions (8.139) we can then write

$$\boldsymbol{P}_{\|}(\mathbf{n}) \cdot \mathbf{p}(\mathbf{x}) = \boldsymbol{P}_{\|}(\mathbf{n}) \cdot \langle \mathbf{D} \rangle + \boldsymbol{S}(\mathbf{x}) \cdot \mathbf{p}(\mathbf{x}) - \boldsymbol{S}_{\text{eff}} \cdot \langle \mathbf{p} \rangle - \varepsilon_0 \boldsymbol{P}_{\|}(\mathbf{n}) \cdot \langle \mathbf{E} \rangle .$$

Define

$$\mathbf{v}(\mathbf{n}, \mathbf{x}) := \boldsymbol{P}_{\|}(\mathbf{n}) \cdot \langle \mathbf{D} \rangle - \boldsymbol{S}_{\text{eff}} \cdot \langle \mathbf{p} \rangle - \varepsilon_0 \boldsymbol{P}_{\|}(\mathbf{n}) \cdot \langle \mathbf{E} \rangle . \tag{8.140}$$

Then we have

$$\boldsymbol{P}_{\|}(\mathbf{n}) \cdot \mathbf{p}(\mathbf{x}) = \boldsymbol{S}(\mathbf{x}) \cdot \mathbf{p}(\mathbf{x}) + \mathbf{v}(\mathbf{n}, \mathbf{x})$$

or

$$\mathbf{v}(\mathbf{n}, \mathbf{x}) = -\left[\boldsymbol{S}(\mathbf{x}) - \boldsymbol{P}_{\|}(\mathbf{n})\right] \cdot \mathbf{p}(\mathbf{x}) . \tag{8.141}$$

Also, from equations (8.140) and (8.138) we have

$$\mathbf{v}(\mathbf{n},\mathbf{x}) = \boldsymbol{P}_{\parallel}(\mathbf{n})\cdot[\langle\mathbf{D}\rangle - \varepsilon_0\cdot\langle\mathbf{E}\rangle] - \boldsymbol{S}_{\text{eff}}\cdot\langle\mathbf{p}\rangle = \boldsymbol{P}_{\parallel}(\mathbf{n})\cdot\langle\mathbf{p}\rangle - \boldsymbol{S}_{\text{eff}}\cdot\langle\mathbf{p}\rangle$$

or

$$\mathbf{v}(\mathbf{n},\mathbf{x}) = -\left[\boldsymbol{S}_{\text{eff}} - \boldsymbol{P}_{\parallel}(\mathbf{n})\right]\cdot\langle\mathbf{p}\rangle\,. \tag{8.142}$$

Inverting (8.141) and (8.142) we have

$$\begin{aligned} \mathbf{p}(\mathbf{x}) &= -\left[\boldsymbol{S}(\mathbf{x}) - \boldsymbol{P}_{\parallel}(\mathbf{n})\right]^{-1}\cdot\mathbf{v}(\mathbf{n},\mathbf{x}) \\ \langle\mathbf{p}\rangle &= -\left[\boldsymbol{S}_{\text{eff}} - \boldsymbol{P}_{\parallel}(\mathbf{n})\right]^{-1}\cdot\mathbf{v}(\mathbf{n},\mathbf{x})\,. \end{aligned} \tag{8.143}$$

Also, taking the volume average of the first of equations (8.143), we have

$$\langle\mathbf{p}\rangle = -\left\langle\left[\boldsymbol{S}(\mathbf{x}) - \boldsymbol{P}_{\parallel}(\mathbf{n})\right]^{-1}\right\rangle\cdot\mathbf{v}(\mathbf{n},\mathbf{x})\,. \tag{8.144}$$

Comparing the second equation in (8.143) with (8.144) and invoking the arbitrariness of ε_0, we have

$$\boxed{\left[\boldsymbol{S}_{\text{eff}} - \boldsymbol{P}_{\parallel}(\mathbf{n})\right]^{-1} = \left\langle\left[\boldsymbol{S}(\mathbf{x}) - \boldsymbol{P}_{\parallel}(\mathbf{n})\right]^{-1}\right\rangle\,.} \tag{8.145}$$

Since ε_0 is arbitrary we can take it to be 1, i.e.,

$$\boxed{\boldsymbol{S}(\mathbf{x}) = [\boldsymbol{1} - \boldsymbol{\varepsilon}(\mathbf{x})]^{-1} \quad \text{and} \quad \boldsymbol{S}_{\text{eff}} := [\boldsymbol{1} - \boldsymbol{\varepsilon}_{\text{eff}}]^{-1}\,.}$$

But it is often more useful if we choose a permittivity that matches one of the components of the laminate. Equation (8.145) provides us with a means of computing the effective permittivity of a layered medium oriented at an arbitrary angle (given by the normal $\mathbf{n}$).

8.6.4 Tartar-Murat-Lurie-Cherkaev formula

Consider the periodic rank-1 laminate shown in Figure 8.11. The layers have permittivities alternating between an anisotropic second-order tensor, $\boldsymbol{\varepsilon}_1$, and an isotropic second-order tensor, $\varepsilon_2\boldsymbol{1}$. The volume fraction of phase 1 is f_1 while that of phase 2 is f_2 such that $f_1 + f_2 = 1$. Recall that

$$\boldsymbol{S}(\mathbf{x}) = \varepsilon_0\left[\varepsilon_0\,\boldsymbol{1} - \boldsymbol{\varepsilon}(\mathbf{x})\right]^{-1}.$$

Let us take the limit as $\varepsilon_0 \to \varepsilon_2$. Since $\varepsilon(\mathbf{x}) = \varepsilon_2\,\boldsymbol{1}$ in phase 2, we have

$$\boldsymbol{S}(\mathbf{x}) \to \infty \qquad \Longrightarrow \qquad \left[\boldsymbol{S}(\mathbf{x}) - \boldsymbol{P}_{\parallel}(\mathbf{n})\right]^{-1} \to 0 \quad \text{in phase 2}\,.$$

Hence, the right-hand side of (8.145) reduces to an average only over phase 1. If we define

$$\boldsymbol{S}_1 := \varepsilon_0[\varepsilon_0\boldsymbol{1} - \boldsymbol{\varepsilon}_1]^{-1} \to \varepsilon_2[\varepsilon_2\boldsymbol{1} - \boldsymbol{\varepsilon}_1]^{-1}$$

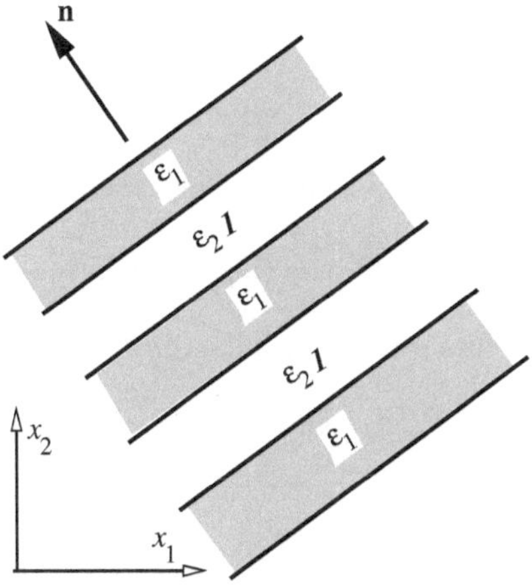

FIGURE 8.11
A rank-1 laminate consisting of alternating layers of permittivity $\boldsymbol{\varepsilon}_1$ and ε_2.

we get

$$\left[\boldsymbol{S}_{\text{eff}} - \boldsymbol{P}_{\|}(\mathbf{n})\right]^{-1} = f_1 \left[\boldsymbol{S}_1 - \boldsymbol{P}_{\|}(\mathbf{n})\right]^{-1}. \tag{8.146}$$

Taking the inverse of both sides of (8.146) gives

$$f_1 \left[\boldsymbol{S}_{\text{eff}} - \boldsymbol{P}_{\|}(\mathbf{n})\right] = \boldsymbol{S}_1 - \boldsymbol{P}_{\|}(\mathbf{n})$$

or

$$f_1 \boldsymbol{S}_{\text{eff}} = \boldsymbol{S}_1 - (1 - f_1)\boldsymbol{P}_{\|}(\mathbf{n}) = \boldsymbol{S}_1 - f_2 \boldsymbol{P}_{\|}(\mathbf{n}).$$

Since

$$\boldsymbol{S}_{\text{eff}} = \varepsilon_0 \left[\varepsilon_0 \boldsymbol{1} - \boldsymbol{\varepsilon}_{\text{eff}}\right]^{-1} \to \varepsilon_2 \left[\varepsilon_2 \boldsymbol{1} - \boldsymbol{\varepsilon}_{\text{eff}}\right]^{-1}$$

we then have

$$\boxed{f_1 \varepsilon_2 [\varepsilon_2 \boldsymbol{1} - \boldsymbol{\varepsilon}_{\text{eff}}]^{-1} = \varepsilon_2 [\varepsilon_2 \boldsymbol{1} - \boldsymbol{\varepsilon}_1]^{-1} - f_2 \boldsymbol{P}_{\|}(\mathbf{n}).} \tag{8.147}$$

This is the formula of Tartar, Murat, Lurie, and Cherkaev (TMLC) and can be shown to be equivalent to the Backus formula. The idea was originally presented Tartar (1976) and later by Lurie and Cherkaev. However, these original works are not readily available. Wider dissemination followed the work by Milton (1990) and the topic is discussed in detail in the monograph by Milton (2002).

In fact, there is a formula as simple as the TMLC formula even when $\boldsymbol{\varepsilon}_1$ and $\boldsymbol{\varepsilon}_2$ are both anisotropic. In that case we use an anisotropic reference material $\boldsymbol{\varepsilon}_0$ and define the polarization as

$$\mathbf{p}(\mathbf{x}) := [\boldsymbol{\varepsilon}(\mathbf{x}) - \boldsymbol{\varepsilon}_0] \cdot \mathbf{E}(\mathbf{x}) = \mathbf{D}(\mathbf{x}) - \boldsymbol{\varepsilon}_0 \cdot \mathbf{E}(\mathbf{x}). \tag{8.148}$$

The volume average of this field is given by

$$\langle \mathbf{p} \rangle := \langle \boldsymbol{\varepsilon} \cdot \mathbf{E} \rangle - \boldsymbol{\varepsilon}_0 \cdot \langle \mathbf{E} \rangle = \langle \mathbf{D} \rangle - \boldsymbol{\varepsilon}_0 \cdot \langle \mathbf{E} \rangle \,.$$

Therefore, the difference between the field and its volume average is

$$\mathbf{p}(\mathbf{x}) - \langle \mathbf{p} \rangle = [\mathbf{D}(\mathbf{x}) - \langle \mathbf{D} \rangle] - \boldsymbol{\varepsilon}_0 \cdot [\mathbf{E}(\mathbf{x}) - \langle \mathbf{E} \rangle] \,. \tag{8.149}$$

Taking the projection of both sides of equation (8.149) we get

$$\boldsymbol{P}_{\|}(\mathbf{n}) \cdot [\mathbf{p}(\mathbf{x}) - \langle \mathbf{p} \rangle] = \boldsymbol{P}_{\|}(\mathbf{n}) \cdot [\mathbf{D}(\mathbf{x}) - \langle \mathbf{D} \rangle] - \boldsymbol{P}_{\|}(\mathbf{n}) \cdot \{\boldsymbol{\varepsilon}_0 \cdot [\mathbf{E}(\mathbf{x}) - \langle \mathbf{E} \rangle]\}$$

or

$$\boldsymbol{P}_{\|}(\mathbf{n}) \cdot [\mathbf{p}(\mathbf{x}) - \langle \mathbf{p} \rangle + \boldsymbol{\varepsilon}_0 \cdot \mathbf{E}(\mathbf{x}) - \boldsymbol{\varepsilon}_0 \cdot \langle \mathbf{E} \rangle] = \boldsymbol{P}_{\|}(\mathbf{n}) \cdot [\mathbf{D}(\mathbf{x}) - \langle \mathbf{D} \rangle] \,.$$

Now continuity of the normal component of $\mathbf{D}$ and the piecewise constant nature of the field implies that the normal component of $\mathbf{D}$ is constant. Therefore,

$$\boldsymbol{P}_{\|}(\mathbf{n}) \cdot [\mathbf{D}(\mathbf{x}) - \langle \mathbf{D} \rangle] = [\mathbf{n} \cdot \mathbf{D}(\mathbf{x}) - \mathbf{n} \cdot \langle \mathbf{D} \rangle] \, \mathbf{n} = \mathbf{0}$$

and we have

$$\boldsymbol{P}_{\|}(\mathbf{n}) \cdot \Big[\mathbf{p}(\mathbf{x}) - \langle \mathbf{p} \rangle - \boldsymbol{\varepsilon}_0 \cdot [\langle \mathbf{E} \rangle - \mathbf{E}(\mathbf{x})] \Big] = \mathbf{0} \,. \tag{8.150}$$

Also, if we rearrange equation (8.136), we have

$$\boldsymbol{P}_{\|}(\mathbf{n}) \cdot [\langle \mathbf{E} \rangle - \mathbf{E}(\mathbf{x})] = \langle \mathbf{E} \rangle - \mathbf{E}(\mathbf{x}) \,. \tag{8.151}$$

Notice that equations (8.150) and (8.151) have the forms

$$\boldsymbol{P}_{\|}(\mathbf{n}) \cdot (\mathbf{a} - \boldsymbol{\varepsilon}_0 \cdot \mathbf{b}) = \mathbf{0} \quad \text{and} \quad \boldsymbol{P}_{\|}(\mathbf{n}) \cdot \mathbf{b} = \mathbf{b}$$

where

$$\mathbf{a} := \mathbf{p}(\mathbf{x}) - \langle \mathbf{p} \rangle \quad \text{and} \quad \mathbf{b} := \langle \mathbf{E} \rangle - \mathbf{E}(\mathbf{x}) \,.$$

We would like to find a relation between $\mathbf{a}$ and $\mathbf{b}$. Let us introduce a new operator $\boldsymbol{P}_{\mathbf{n}}$ defined through its action on a vector $\mathbf{a}$ (we could alternatively define the operator by its action on a tensor as discussed in Milton (2002), p. 171). Let the operation produce a vector $\mathbf{b}$ ($\mathbf{b} = \boldsymbol{P}_{\mathbf{n}} \cdot \mathbf{a}$) which has the properties

$$\boldsymbol{P}_{\|}(\mathbf{n}) \cdot \mathbf{b} = \mathbf{b} \qquad \text{and} \quad \boldsymbol{P}_{\|}(\mathbf{n}) \cdot (\mathbf{a} - \boldsymbol{\varepsilon}_0 \cdot \mathbf{b}) = \mathbf{0} \tag{8.152}$$

where $\boldsymbol{P}_{\|}(\mathbf{n}) = \mathbf{n} \otimes \mathbf{n}$ and projects parallel to $\mathbf{n}$. Now,

$$(\mathbf{n} \otimes \mathbf{n}) \cdot \mathbf{b} = (\mathbf{b} \cdot \mathbf{n})\mathbf{n} =: \alpha \mathbf{n}$$

and

$$(\mathbf{n} \otimes \mathbf{n}) \cdot (\mathbf{a} - \boldsymbol{\varepsilon}_0 \cdot \mathbf{b}) = (\mathbf{a} \cdot \mathbf{n})\mathbf{n} - \alpha (\mathbf{n} \otimes \mathbf{n})(\boldsymbol{\varepsilon}_0 \cdot \mathbf{n}) = (\mathbf{a} \cdot \mathbf{n} - \alpha \mathbf{n} \cdot \boldsymbol{\varepsilon}_0 \cdot \mathbf{n})\mathbf{n}.$$

Therefore, from equations (8.152), we have

$$\alpha\mathbf{n} = \mathbf{b} \qquad \text{and} \quad (\mathbf{a}\cdot\mathbf{n} - \alpha\mathbf{n}\cdot\boldsymbol{\varepsilon}_0\cdot\mathbf{n})\mathbf{n} = \mathbf{0}.$$

Solving for α gives

$$\alpha = \frac{\mathbf{a}\cdot\mathbf{n}}{\mathbf{n}\cdot\boldsymbol{\varepsilon}_0\cdot\mathbf{n}}.$$

We can now find the relation between $\mathbf{b}$ and $\mathbf{a}$ because

$$\mathbf{b} = \alpha\mathbf{n} = \left(\frac{\mathbf{a}\cdot\mathbf{n}}{\mathbf{n}\cdot\boldsymbol{\varepsilon}_0\cdot\mathbf{n}}\right)\mathbf{n} = \frac{(\mathbf{n}\otimes\mathbf{n})\cdot\mathbf{a}}{\mathbf{n}\cdot\boldsymbol{\varepsilon}_0\cdot\mathbf{n}} = \left(\frac{\boldsymbol{P}_{\parallel}(\mathbf{n})}{\mathbf{n}\cdot\boldsymbol{\varepsilon}_0\cdot\mathbf{n}}\right)\cdot\mathbf{a}.$$

From the definition of $\boldsymbol{P}_{\mathbf{n}}$ we then have

$$\boxed{\boldsymbol{P}_{\mathbf{n}} = \frac{\boldsymbol{P}_{\parallel}(\mathbf{n})}{\mathbf{n}\cdot\boldsymbol{\varepsilon}_0\cdot\mathbf{n}}.} \tag{8.153}$$

Now that we have found the relation between $\mathbf{a}$ and $\mathbf{b}$, we can return to our problem for which $\mathbf{a} = \mathbf{p}(\mathbf{x}) - \langle\mathbf{p}\rangle$ and $\mathbf{b} = \langle\mathbf{E}\rangle - \mathbf{E}(\mathbf{x})$. Then, from the definition of $\boldsymbol{P}_{\mathbf{n}}$,

$$\langle\mathbf{E}\rangle - \mathbf{E}(\mathbf{x}) = \boldsymbol{P}_{\mathbf{n}}\cdot[\mathbf{p}(\mathbf{x}) - \langle\mathbf{p}\rangle]. \tag{8.154}$$

Recall that from equation (8.148) we have

$$\mathbf{E}(\mathbf{x}) = [\boldsymbol{\varepsilon}(\mathbf{x}) - \boldsymbol{\varepsilon}_0]^{-1}\cdot\mathbf{p}(\mathbf{x}).$$

Plugging this into (8.154) and rearranging gives

$$\left[[\boldsymbol{\varepsilon}(\mathbf{x}) - \boldsymbol{\varepsilon}_0]^{-1} + \boldsymbol{P}_{\mathbf{n}}\right]\cdot\mathbf{p}(\mathbf{x}) = \langle\mathbf{E}\rangle + \boldsymbol{P}_{\mathbf{n}}\cdot\langle\mathbf{p}\rangle =: \mathbf{v}.$$

Note that $\mathbf{v} = \langle\mathbf{E}\rangle + \boldsymbol{P}_{\mathbf{n}}\cdot\langle\mathbf{p}\rangle$ is constant throughout the laminate. Therefore we have

$$\mathbf{p}(\mathbf{x}) = \left[[\boldsymbol{\varepsilon}(\mathbf{x}) - \boldsymbol{\varepsilon}_0]^{-1} + \boldsymbol{P}_{\mathbf{n}}\right]^{-1}\cdot\mathbf{v}.$$

If we now take a volume average, we get

$$\langle\mathbf{p}\rangle = \left\langle\left[[\boldsymbol{\varepsilon}(\mathbf{x}) - \boldsymbol{\varepsilon}_0]^{-1} + \boldsymbol{P}_{\mathbf{n}}\right]^{-1}\right\rangle\cdot\mathbf{v}. \tag{8.155}$$

Also, from the definition of $\mathbf{p}(\mathbf{x})$ in equation (8.148), we have

$$\langle\mathbf{E}\rangle = (\boldsymbol{\varepsilon}_{\text{eff}} - \boldsymbol{\varepsilon}_0)^{-1}\cdot\langle\mathbf{p}\rangle.$$

Therefore,

$$\mathbf{v} = \langle\mathbf{E}\rangle + \boldsymbol{P}_{\mathbf{n}}\cdot\langle\mathbf{p}\rangle = \left[(\boldsymbol{\varepsilon}_{\text{eff}} - \boldsymbol{\varepsilon}_0)^{-1} + \boldsymbol{P}_{\mathbf{n}}\right]\cdot\langle\mathbf{p}\rangle$$

or

$$\langle\mathbf{p}\rangle = \left[(\boldsymbol{\varepsilon}_{\text{eff}} - \boldsymbol{\varepsilon}_0)^{-1} + \boldsymbol{P}_{\mathbf{n}}\right]^{-1}\cdot\mathbf{v}. \tag{8.156}$$

Comparing equations (8.155) and (8.156) and invoking the arbitrariness of $\boldsymbol{\varepsilon}_0$, we get

$$\boxed{\left[(\boldsymbol{\varepsilon}_{\text{eff}} - \boldsymbol{\varepsilon}_0)^{-1} + \boldsymbol{P}_{\mathbf{n}}\right]^{-1} = \left\langle \left[[\boldsymbol{\varepsilon}(\mathbf{x}) - \boldsymbol{\varepsilon}_0]^{-1} + \boldsymbol{P}_{\mathbf{n}}\right]^{-1} \right\rangle.} \tag{8.157}$$

This is Milton's relation, a generalization of the TMLC formula which has a simple form and can be used when the phases are anisotropic. For a rank-1 laminate where $\boldsymbol{\varepsilon}_0 = \boldsymbol{\varepsilon}_2$, equation (8.157) reduces to

$$f_1 (\boldsymbol{\varepsilon}_{\text{eff}} - \boldsymbol{\varepsilon}_2)^{-1} = (\boldsymbol{\varepsilon}_1 - \boldsymbol{\varepsilon}_2)^{-1} + f_2 \boldsymbol{P}_{\mathbf{n}} \quad \text{where} \quad \boldsymbol{P}_{\mathbf{n}} = \frac{\mathbf{n} \otimes \mathbf{n}}{\mathbf{n} \cdot \boldsymbol{\varepsilon}_2 \cdot \mathbf{n}}.$$

Linear elastic laminates

The same analysis can be applied for elastic laminates with anisotropic layers and arbitrary direction of lamination. In this case we introduce a reference stiffness tensor $\mathbf{C}_0$ and define a second-order polarization tensor as

$$\boldsymbol{\pi}(\mathbf{x}) := [\mathbf{C}(\mathbf{x}) - \mathbf{C}_0] : \boldsymbol{\varepsilon}(\mathbf{x}) \tag{8.158}$$

where $\boldsymbol{\varepsilon}(\mathbf{x})$ is the strain tensor. Because we are dealing with fourth-order tensors, the appropriate projection operators for elasticity are defined using

$$\mathbf{P}_{\parallel}(\mathbf{n}) : \boldsymbol{A} = (\boldsymbol{A} \cdot \mathbf{n}) \otimes \mathbf{n} + \mathbf{n} \otimes (\boldsymbol{A} \cdot \mathbf{n}) - (\mathbf{n} \cdot \boldsymbol{A} \cdot \mathbf{n})(\mathbf{n} \otimes \mathbf{n})$$

and

$$\mathbf{P}_{\perp}(\mathbf{n}) : \boldsymbol{A} = [\mathbf{1} - \mathbf{P}_{\parallel}(\mathbf{n})] : \boldsymbol{A}$$

where $\boldsymbol{A}$ is a second-order tensor and $\mathbf{1}$ is the symmetric part of the fourth-order identity tensor. Clearly these operators are also fourth-order tensors. Expressed in components with respect to a Cartesian basis, these projection tensors are given by

$$\mathbf{P}_{\parallel}(\mathbf{n}) = \left[\tfrac{1}{2}(n_i n_\ell \delta_{jk} + n_i n_k \delta_{j\ell} + n_j n_\ell \delta_{ik} + n_j n_k \delta_{i\ell}) - n_j n_j n_k n_\ell\right] \mathbf{e}_i \otimes \mathbf{e}_j \otimes \mathbf{e}_k \otimes \mathbf{e}_\ell$$

and

$$\mathbf{P}_{\perp}(\mathbf{n}) = \tfrac{1}{2}(\delta_{ik}\delta_{j\ell} + \delta_{i\ell}\delta_{jk}) \mathbf{e}_i \otimes \mathbf{e}_j \otimes \mathbf{e}_k \otimes \mathbf{e}_\ell - \mathbf{P}_{\parallel}(\mathbf{n}).$$

In this case we define two second-order tensors $\boldsymbol{A}$ and $\boldsymbol{B}$ such that

$$\boldsymbol{A} := \boldsymbol{\pi}(\mathbf{x}) - \langle \boldsymbol{\pi} \rangle \quad \text{and} \quad \boldsymbol{B} := \langle \boldsymbol{\varepsilon} \rangle - \boldsymbol{\varepsilon}(\mathbf{x}).$$

The fourth-order operator $\mathbf{P}_{\mathbf{n}}$ is defined in terms of these second-order tensors such that

$$\boldsymbol{B} = \mathbf{P}_{\mathbf{n}} : \boldsymbol{A}.$$

The condition $\mathbf{P}_{\parallel}(\mathbf{n}) : \boldsymbol{B} = \boldsymbol{B}$ is satisfied if there exists a vector $\mathbf{b}$ such that

$$\boldsymbol{B} = \mathbf{n} \otimes \mathbf{b} + \mathbf{b} \otimes \mathbf{n} \tag{8.159}$$

and the condition $\mathsf{P}_{\|}(\mathbf{n}) : (\boldsymbol{A} - \mathsf{C}_0 : \boldsymbol{B}) = \mathbf{0}$ is satisfied if

$$\mathbf{n} \cdot (\boldsymbol{A} - \mathsf{C}_0 : \boldsymbol{B}) = \mathbf{0}. \tag{8.160}$$

Substituting (8.159) into (8.160), and using the symmetry of C_0, gives

$$\mathbf{n} \cdot \boldsymbol{A} - 2(\mathbf{n} \cdot \mathsf{C}_0 \cdot \mathbf{n}) \cdot \mathbf{b} = \mathbf{0} \quad \text{or} \quad \mathbf{n} \cdot \boldsymbol{A} - 2\boldsymbol{C}(\mathbf{n}) \cdot \mathbf{b} = \mathbf{0}.$$

Notice that the tensor $\boldsymbol{C}(\mathbf{n}) := \mathbf{n} \cdot \mathsf{C}_0 \cdot \mathbf{n}$ is the second-order acoustic tensor. The symmetry of the stress tensor implies that the polarization tensor is also symmetric and we have

$$\mathbf{b} = \tfrac{1}{4}\boldsymbol{C}^{-1}(\mathbf{n}) \cdot \left(\boldsymbol{A} \cdot \mathbf{n} + \boldsymbol{A}^T \cdot \mathbf{n}\right).$$

Substituting this expression for $\mathbf{b}$ into (8.159) and expressing all quantities in terms of components with respect to a Cartesian basis, we have

$$B_{ij} = \tfrac{1}{4}\left(C^{-1}_{j\ell} n_i n_k + C^{-1}_{jk} n_i n_\ell + C^{-1}_{i\ell} n_j n_k + C^{-1}_{ik} n_j n_\ell\right) A_{k\ell}.$$

Therefore the operator $\mathsf{P}_{\mathbf{n}}$ has the form

$$\boxed{\mathsf{P}_{\mathbf{n}} = \tfrac{1}{4}\left(C^{-1}_{j\ell} n_i n_k + C^{-1}_{jk} n_i n_\ell + C^{-1}_{i\ell} n_j n_k + C^{-1}_{ik} n_j n_\ell\right) \mathbf{e}_i \otimes \mathbf{e}_j \otimes \mathbf{e}_k \otimes \mathbf{e}_l.}$$

Following the process that we used to determine the effective permittivity, we can show that the effective elastic stiffness of a laminate can be determined from the formula

$$\boxed{\left[(\mathsf{C}_{\text{eff}} - \mathsf{C}_0)^{-1} + \mathsf{P}_{\mathbf{n}}\right]^{-1} = \left\langle \left[[\mathsf{C}(\mathbf{x}) - \mathsf{C}_0]^{-1} + \mathsf{P}_{\mathbf{n}}\right]^{-1} \right\rangle.} \tag{8.161}$$

If C_0 is isotropic, i.e., $\mathsf{C}_0 = \lambda_0 \boldsymbol{1} \otimes \boldsymbol{1} + 2\mu_0 \mathsf{1}$, where λ_0 is a reference Lamé modulus and μ_0 is a reference shear modulus, $\mathsf{P}_{\mathbf{n}}$ simplifies to

$$[\mathsf{P}_{\mathbf{n}}]_{ijk\ell} = \left(\tfrac{1}{\lambda_0 + 2\mu_0} - \tfrac{1}{\mu_0}\right) n_i n_j n_k n_\ell + \tfrac{1}{4\mu_0}(n_i n_\ell \delta_{jk} + n_i n_k \delta_{j\ell} + n_j n_\ell \delta_{ik} + n_j n_k \delta_{i\ell}).$$

8.6.5 Hierarchical laminates

We have found that the effective material properties of rank-1 laminates can be calculated using the formula of Tartar-Murat-Lurie-Cherkaev and its extensions. Let us now explore laminates made of laminates. Such hierarchical structures have remarkable properties and a detailed exposition is beyond the scope of this book. Much more can be found in Milton (2002) and the references therein. For diverse examples of such hierarchical structures and their uses see Lloyd and Molina-Aldareguia (2003) and Shawkey et al. (2009).

An example of a hierarchical laminate is shown in Figure 8.12. The idea of such materials goes back to Maxwell (see Maxwell (1954)). In the rank-2 laminate shown

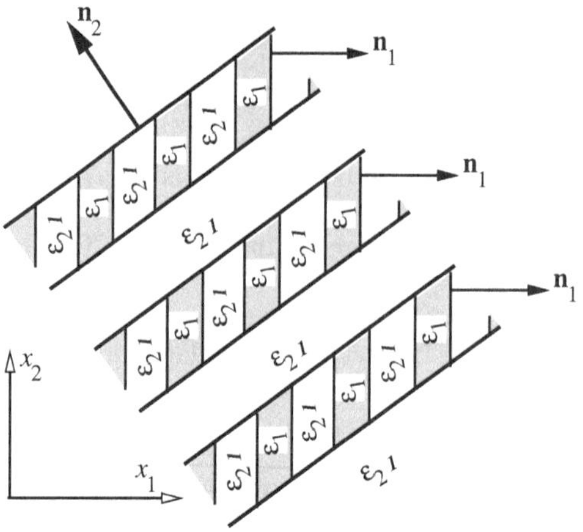

FIGURE 8.12
A rank-2 hierarchical laminate.

in the figure there are two length scales which are assumed to be sufficiently separated so that the ideas in the previous sections can be exploited and the layered material at each length scale can be replaced by its effective tensor. Recall the Tartar-Murat-Lurie-Cherkaev formula in equation (8.147) for the effective permittivity of a rank-1 laminate:

$$\begin{aligned} f_1^{(1)}[\varepsilon_2 \mathbf{1} - \boldsymbol{\varepsilon}_{\text{eff}}^{(1)}]^{-1} &= [\varepsilon_2 \mathbf{1} - \boldsymbol{\varepsilon}_1]^{-1} - (1 - f_1^{(1)})\varepsilon_2^{-1}\boldsymbol{P}_{\|}(\mathbf{n}_1) \\ &= f_1^{(0)}[\varepsilon_2 \mathbf{1} - \boldsymbol{\varepsilon}_1]^{-1} - (f_1^{(0)} - f_1^{(1)})\varepsilon_2^{-1}\boldsymbol{P}_{\|}(\mathbf{n}_1) \end{aligned}$$

where the superscript denotes the rank of the laminate and we have used $f_2^{(1)} = 1 - f_1^{(1)}$. The volume fraction of phase 1 in the rank-0 laminate is 1 because we are creating the laminate by adding the phase 2 material to it. To find the effective permittivity of the rank-2 laminate in Figure 8.12 we just adjust the volume fractions of the two phases in the laminate and replace $\boldsymbol{\varepsilon}_1$ with $\boldsymbol{\varepsilon}_{\text{eff}}$ for the rank-1 laminate. Then we get

$$f_1^{(2)}[\varepsilon_2 \mathbf{1} - \boldsymbol{\varepsilon}_{\text{eff}}^{(2)}]^{-1} = f_1^{(1)}[\varepsilon_2 \mathbf{1} - \boldsymbol{\varepsilon}_{\text{eff}}^{(1)}]^{-1} - (f_1^{(1)} - f_1^{(2)})\varepsilon_2^{-1}\boldsymbol{P}_{\|}(\mathbf{n}_2)$$

or

$$\begin{aligned} & f_1^{(2)}[\varepsilon_2 \mathbf{1} - \boldsymbol{\varepsilon}_{\text{eff}}^{(2)}]^{-1} = \\ & \qquad [\varepsilon_2 \mathbf{1} - \boldsymbol{\varepsilon}_1]^{-1} - (f_1^{(0)} - f_1^{(1)})\varepsilon_2^{-1}\boldsymbol{P}_{\|}(\mathbf{n}_1) - (f_1^{(1)} - f_1^{(2)})\varepsilon_2^{-1}\boldsymbol{P}_{\|}(\mathbf{n}_2). \end{aligned}$$

Since $f_1^{(2)} = f_1$ is the volume fraction of phase 1 in the final composite, $\boldsymbol{\varepsilon}_{\text{eff}}^{(2)}$ is the effective modulus of the rank-2 laminate, and if $f_2 = 1 - f_1$ is the final volume fraction of the second phase, then we can write the above formula as

$$f_1[\varepsilon_2 \mathbf{1} - \boldsymbol{\varepsilon}_{\text{eff}}]^{-1} = [\varepsilon_2 \mathbf{1} - \boldsymbol{\varepsilon}_1]^{-1} - f_2\varepsilon_2^{-1}\left[c_1\boldsymbol{P}_{\|}(\mathbf{n}_1) + c_2\boldsymbol{P}_{\|}(\mathbf{n}_2)\right] \tag{8.162}$$

where

$$c_1 = \frac{f_1^{(0)} - f_1^{(1)}}{f_2} \quad \text{and} \quad c_2 = \frac{f_1^{(1)} - f_1^{(2)}}{f_2}.$$

By iterating (8.162) one gets, for a rank-m laminate, the Tartar formula

$$f_1 [\varepsilon_2 \mathbf{1} - \boldsymbol{\varepsilon}_{\text{eff}}]^{-1} = [\varepsilon_2 \mathbf{1} - \boldsymbol{\varepsilon}_1]^{-1} - f_2 \varepsilon_2^{-1} \boldsymbol{M} \tag{8.163}$$

where $f_1 = f_1^{(m)}$, $f_1^{(0)} = 1$,

$$\boldsymbol{M} := \sum_{j=1}^{m} c_j \, \boldsymbol{P}_{\parallel}(\mathbf{n}_j) \; ; \; c_j = \frac{f^{(j-1)} - f^{(j)}}{1 - f_1} \geq 0 \; ; \; f^{(j-1)} > f^{(j)}$$

and m is the number of laminates in the hierarchy, $f^{(j)}$ is the proportion of phase 1 in a rank-j laminate, and $\mathbf{n}_j$ is the orientation of the j-th laminate. Also observe that

$$\sum_{j=1}^{m} c_j = \frac{f^{(0)} - f^{(m)}}{1 - f_1} = 1\,.$$

For a rank-3 laminate, if the normals $\mathbf{n}_1$, $\mathbf{n}_2$, and $\mathbf{n}_3$ are three orthogonal vectors, then

$$\boldsymbol{M} = c_1 \mathbf{n}_1 \otimes \mathbf{n}_1 + c_2 \mathbf{n}_2 \otimes \mathbf{n}_2 + c_3 \mathbf{n}_3 \otimes \mathbf{n}_3\,.$$

If we choose the $f^{(j)}$s such that $c_1 = c_2 = c_3 = 1/2$, then

$$\boldsymbol{M} = \tfrac{1}{3}(\mathbf{n}_1 \otimes \mathbf{n}_1 + \mathbf{n}_2 \otimes \mathbf{n}_2 + \mathbf{n}_3 \otimes \mathbf{n}_3) = \mathbf{1}.$$

In this case, equation (8.163) coincides with the solution for the Hashin sphere assemblage! This implies that different geometries can have the same $\boldsymbol{\varepsilon}_{\text{eff}}(\varepsilon_2, \boldsymbol{\varepsilon}_1)$. This observation has led to the discovery that a large range of material properties can be realized using laminated structures. Milton (2010) examines some of the possibilities in the context of electrodynamics. Many of these ideas can be extended to acoustics and elastodynamics and a detailed discussion can be found in Milton (2002).

Exercises

Problem 8.1 A plane wave propagating through a layered medium can be expressed in the form

$$u_2(x_1, x_3, t) = (A_{1j} e^{k_j x_3} + A_{2j} e^{-k_j x_3})\, e^{i(k_1 x_1 - \omega t)}\,.$$

Show that if $\text{Re}(k_j) \geq 0$ we must have $\text{Im}(k_j) \leq 0$.

Problem 8.2 Verify that the generalized reflection and transmission coefficients in Section 8.2 can also be applied without change to acoustic waves. What are the equivalent expressions for $R_{j,j+1}$ and $T_{j,j+1}$ for acoustic waves?

Problem 8.3 Find the matrix $\underline{\underline{\mathsf{H}}}$ and the associated state vector $\underline{\underline{\mathsf{v}}}$ for the propagation of P-SV waves in a layered medium.

Problem 8.4 Use the WKBJ method to find the transmission coefficient for a plane wave propagating through the graded slab shown in the figure below.

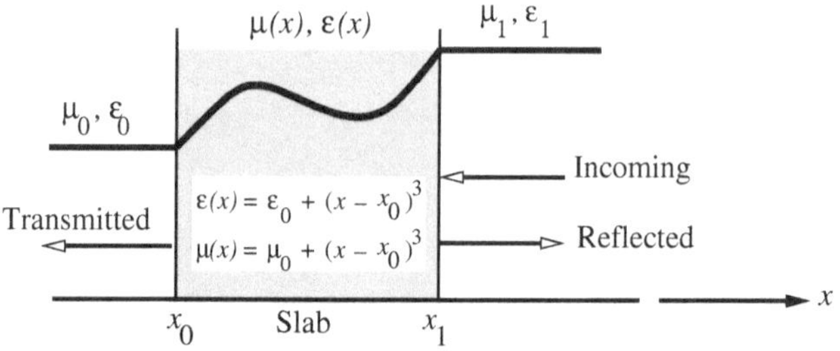

Problem 8.5 Let z_0 and z be the locations of the top and bottom of an isotropic elastic layer, respectively. Show that the propagator matrix for antiplane shear waves in the layer can be expressed as

$$\underline{\underline{\mathsf{P}}}(z,z_0) = \begin{bmatrix} \cos k_z(z-z_0) & (\mu k_z)^{-1}\sin k_z(z-z_0) \\ -\mu k_z \sin k_z(z-z_0) & \cos k_z(z-z_0) \end{bmatrix}.$$

Use equation (8.78) to calculate the propagator matrix and then verify your result using the matrix exponential solution in (8.80) and Sylvester's formula. Note that the eigenvalues of a square matrix of the form $\alpha\underline{\underline{\mathsf{A}}}$ where α is a scalar are given by $\alpha\lambda_j$ where λ_j are the eigenvalues of $\underline{\underline{\mathsf{A}}}$.

Problem 8.6 Derive the relations between the $\mathbf{E}_z, \mathbf{H}_z$ and $\mathbf{E}_s, \mathbf{H}_s$ given in equation (8.92). Also, find the explicit form of the matrix $\underline{\underline{\mathsf{H}}}$ in equation (8.93).

Problem 8.7 In the Schoenberg-Sen model of a periodic layered medium, the slowness component s_z is related to the angle of incidence θ_i by $s_z = \sin\theta_i / c_0$ where c_0 is the phase speed in the medium of incidence. Show that, in the low-frequency limit, the effective angle of transmission (θ_t) into the layered medium is given by

$$\theta = \tan^{-1}\left\{ \sin\theta_i \left[\langle\rho\rangle^{\frac{1}{2}} \langle 1/\rho\rangle^{\frac{1}{2}} \left(c_0^2 \frac{\langle 1/\kappa\rangle}{\langle 1/\rho\rangle} - \sin^2\theta_i \right)^{\frac{1}{2}} \right]^{-1} \right\}.$$

Show that the same expression is obtained for the angle of transmission in a homogeneous medium with

$$\kappa = \frac{1}{\langle 1/\kappa\rangle} \quad \text{and} \quad \boldsymbol{\rho} \equiv \begin{bmatrix} \langle\rho\rangle & 0 & 0 \\ 0 & 1/\langle 1/\rho\rangle & 0 \\ 0 & 0 & 1/\langle 1/\rho\rangle \end{bmatrix}.$$

Problem 8.8 Verify that the effective stiffness of an elastic laminate can be expressed as

$$\underline{\underline{\mathbf{C}}}^{nn}_{\text{eff}} = \left\langle (\underline{\underline{\mathbf{C}}}^{nn})^{-1} \right\rangle^{-1}$$

$$\underline{\underline{\mathbf{C}}}^{nt}_{\text{eff}} = \left\langle (\underline{\underline{\mathbf{C}}}^{nn})^{-1} \right\rangle^{-1} \cdot \left\langle (\underline{\underline{\mathbf{C}}}^{nn})^{-1} \cdot \underline{\underline{\mathbf{C}}}^{nt} \right\rangle$$

$$\underline{\underline{\mathbf{C}}}^{tt}_{\text{eff}} = \left\langle \underline{\underline{\mathbf{C}}}^{tt} - \underline{\underline{\mathbf{C}}}^{tn} \cdot (\underline{\underline{\mathbf{C}}}^{nn})^{-1} \cdot \underline{\underline{\mathbf{C}}}^{nt} \right\rangle +$$
$$\left\langle \underline{\underline{\mathbf{C}}}^{tn} \cdot (\underline{\underline{\mathbf{C}}}^{nn})^{-1} \right\rangle \cdot \left\langle (\underline{\underline{\mathbf{C}}}^{nn})^{-1} \right\rangle^{-1} \cdot \left\langle (\underline{\underline{\mathbf{C}}}^{nn})^{-1} \cdot \underline{\underline{\mathbf{C}}}^{nt} \right\rangle .$$

Consider the three-layered laminate shown in the figure below and assume that the x-direction is the direction of lamination and the y-, z-directions are in the plane of the laminate. The stiffness of layer i is labeled $\mathbf{C}_i$ in the figure.

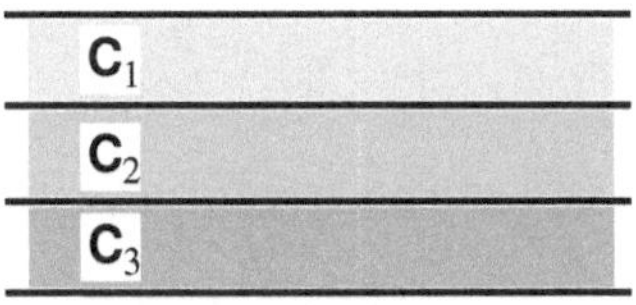

Assume that layers 1 and 3 are made of a transversely isotropic AS/3501 (carbon fiber/epoxy) composite with Young's moduli $E_x = 9$ GPa, $E_y = E_z = 140$ GPa, Poisson's ratios $\nu_{xy} = \nu_{yz} = \nu_{xz} = 0.3$, and shear moduli $G_{xy} = G_{yz} = 7$ GPa. Assume that layer 2 is isotropic with Young's modulus $E = 70$ GPa and Poisson's ratio $\nu = 0.2$. What are the quasistatic effective elastic stiffnesses of the three layer composite if the thicknesses of layers 1 and 3 are 2 mm and that of layer 2 is 5 mm?

Problem 8.9 Show that for a laminate made of isotropic linear elastic layers, the effective stiffness tensor has components

$$C^{\text{eff}}_{1111} = \left\langle \frac{1}{\lambda+2\mu} \right\rangle^{-1} \; ; \; C^{\text{eff}}_{1122} = C^{\text{eff}}_{1133} = \left\langle \frac{\lambda}{\lambda+2\mu} \right\rangle \left\langle \frac{1}{\lambda+2\mu} \right\rangle^{-1}$$

$$C^{\text{eff}}_{1212} = C^{\text{eff}}_{1313} = \left\langle \frac{1}{\mu} \right\rangle^{-1} \; ; \; C^{\text{eff}}_{2323} = \langle \mu \rangle \; ; \; C^{\text{eff}}_{1112} = C^{\text{eff}}_{1113} = C^{\text{eff}}_{1123} = 0$$

$$C^{\text{eff}}_{2222} = C^{\text{eff}}_{3333} = \left\langle \frac{4\mu\lambda(\lambda+\mu)}{(\lambda+2\mu)^2} \right\rangle + \left\langle \frac{\lambda}{\lambda+2\mu} \right\rangle \left\langle \frac{1}{\lambda+2\mu} \right\rangle^{-1}$$

$$C^{\text{eff}}_{2233} = \left\langle \frac{2\mu\lambda}{\lambda+2\mu} \right\rangle + \left\langle \frac{\lambda}{\lambda+2\mu} \right\rangle^{2} \left\langle \frac{1}{\lambda+2\mu} \right\rangle^{-1} .$$

Problem 8.10 Verify that the quasistatic effective permittivity of a laminate is given by the relations

$$\varepsilon^{\text{eff}}_{11} = \left\langle \frac{1}{\varepsilon_{11}} \right\rangle^{-1}$$

$$\varepsilon^{\text{eff}}_{1j} = \left\langle \frac{1}{\varepsilon_{11}} \right\rangle^{-1} \left\langle \frac{\varepsilon_{1j}}{\varepsilon_{11}} \right\rangle$$

$$\varepsilon^{\text{eff}}_{ij} = \left\langle \varepsilon_{ij} - \frac{\varepsilon_{i1}\,\varepsilon_{1j}}{\varepsilon_{11}} \right\rangle + \left\langle \frac{1}{\varepsilon_{11}} \right\rangle^{-1} \left\langle \frac{\varepsilon_{i1}}{\varepsilon_{11}} \right\rangle \left\langle \frac{\varepsilon_{1j}}{\varepsilon_{11}} \right\rangle \; , i, j \neq 1 .$$

Problem 8.11 Use the Backus approach to find expressions for the effective magnetic permeability of a laminate with lamination in the x_1-direction.

Problem 8.12 Use the relation

$$\left[\boldsymbol{S}_{\text{eff}} - \boldsymbol{P}_{\|}(\mathbf{n})\right]^{-1} = \left\langle \left[\boldsymbol{S}(\mathbf{x}) - \boldsymbol{P}_{\|}(\mathbf{n})\right]^{-1} \right\rangle$$

to find the effective permittivity of a laminate that is oriented in the x_2-direction.

Problem 8.13 Show that the Tartar-Murat-Lurie-Cherkaev formula for the effective permittivity of a rank-1 laminate

$$f_1 \varepsilon_2 [\varepsilon_2 \boldsymbol{1} - \boldsymbol{\varepsilon}_{\text{eff}}]^{-1} = \varepsilon_2 [\varepsilon_2 \boldsymbol{1} - \boldsymbol{\varepsilon}_1]^{-1} - f_2 \boldsymbol{P}_{\|}(\mathbf{n})$$

is equivalent to the formula derived using the Backus method.

Problem 8.14 Show that, for a rank-1 laminate with layer permittivities $\boldsymbol{\varepsilon}_1$ and $\boldsymbol{\varepsilon}_2$ and volume fractions f_1 and f_2, Milton's relation reduces to

$$f_1 (\boldsymbol{\varepsilon}_{\text{eff}} - \boldsymbol{\varepsilon}_2)^{-1} = (\boldsymbol{\varepsilon}_1 - \boldsymbol{\varepsilon}_2)^{-1} + f_2 \boldsymbol{P}_{\mathbf{n}} \quad \text{where} \quad \boldsymbol{P}_{\mathbf{n}} = \frac{\mathbf{n} \otimes \mathbf{n}}{\mathbf{n} \cdot \boldsymbol{\varepsilon}_2 \cdot \mathbf{n}}.$$

Problem 8.15 Show that for an isotropic reference material with

$$\mathsf{C}_0 = \lambda_0 \boldsymbol{1} \otimes \boldsymbol{1} + 2\mu_0 \mathbf{1}$$

the Cartesian components of the operator $\mathsf{P}_{\mathbf{n}}$ have the form

$$(\mathsf{P}_{\mathbf{n}})_{ijk\ell} = \left(\tfrac{1}{\lambda_0 + 2\mu_0} - \tfrac{1}{\mu_0}\right) n_i n_j n_k n_\ell + \tfrac{1}{4\mu_0} \left(n_i n_\ell \delta_{jk} + n_i n_k \delta_{j\ell} + n_j n_\ell \delta_{ik} + n_j n_k \delta_{i\ell}\right).$$

Epilogue

> In any case, the triple pedagogical goals of understanding, skills, and design can be achieved only via independent practice, hard work, and creative thinking.
>
> C. KLEINSTREUER, *Two-phase flow: Theory and applications*, 2003.

There are many important topics in wave propagation that have not been covered in this book. Some topics of practical and pedagogical significance are:

- *Numerical techniques for the simulation of wave propagation.* A good rule of thumb for full wave simulations with finite elements is to use 15–20 elements per wavelength. Clearly, such a high resolution makes the computation of all but very simple problems prohibitively expensive. Efficient techniques typically involve simplifying the problem using some knowledge of the structure. An example is the efficient and stable method of Rokhlin and co-workers for problems that can be described using propagator matrices.

- *Design paradigms and techniques.* A design paradigm that can be easily used for periodic structures is to take the vast library of known results on elastostatic composites and extend them to dynamic, complex-valued problems as we have discussed in Chapter 7. Numerous elastostatic and electrostatic results based on duality relations can be found in the monograph by Milton. Layered media are usually easier to manufacture than periodic media with complex geometries. Hierarchical laminates and other layered geometries appear to be the most straightforward path toward realizing metamaterials. Another design paradigm is AC circuit analogy and the related transmission line theory which has been used extensively in the design of acoustic resonators by Fang and co-workers.

- *Homogenization methods.* Though we have woven homogenization methods into most of the chapters in this book, more can be said about the subject. Homogenization in the presence of resonances, and the homogenization of dynamics in general, can be quite involved and no single technique has gained wide acceptance yet. Tartar's recently published monograph on homogenization gives a flavor of the complexities involved.

- *Specific designs.* There has been a proliferation of designs of metamaterial structures and transformation-based coatings. We have discussed a few in the book but the reader is urged to explore the literature for other designs.

- *Manufacturing methods and experimental procedures.* There has been an explosion of techniques for rapidly manufacturing small structures. Knowledge of manufacturing techniques and experimental methods are crucial for the future researcher in metamaterials and photonic/phononic devices.

References

Abramowitz, M. and Stegun, I. A. (1972). Airy functions. In *Handbook of mathematical functions with formulas, graphs, and mathematical tables*, pages 446–452. Dover, New York.

Achenbach, J. (1973). *Wave propagation in elastic solids*. North-Holland, Amsterdam.

Adam, J. A. (2002). The mathematical physics of rainbows and glories. *Physics Reports*, 356:229–365.

Aki, K. and Richards, P. G. (1980). *Quantitative seismology: Theory and methods*. Freeman, San Francisco.

Alshits, V. I. and Maugin, G. A. (2005). Dynamics of multilayers: Elastic waves in an anisotropic graded or stratified plate. *Wave Motion*, 41(4):357–394.

Ament, W. S. (1953). Sound propagation in gross mixtures. *Journal of the Acoustical Society of America*, 25:638.

Arfken, G. B. and Weber, H. J. (2005). *Mathematical methods for physicists: 6th edition*. Elsevier Academic Press, Burlington, MA.

Ashcroft, N. W. and Mermin, N. D. (1976). *Solid state physics*. Saunders, New York.

Atkin, R. J. and Fox, N. (1980). *An introduction to the theory of elasticity*. Longman, New York.

Avila, A., Griso, G., and Miara, B. (2005). Bandes photoniques interdies en elasticite linearisee. *Comptes Rendus de l'Académie des Sciences Series I*, 340:933–938.

Axmann, W. and Kuchment, P. (1999). An efficient finite element method for computing spectra of photonic and acoustic band-gap materials I. Scalar case. *Journal of Computational Physics*, 150(2):468–481.

Backus, G. E. (1962). Long-wave elastic anisotropy produced by horizontal layering. *Journal of Geophysical Research*, 67:4427–4440.

Bathe, K. J. (1997). *Finite element procedures*. Prentice-Hall, Upper Saddle River, NJ.

Bernstein, S. (1928). Sur les fonctions absolument monotones. *Acta Mathematica*, 52:1–66.

Berryman, J. G. (1980). Long-wavelength propagation in composite elastic media I. Spherical inclusions. *Journal of the Acoustical Society of America*, 68:1809–1819.

Bertoldi, K., Reis, P. M., Willshaw, S., and Mullin, T. (2010). Negative Poisson's ratio behavior induced by an elastic instability. *Advanced Materials*, 22(3):361–366.

Bloch, F. (1929). Über die Quantenmechanik der Elektronen in Kristallgittern. *Zeitschrift für Physik A Hadrons and Nuclei*, 52(7):555–600.

Bouchitté, G. and Schweizer, B. (2010). Homogenization of Maxwell's equations in a split ring geometry. *Multiscale Modeling & Simulation*, 8:717.

Bower, A. F. (2010). *Applied mechanics of solids*. CRC Press, Boca Raton, FL.

Bowman, J. J., Senior, T. B. A., and Uslenghi, P. L. E. (1969). *Electromagnetic and acoustic scattering by simple shapes*. North-Holland, Amsterdam.

Brekhovskikh, L. M. (1960). *Waves in layered media*. Transl. by David Lieberman and Robert T. Beyer. Applied Mathematics and Mechanics. Academic Press, New York.

Brill, D. and Gaunaurd, G. (1987). Resonance theory of elastic waves ultrasonically scattered from an elastic sphere. *Journal of the Acoustical Society of America*, 81:1.

Brillouin, L. (2003). *Wave propagation in periodic structures*. Dover, New York.

Brocato, M. and Capriz, G. (2001). Gyrocontinua. *International Journal of Solids and Structures*, 38(6-7):1089–1103.

Brun, M., Guenneau, S., and Movchan, A. B. (2009). Achieving control of in-plane elastic waves. *Applied Physics Letters*, 94(6):061903.

Busch, K., von Freymann, G., Linden, S., Mingaleev, S. F., Tkeshelashvili, L., and Wegener, M. (2007). Periodic nanostructures for photonics. *Physics Reports*, 444(3–6):101–202.

Cai, L. W. and Sánchez-Dehesa, J. (2007). Analysis of Cummer–Schurig acoustic cloaking. *New Journal of Physics*, 9:450.

Cai, W. and Shalaev, V. (2009). *Optical metamaterials: Fundamentals and applications*. Springer-Verlag, New York.

Carcione, J. M. (2007). *Wave fields in real media: Wave propagation in anisotropic, anelastic, porous and electromagnetic media*. Elsevier Science, Oxford.

Chen, H. and Chan, C. T. (2007a). Acoustic cloaking in three dimensions using acoustic metamaterials. *Applied Physics Letters*, 91:183518.

Chen, H. and Chan, C. T. (2007b). Transformation media that rotate electromagnetic fields. *Applied Physics Letters*, 90:241105.

Chen, H. and Chan, C. T. (2010). Acoustic cloaking and transformation acoustics. *Journal of Physics D: Applied Physics*, 43:113001.

Chen, H., Chan, C. T., and Sheng, P. (2010). Transformation optics and metamaterials. *Nature Materials*, 9(5):387–396.

Cheng, Q., Cui, T. J., Jiang, W. X., and Cai, B. G. (2009). An electromagnetic black hole made of metamaterials. *ArXiv preprint arXiv:0910.2159*.

Cheng, Q., Cui, T. J., Jiang, W. X., and Cai, B. G. (2010). An omnidirectional electromagnetic absorber made of metamaterials. *New Journal of Physics*, 12:063006.

Chew, W. C. (1995). *Waves and fields in inhomogeneous media*. IEEE Press, New York.

Christensen, R. M. (1984). *Mechanics of Composite Materials*. John Wiley & Sons, New York.

Christensen, R. M. (2003). *Theory of viscoelasticity: 2nd edition*. Courier Dover Publications, London.

Craster, R. V., Kaplunov, J., and Pichugin, A. V. (2010). High-frequency homogenization for periodic media. *Proceeding of the Royal Society A*, 466:2341–2362.

Cummer, S. A., Popa, B. I., Schurig, D., Smith, D. R., and Pendry, J. (2006). Full-wave simulations of electromagnetic cloaking structures. *Physical Review E*, 74(3):36621.

Cummer, S. A., Popa, B. I., Schurig, D., Smith, D. R., Pendry, J., Rahm, M., and Starr, A. (2008). Scattering theory derivation of a 3D acoustic cloaking shell. *Physical Review Letters*, 100(2):24301.

Cummer, S. A. and Schurig, D. (2007). One path to acoustic cloaking. *New Journal of Physics*, 9:45.

D'Eleuterio, G. M. T. and Hughes, P. C. (1984). Dynamics of gyroelastic continua. *Journal of Applied Mechanics*, 51:415.

Delgado, V., Sydoruk, O., Tatartschuk, E., Marqués, R., Freire, M. J., and Jelinek, L. (2009). Analytical circuit model for split ring resonators in the far infrared and optical frequency range. *Metamaterials*, 3(2):57–62.

Ding, Y., Liu, Z., Qiu, C., and Shi, J. (2007). Metamaterial with simultaneously negative bulk modulus and mass density. *Physical Review Letters*, 99(9):93904.

Dobson, D. C. (1999). An efficient method for band structure calculations in 2D photonic crystals. *Journal of Computational Physics*, 149(2):363–376.

Ergin, T., Stenger, N., Brenner, P., Pendry, J. B., and Wegener, M. (2010). Three-dimensional invisibility cloak at optical wavelengths. *Science*, 328(5976):337.

Erofeyev, V. I. (2003). *Wave processes in solids with microstructure*. World Scientific, Singapore.

Fang, N., Lee, H., Sun, C., and Zhang, X. (2005). Sub-diffraction-limited optical imaging with a silver superlens. *Science*, 308(5721):534.

Fang, N., Xi, D., Xu, J., Ambati, M., Srituravanich, W., Sun, C., and Zhang, X. (2006). Ultrasonic metamaterials with negative modulus. *Nature Materials*, 5:452–456.

Farhat, M., Enoch, S., Guenneau, S., and Movchan, A. B. (2008). Broadband cylindrical acoustic cloak for linear surface waves in a fluid. *Physical Review Letters*, 101(13):134501.

Farhat, M., Guenneau, S., Enoch, S., and Movchan, A. B. (2009). Negative refraction, surface modes, and superlensing effect via homogenization near resonances for a finite array of split ring resonators. *Physical Review E*, 80:046309.

Felbacq, D. and Bouchitté, G. (1997). Homogenization of a set of parallel fibres. *Waves in Random and Complex Media*, 7(2):245–256.

Felbacq, D. and Bouchitté, G. (2005). Left-handed media and homogenization of photonic crystals. *Optics Letters*, 30(10):1189–1191.

Feynman, R. P., Leighton, R. B., and Sands, M. (1964). *The Feynman lectures on physics: Mainly electromagnetism and matter.* Addison-Wesley, Reading, MA.

Figotin, A. and Kuchment, P. (1996). Band-gap structure of spectra of periodic dielectric and acoustic media. I. Scalar model. *SIAM Journal on Applied Mathematics*, 56(1):68–88.

Floquet, G. (1880). Sur les équations différentielles linéairesa coefficients périodiques. *Comptes Rendu de l'Académíe des sciences Paris*, 91:880–882.

Foldy, L. L. (1945). The multiple scattering of waves. I. General theory of isotropic scattering by randomly distributed scatterers. *Physical Review*, 67(3-4):107–119.

Galilei, G. (1638). *Dialogues concerning two new sciences*, Dover Publications (1954).

Geertsma, J. and Smit, D. C. (1961). Some aspects of elastic wave propagation in fluid-saturated porous solids. *Geophysics*, 26(2):169–181.

Gilbert, F. and Backus, G. E. (1966). Propagator matrices in elastic wave and vibration problems. *Geophysics*, 31(2):326–332.

Greenleaf, A., Kurylev, Y., Lassas, M., and Uhlmann, G. (2007). Full-wave invisibility of active devices at all frequencies. *Communications in Mathematical Physics*, 275(3):749–789.

Greenleaf, A., Kurylev, Y., Lassas, M., and Uhlmann, G. (2008). Comment on "Scattering Theory Derivation of a 3D Acoustic Cloaking Shell." *Arxiv preprint arXiv:0801.3279.*

Greenleaf, A., Lassas, M., and Uhlmann, G. (2003a). Anisotropic conductivities that cannot be detected by EIT. *Physiological Measurement*, 24:413–419.

Greenleaf, A., Lassas, M., and Uhlmann, G. (2003b). On non-uniqueness for Calderon's inverse problem. *Mathematical Research Letters*, 10:685–693.

Grekova, E. F. and Maugin, G. A. (2005). Modelling of complex elastic crystals by means of multi-spin micromorphic media. *International Journal of Engineering Science*, 43(5-6):494–519.

Guenneau, S., Movchan, A., Petursson, G., and Ramakrishna, S. A. (2007). Acoustic metamaterials for sound focusing and confinement. *New Journal of Physics*, 9:399.

Hall, A. J., Calius, E. P., Dodd, G., and Wester, E. (2010). Modelling and experimental validation of complex locally resonant structures. In *Proceedings of 20th International Conference on Acoustics, ICA 2010*, pages 1–9, August 2010, Sydney, Australia.

Harris, J. G. (2001). *Linear elastic waves*. Cambridge Texts in Applied Mathematics. Cambridge University Press, Cambridge.

Hashin, Z. (1962). The elastic moduli of heterogeneous materials. *Journal of Applied Mechanics*, 29:143–150.

Hashin, Z. and Shtrikman, S. (1962). A variational approach to the theory of the effective magnetic permeability of multiphase materials. *J. Appl. Phys.*, 33(10):3125–3131.

Heaviside, O. (1885). On the electromagnetic wave-surface. *Philosophical Magazine Series 5*, 19(21):397–419.

Helmholtz, H. von. (1869). On the aim and progress of physical science. In Cahan, D., editor, *Science and culture: Popular and philosophical essays*, pages 266–278. University of Chicago Press, Chicago, 1995.

Hill, R. (1963). Elastic properties of reinforced solids: Some theoretical principles. *Journal of the Mechanics and Physics of Solids*, 11:357–372.

Hill, R. (1964). Theory of mechanical properties of fibre-strengthened materials: I. Elastic behavior. *Journal of the Mechanics and Physics of Solids*, 12:199–212.

Ho, K. M., Yang, Z., Zhang, X. X., and Sheng, P. (2005). Measurements of sound transmission through panels of locally resonant materials between impedance tubes. *Applied Acoustics*, 66(7):751–765.

Huang, H. H., Sun, C. T., and Huang, G. L. (2009). On the negative effective mass density in acoustic metamaterials. *International Journal of Engineering Science*, 47:610–617.

Hughes, T. J. R. (2000). *The finite element method: Linear static and dynamic finite element analysis*, volume 682. Dover, New York.

Hussein, M. I. (2009). Reduced Bloch mode expansion for periodic media band structure calculations. *Proceedings of the Royal Society A*, 465(2109):2825.

Hussein, M. I. and Frazier, M. J. (2010). Band structure of phononic crystals with general damping. *Journal of Applied Physics*, 108(9):093506.

Ishimaru, A. (1978). *Wave propagation and scattering in random media*. Academic Press, New York.

Itskov, M. (2000). On the theory of fourth-order tensors and their applications in computational mechanics. *Computer Methods in Applied Mechanics and Engineering*, 189(2):419–438.

Jackson, J. D. (1999). *Classical electrodynamics: 3rd edition*. John Wiley & Sons, New York.

Jaglinski, T., Kochmann, D., Stone, D., and Lakes, R. S. (2007). Composite materials with viscoelastic stiffness greater than diamond. *Science*, 315(5812):620.

Jensen, J. S. (2003). Phononic band gaps and vibrations in one-and two-dimensional mass-spring structures. *Journal of Sound and Vibration*, 266(5):1053–1078.

Joannopoulos, J. D. and Winn, J. N. (2008). *Photonic crystals: Molding the flow of light*. Princeton University Press, Princeton, NJ.

John, S. (1987). Strong localization of photons in certain disordered dielectric superlattices. *Physical Review Letters*, 58(23):2486–2489.

Johnson, S. G. and Joannopoulos, J. D. (2002). *Photonic crystals: The road from theory to practice*. Springer Science + Business Media, New York.

Kerker, M. (1969). *The scattering of light*. Academic Press, New York.

Kleinstreuer, C. (2003). *Two-phase flow: Theory and applications*. Taylor & Francis, New York.

Kohn, R. V. and Shipman, S. P. (2008). Magnetism and homogenization of microresonators. *Multiscale Modeling & Simulation*, 7:62.

Kohn, W., Krumhansl, J. A., and Lee, E. H. (1972). Variational methods for dispersion relations and elastic properties of composite materials. *Journal of Applied Mechanics*, 39:327.

Korneev, V. A. and Johnson, L. (1993). Scattering of elastic waves by a spherical inclusion. I. Theory and numerical results. *Geophysical Journal International*, 115(1):230–250.

Kuchment, P. (1993). *Floquet theory for partial differential equations*. Birkhäuser, Basel.

Kurter, C., Abrahams, J., and Anlage, S. (2010). Miniaturized superconducting metamaterials for radio frequencies. *Applied Physics Letters*, 96(25):253504.

Kushwaha, M. S., Halevi, P., Dobrzynski, L., and Djafari-Rouhani, B. (1993). Acoustic band structure of periodic elastic composites. *Physical Review Letters*, 71(13):2022–2025.

Kuster, G. T. and Toksöz, M. N. (1974). Velocity and attenuation of seismic waves in two-phase media: Part I. Theoretical formulations. *Geophysics*, 39:587.

Lakes, R. (1987). Foam structures with a negative Poisson's ratio. *Science*, 235:1038–1040.

Lakes, R. (1993). Advances in negative Poisson's ratio materials. *Advanced Materials (Weinheim)*, 5(4):293–296.

Lakes, R., Lee, T., Bersie, A., and Wang, Y. C. (2001). Extreme damping in composite materials with negative-stiffness inclusions. *Nature*, 410:565–567.

Larsen, U. D., Sigmund, O., and Bouwstra, S. (2002). Design and fabrication of compliant micromechanisms and structures with negative Poisson's ratio. In *Micro Electro Mechanical Systems, 1996, MEMS '96, Proceedings. An Investigation of Micro Structures, Sensors, Actuators, Machines and Systems. IEEE, The Ninth Annual International Workshop on*, pages 365–371. IEEE, Piscataway, NJ.

Lax, M. (1951). Multiple scattering of waves. *Reviews of Modern Physics*, 23(4):287–310.

Leonhardt, U. (2006a). Notes on conformal invisibility devices. *New Journal of Physics*, 8:118.

Leonhardt, U. (2006b). Optical conformal mapping. *Science*, 23:1777–1780.

Leonhardt, U. and Philbin, T. G. (2006). General relativity in electrical engineering. *New Journal of Physics*, 8:247.

Li, J. and Chan, C. T. (2004). Double-negative acoustic metamaterial. *Physical Review E*, 70(5):55602.

Li, J., Fok, L., Yin, X., Bartal, G., and Zhang, X. (2009). Experimental demonstration of an acoustic magnifying hyperlens. *Nature Materials*. 8(12): 931–934.

Li, J. and Pendry, J. B. (2008). Hiding under the carpet: A new strategy for cloaking. *Physical Review Letters*, 101(20):203901.

Liu, R., Ji, C., Mock, J. J., Chin, J. Y., Cui, T. J., and Smith, D. R. (2009). Broadband ground-plane cloak. *Science*, 323(5912):366.

Liu, Z., Chan, C. T., and Sheng, P. (2002). Three-component elastic wave band-gap material. *Physical Review B*, 65(16):165116.

Liu, Z., Chan, C. T., and Sheng, P. (2005). Analytic model of phononic crystals with local resonances. *Physical Review B. Solid State*, 71:014103.

Liu, Z., Zhang, X., Mao, Y., Zhu, Y. Y., Yang, Z., Chan, C. T., and Sheng, P. (2000). Locally resonant sonic materials. *Science*, 289(5485):1734.

Lloyd, S. J. and Molina-Aldareguia, J. M. (2003). Multilayered materials: A palette for the materials artist. *Philosophical Transactions of the Royal Society of London. Series A: Mathematical, Physical and Engineering Sciences*, 361(1813):2931.

Lorenz, L. (1867). On the identity of the vibrations of light with electrical currents. *Philosophical Magazine*, 34:287–301.

Lorrain, P., Corson, D. R., and Lorrain, F. (1988). *Electromagnetic fields and waves: Including electric circuits*. Freeman, New York.

Lourtioz, J. M., Benisty, H., Berger, V., Pagnoux, D., Gerard, J. M., Maystre, D., and Tchelnokov, A. (2008). *Photonic crystals: Towards nanoscale photonic devices*. Springer-Verlag, Berlin.

Luo, Y., Chen, H., Zhang, J., Ran, L., and Kong, J. A. (2008). Design and analytical full-wave validation of the invisibility cloaks, concentrators, and field rotators created with a general class of transformations. *Physical Review B*, 77(12):125127.

Marqués, R., Martín, F., and Sorolla, M. (2008). *Metamaterials with negative parameters: Theory, design, and microwave applications*. Wiley-Interscience, Hoboken, NJ.

Marsden, J. E. and Hughes, T. J. R. (1993). *The mathematical foundations of elasticity*. Dover, New York.

Marsden, J. E. and Ratiu, T. S. (1999). *Introduction to mechanics and symmetry: A basic exposition of classical mechanical systems*. Springer-Verlag, New York.

Martin, P. A. (2006). *Multiple scattering: Interaction of time-harmonic waves with N obstacles*. Cambridge University Press, Cambridge.

Martin, P. A., Maurel, A., and Parnell, W. J. (2010). Estimating the dynamic effective mass density of random composites. *The Journal of the Acoustical Society of America*, 128:571.

Martinsson, P. G. and Movchan, A. B. (2003). Vibrations of lattice structures and phononic band gaps. *Quarterly Journal of Mechanics and Applied Mathematics*, 56(1):45.

Maxwell, J. C. (1954). *A treatise on electricity and magnetism*, vols. 1 and 2. Dover, New York.

Maysenhölder, W. (2003). Transmission loss of plates with internal resonators modelled by harmonic oscillators with frequency dependent complex mass and spring stiffness. Euronoise 2003, Naples, Italy.

Mei, J., Liu, Z., Wen, W., and Sheng, P. (2006). Effective mass density of fluid-solid composites. *Physical Review Letters*, 96(2):24301.

Mei, J., Liu, Z., Wen, W., and Sheng, P. (2007). Effective dynamic mass density of composites. *Physical Review B*, 76(13):134205.

Meng, F. Y., Liang, Y., Wu, Q., and Li, L. W. (2009). Invisibility of a metamaterial cloak illuminated by spherical electromagnetic wave. *Applied Physics A: Materials Science and Processing*, 95(3):881–888.

Merlin, R. (2009). Metamaterials and the Landau–Lifshitz permeability argument: Large permittivity begets high-frequency magnetism. *Proceedings of the National Academy of Sciences*, 106(6):1693.

Mie, G. (1908). Beiträge zur Optik trüber Medien, speziell kolloidaler Metallösungen. *Annalen der Physik*, 25:377–445.

Miller, D. A. B. (2006). On perfect cloaking. *Optics Express*, 14:12457–12466.

Milton, G. W. (1990). On characterizing the set of possible effective tensors of composites: the variational method and the translation method. *Communications on Pure and Applied Mathematics*, 43(1):63–125.

Milton, G. W. (2002). *Theory of composites*. Cambridge University Press, New York.

Milton, G. W. (2007). New metamaterials with macroscopic behavior outside that of continuum elastodynamics. *New Journal of Physics*, 9:359.

Milton, G. W. (2010). Realizability of metamaterials with prescribed electric permittivity and magnetic permeability tensors. *New Journal of Physics*, 12:033035.

Milton, G. W., Briane, M., and Willis, J. R. (2006). On cloaking for elasticity and physical equations with a transformation invariant form. *New Journal of Physics*, 8:248.

Milton, G. W. and Cherkaev, A. V. (1995). Which elasticity tensors are realizable? *Journal of Engineering Materials and Technology*, 117:483.

Milton, G. W. and Nicorovici, N.-A. P. (2006). On the cloaking effects associated with anomalous localized resonance. *Proceedings of the Royal Society of London A*, 462:3027–3059.

Milton, G. W., Nicorovici, N. A. P., McPhedran, R. C., Cherednichenko, K., and Jacob, Z. (2008). Solutions in folded geometries, and associated cloaking due to anomalous resonance. *New Journal of Physics*, 10:115021.

Milton, G. W., Nicorovici, N. A. P., McPhedran, R. C., and Podolskiy, V. A. (2005). A proof of superlensing in the quasistatic regime, and limitations of superlenses in this regime due to anomalous localized resonance. *Proceedings of the Royal Society A: Mathematical, Physical and Engineering Science*, 461(2064):3999.

Milton, G. W. and Willis, J. R. (2007). On modifications of Newton's second law and linear continuum elastodynamics. *Proceeding of the Royal Society of London A*, 463:855–880.

Minagawa, S., Nemat-Nasser, S., and Yamada, M. (1984). Dispersion of waves in two-dimensional layered, fiber-reinforced, and other elastic composites. *Computers and Structures*, 19(1-2):119–128.

Moakher, M. (2008). Fourth-order cartesian tensors: Old and new facts, notions and applications. *Quarterly Journal of Mechanics and Applied Mathematics*, 61(2):181.

Morse, P. M. (1977). *In at the beginnings: A physicist's life*. MIT Press, Cambridge, MA.

Morse, P. M. C. and Ingard, K. U. (1986). *Theoretical acoustics*. Princeton University Press, Princeton, NJ.

Movchan, A. B. and Guenneau, S. (2004). Split-ring resonators and localized modes. *Physical Review B*, 70:125116.

Movchan, A. B., Nicorovici, N. A., and McPhedran, R. C. (1997). Green's tensors and lattice sums for elastostatics and elastodynamics. *Proceedings: Mathematical, Physical and Engineering Sciences*, 453(1958):643–662.

Musgrave, M. J. P. (1970). *Crystal acoustics: Introduction to the study of elastic waves and vibrations in crystals*. Holden-Day, San Francisco.

Nemat-Nasser, S. and Hori, M. (1993). *Micromechanics: Overall properties of heterogeneous materials*. North-Holland, Amsterdam.

Ng, J., Chen, H., and Chan, C. T. (2009). Metamaterial frequency-selective superabsorber. *Optics Letters*, 34(4):644–646.

NIST (2010). *Digital Libray of Mathematical Functions (Release date: 2010-05-07)*. National Institute of Standards and Technology, Washington, DC.

Norris, A. N. (2008). Acoustic cloaking theory. *Proceedings of the Royal Society A*, 464(2097):2411.

Norris, A. N. (2009). Acoustic metafluids. *Journal of the Acoustical Society of America*, 125(2):839–849.

Norris, A. N. and Nagy, A. J. (2009). Acoustic metafluids made from three acoustic fluids. *Journal of the Acoustical Society of America*, 128(4):1606–1616.

Ogden, R. W. (1997). *Non-linear elastic deformations*. Dover, New York.

Onsager, L. (1931). Reciprocal relations in irreversible processes. I. *Physical Review*, 37(4):405–426.

Ostoja-Starzewski, M. (2002). Lattice models in micromechanics. *Applied Mechanics Reviews*, 55(1):35–60.

Peck, M. A. and Cavender, A. R. (2004). Practicable gyroelastic technology. *Advances in the Astronautical Sciences*, 118:1–15.

Pendry, J. B. (2000). Negative refraction makes a perfect lens. *Physical Review Letters*, 85(18):3966–3969.

Pendry, J. B., Holden, A. J., Robbins, D. J., and Stewart, W. J. (1998). Low frequency plasmons in thin-wire structures. *Journal of Physics: Condensed Matter*, 10:4785.

Pendry, J. B., Holden, A. J., Robbins, D. J., and Stewart, W. J. (1999). Magnetism from conductors, and enhanced non-linear phenomena. *IEEE Transactions on Microwave Theory and Techniques*, 47(11):2075–2084.

Pendry, J. B. and Li, J. (2008). An acoustic metafluid: Realizing a broadband acoustic cloak. *New Journal of Physics*, 10:115032.

Pendry, J. B., Schurig, D., and Smith, D. R. (2006). Controlling electromagnetic fields. *Science*, 312:1780–1782.

Postma, G. W. (1955). Wave propagation in a stratified medium. *Geophysics*, 20:780–806.

Poulton, C. G., Movchan, A. B., McPhedran, R. C., Nicorovici, N. A., and Antipov, Y. A. (2000). Eigenvalue problems for doubly periodic elastic structures and phononic band gaps. *Proceedings of the Royal Society of London. Series A: Mathematical, Physical and Engineering Sciences*, 456(2002):2543.

Rahm, M., Schurig, D., Roberts, D. A., Cummer, S. A., Smith, D. R., and Pendry, J. B. (2008). Design of electromagnetic cloaks and concentrators using form-invariant coordinate transformations of Maxwell's equations. *Photonics and Nanostructures—Fundamentals and Applications*, 6(1):87–95.

Ramakrishna, S. A. (2005). Physics of negative refractive index materials. *Reports on Progress in Physics*, 68:449.

Ramakrishna, S. A. and Grzegorczyk, T. M. (2009). *Physics and applications of negative refractive index materials*. CRC Press, Boca Raton, FL.

Rayleigh, L. (1887). On the maintenance of vibrations by forces of double frequency, and on the propagation of waves through a medium endowed with a periodic structure. *Philosophical Magazine*, 24(147):145–159.

Rayleigh, L. (1892). On the influence of obstacles arranged in rectangular order upon the properties of a medium. *Philosophical Magazine*, 34(481-502):205.

Reddy, B. D. (1998). *Introductory functional analysis*. Springer, New York.

Reynolds, D. D. (1981). *Engineering principles of acoustics: Noise and vibration control*. Allyn & Bacon, Boston.

Rytov, S. M. (1956). Acoustical properties of a thinly laminated medium. *Soviet Physics—Acoustics*, 2:68–80.

Sarychev, A. K., McPhedran, R. C., and Shalaev, V. M. (2000). Electrodynamics of metal-dielectric composites and electromagnetic crystals. *Physical Review B*, 62(12):8531–8539.

Sarychev, A. K. and Shalaev, V. M. (2007). *Electrodynamics of metamaterials*. World Scientific, Singapore.

Schoenberg, M. and Sen, P. N. (1983). Properties of a periodically stratified acoustic half-space and its relation to a Biot fluid. *Journal of the Acoustical Society of America*, 73:61.

Schurig, D., Mock, J. J., Justice, B. J., Cummer, S. A., Pendry, J. B., Starr, A. F., and Smith, D. R. (2006a). Metamaterial electromagnetic cloak at microwave frequencies. *Science*, 314(5801):977.

Schurig, D., Mock, J. J., and Smith, D. R. (2006b). Electric-field-coupled resonators for negative permittivity metamaterials. *Applied Physics Letters*, 88:041109.

Schurig, D., Pendry, J. B., and Smith, D. R. (2007). Transformation-designed optical elements. *Optics Express*, 15(22):14772–14782.

Shalaev, V. M. (2007). Optical negative-index metamaterials. *Nature Photonics*, 1(1):41–48.

Shawkey, M. D., Morehouse, N. I., and Vukusic, P. (2009). A protean palette: Colour materials and mixing in birds and butterflies. *Journal of The Royal Society Interface*, 6(Suppl. 2):S221.

Shelkunoff, S. A. and Friis, H. T. (1952). *Antennas: Theory and practice*. Wiley, New York.

Sheng, P. (2006). *Introduction to wave scattering, localization, and mesoscopic phenomena*. Springer-Verlag, Berlin.

Sheng, P., Zhang, X. X., Liu, Z., and Chan, C. T. (2003). Locally resonant sonic materials. *Physica B. Condensed Matter*, 338:201–205.

Shuvalov, A., Kutsenko, A., and Norris, A. (2010). Divergence of logarithm of a unimodular monodromy matrix near the edges of the Brillouin zone. *Wave Motion*, 47(6):370–382.

Sigalas, M., Kushwaha, M. S., Economou, E. N., Kafesaki, M., Psarobas, I. E., and Steurer, W. (2005). Classical vibrational modes in phononic lattices: Theory and experiment. *Zeitschrift für Kristallögraphie*, 220:765–809.

Sigalas, M. M. and Economou, E. N. (1992). Elastic and acoustic wave band structure. *Journal of Sound Vibration*, 158:377–382.

Sigalas, M. M. and Garcia, N. (2000). Theoretical study of three dimensional elastic band gaps with the finite-difference time-domain method. *Journal of Applied Physics*, 87:3122.

Sigmund, O. (2000). A new class of extremal composites. *Journal of the Mechanics and Physics of Solids*, 48(2):397–428.

Sigmund, O. and Jensen, J. S. (2003). Systematic design of phononic band–gap materials and structures by topology optimization. *Philosophical Transactions of the Royal Society of London. Series A: Mathematical, Physical and Engineering Sciences*, 361(1806):1001–1019.

Skudrzyk, E. J. (1971). *The foundations of acoustics: Basic mathematics and basic acoustics*. Springer-Verlag, Vienna.

Slaughter, W. S. (2002). *The linearized theory of elasticity*. Birkhäuser, Boston.

Smith, D. R., Padilla, W. J., Vier, D. C., Nemat-Nasser, S. C., and Schultz, S. (2000). Composite medium with simultaneously negative permeability and permittivity. *Physical Review Letters*, 84(18):4184–4187.

Smith, D. R., Pendry, J. B., and Wiltshire, M. C. K. (2004). Metamaterials and negative refractive index. *Science*, 305(5685):788.

Smith, D. R., Schurig, D., Rosenbluth, M., Schultz, S., Ramakrishna, S. A., and Pendry, J. B. (2003). Limitations on subdiffraction imaging with a negative refractive index slab. *Applied Physics Letters*, 82:1506.

Stakgold, I. (2000). *Boundary value problems of mathematical physics*. Society for Industrial Mathematics, Philadelphia.

Suzuki, T. and Yu, P. K. L. (1998). Complex elastic wave band structures in three-dimensional periodic elastic media. *Journal of the Mechanics and Physics of Solids*, 46(1):115–138.

Tartar, L. (1976). Estimation de coefficients homogeneises. In Glowinski, R. and Lions, J. L., editors, *Computer methods in applied sciences and engineering*, pages 136–212. Springer-Verlag, Berlin.

Tartar, L. (2009). *The general theory of homogenization: A personalized introduction*. Springer-Verlag, Berlin.

Ting, T. C. T. (1996). *Anisotropic elasticity—Theory and applications*. Oxford University Press, New York.

Torquato, S. (2001). *Random heterogeneous materials: Microstructure and macroscopic properties*. Springer-Verlag, New York.

Torrent, D. and Sánchez-Dehesa, J. (2006). Effective parameters of clusters of cylinders embedded in a nonviscous fluid or gas. *Physical Review B*, 74(22):224305.

Torrent, D. and Sánchez-Dehesa, J. (2008a). Acoustic cloaking in two dimensions: A feasible approach. *New Journal of Physics*, 10:063015.

Torrent, D. and Sánchez-Dehesa, J. (2008b). Anisotropic mass density by two-dimensional acoustic metamaterials. *New Journal of Physics*, 10:023004.

Tsang, M. and Psaltis, D. (2008). Magnifying perfect lens and superlens design by coordinate transformation. *Physical Review B*, 77(3):35122.

Urzhumov, Y., Ghezzo, F., Hunt, J., and Smith, D. R. (2010). Acoustic cloaking transformations from attainable material properties. *New Journal of Physics*, 12:073014.

Valentine, J., Zhang, S., Zentgraf, T., Ulin-Avila, E., Genov, D. A., Bartal, G., and Zhang, X. (2008). Three-dimensional optical metamaterial with a negative refractive index. *Nature*, 455(7211):376–379.

Vasquez, F. G., Milton, G. W., and Onofrei, D. (2009). Active exterior cloaking for the 2D Laplace and Helmholtz equations. *Physical Review Letters*, 103(7):73901.

Veselago, V. G. (1968). The electrodynamics of substances with simultaneous negative values of ε and μ. *Soviet Physics Uspekhi*, 10:509–514.

Volterra, E. and Zachmanoglou, E. C. (1965). *Dynamics of vibrations*. C. E. Merrill Books, Columbus, OH.

Waterman, P. C. and Truell, R. (1961). Multiple scattering of waves. *Journal of Mathematical Physics*, 2:512.

Weyl, H. (1919). Ausbreitung electromagnetischer Wellen über einem ebenen Leiter. *Annalen der Physik*, 60:481–500.

Wilcox, C. H. (1957). Debye potentials. *Journal of Mathematics and Mechanics*, 6:167–201.

Willis, J. R. (1980). Polarization approach to the scattering of elastic waves–I. Scattering by a single inclusion. *Journal of the Mechanics and Physics of Solids*, 28(5-6):287–305.

Willis, J. R. (1981a). Variational and related methods for the overall properties of composites. *Advances in Applied Mechanics*, 21:1–78.

Willis, J. R. (1981b). Variational principles for dynamics problems in inhomogenous elastic media. *Wave Motion*, 3:1–11.

Willis, J. R. (1997). Dynamics of composites. In Suquet, P. editor, *Continuum micromechanics: CISM courses and lectures no. 377*, pages 265–290. Springer-Verlag, New York.

Yablonovitch, E. (1987). Inhibited spontaneous emission in solid-state physics and electronics. *Physical Review Letters*, 58(20):2059–2062.

Yang, S., Page, J. H., Liu, Z., Cowan, M. L., Chan, C. T., and Sheng, P. (2004). Focusing of sound in a 3D phononic crystal. *Physical Review Letters*, 93(2):24301.

Ying, C. F. and Truell, R. (1956). Scattering of a plane longitudinal wave by a spherical obstacle in an isotropically elastic solid. *Journal of Applied Physics*, 27:1086–1097.

Zhang, S. (2010). *Acoustic metamaterial design and applications*. PhD thesis, University of Illinois at Urbana-Champaign.

Zhang, S., Xia, C., and Fang, N. (2011). Broadband acoustic cloak for ultrasound waves. *Physical Review Letters*, 106(2):024301.

Zhang, S., Yin, L., and Fang, N. (2009). Focusing ultrasound with an acoustic metamaterial network. *Physical Review Letters*, 102(19):194301.

Zhang, X. and Liu, Z. (2004). Negative refraction of acoustic waves in two-dimensional phononic crystals. *Applied Physics Letters*, 85:341.

Zhang, X. and Liu, Z. (2008). Superlenses to overcome the diffraction limit. *Nature Materials*, 7(6):435–441.

Zharov, A. A., Shadrivov, I. V., and Kivshar, Y. S. (2005). Suppression of left-handed properties in disordered metamaterials. *Journal of Applied Physics*, 97:113906.

Zhikov, V. V., Kozlov, S. M., and Oleinik, O. A. (1994). *Homogenization of differential operators and integral functionals*. Springer-Verlag, Berlin.

Zhu, J., Christensen, J., Jung, J., Martin-Moreno, L., Yin, X., Fok, L., Zhang, X., and Garcia-Vidal, F. J. (2010). A holey-structured metamaterial for acoustic deep-subwavelength imaging. *Nature Physics*, 7(1), 52–55.

Zohdi, T. I. and Wriggers, P. (2008). *An introduction to computational micromechanics*. Springer-Verlag, Berlin.

Index

acoustic cloaking, 218, 234
  inertial, 246
acoustic intensity level, 23
acoustic metafluid, 214, 245, 246
acoustic metamaterial, 185
acoustic mode, 270
acoustic power, 23
acoustic tensor, 50, 331
  effective, 261
acoustics, 16, 61, 62, 105, 253
adiabatic, 17
adiabatic index, 17
adjoint, 193
Airy equation, 296
Airy function, 296
Ament's relation, 147
Ampere's law, 26
angular momentum, 4, 171
anisotropic mass, 161
anisotropic mass density, 40, 161, 310
anisotropy, 308, 309
anomalous cloaking, 233
antiplane shear, 13, 37, 38, 52, 184
artificial magnetism, 133
attenuation length, 268
auxetic material, 208
axial vector, 169

Backus upscaling, 318
band gap, 184, 272
Bessel function, 95, 120, 132, 137, 296
  spherical, 105
Bessel's equation, 95, 116
bi-anisotropic medium, 196
Biot-Savart law, 26
Bloch condition, 255, 278
Bloch wave, 202, 255, 258, 265
Bloch wave mode, 267
Bloch wave vector, 255, 258
Bloch-Floquet theorem, 199, 255, 272, 275
body force, 13, 190
bound electron, 126
Brewster's angle, 64, 74
Brillouin zone, 270
bubbly liquid, 261
bulk modulus, 17, 253, 261
  complex, 175
bulk viscosity, 18
  effective, 264

capacitance, 137
Cartesian coordinates
  acoustics, 19
  elastodynamics, 7
  electrodynamics, 31
Cauchy principal value, 165
Cauchy stress, 4, 5, 7, 8, 11, 14, 17, 18, 54, 60, 67, 112, 176, 180, 183, 190, 197, 205, 242, 257, 258, 263, 285, 309, 319, 320, 331
  unsymmetric, 207
Cauchy's integral formula, 163
Cauchy-Goursat theorem, 165
Cauchy-Riemann equations, 163
causality, 5, 17, 20, 34, 41, 136, 163, 178, 231
chiral medium, 197
Christoffel symbols, 9
cloaking, 217
  acoustic, 234

elastodynamics, 241
electromagnetism, 225
cloaking transformation, 228
coated sphere, 261
coherent potential approximation *see also* CPA method, 117, 186
conductivity
electrical, 221
conservation
angular momentum, 4
linear momentum, 4, 18
mass, 16
constitutive relation, 5, 190
convolution, 10, 191
convolution theorem, 163
Coulomb's law, 25
curl, 31
curvilinear coordinates, 32, 228
acoustics, 19
elastodynamics, 8
electrodynamics, 31
cylindrical coordinates, 94, 120, 122, 130, 137

damping constant, 156
damping factor, 176
dashpot, 154
Debye potential, 111, 114
deformation gradient, 220
deviatoric stress, 18
dielectric constant, 27, 127
diffraction, 141
dipole moment, 27, 126, 127
dipole source, 97
Dirac delta function, 94, 96
Dirichlet to Neumann map, 223
dispersion, 49, 108, 277
dispersion relation, 49, 129, 200, 259, 261, 267, 268, 270, 275
displacement, 51, 191
displacement gradient, 6
dissipation, 13, 23, 41, 42
divergence theorem, 14, 42, 238
Drude model, 129
dynamic mass, 155

easy strain, 211
effective bulk modulus, 263
effective density, 317
effective dynamic mass, 153
effective field, 117
effective mass, 157
effective permeability, 184, 266
effective permittivity, 266, 323
effective phase velocity, 317
effective property, 318
eigen decomposition, 209, 312
eigenvalue, 160, 200, 209, 304
eigenvalue problem, 253, 258, 265, 267, 303
eigenvector, 160, 304
Eikonal equation, 299
elastic wave equation, 40
elastodynamic power, 13
elastodynamic wave, 256
elastodynamics, 49, 110, 277
governing equations, 3
periodic, 256
quasistatic limit, 256
Willis equations, 188
electric current, 26
electric dipole, 104
electric displacement, 27
electric field, 25
electric permittivity, 27
electric polarization, 27
electric susceptibility, 27
electrical conductivity, 28, 41
electrodynamic metamaterial, 125
electrodynamics, 24, 69, 70, 85, 102, 113
periodic, 264
quasistatic limit, 264
electromagnetic impedance
characteristic, 77
electromotive force, 27
electrostatics, 25
ensemble averaging, 188
ergodicity, 188

Euler-Lagrange equations, 157–159, 199
Eulerian description, 16
evanescent wave, 87, 97, 129, 143, 313
extremal material, 208

fading memory, 179, 182
far-field pattern, 121
Faraday's law, 27
folding transformation, 225
Foldy-Lax method, 117
Fourier transform, 10, 11, 96, 99
free-body diagram, 152
frequency, 62
Fresnel equations, 73, 87
function
    completely monotonic, 182

gauge transformation, 30
Gauss's law, 25
Graf's addition theorem, 122
Green's function, 94, 95, 193
Greenleaf-Leonhardt-Pendry, 217
group velocity, 49, 143, 268, 270, 272
gyrocontinua, 167

Hamiltonian, 157
Hankel function, 95, 97
    spherical, 106
harmonic plane wave, 48
Hashin's relations, 187, 263
    elastostatics, 261, 280
    electrostatics, 280, 333
Helmholtz equation, 47
Helmholtz potential
    body force, 6
Helmholtz resonator, 173, 186
Hertz vector potentials, 30, 114
hierarchical laminates, 208
homogeneous material, 6
homogenization, 132, 135, 139, 163, 184, 252, 260, 283, 292, 318
Hooke's law, 152

ideal gas, 17
image resolution, 141
impedance, 137, 161
    acoustic, 65, 81, 157, 314
    characteristic, 65
    effective, 315
    elastic, 56, 159
    electrical, 221
    electromagnetic, 77, 231
    normal, 77
impedance matching, 66, 78, 144, 231
impedance tube, 82, 83, 89, 157
incident wave, 62, 70
independent scattering, 117
inductance, 137
inertia
    frequency dependent, 167
inertial cloak, 234
infinite mass, 239
infinitesimal strain, 6
inhomogeneous wave, 102
integral matrix, 305
intensity, 23
irrotational flow, 18
isotropic material, 6

Jacobian, 219, 220
Jacobian matrix, 101

Kelvin moduli, 209
Korringa-Kohn-Rostoker method, 117
Kramers-Kronig relation, 163

Lagrangian, 156, 159, 198
Lagrangian description, 16
Lamé moduli, 6
Lamé's theorem, 6
Laplace-Beltrami operator, 9, 250
lattice spacing, 257
layered media, 283
    periodic, 310

Legendre polynomial, 105, 109
associated, 115
Legendre's equation, 105
Leibniz rule, 42
linear elastic, 190
micromorphic, 167
linear momentum, 4, 150, 170
linear operator, 193, 253, 258
linear superposition, 25
linear viscoelasticity, 178
locally resonant sonic material, *see* LRSM
Lorenz condition, 30
Love wave, 284
LRSM, 84, 148, 277
lumped-mass model, *see* spring-mass system

magnetic field, 28
magnetic flux density, 25
magnetic induction, 25
magnetic moment density, 28
magnetic monopole, 25
magnetic permeability, 26, 28
effective, 135
magnetization vector, 28
magnetostatics, 25
magnification, 233
mass
frequency dependent, 148, 153, 155, 163
mass density, 191, 253, 257
anisotropic, 148
dynamic, 148
effective, 147
negative, 148
static, 147
transversely isotropic, 318
matricant, 310
Maxwell model, 154
generalized, 178, 183
Maxwell's equations, 24, 28, 31–34, 37, 39, 69, 70, 102, 115, 130, 217, 225, 228, 244, 256, 265, 266, 308
Mie solution, 113
Milton's relation, 330
molecule, 147
momentum, 190, 204
balance of, 190
linear, 203
monochromatic plane wave, 48
monodromy matrix, 310, 312
multilayered cloak, 237
multiple scales analysis, 259
multiple scattering, 117, 187
multipole, 119

Nanson's formula, 238
Navier-Stokes equations, 16
negative density, 60
negative index, 141
negative mass, 40, 153
negative modulus, 149, 173
negative refractive index, 60, 65, 68, 78, 125
mirror, 78
negative shear modulus, 60
Newton's second law, 152
Newtonian fluid, 18
nonlocal, 5
normal strain, 214
null space, 254

Ohm's law, 28
Onsager reciprocity relations, 41
optical mode, 272

p-polarization, 38
P-wave, 6, 9, 53, 54, 58
pentamode material, 208, 211, 244–246
perfect lens, 139, 231
periodic media
one dimensional, 310
periodic medium, 257
permittivity, 25, 126, 127, 155
negative, 133
perturbation, 259
phase, 49

phase speed, 70
phase velocity, 36, 49, 71, 143, 268
phononic crystals, 251
photonic crystals, 251
plane wave, 48
acoustics, 61
elastodynamics, 49
electrodynamics, 69
plane wave expansion, 93
plasma frequency, 129
magnetic, 139
plasmon, 129
Poisson's ratio
negative, 208
polar decomposition theorem, 239
polarization, 52, 71
polarization field, 192, 325
porous media, 147
Poynting vector, 143
principle of minimum dissipation, 42
projection operator, 240, 325, 330
propagating wave, 97
propagator matrix, 286, 302, 304, 310
pseudo-pressure, 214, 246
pull-back, 226, 242

quasistatic limit, 259

radial map, 230
rank-1 laminate, 326
ray parameter, 56
ray theory, 299
Rayleigh multipole method, 118
reciprocal lattice vector, 255
recursion relation, 290
reflected wave, 63, 70
reflection coefficient, 55, 63, 73, 82, 87, 143, 290, 306, 315
apparent, 87
generalized, 292
reflectivity, 85
refractive index, 60, 68, 71
negative, 185
relaxation spectrum, 179, 182
relaxation time, 179
representative volume element, *see* RVE
residue, 100
resistance, 137
resonance, 27, 85, 91, 124, 125, 127, 129, 144, 150, 153, 156, 163, 177, 186, 208
anomalous, 232
frequency, 153, 157
Riccati equation, 294
rotating ring, 167
Runge-Kutta method, 294
RVE, 132, 133, 135, 150, 197, 199, 201, 202, 204–206, 252, 257, 259, 260, 263, 266, 268, 269, 272, 273, 278

s-polarized wave, 37
S-wave, 6
scalar potential, 41
acoustic, 19
elastodynamics, 6
electrodynamics, 29
scattering, 105, 251
acoustic, 105, 108, 120
cross-section, 108
elastodynamic, 110
electrodynamic, 113
scattering cross-section, 113
Schoenberg-Sen model, 186, 310
self-consistent method, 117
SH-wave, 13, 52, 59, 60, 183, 184, 237, 279
shear modulus, 6, 184, 261, 284
complex, 264
negative, 185
shear strain, 214
shear viscosity, 18, 154, 264
single scattering, 105, 117, 231
slab, 80
slowness surface, 214
slowness vector, 50, 51, 53, 54, 316, 317
Snell's law, 56, 64, 71

spectral decomposition, 208
spherical coordinates, 94, 98, 101, 106, 107, 111, 114, 116
spherical harmonics, 115
split-ring resonator, 133, 134, 136, 138, 139, 183, 184
spring constant
  complex, 154
  effective, 179
  real, 154
spring-mass system, 152, 153, 156, 159, 161, 173, 183, 184, 186, 197, 198, 251, 269, 276
SRR, *see* split-ring resonator
state equation, 283, 286, 303, 309, 310
state vector, 286, 308, 309
stiffness tensor, 5, 6, 190, 191, 256, 257, 277
strain, *see* strain tensor
strain energy, 213, 244, 245, 247, 248
strain tensor, 5–7, 11–14, 176, 180, 183, 191, 192, 196, 210–212, 257, 258, 260, 263, 285, 319, 320, 330
stress, *see* Cauchy stress
Stroh formalism, 309
Stroh matrix, 309
superpotentials, 30
supporting stress, 211
surface traction, 4, 13
SV-wave, 52, 54–57, 59, 66, 67, 110, 124
Sylvester's formula, 305

T-matrix, 121
Tartar formula, 333
Tartar-Murat-Lurie-Cherkaev formula, *see* TMLC formula
TE-wave, 37, 39, 75, 86, 94, 143, 183, 185, 217, 234, 287–293, 303, 308
TM-wave, 37–39, 72, 94, 185, 288, 292
TMLC formula, 327, 330
traction vector, 4, 203
transfer matrix, 302, 310
transformation optics, 225
transformation rule
  acoustics, 234, 236
  Chen-Chan, 236
  conductivity, 222
  Cummer-Schurig, 234
  current, 223
  elastodynamics, 244
  electromagnetism, 227
  tensor, 219
  vector, 218
transition matrix, 302, 304
translation operator, 254
transmission coefficient, 63, 73, 82, 87, 143, 307, 315
  generalized, 292
transmission line model, 186
transmission loss, 158
transmissivity, 85
transmitted wave, 63, 70
transverse electric field, *see* TE wave
transverse magnetic field, *see* TM wave

unit cell, 132, 133, 135, 150, 197, 199, 201, 202, 204–206, 252, 257, 259, 260, 263, 266, 268, 269, 272, 273, 278

variation, 42
vector potential
  elastodynamics, 6
  electrodynamics, 29
  Hertz, 30
Voigt notation, 209
volume average, 188, 319, 321, 323
volumetric strain, 6
volumetric stress, 18

wave equation

acoustic, 19
elastodynamics, 12
pentamode, 214
wave number, 48
wave vector, 48, 62, 70, 202
wavenumber vector, *see* wave vector
weak form, 278
Wentzel-Kramers-Brillouin-Jeffreys method, *see* WKBJ method
Weyl identity, 101
Weyl integral, 97
Willis equations, 188, 191, 244, 310
Willis materials, 188
WKBJ method, 297, 300
Wronskian, 95

Young's modulus, 176
frequency dependent, 179